AMA;

Erbium-Doped Fiber Amplifiers
Fundamentals and Technology

OPTICS AND PHOTONICS
(formerly Quantum Electronics)

EDITED BY

PAUL L. KELLY
Tufts University
Medford, Massachusetts

IVAN KAMINOW
AT&T Bell Laboratories
Holmdel, New Jersey

GOVIND AGRAWAL
University of Rochester
Rochester, New York

A complete list of titles in this series appears at the end of this volume.

Erbium-Doped Fiber Amplifiers
Fundamentals and Technology

P. C. BECKER
Passive Optical Networks Group
Switching and Access Group
Lucent Technologies
Tokyo, Japan

N. A. OLSSON
Passive Optical Networks Group
Switching and Access Group
Lucent Technologies
Murray Hill, New Jersey

J. R. SIMPSON
Ciena Corporation
Linthicum, Maryland

ACADEMIC PRESS

San Diego London Boston
New York Sydney Tokyo Toronto

This book is printed on acid-free paper. ∞
Copyright © 1999 by Lucent Technologies

All rights reserved.
No part of this publication may be reproduced or
transmitted in any form or by any means, electronic
or mechanical, including photocopy, recording, or
any information storage and retrieval system, without
permission in writing from the publisher.

All brand names and product names mentioned in this book
are trademarks or registered trademarks of their respective companies.

Academic Press
a division of Harcourt Brace & Company
525 B Street, Suite 1900, San Diego, CA 92101-4495, USA
http://www.apnet.com

Academic Press
24-28 Oval Road, London NW1 7DX, UK
http://www.hbuk.co.uk/ao/

Library of Congress Cataloging-in-Publication Data
Becker, P. C.
  Erbium-Doped Fiber Amplifiers: Fundamentals and Technology / P. C. Becker,
N. A. Olsson, J. R. Simpson.
    p. cm. — (Optics and Photonics)
  Includes bibliographical references and index.
  ISBN 0-12-084590-3—ISBN 0-12-084591-1 (Diskette)
  1. Optical communications — Equipment and supplies 2. Optical amplifiers.
3. Optical fibers   I. Olsson, N. A. II. Simpson, J. R. III. Title IV. Series.
TK5103.59.B43   1997
621.382'75 — DC21   97-29070
                                                                       CIP

Printed in the United States of America
99 00 01 02 03 IP 9 8 7 6 5 4 3 2 1

Contents

	Foreword	x
	Preface	xiii
	Acknowledgements	xv
1	**INTRODUCTION**	**1**
1.1	Long Haul Fiber Networks	1
1.2	Historical Development of Erbium-Doped Fiber Amplifiers	5
1.3	From Glass to Systems — Outline	9
2	**OPTICAL FIBER FABRICATION**	**13**
2.1	Introduction	13
2.2	Conventional Communication Fiber	14
2.3	Rare Earth Doped Fibers	16
	2.3.1 Rare Earth Vapor Phase Delivery Methods	16
	2.3.2 Rare Earth Solution-Doping Methods	21
	2.3.3 Rod and Tube Methods	23
2.4	Pump-Signal Interaction Methods	25
	2.4.1 Evanescent Field	25
	2.4.2 Double Clad Fiber Design	26
2.5	Compositions	27
2.6	Physical Properties	29
	2.6.1 Fiber Refractive Index and Composition Profile	29
	2.6.2 Strength and Reliability	30
	2.6.3 Alternate Glass Host Fabrication	30
3	**COMPONENTS AND INTEGRATION**	**43**
3.1	Introduction	43
3.2	Fiber Connectors	43
3.3	Fusion Splicing	48
3.4	Pump and Signal Combiners	50
3.5	Isolators	52

	3.6	Circulators	53
	3.7	Filters	55
	3.8	Fiber Gratings	55
		3.8.1 Introduction	55
		3.8.2 Applications of Bragg Gratings	57
		3.8.3 Long Period Gratings	59
	3.9	Signal Multiplexers and Demultiplexers	61
	3.10	Signal Add/Drop Components	62
	3.11	Dispersion Compensation Components	63
	3.12	Integrated Components	66
	3.13	Pump Lasers	66

4 RARE EARTH IONS — INTRODUCTORY SURVEY 87

	4.1	Introduction	87
	4.2	Atomic Physics of the Rare Earths	87
		4.2.1 Introduction — The 4f Electron Shell	87
		4.2.2 The "Puzzle" of 4f Electron Optical Spectra	91
		4.2.3 Semiempirical Atomic and Crystal Field Hamiltonians	92
		4.2.4 Energy Level Fitting	94
	4.3	Optical Spectra of Rare Earth Ions	95
		4.3.1 The Character of $4f^N \to 4f^N$ Optical Transitions	95
		4.3.2 Intensities of One-Photon Transitions — Judd-Ofelt Theory	96
	4.4	Fundamental Properties	99
		4.4.1 Transition Cross Sections	99
		4.4.2 Lifetimes	105
		4.4.3 Linewidths and Broadening	108
	4.5	Spectroscopy of the Er^{3+} Ion	110
		4.5.1 Lifetimes	111
		4.5.2 Er^{3+} Spectra, Cross Sections, and Linewidths	114
	4.6	Er^{3+}-Er^{3+} Interaction Effects	120

5 ERBIUM-DOPED FIBER AMPLIFIERS — AMPLIFIER BASICS 131

	5.1	Introduction	131
	5.2	Amplification in Three-Level Systems — Basics	131
		5.2.1 Three-Level Rate Equations	131
		5.2.2 The Overlap Factor	140
	5.3	Reduction of the Three-Level System to the Two-Level System	144
		5.3.1 Validity of the Two-Level Approach	144
		5.3.2 Generalized Rate Equations	146
	5.4	Amplified Spontaneous Emission	147
	5.5	Analytical Solutions to the Two-Level System	149

6 ERBIUM-DOPED FIBER AMPLIFIERS – MODELING AND COMPLEX EFFECTS 153

	6.1	Introduction	153
	6.2	Absorption and Emission Cross Sections	153

6.3	Gain and ASE Modeling			156
	6.3.1	Model Equations – Homogeneous Broadening		156
	6.3.2	Average Inversion Relationship		158
	6.3.3	Inhomogeneous Broadening		159
6.4	Amplifier Simulations			161
	6.4.1	Signal Gain, ASE Generation, and Population Inversion		161
	6.4.2	Gain as a Function of Fiber Length		169
	6.4.3	Spectral Profile of the ASE		169
	6.4.4	Small Signal Spectral Gain and Noise Modeling		171
	6.4.5	Saturation Modeling — Signal Gain and Noise Figure		173
	6.4.6	Power Amplifier Modeling		175
	6.4.7	Effective Parameter Modeling		178
6.5	Transverse Mode Models — Erbium Confinement Effect			180
6.6	Excited State Absorption Effects			186
	6.6.1	Model Equations		186
	6.6.2	Modeling Results in the Presence of ESA		188
	6.6.3	800 nm Band Pumping		188
6.7	Er^3 - Er^3 Interaction Effects			191
	6.7.1	Upconversion Effects on Amplifier Performance		193
	6.7.2	Pair Induced Quenching		195

7 OPTICAL AMPLIFIERS IN FIBER OPTIC COMMUNICATION SYSTEMS – THEORY   201

7.1	Introduction		201
7.2	Optical Noise: Device Aspects		202
	7.2.1	Classical Derivation of Optical Amplifier Noise	202
	7.2.2	Noise at the Output of an Optical Amplifier	205
	7.2.3	Comparison of Optical Amplifier Devices	210
7.3	Optical Noise: System Aspects		212
	7.3.1	Receivers	213
	7.3.2	Bit Error Rate Calculations - Direct Detection	214
	7.3.3	Optical Preamplifiers – Noise Figure and Sensitivity	220
	7.3.4	Optical Inline Amplifiers - Amplifier Chains	226
	7.3.5	Noise in Optical Power Amplifiers	235
	7.3.6	Nonlinearity Issues	236
	7.3.7	Analog Applications	240

8 AMPLIFIER CHARACTERIZATION AND DESIGN ISSUES   251

8.1	Introduction		251
8.2	Basic Amplifier Measurement Techniques		251
	8.2.1	Gain Measurements	251
	8.2.2	Power Conversion Efficiency	257
	8.2.3	Noise Figure Measurements	258
8.3	Amplifier Design Issues		263
	8.3.1	Copropagating and Counterpropagating Pumping Issues	265

	8.3.2	Choice of Fiber Lengths and Geometries for Various Applications	268
	8.3.3	Multistage Amplifiers	273
	8.3.4	Bidirectional Amplifiers	277
	8.3.5	Power Amplifiers	280
	8.3.6	WDM Amplifier Design Issues	284
	8.3.7	Distributed Amplifiers	295
	8.3.8	Waveguide Amplifiers	302

9 SYSTEM IMPLEMENTATIONS OF AMPLIFIERS   321
9.1 Introduction   321
9.2 System Demonstrations and Issues   323
- 9.2.1 Preamplifiers   323
- 9.2.2 Inline Amplifiers - Single Channel Transmission   327
- 9.2.3 Inline Amplifiers - WDM Transmission   335
- 9.2.4 Repeaterless Systems   345
- 9.2.5 Remote Pumping   346
- 9.2.6 Analog Applications   351
- 9.2.7 Gain Peaking and Self-Filtering   354
- 9.2.8 Polarization Issues   359
- 9.2.9 Transient Effects   363
9.3 Soliton Systems   367
- 9.3.1 Principles   367
- 9.3.2 System Results and Milestones   374

10 FOUR LEVEL FIBER AMPLIFIERS FOR 1.3 M AMPLIFICATION   401
10.1 Introduction   401
- 10.1.1 Gain in a Four-Level System   401
10.2 Pr^3-doped Fiber Amplifiers   404
- 10.2.1 Introduction   404
- 10.2.2 Spectroscopic Properties   405
- 10.2.3 Gain Results for Pr^3-doped Fiber Amplifiers   406
- 10.2.4 Modeling of the Pr^3-doped Fiber Amplifier Gain   412
- 10.2.5 System Results   416
10.3 Nd^3-Doped Fiber Amplifiers   418
- 10.3.1 Introduction   418
- 10.3.2 Gain Results for Nd^3-Doped Fiber Amplifiers   419
- 10.3.3 Modeling of the Nd^3-Doped Fiber Amplifier Gain   420

Appendix A   429
A.1 OASIX[R R] Amplifier Simulation Software   429
A.2 Introduction   429
- A.2.1 System Requirements   429
- A.2.2 Installing OASIX[R]   430
- A.2.3 Starting OASIX[R]   430
- A.2.4 What to do next   430

A.3	A Quick Overview and Tour		430
	A.3.1	Fibers and Modeling Parameters	430
	A.3.2	Saving a Simulation Configuration	431
	A.3.3	Device Types Simulated	431
	A.3.4	Data Entry and Device Conventions	432
	A.3.5	Screens and Menus	432
	A.3.6	Simulation Looping and Output Modes	433
A.4	Screen Contents and Simulation Methodology		434
	A.4.1	Main/Entry Screen	434
	A.4.2	Single-Stage Setup Screen	435
	A.4.3	Additional Signals Screen	435
	A.4.4	Output Setup Screen	436
	A.4.5	Simulation Status Box	437
A.5	Simulation Looping Structure		438
	A.5.1	Specifying Loop Parameters	438
	A.5.2	Choosing Loop Order	438
	A.5.3	Linear or Logarithmic Looping	439
	A.5.4	Multiple Parameters Varied in a Loop	439
	A.5.5	Influence on Output Format	440
	A.5.6	Output Modes	440
A.6	Sample Simulations		442
	A.6.1	Single-Run, Single-Stage EDFA	442
	A.6.2	Multiple-Run, Single-Stage EDFA	443
	A.6.3	Other simulations to try	443
A.7	Computation of Signal Related Quantities		443
A.8	Computation of ASE Related Quantities		444
A.9	Basic Operating Principles		445
	A.9.1	Simulation Speed and the Number of Waves	446
	A.9.2	Causes and Remedies for Convergence Failure	447
A.10	Comment on the treatment of losses		448

INDEX   **451**

Foreword

The telecommunications industry has been in a constant state of agitation in recent years, driven by wider competition and consumer demand. Innovations in information technology and government regulatory relief are largely responsible for much of the activity both by individual users and by service providers. For example, increased competition for global telecommunications markets has increased equipment sales and reduced consumer costs to the point that international fax and Internet communication is commonplace. At the same time, photonic technology has revolutionized long distance, and now local access, capabilities, thereby helping to sustain the boil in the information marketplace.

The first generation lightwave systems were made possible by the development of low-loss, single-mode silica fiber and efficient, double-heterostructure, single-mode injection lasers in the 1970's. The new lightwave generation, with vastly improved capacity and cost, is based on the recent development of erbium-doped fiber amplifiers (EDFA's). Undersea systems were the early beneficiaries, as EDFA repeaters replaced expensive and intrinsically unreliable electronic regenerators. Indeed, early EDFA technology was driven by the submarine system developers who were quick to recognize its advantages, soon after the first diode-pumped EDFA was demonstrated in 1989. Terrestrial telecommunications systems have also adopted EDFA technology in order to avoid electronic regeneration. And hybrid fiber/coax cable television networks employ EDFA's to extend the number of homes served. An equally attractive feature of the EDFA is its wide gain bandwidth. Along with providing gain at 1550 nm, in the low-loss window of silica fiber, it can provide gain over a band that is more than 4000 GHz wide. With available wavelength division multiplexing (WDM) components, commercial systems transport more than 16 channels on a single fiber; and the number is expected to reach 100. Hence, installed systems can be upgraded many fold without adding new fiber, and new WDM systems can be built inexpensively with much greater capacity.

As the EDFA technology matures, more applications, some outside the telecommunications field, become feasible. Commercial amplified soliton transmission now looks more promising; and high power rare-earth-doped fiber amplifiers and lasers have been demonstrated. The latter devices have a wide range of applications in printing and machining.

Foreword

The present book is a much-needed and authoritative exposition of the EDFA by three researchers who have been early contributors to its development. No other book provides an up-to-date engineering account of the basics of operation, methods of doped fiber fabrication, amplifier design, and system performance considerations. The Becker, Olsson, Simpson book focuses on the technology through 1998 in a thorough but concise format. The authors cover work at AT&T Bell Labs, now Lucent Bell Labs, along with developments world-wide. The contents of each chapter are surveyed in the following paragraphs.

A short historical introduction is given in Chapter 1. The methods of fabricating rare-earth-doped fiber, including the double-clad fiber used in some diode-pumped devices, are reviewed in Chapter 2. The physical properties of the doped glass are also discussed briefly.

Chapter 3 provides background on the passive fiber components that are required to build an amplifier and use it in a WDM system. Here we learn about those properties of commercial transmission fiber that are needed to couple them to doped fiber, and the operation of such components as wavelength division multiplexers for coupling the diode pump laser, isolators for blocking reflections, circulators for separating incident and reflected signals, fiber grating filters, and add/drop multiplexers for system applications.

The rare earth group of ions have the special property that their atomic spectra are only moderately influenced by chemical bonds to the host glass matrix; the reason being that the electrons responsible for the spectra are in incomplete shells deep inside the atom. The physical properties of the rare earths that bear on their behavior in lasers and amplifiers is summarized in Chapter 4. The energy levels, spectra, line shapes, and lifetimes are covered along with the small but significant influences of the host glass and doping concentration on these parameters. The detailed emphasis is on the erbium ion, Er^3, found in EDFA's.

The erbium amplifier is a three-level system, as opposed to neodymium, which is a four-level system; the difference being that a good deal more pump power is required to invert the three-level system. Hence, neodymium was the earliest successful rare-earth laser ion. The rate equations that model the gain process in erbium amplifiers are introduced in Chapter 5. With this model, one can optimize the EDFA design in terms of fiber length and index profile. Amplified spontaneous emission (ASE), which is critical in defining the noise figure of EDFA's, is included in the model.

The study of the gain process, including saturation, continues in more detail in Chapter 6, where the spectral gain functions are modeled along with ASE. The spectral shape of the gain curve is crucial to WDM applications in that it determines the number of channels allowed.

The authors advise that a reader interested only in technical amplifier design can skip the background material provided in Chapters 2 to 6, and go directly to Chapter 7. At the outset of this chapter, a fundamental formula for ASE power is derived in terms of the spontaneous emission, or inversion, parameter, nsp. The electrical noise in a receiver due to mixing of optical signal and ASE in the square-law photodetector can then be determined. With this information one can design systems with optimized performance, depending upon system requirements. For example, a cascade of in-line amplifiers can be employed on a very long submarine system to overcome fiber loss, or a preamplifier can be placed before the output receiver to optimize signal-to-noise ratio (SNR), or a power amplifier can be placed after the transmitter laser to boost output power and transmission distance in unrepeatered systems. In addition to these design features, nonlinear effect limitations, such as four-wave mixing and cross phase modulation are considered. A final section covers the use of amplifiers for analog applications, such as cable television.

Chapter 8 gets into the practical considerations of amplifier characterization and design for particular applications. The methods used for measurements of gain, noise figure, and pump power efficiency are detailed. Pumping configurations are compared, as are multi-stage configurations. Other issues are covered, including high power booster amplifiers realized by employing ytterbium/erbium co-doping along with a 1060 nm neodymium ion pump laser, and techniques for flattening the gain spectrum to meet the needs of WDM systems.

Chapter 9 brings us to the core of EDFA fever: the system implementations. Some 300 references recall the record-setting system experiments achieved with EDFA's in their roles as preamplifiers, in-line, and power amplifiers. Before the advent of the EDFA, the only candidates for high-sensitivity receivers were coherent detectors or avalanche photodiodes. Nowadays, the only commercially viable means to realize high sensitivity, measured in terms of minimal photons/bit needed to achieve a given bit-error-rate, is with an EDFA preamplifier. In-line amplifiers have been employed in the lab and under the sea to span 10,000 km distances at 5 Gbs and beyond. WDM experiments and their constraints are also recounted; and repeaterless and remotely-pumped systems are discussed. System requirements and the performance of analog systems are extended beyond the treatment in Chapter 7. Finally, the remarkable soliton experiments and their extreme performance are reported. The reader having reached this point will be conversant with all the terminology needed to get started on his own system work.

The book closes with a chapter on rare-earth-doped amplifiers for the 1300 nm band, corresponding to the other important low-loss window in silica, which was the first to be exploited commercially. The ions of choice are the four-level systems neodymium and praseodymium in non-silica host glasses. Although several experiments have been reported, their performance is not yet competitive with EDFA's.

In summary, Erbium-Doped Fiber Amplifiers: Fundamentals and Technology is an excellent place for graduate students, device developers and system designers to enter the field of amplified systems and components. They will learn the language, the achievements, and the remaining problems in as brief a time as is reasonable.

Ivan P. Kaminow

Preface

With the development of low loss fibers as the communications medium, efficient compact laser diodes as the modulated light sources, fast detectors and the auxiliary equipment necessary to connect these components, fiber optics became a competitive alternative technology to electrical systems for telecommunication. However, the optical amplifier was the essential missing link that now makes fiber optic systems so compelling. The stunning success of the erbium-doped fiber amplifier has inspired thousands of papers and continues to motivate research on the many diverse components that are required in these systems.

There continues to be many hundreds of publications per year on various aspects of erbium spectroscopy, fiber design, fabrication methods, systems and applications. The compelling features of this book are not only that it brings up to date a report on the technology of erbium-doped fiber amplifiers, but for completeness it deals with introductory material on spectroscopy, practical amplifier design and systems, so as to provide a complete self contained volume.

Chapter 1 opens with a convincing enumeration of the applications to long haul networks. Especially impressive are the undersea connections from the Americas to Europe and Asia. This is followed by a selected history of significant achievements in rare earth doped fiber lasers. In Chapter 2 the authors restate some of the fundamental concepts of various chemical vapor deposition procedures and bring us up to date with an evaluation of how these are utilized to dope the fiber cores with rare earths. They also deal with sol gel preparation methods. Of particular interest is their description of novel fiber designs to facilitate optical pumping of the core, such as the double clad configuration. Much of the recent work on fabrication is directed towards composition effects on the optical and physical properties of the fiber. Chapter 3 is a delight. It deals in a direct and relatively simple manner with the many issues of connectors, couplers, splices, optical isolators, circulatory and filters. In the latter category they describe the important recent developments in fiber gratings, both short and long period, and their applications to add/drop components in WDM systems, dispersion compensation, and gain flattening.

Chapter 4 covers the usual items of rare earth spectroscopy, Judd-Ofelt computations and non-radiative phonon relaxation. A nice feature is the way in which they describe the influence of various glass hosts on the spectroscopic properties, such as fluorescent lifetime, absorption and fluorescence spectra, transition gain cross sections, and line shapes. There is also a brief but useful discussion of upconversion (useful for the Tm laser in the blue but no so useful for an Er amplifier at 1.5 m). For completeness, in order to make the book a self contained document, Chapter 5 reviews basic concepts in amplifiers. This is augmented in Chapter 6 with more complex modeling and nicely presented data on ASE backward and forward gain as a function of pump power and fiber length, and the difference in behavior for pumping at 980 nm versus 1480 nm. Chapter 6 concludes with a caution that too much erbium can be too much of a good thing because of clustering and cooperative up-conversion quenching of the excited state. Chapters 7, 8, and 9 deal with the many issues involved in system applications. They cover the theory of noise in optically amplified systems and then review the many systems experiments performed with erbium-doped fiber amplifiers. Of particular interest are the sections on WDM systems and gain flattening of amplifiers. The final chapter is entitled "Four level fiber amplifiers for 1.3 m amplification". While it takes up the use of Nd^3 in selected hosts, its primary emphasis is on Pr^3 in fluoride hosts, a leading candidate for 1.3 m amplifiers. The contrast between it and erbium-doped fiber amplifiers illustrates how much of a gift Nature has made to us with erbium!

The continuing research on erbium-doped fiber amplifiers and their applications justifies the need for a book such as the present one. An intuitive and understandable monograph, it guides the reader through many aspects of the fiber amplifier field. It is an authoritative and comprehensive review of many of the necessary building blocks for understanding erbium fiber amplifiers and optically amplified systems.

Elias Snitzer
Professor Emeritus
Department of Materials Science and Ceramics
Rutgers University
Piscataway, NJ 08854

Acknowledgements

Covering a subject as large, and growing as fast, as that of erbium-doped fiber amplifiers, is a very daunting task. By diving the work among us according to our specializations has made the task easier. Nevertheless, this work would not have been possible without the support and help of a number of individuals. Miriam Barros was instrumental in supporting the effort with her research. We are indebted to Bell Laboratories for providing the environment which enabled and provided the intellectual stimulation for our research in the field of erbium-doped fiber amplifiers and their applications. We are grateful for our collaborators over the years, E. Desurvire, D. DiGiovanni, C. Giles, S. Kramer, and G. Nykolak. We would also like to convey our appreciation to our colleagues who assisted in the reviewing of the manuscript, N. Bergano, A. Chraplyvy, T. Cline, J.-M. Delavaux, P. Hansen, C. Headley, G. Jacobovitz-Veselka, F. Kerfoot, A. Lucero, L. Lunardi, L. Mollenauer, T. Nielsen, Y. Park, D. Piehler, A. Stentz, R. Tench, R. Tkach, J. Wagener, and P. Wysocki. Many thanks to our Academic Press colleagues, our editor, Zvi Ruder, and our production editor, Diane Grossman, for their support. Thank you also to our LaTeX consultant, Amy Hendrickson. We would also like to thank the authors of our preface and foreword, E. Snitzer and I. Kaminow, for their kindness in providing their gracious remarks for the present text.

Philippe C. Becker
N. Anders Olsson
Jay R. Simpson

December, 1998

*Dedicated to our parents,
Jean-Philippe and Lissy,
Nils-Petter and Inga,
Harold and Edith,
and to our wives,
Tomomi, Lana, Carol,
and to our children,
Fumiyuki, Nicolas, Anna,
Julie and Katie,
for their support during this
long project.*

Chapter 1
Introduction

The erbium-doped fiber amplifier is emerging as a major enabler in the development of worldwide fiber-optic networks. The purpose of this chapter is to present an introduction to the history of the erbium-doped fiber amplifier, as well as the context within which fiber amplifiers are having a very significant commercial impact. The emergence of the fiber amplifier foreshadows the invention and development of further guided wave devices that should play a major role in the continuing increase in transmission capacity and functionality of fiber networks.

1.1 LONG HAUL FIBER NETWORKS

Recent years have witnessed an explosive and exponential growth in worldwide fiber networks. As of the end of 1997, the embedded fiber base was 69 million km in North America, 35 million km in Europe, 59 million km in Asia-Pacific, and 8 million km elsewhere, for a total of 171 million km, according to KMI Corporation, Newport, RI. In 1997 alone, 38 million km of fiber were added worldwide. Additionally, by 1997, over 366,000 cable-km of fiber-optic undersea cable had been installed, up from 321,000 cable-km as of year-end 1996.[1]

Currently fiber networks are used predominantly in long distance telephone networks, high-density metropolitan areas, and in cable television trunk lines. The next decade should witness a large increase in fiber networks for access applications, if the economics warrant it. Given the current high price of erbium-doped fiber amplifiers (US $10,000 and up at the time of this writing), they are used primarily in high-capacity backbone routes and are not yet slated for high-volume applications in the local loop.

The most vivid illustrations of fiber-based transmission systems have been in undersea transcontinental cables. Figure 1.1 shows the long distance networks (existing and planned) that form the worldwide undersea links as of late 1998, regenerator based cables as well as optically amplified cables with inline amplifiers. Proposed fiber-optic cables should allow modern digital transmission techniques to become implanted in most corners of the world. Quite often, sea-based cables (known as offshore trunk routes) are a convenient way to connect the major hubs of a region. One example is the FLAG (Fiber Loop Around the Globe) cable that connects Europe and Asia and has a

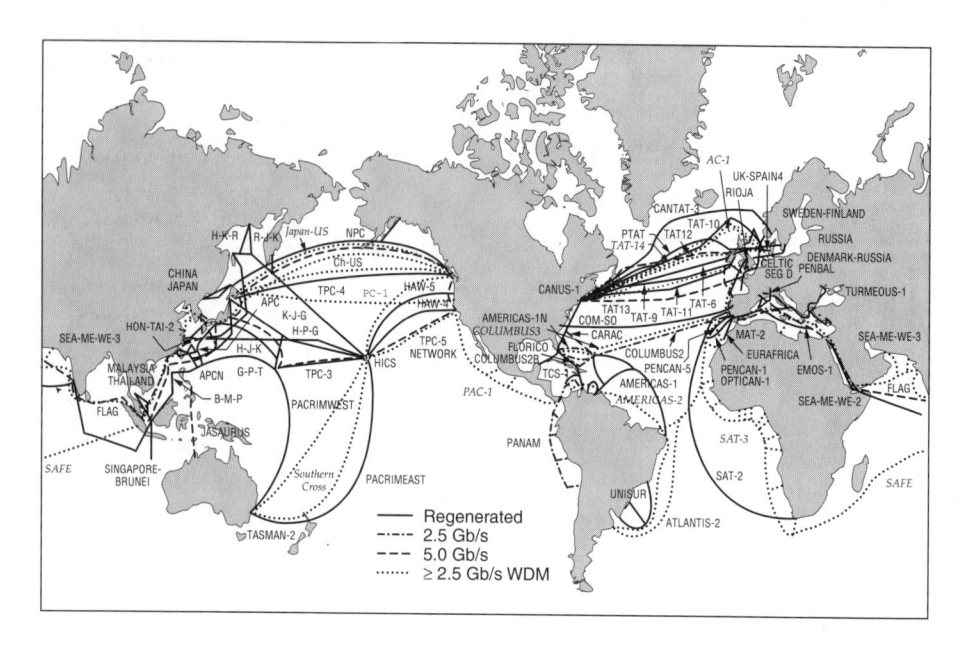

Figure 1.1: Global undersea fiber-optic cable network existing and planned as of late 1998. Planned cables are labeled in italics. Adapted from reference [2]. Original map copyright ©1995, *The AT&T Technical Journal*. All rights reserved. Reprinted with permission. Updates courtesy W. Marra, Tyco Submarine Systems Limited, Holmdel, NJ.

number of festoons for local connections, in particular in Southeast Asia. There is currently a significant number of new cables being planned, based on WDM technology.

Optical amplifiers play an exceptionally important role in long haul networks. Prior to the advent of optical amplifiers, the standard way of coping with the attenuation of light signals along a fiber span was to periodically space electronic regenerators along the line. Such regenerators consist of a photodetector to detect the weak incoming light, electronic amplifiers, timing circuitry to maintain the timing of the signals, and a laser along with its driver to launch the signal along the next span. Such regenerators are limited by the speed of their electronic components. Thus, even though fiber systems have inherently large transmission capacity and bandwidth, due to their optical nature, they are limited by electronic regenerators in the event such regenerators are employed. Optical fiber amplifiers, on the other hand, are purely optical in nature and require no high-speed circuitry. The signal is not detected then regenerated; rather, it is very simply optically amplified in strength by several orders of magnitude as it traverses the amplifier, without being limited by any electronic bandwidth. The shift from regenerators to amplifiers thus permits a dramatic increase in capacity of the transmission system. In addition, well-engineered amplified links can be upgraded in terms of bit rate from the terminal end alone, reusing the undersea cable and amplifiers. Since the introduction of optical amplifiers, rapid progress has been made in increasing the capacity of systems using such amplifiers. Table 1.1 traces the evolution of transatlantic

1.1. LONG HAUL FIBER NETWORKS

Year Installed	System	Bandwidth or Bit Rate	Number of Basic Channels	Capacity in Voice Channels	Technology
1963	TAT-3	1.1 MHz	140	315	Copper coax; analog; vacuum tubes
1965	TAT-4	1.1 MHz	140	315	"
1970	TAT-5	6 MHz	840	1900	Ge transistors
1976	TAT-6	30 MHz	4200	9450	Si transistors
1983	TAT-7	30 MHz	4200	9450	"
1988	TAT-8	280 Mb/s	8000	40000	Digital; optical fiber; $\lambda_0 = 1.3\ \mu m$
1992	TAT-9	560 Mb/s	16000	80000	Digital; optical fiber; $\lambda_0 = 1.5\ \mu m$
1992	TAT-10	560 Mb/s	24000	120000	"
1993	TAT-11	560 Mb/s	24000	120000	"
1996	TAT-12,13	5 Gb/s	122880	614400	Optical amplifiers

Table 1.1: Transatlantic cable systems and capacity in simultaneous calls. From reference [3] (©1993 IEEE). The capacity in voice channels is larger than that in basic channels (which itself makes use of compression techniques) as a result of the use of statistical multiplexing techniques, such as DCMS (digital circuit multiplication system) for digital transmission systems.

cables and their capacity. The shift from analog to digital occurred in the late 1980s, and the capacity of digital systems has grown rapidly since then.

The first implementation of erbium-doped fiber amplifiers has been in long haul systems, such as the TAT-12,13 fiber cable that AT&T and its European partners installed across the Atlantic in 1996. This cable, the first transoceanic cable to use fiber amplifiers, provides a near tenfold increase in voice and data transmission capacity over the previous transatlantic cable. A similar cable, TPC-5, was also installed in 1996 and links the United States and Japan. These cables operate at 5 Gb/s. The approximate length and optical amplifier spacing for several commercially deployed undersea systems are shown in Table 1.2.

Future long haul systems will operate at higher bit rates, in the 5 to 10 Gb/s range. They will also have multiple wavelength channels and make use of WDM (Wavelength Division Multiplexing) technology. Recent experiments using optical amplifiers and dense WDM (50 to 132 channels) have crossed the Tb/s barrier for information transmission, over distances in some cases as long as 600 km.[4, 5, 6, 7] Even higher bit rate systems (100 Gb/s per wavelength channel is possible) are promised by using soliton pulses, which make use of many of the fiber nonlinearities that limit conventional

System	Landing Points	Approximate Length (km)	Amplifier Spacing (km)
Americas 1	Vero Beach, FL - St. Thomas	2000	80
TPC-5G	San Luis Obispo, CA - Keawaula, HI	4200	68
TAT-12	Green Hill, NY - Lands End, UK	5900	45
TPC-5J	Coos Bay, OR - Ninomiya, Japan	8600	33

Table 1.2: Length and amplifier spacing of several representative commercial undersea cable systems (courtesy W. Marra, Tyco Submarine Systems Limited, Holmdel, NJ).

Figure 1.2: Albert Gore, vice president of the United States, examines an erbium-doped fiber amplifier during a 1993 visit to AT&T Bell Laboratories, in the presence of researchers Miriam Barros and Gerald Nykolak. Photograph property of AT&T Archives. Reprinted with permission of AT&T.

transmission systems. Erbium-doped fiber amplifiers are key enablers for the development of all optical networks under study in the United States (MONET program) and in Europe (ACTS program). As such, they have attracted high level political attention, as witnessed by the photo of Figure 1.2.

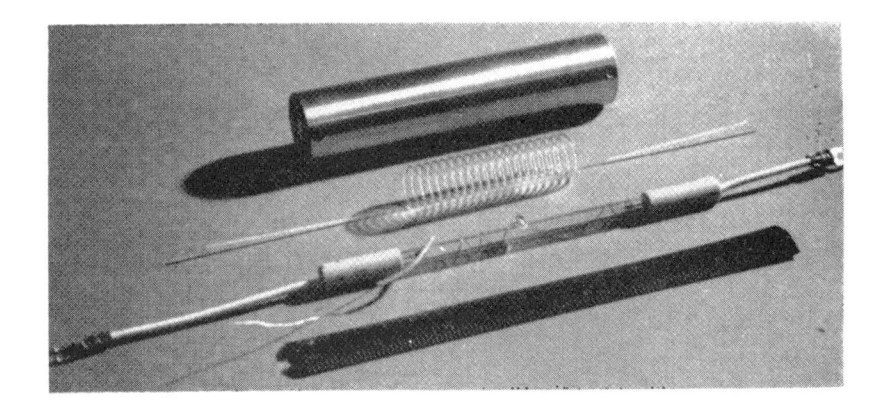

Figure 1.3: Optical components used in the first rare earth ion doped fiber amplifier demonstration. From top to bottom, the elements are the laser cavity, the fiber laser (fabricated in the form of a helix so as to be wrapped around the flashtube), a flashtube, and an 18 cm scale. From reference [9].

1.2 HISTORICAL DEVELOPMENT OF ERBIUM-DOPED FIBER AMPLIFIERS

The basic concept of a traveling wave optical amplifier was first introduced in 1962 by Geusic and Scovil.[8] Shortly thereafter, optical fiber amplifiers were invented in 1964 by E. Snitzer, then at the American Optical Company. He demonstrated a neodymium doped fiber amplifier at 1.06 μm. The fiber had a core of 10 μm with a 0.75 to 1.5 mm cladding, a typical length of 1 m, and was wrapped around a flashlamp that excited the neodymium ions.[9] Figure 1.3 shows the components used in the 1964 experiment. The fiber ends were polished at an angle to prevent laser oscillation, a technique that was used again by workers in the field more than twenty years later. Application to communications, and the appearance of noise from spontaneous emission, was mentioned by Snitzer in the conclusion of his paper. This work lay dormant for many years thereafter. It emerged as an exceedingly relevant technological innovation after the advent of silica glass fibers for telecommunications. Snitzer also demonstrated the first erbium-doped glass laser.[10]

Interestingly, rare earth doped lasers in a small diameter crystal fiber form were investigated during the early 1970s as potential devices for fiber transmission systems. This work was done by Stone and Burrus at Bell Telephone Laboratories.[11, 12, 13] The crystal fibers had cores as small as 15 μm in diameter, with typical values in the 25 μm to 70 μm range. The cores were doped with neodymium, with a surrounding fused silica cladding. Lasing of this device was achieved for a laser wavelength of 1.06 μm. A laser was typically fabricated by polishing the end faces of the laser and coating them with dielectric coatings. The fiber was then aligned to a pump laser, as shown in Figure 1.4 in the case of a laser diode pump.[12] In the case of a fiber with a core diameter of 35 μm, the laser pump threshold was as low as 0.6 mW of launched pump power at 890 nm. Lasing was even demonstrated with an LED pump.[13] Since

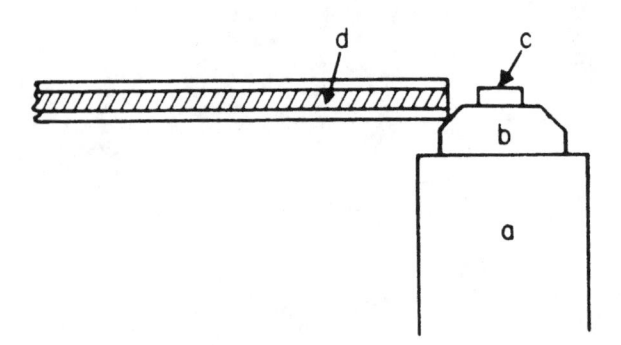

Figure 1.4: Fiber laser pumped by a diode laser. From reference [12]. (a) Copper support; (b) diamond heat sink; (c) laser chip; (d) fiber laser (not drawn to scale).

commercial fiber-optic transmission systems did not adopt the 1.06 μm wavelength as a signal wavelength, these lasers did not make their way into today's fiber communication systems.

The first demonstration of rare earth doping of single-mode fibers occurred in 1983. Performed by Broer and Simpson and coworkers at Bell Telephone Laboratories, the purpose of the work was to study of the physics of fundamental relaxation mechanisms of rare earth ions in amorphous hosts.[14, 15] The fiber, fabricated by the MCVD method, had a 6 μm core of pure silica (SiO$_2$), doped with 10 ppm of Nd^{3+}, surrounded by a depressed index fluorine doped silica cladding. The background loss of the fiber, away from any Nd^{3+} absorption peak, was relatively high (8 dB/km at 1.38 μm).[15] A few years later, further improvements in using the MCVD technique to fabricate rare earth doped single-mode fibers were achieved by Poole and coworkers at the University of Southampton, UK.[16, 17] A schematic of the MCVD setup for rare earth doped single-mode fiber fabrication used by the Southampton group is shown in Figure 1.5. This resulted in rare earth doped fibers with low background loss. An Nd^{3+} doped single-mode fiber laser, pumped by a GaAlAs laser diode, was demonstrated for the first time, at the University of Southampton, in 1985.[18] The laser was 2 m in length, with the cleaved fiber ends butted directly to mirrors highly reflective at the lasing wavelength, and transmissive at the pump wavelength, as the pump light was injected through one of the ends of the fiber.

All the necessary ingredients now being in place, the development of low-loss single-mode fiber lasers was followed shortly thereafter by that of fiber amplifiers. Erbium-doped single-mode fiber amplifiers for traveling wave amplification of 1.5 μm signals were simultaneously developed in 1987 at the University of Southampton and at AT&T Bell Laboratories.[19, 20, 21] Apart from the technical refinements that reflected the advance of state of the art in fiber optics and optical engineering in the 1980s versus the 1960s, these experiments were a restatement of Snitzer's original discovery in 1964 and a vindication of his prediction regarding the use of fiber amplifiers for communications. A key advance was the recognition that the Er^{3+} ion, with its pro-

1.2. HISTORICAL DEVELOPMENT OF ERBIUM-DOPED FIBER AMPLIFIERS   7

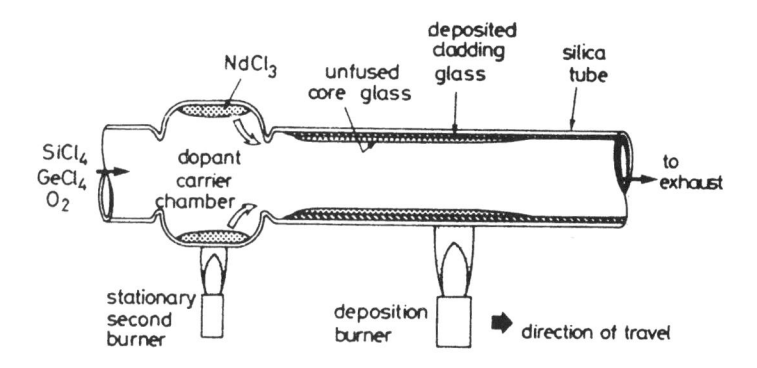

Figure 1.5: Experimental setup for MCVD fabrication of low-loss rare earth-doped single-mode fibers. From reference [16].

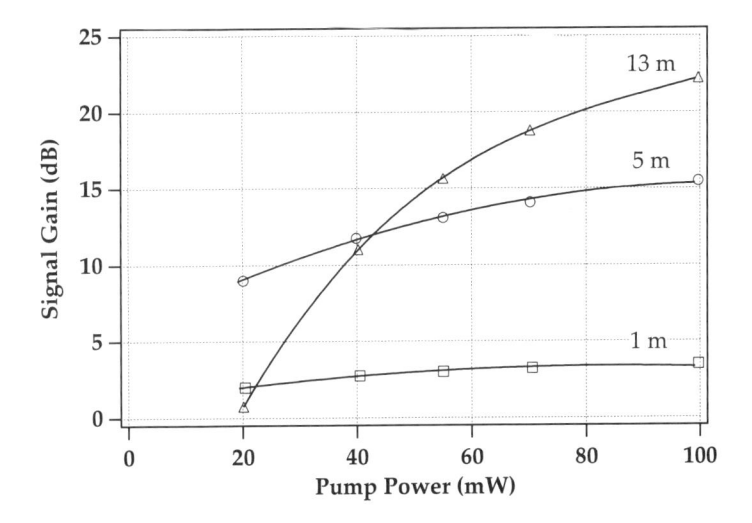

Figure 1.6: Early demonstration of gain at 1.53 μm in a single-mode erbium-doped fiber amplifier pumped by a 514 nm argon ion laser, in fibers of length 1 m, 5 m, and 13 m. From reference [21].

pitious transition at 1.5 μm, was ideally suited as an amplifying medium for modern fiber-optic transmission systems at 1.5 μm. Both of the demonstrations involved large frame lasers; an argon laser-pumped dye laser operating at 650 nm for the Southampton group, and an argon laser operating at 514 nm for the AT&T Bell Laboratories group. The high signal gains obtained with these erbium-doped fibers, shown in Figure 1.6, immediately attracted worldwide attention. In these early experiments, the ends of the fibers were immersed in cells containing index matching-fluid to prevent laser oscillation. Today's erbium-doped fiber amplifiers are fusion spliced to standard single-mode fiber, and fiber isolators placed after these splices prevent the laser oscillation.

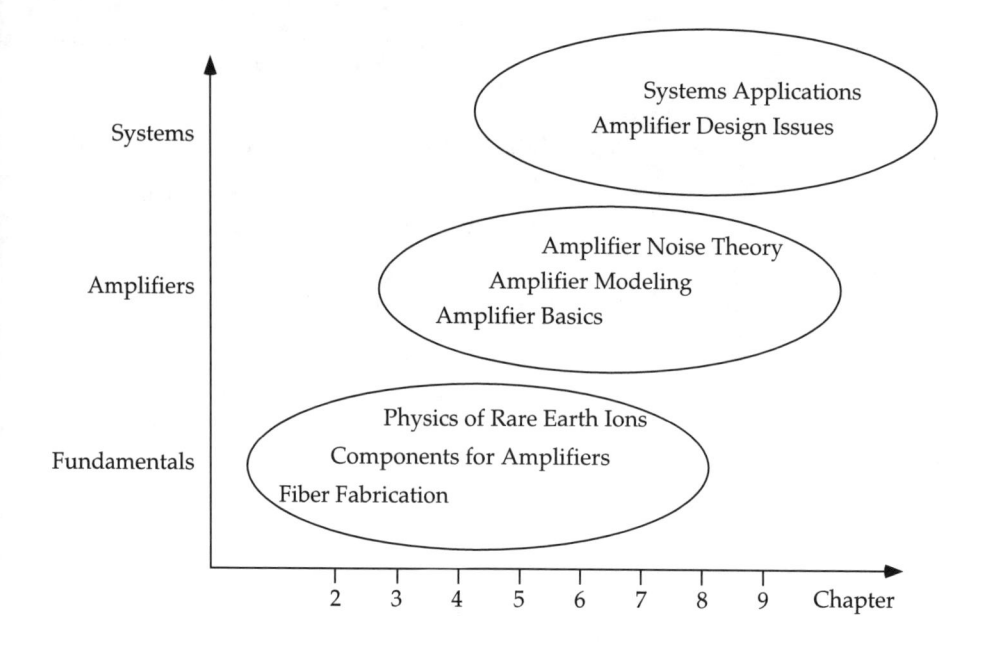

Figure 1.7: Outline of the book.

Given that the previously mentioned amplifier demonstrations used large frame laser pumps, one last remaining hurdle was to demonstrate an effective erbium-doped fiber amplifier pumped by a laser diode. This was achieved in 1989 by Nakazawa and coworkers, after the demonstration by Snitzer that 1.48 μm was a suitable pump wavelength for erbium amplification in the 1.53 μm to 1.55 μm range.[22] Nakazawa was able to use high-power 1.48 μm laser diode pumps previously developed for fiber Raman amplifiers.[23] This demonstration opened the way to serious consideration of amplifiers for systems application. Previous work, exploring optical amplification with semiconductor amplifiers, provided a foundation for understanding signal and noise issues in optically amplified transmission systems.[24]

It is safe to say that, starting in 1989, erbium-doped fiber amplifiers were the catalyst for an entirely new generation of high-capacity undersea and terrestrial fiber-optic links and networks. The first undersea test of erbium-doped fiber amplifiers in a fiber-optic transmission cable occurred in 1989.[25] A few years later, commercial amplifiers were for sale and were being installed by major telecommunications companies. MCI, for example, purchased and began the installation of 500 optical amplifiers in 1993. By 1996, erbium-doped fiber amplifiers were in commercial use in a number of undersea links, in particular TPC-5 and TAT-12,13, increasing the capacity near tenfold over the previous generation of cables. The erbium-doped fiber amplifier also reinvigorated the study of optical solitons for fiber-optic transmission, since it now made practical the long distance transmission of solitons. In conjunction with recent advances made throughout the 1990s in a number of optical transmission technologies, be it lasers or novel components such as fiber-grating devices or signal-processing fiber devices, the optical amplifier offers a solution to the high-capacity needs of today's voice and data transmission applications.

1.3 FROM GLASS TO SYSTEMS — OUTLINE

This book is organized so as to provide a basis for understanding the underlying materials and physics fundamentals of erbium-doped fiber amplifiers, which then leads into amplifier design issues and system applications, as shown in Figure 1.7. Because a deep understanding of the materials and physics fundamentals is not necessary to understand the design and systems implementation issues, the beginning chapters—Chapters 2, 3, 4, 5, and 6—can be used for reference as needed. The noise theory chapter—Chapter 7—is used frequently in the chapters on amplifier design and system applications—Chapters 8 and 9. Chapter 10 is included as an introduction to 1.3 μm amplifiers.

Bibliography

[1] Courtesy KMI Corporation, Newport, RI.

[2] J. M. Sipress, AT&T Technical Journal, January/February 1995, p. 5, with updates courtesy W. Marra, Tyco Submarine Systems Limited.

[3] T. Li, *Proc. of the IEEE* **81**, 1568 (1993).

[4] H. Onaka, H. Miyata, G. Ishikawa, K. Otsuka, H. Ooi, Y. Kai, S. Kinoshita, M. Seino, H. Nishimoto, and T. Chikama, "1.1 Tb/s WDM transmission over a 150 km 1.3 μm zero-dispersion single-mode fiber," in *Optical Fiber Communication Conference*, Vol. 2, 1996 OSA Technical Digest Series (Optical Society of America, Washington D.C.,1996), pp. 403–406.

[5] A. H. Gnauck, F. Forghieri, R. M. Derosier, A. R. McCormick, A. R. Chraplyvy, J. L Zyskind, J. W. Sulhoff, A. J. Lucero, Y. Sun, R. M. Jopson, and C. Wolf, "One terabit/s transmission experiment," in *Optical Fiber Communication Conference*, Vol. 2, 1996 OSA Technical Digest Series (Optical Society of America, Washington D.C.,1996), pp. 407–410.

[6] Y. Yano, T. Ono, K. Fukuchi, T. Ito, H. Yamazaki, M. Yamaguchi, and K. Emura, in *22nd European Conference on Optical Communication*, Proceedings Vol. 5, pp. 5.3-5.6 (1996).

[7] S. Aisawa, T. Sakamoto, M. Fukui, J. Kani, M. Jinno, and K. Oguchi, "Ultra-wide band, long distance WDM transmission demonstration:1 Tb/s (50 × 20 Gb/s, 600 km transmission using 1550 and 1580 nm wavelength bands," in *Optical Fiber Communication Conference*, Vol. 2, 1998 OSA Technical Digest Series (Optical Society of America, Washington D.C., 1998), pp. 468-471.

[8] J. E. Geusic and H. E. D. Scovil, *Bell Syst. Tech. J.* **41**, 1371 (1962).

[9] C. J. Koester and E. Snitzer, *Appl. Opt.* **3**, 1182 (1964).

[10] E. Snitzer and R. Woodcock, *Appl. Phys. Lett.* **6**, 45 (1965).

[11] J. Stone and C. A. Burrus, *Appl. Phys. Lett.* **23**, 388 (1973).

[12] J. Stone and C. A. Burrus, *Appl. Opt.* **13**, 1256 (1974).

[13] J. Stone, C. A. Burrus, A. G. Dentai, and B. I. Miller, *Appl. Phys. Lett.* **29**, 37 (1976).

[14] J. Hegarty, M. M. Broer, B. Golding, J. R. Simpson, and J. B. MacChesney, *Phys. Rev. Lett.* **51**, 2033 (1983).

[15] M. M. Broer, B. Golding, W. H. Heammerle, and J. R. Simpson, *Phys. Rev. B* **33**, 4160 (1986).

[16] S. B. Poole, D.N. Payne, and M.E. Fermann, *Elect. Lett.* **21**, 737 (1985).

[17] S.B. Poole, D. N. Payne, R. J. Mears, M. E. Fermann,and R. I. Laming, *J. Light. Tech.* **21**, 737 (1985).

[18] R. J. Mears, L. Reekie, S. B. Poole, and D. N. Payne, *Elect. Lett.* **LT-4**, 870 (1986).

[19] R. J. Mears, L. Reekie, I. M. Jauncie, and D. N. Payne, "High-gain rare-earth doped fiber amplifier at 1.54 μm," in *Optical Fiber Communication Conference*, Vol. 3, 1987 OSA Technical Digest Series, (Optical Society of America, Washington, DC., 1987) p. 167.

[20] R. J. Mears, L. Reekie, I. M. Jauncie, and D. N. Payne, *Elect. Lett.* **23**, 1026 (1987).

[21] E. Desurvire, J. R. Simpson, and P. C. Becker, *Opt. Lett.* **12**, 888 (1987).

[22] E. Snitzer, H. Po, F. Hakimi, R. Tuminelli, and B. C. MaCollum, "Erbium fiber laser amplifier at 1.55 μm with pump at 1.49 μm and Yb sensitized Er oscillator," in *Optical Fiber Communication Conference*, Vol. 1, 1988 OSA Technical Digest Series (Optical Society of America, Washington, D.C., 1988), pp. 218–221.

[23] M. Nakazawa, Y. Kimura, and K. Suzuki, *Appl. Phys. Lett.* **54**, 295 (1989).

[24] N. A. Olsson, *J. Light. Tech.* **7**, 1071 (1989).

[25] N. Edagawa, K. Mochizuki, and H. Wakabayashi, *Elect. Lett.* **25**, 363 (1989).

Chapter 2
Optical Fiber Fabrication

Fabrication of suitable erbium-doped fiber is one of the keys to creating an appropriate amplifier for a particular application. Fortunately, many of the methods used in fabricating low-loss silica transmission fiber can be used in this context. In most cases the concentration of erbium is low enough that the fabrication methods do not entail a significant change in the fundamental structure of the underlying glass host. This chapter will mainly focus on describing the methods developed for fabricating rare earth doped silica-based fibers. We will emphasize the MCVD, OVD, and VAD fabrication techniques. Fluoride fiber fabrication and fiber structures for specific amplifier designs, such as double clad fibers, will also be discussed.

2.1 INTRODUCTION

Rare earth doped fibers can be fabricated by a wide variety of methods, each suited for different amplifier design needs. The concentration of rare earth dopant ranges from very high (thousands of ppm) in multicomponent glasses, to less than 1 ppm in distributed erbium-doped fibers. The background losses are comparable to state-of-the-art transmission grade fiber. The methods used to fabricate rare earth doped optical fiber are, in general, variations on the techniques used to produce low-loss communications grade fiber.[1, 2, 3, 4, 5] New compositions that offer improved amplifier performance, be it from a geometry or a host composition perspective, will continue to challenge the techniques of fabrication. Commercialization of erbium-doped fiber amplifiers has required greater attention to reproducibility of core and fiber geometry, as well as dopant control to assure uniformity of doping along the longitudinal and transverse fiber axes.

There is a strong incentive to maintain compatibility between standard low-attenuation silica-based fiber and rare earth doped fiber. Connectivity of rare earth fiber components to doped silica telecommunications fiber by fusion splicing results in the low insertion loss and low reflectivity necessary for stable, low-noise, high-gain amplifiers. Rare earth doped fibers based on traditional silica processing have therefore become the media of most interest, due to ease of fusion splicing. Less compatible glass host compositions including compound (e.g., SiO_2-Al_2O_3-NaO_2-CaO), phosphate, fluorozirconate (e.g., ZrF_4-BaF_2-LaF_3-AlF_3-NaF, also called ZBLAN), tellurite, sulfide, and others may offer benefits such as higher gain, higher output power, or broader band operation.

2.2 CONVENTIONAL COMMUNICATION FIBER

Before discussing the challenges of rare earth doping, we first review the traditional methods of making low-impurity fiber materials. These techniques may be divided into three general categories:

- Hydrolysis (reaction with H_2O)
- Oxidation (reaction with O_2)
- Sol-gel (reactions with a suspension of silica)

The hydrolysis method is accomplished by flowing $SiCl_4$ vapor into a hydrogen flame with the resulting "fumed" silica submicron particles collected on a rotating target. The chemistry of this flame hydrolysis is dominated by the reaction of the halide with the water of reaction within the flame as indicated by the hydrolysis reaction.

$$SiCl_4 + 2H_2O = SiO_2 + 4HCl \tag{2.1}$$

Other halide dopants that may be added to the flame (e.g., $GeCl_4$, $POCl_3$) will likewise react to form their respective oxides. The large amount of hydrogen created by this reaction results in substantial OH incorporation in the glass particles. The resulting porous cylinder is then treated at a temperature near 800°C with an atmosphere of $SOCl_2$ to reduce the OH content of the glass. Following this, a transparent glass preform is made by fusing the particles at a temperature of 1500°C, a process referred to as sintering. The resulting glass preform is then drawn into fiber. Processes based on this chemistry are commonly referred to as vapor axial deposition, VAD, and outside vapor deposition, OVD.[6, 7] In the VAD process, the rotation target that collects the submicron particles or "soot" is a rotating pedestal that slowly recedes from the flame. In the OVD process, the rotating target is a rod with the torch traversing back and forth, depositing soot, layer by layer, as shown in Figure 2.1.

The oxidation method reacts the chlorides with oxygen inside a substrate tube. The reaction (as written in equation 2.1) takes place in a region of the substrate tube that is heated from the outside. The tube is typically heated to 1200°C using an oxy-hydrogen torch that traverses slowly along the rotating substrate tube. As the torch traverses the substrate tube, the gases flowing inside the tube are simultaneously reacted, deposited, and sintered into a clear glass layer. The dominant reaction chemistry here is that of oxidation as written for the primary halide constituent $SiCl_4$.

$$SiCl_4 + O_2 = SiO_2 + 2Cl_2 \tag{2.2}$$

Processes based on this method are commonly referred to as modified chemical vapor deposition (MCVD), plasma chemical vapor deposition (PCVD), intrinsic microwave chemical vapor deposition (IMCVD), and surface plasma chemical vapor deposition (SPCVD).[8, 9, 10, 11] For all of these processes, the reaction of halides takes

2.2. CONVENTIONAL COMMUNICATION FIBER 15

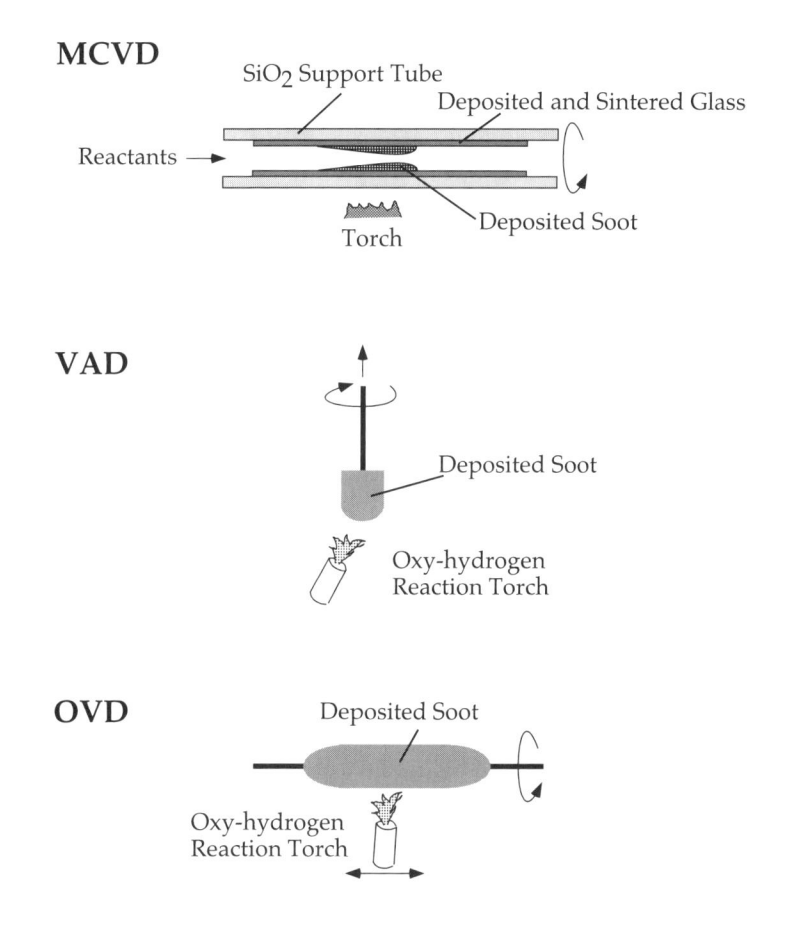

Figure 2.1: Fiber fabrication methods: Modified chemical vapor deposition (MCVD); vapor axial deposition (VAD); and outside vapor deposition (OVD).

place inside a silica support tube. For MCVD, the reaction occurs at a temperature in excess of 1000°C and a pressure near 1 atmosphere. For the PCVD, SPCVD, and IM-CVD processes, the reaction is initiated by a low pressure plasma. All of these methods create a preform, or large geometry equivalent of what is desired in the fiber.

Sol-gel processing for optical fiber has been investigated primarily for the production of silica tubes used to overclad the higher purity core and inner cladding regions of a preform.[12] This outer cladding region accounts for a large fraction of the fiber volume and therefore significantly influences the cost of fiber fabrication. These tubes are fashioned by first creating the sol, a suspension of low-surface-area fumed silica in a basic aqueous solution. This sol is then spun to remove a large fraction of the water and then cast into the shape of a tube. The cast sol then gels to retain the form of the tube mold. Finally the gelled tube is removed from the mold, heat-treated to remove OH and impurities, and then sintered at 1500°C to form a transparent tube. Overcladding is then accomplished by shrinking this sol-gel tube onto a preform rod containing the core and inner cladding. This high-temperature heating allows the surface tension to draw the outer tube onto the inner rod. A preform, made by any of the previously described methods, is then drawn into an optical fiber by heating one end to the softening temperature and pulling it into a fiber at rates as high as 20 meters/second. Details of these methods may be seen in publications edited by Miller and Kaminow, and Li.[13, 14]

It is necessary to add dopants to the primary glass constituent, SiO_2, to change its refractive index, thus allowing control of the fiber waveguide designs. Index-raising dopant ions—such as germanium, phosphorus, and aluminium,—and index-lowering dopants—such as boron and fluorine,—are introduced into the reaction stream as halide vapors carried by oxygen or an inert gas at a temperature near 30°C. The incorporation of the dopant ions in either the hydrolysis or oxidation processes is controlled by the equilibria established during dopant reaction, deposition, and sintering.[15, 16] These equilibria are established between the reactant halides and the resulting oxides during deposition and between the oxides and any reduced state of the oxides at higher temperatures. The equilibria may involve a number of other species as shown below for the case of the reaction of $GeCl_4$ to form GeO_2.[17]

$$GeCl_4 + O_2 = GeO_2 + Cl_2 \quad (2.3)$$
$$GeO_2 + GeCl_4 = 2GeOCl_2 \quad (2.4)$$
$$GeO_2 + 3GeCl_4 = 2Ge_2OCl_6 \quad (2.5)$$
$$GeO_2 + 1/2Cl_2 = GeCl + O_2 \quad (2.6)$$
$$GeO_2 = GeO + 1/2O_2 \quad (2.7)$$

Examples of the refractive index profiles typical of these processes are shown in Figure 2.2. Variations in the refractive index due to the equilibria are evident in the sawtooth pattern in the cladding of the depressed clad design, and the depression in the refractive index in the center (r = 0) of the matched clad design.

The difficulty in delivering rare earth dopants to the reaction zones in conventional fiber preform fabrication methods is a fundamental result of the chemistry of the rare earth compounds. These halide compounds of rare earth ions are generally less volatile than the commonly used chlorides and fluorides of the index modifying ions (Ge, P, Al, and F). The rare earth halide materials therefore require volatilizing and delivery temperatures of a few hundred °C (see Figure 2.3).[18, 19, 20] This requirement has stimulated the vapor and liquid phase handling methods to be discussed below.

2.3 RARE EARTH DOPED FIBERS

2.3.1 *Rare Earth Vapor Phase Delivery Methods*

Methods to deliver rare earth vapor species to the reaction/deposition zone of a preform process have been devised for MCVD, VAD, and OVD techniques. The fabrication configurations employed for MCVD are shown in Figure 2.4. Rare earth dopants are delivered to an oxidation reaction region along with other index controlling dopants. The low vapor pressure rare earth reactant is accommodated either by placing the vapor source close to the reaction zone and immediately diluting it with other reactants

2.3. RARE EARTH DOPED FIBERS   17

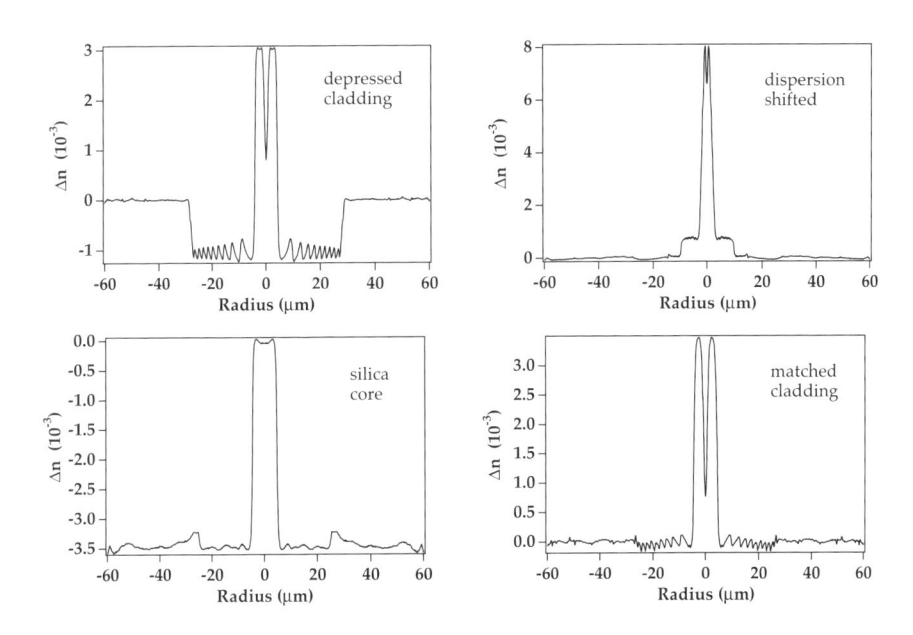

Figure 2.2: Refractive index profiles of conventional communications fiber, with the refractive index difference shown relative to undoped silica (the fiber types are indicated on the graphs). Data courtesy of W. Reed, Lucent Technologies, Murray Hill, NJ.

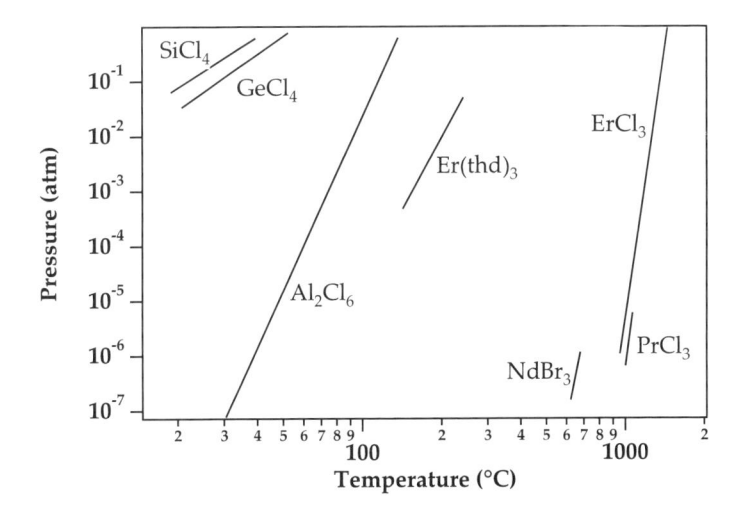

Figure 2.3: Vapor pressures of reactant halides (excepting Er(thd)$_3$, an organic compound) which incorporate the index-raising elements Ge and Al as well as the representative rare earth elements Er, Nd, and Pr. From reference [3].

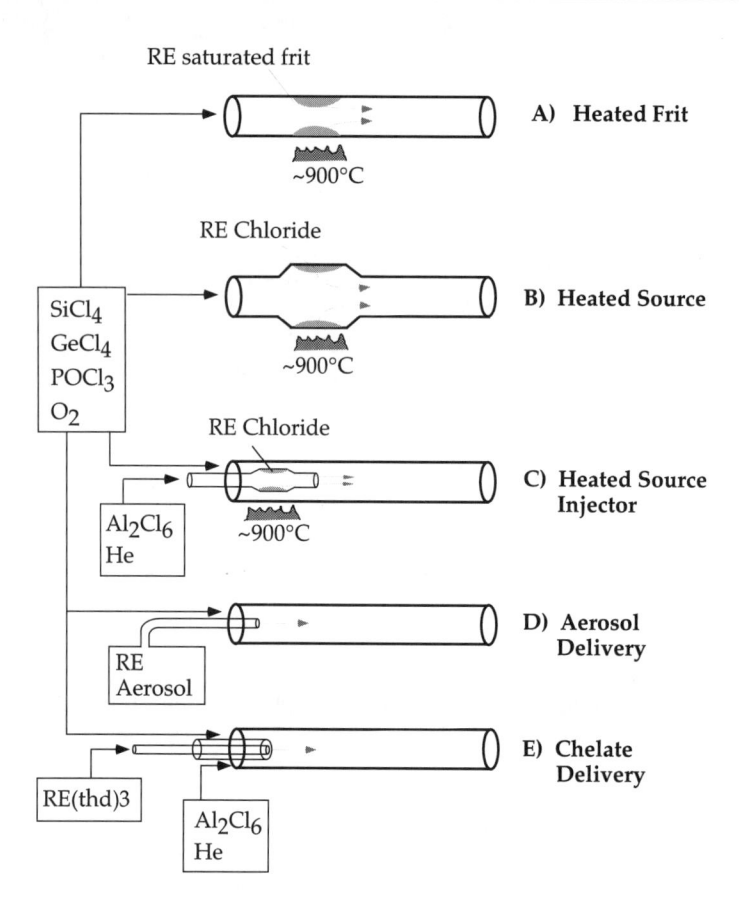

Figure 2.4: Low vapor pressure dopant delivery methods for MCVD. From reference [3].

(Figure 2.4, A–C), or by delivering the material as an aerosol or higher vapor pressure organic (Figure 2.4, D–E).

The heated frit source (Figure 2.4, A) is made by soaking a region of porous soot, previously deposited on the upstream inner wall of an MCVD tube, with a rare earth chloride-ethanol solution.[21] Having been heated to 900°C and allowed to dry, the sponge becomes a vapor source. Two other source methods (Figures 2.4, B and C) use the heated chloride directly as a source after dehydration.[1, 16, 22, 23] The dehydration is necessary in that most rare earth chlorides are in fact hydrated. The dehydration process may be accomplished by heating the material to near 900°C with a flow of Cl_2, $SOCl_2$, or SF_6. The attraction of the heated source injector method is that the rare earth reactant source is isolated from potentially unwanted reactions with the $SiCl_4$, $GeCl_4$, or $POCl_3$ index-raising reactants.

A variation of the heated chloride source method requires a two-step process referred to as transport-and-oxidation.[24] Using this material, the rare earth chloride is first transported to the downstream inner wall by evaporation and condensation, fol-

2.3. RARE EARTH DOPED FIBERS

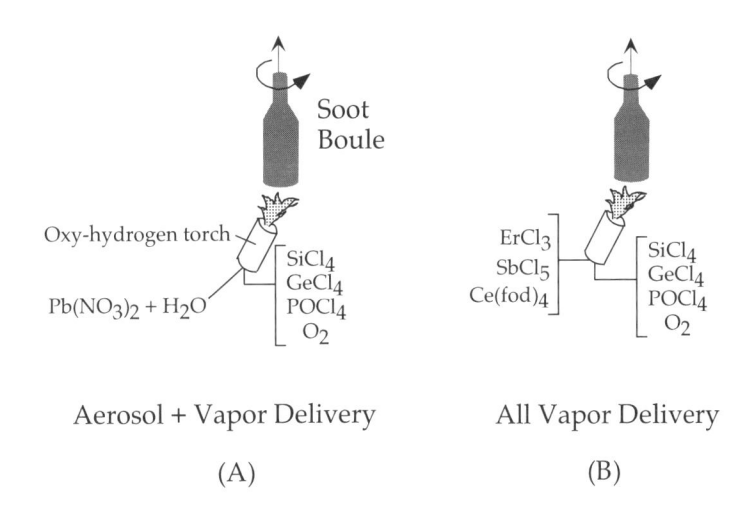

Figure 2.5: Low vapor pressure dopant delivery methods for VAD. From reference [3].

lowed by a separate oxidation step at higher temperatures. The resulting single-mode fiber structure of a P_2O_5-SiO_2 cladding and a Yb_2O_3-SiO_2 core is one of the few reported uses of a rare earth dopant as an index-raising constituent. A 1 mole % Yb_2O_3-SiO_2 core provided the 0.29 % increase in refractive index over the near silica index cladding.

The aerosol delivery method (Figure 2.4, D) overcomes the need for heated source compounds by generating a vapor at the reaction site.[25, 26, 27, 28] A feature of this method is the ability to create an aerosol at a remote location and pipe the resulting suspension of liquid droplets of rare earth dopant into the reaction region of the MCVD substrate tube with a carrier gas. The aerosols delivered this way were generated by a 1.5 MHz ultrasonic nebulizer commonly used in room humidifiers. Both aqueous and organic liquids have been delivered by this technique, allowing the incorporation of lead, sodium, and gallium as well as several rare earths. Given that most of the aerosol fluid materials contain hydrogen, dehydration after deposition is required for low OH content.

Vapor transport of rare earth dopants may also be achieved by using organic compounds that have higher vapor pressures than the chlorides, bromides, or iodides, as shown in Figure 2.3.[20] These materials can be delivered to the reaction in tubing heated to 200°C, rather than the several hundred °C requirements for chlorides. The application of this source to MCVD has been reported using three concentric input delivery lines (Figure 2.4, E).[29] Multiple rare earth doping and high dopant levels are reported with this method, along with background losses of 10 dB/km and moderate OH levels of near 20 ppm.

Rare earth vapor, aerosol, and solution transport may also be used to dope preforms fabricated by the OVD or VAD hydrolysis processes. Such doping may be achieved either during the soot deposition (see Figure 2.5) or after the soot boule has been created (see Figures 2.6 and 2.7).

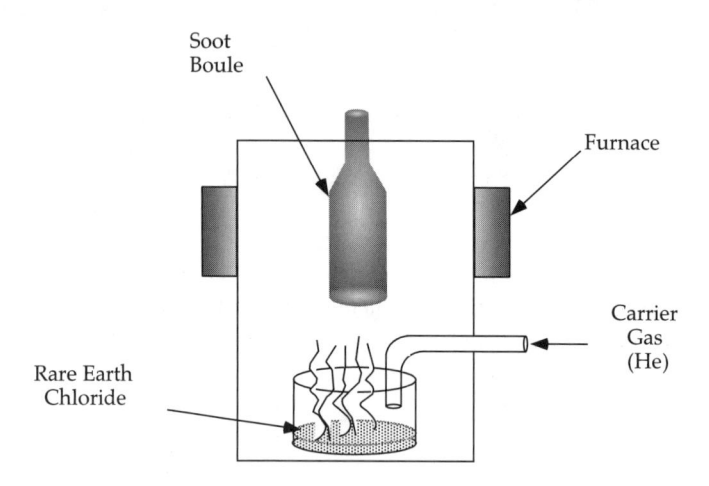

Figure 2.6: Postdeposition low vapor pressure dopant incorporation for VAD or OVD by vapor impregnation of a soot boule. From reference [3].

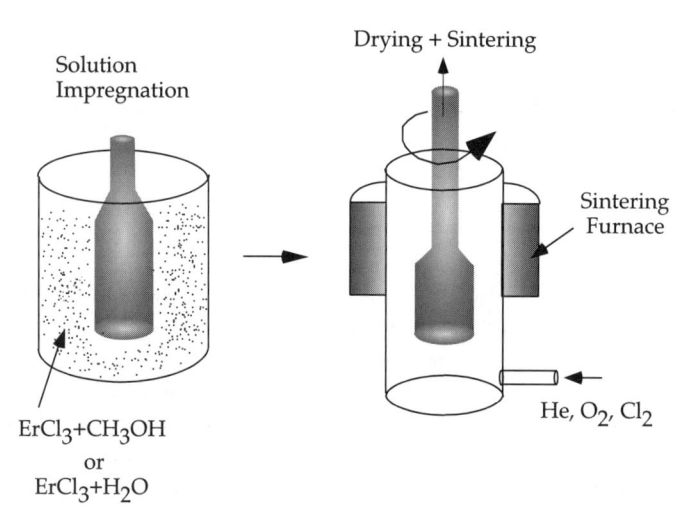

Figure 2.7: Postdeposition low vapor pressure dopant incorporation for VAD or OVD by solution impregnation followed by drying and sintering. From reference [3].

The introduction of low vapor pressure dopants to VAD was initially reported using a combination of aerosol and vapor delivery (see Figure 2.5, A).[30] The incorporation of cerium, neodymium, and erbium has been accomplished in the OVD method by introducing rare earth organic vapors into the reaction flame, as shown in Figure 2.5, B.[31, 32, 33] Cerium, for example, has been introduced as an organic source, cerium beta-diketonate (Ce(fod)$_4$). The high vapor pressure of this compound has allowed delivery to the reaction flame by a more traditional bubbler carrier system with heated delivery lines.[32] Another high vapor pressure organic compound used is the rare earth

2.3. RARE EARTH DOPED FIBERS

chelate RE(thd)$_3$ (2,2,6,6-tetramethyl-3,5-heptanedion).[29] Here a 1.0 wt. % Nd$_2$O$_3$ double clad fiber was fabricated for high output powers with background losses of 10 dB/km. Concentrations of Yb$_2$O$_3$ as high as 11 wt. %, as required for the double clad laser, were also achieved by this method.

Although no rare earth solution aerosol flame demonstrations have been reported, delivery of a nebulized aqueous solution of lead nitrate has been reported, showing the feasibility of this technique.[30] Likewise, there appears to be no report mentioning the delivery of rare earth chloride vapor to OVD or VAD flame reactions, although it is a likely method. The soot boules generated by OVD and VAD undergo a secondary drying and sintering process, which provides another opportunity for dopant incorporation. Rare earth dopant vapors have been incorporated in the glass by this postdeposition diffusion process during sintering, as shown in Figure 2.7.[34] Control of the incorporated dopant is achieved by a combination of dopant concentration in the sintering atmosphere and the pore size or density of the soot preform. Other dopants such as AlCl$_3$ and fluorine have been introduced this way as well.[34, 35]

In the VAD method, core rods of doped materials may be formed using a variety of methods. The cladding may then be deposited onto the core rod as a second operation, followed by sintering to form a preform. Core rods of rare earth doped materials have been fabricated using an RF plasma technique, as shown in Figure 2.8.[36] This technique was used to examine high-concentration doping for a small coil fiber amplifier. Using this method, a core glass composition with 1820 wt-ppm of Er and 5.0 wt. % of Al was fabricated while retaining an Er^{3+} fluoresence life time of 9.5 ms. An amplifier made with this glass achieved a power conversion efficiency of 75% at 1.5 μm.[36] A similar technique was used to fabricate bulk glasses of neodymium doped silica co-doped with aluminum and phosphorus.[37, 38]

2.3.2 Rare Earth Solution-Doping Methods

One of the first reported means for incorporating low-volatility halide ions into high-purity fiber preforms used a liquid phase "soot impregnation method."[39] A pure silica soot boule was first fabricated by flame hydrolysis with a porosity of 60% to 90% (pore diameter of 0.001 μm to 10 μm). The boule was immersed in a methanol solution of the dopant salt for one hour and then allowed to dry for 24 hours, after which the boule was sintered in a He-O$_2$-Cl$_2$ atmosphere to a bubble-free glass rod (see Figure 2.7). The dopant concentration was controlled by the ion concentration in the solution. This general technique, later referred to as molecular stuffing, has been used to incorporate Nd and Ca in silica.[40, 41]

A variation of this method combining MCVD and the solution-doping technique has more recently been reported (see Figure 2.9).[42] This method begins with the deposition of an unsintered (porous) layer of silica inside a silica tube by the MCVD process. The porous layer is then doped by filling the tube with an aqueous rare earth chloride solution; this solution is allowed to soak for nearly an hour, and then the solution is drained from the tube. The impregnated layer is dried at high temperatures in the presence of a flowing chlorine-oxygen mixture. Index-raising dopants such as aluminum have also been incorporated by this method.[43] Although this process would seem to be inherently less pure, it has produced doped fibers with background losses of

SiCl$_4$, AlCl$_3$, ErCl$_3$, O$_2$, Ar

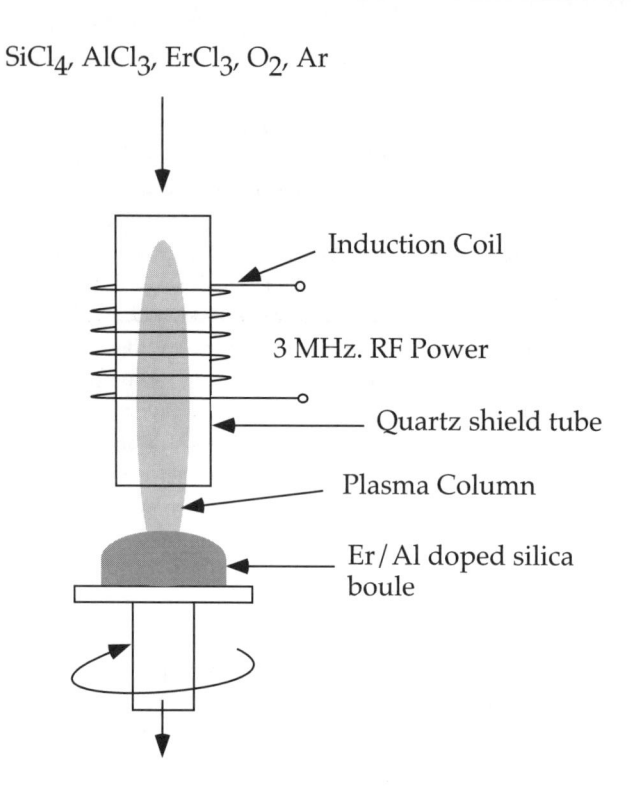

Figure 2.8: RF plasma method of making rare earth doped bulk glass for VAD core rods. From reference [36].

0.3 dB/km.[44] This general method has also been extended by replacing aqueous solutions with ethyl alcohol, ethyl ether, or acetone solvents for Al^{3+} and rare earth halide. Solubilities vary widely between the rare earth nitrates, bromides, and chlorides, and all are useful. Fibers made with these nonaqueous solvents contained a relatively low OH impurity level as evidenced by the less than 10 dB/m absorption at 1.38 μm.[45] Aqueous solution methods may also produce low OH fibers with proper dehydration techniques. A variation of the solution-doping method has been described that allows incorporation of up to 33 wt.% P$_2$O$_5$ as required for the Er-Yb co-doped materials.[46] The high concentration of P$_2$O$_5$ is accomplished using a pure acid melt of phosphoric acid (H$_3$PO$_4$) in combination with the rare earth metal ions in place of the aqueous solution to saturate the porous layer. The saturated porous layer is then "flash" heated to 1000°C in the presence of Cl$_2$ and O$_2$ to complete the reaction. Rare earth dopant levels of as high as 3 mole % have been achieved with this method.

As erbium-doped silica amplifiers were developed, it became clear that confinement of the dopant to the central region of the core was important for low threshold applications. In addition, the uniformity and homogeneity of the deposit were important. To improve these properties, another MCVD dopant method was developed, referred to as sol-gel dipcoating.[47] The process coats the inside of an MCVD sub-

2.3. RARE EARTH DOPED FIBERS

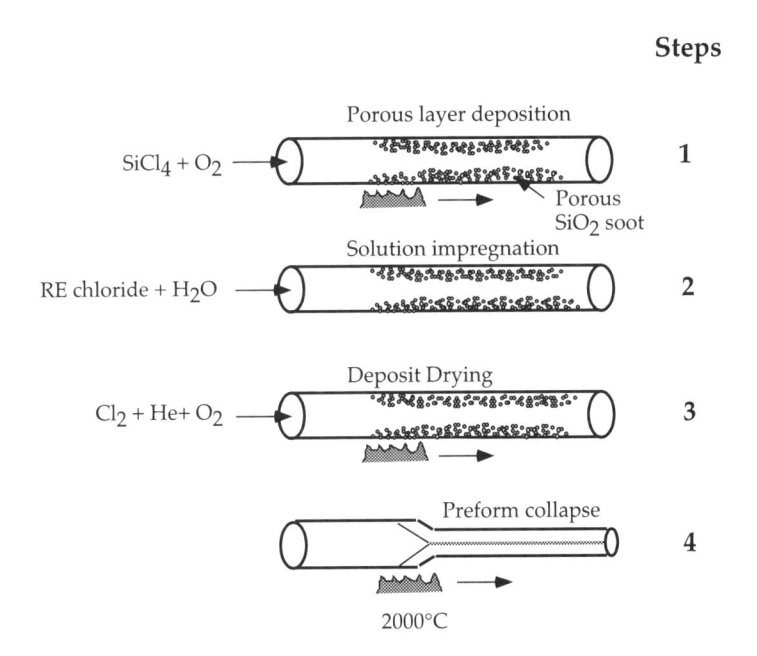

Figure 2.9: The MCVD solution doping method. Steps include (1) deposition of a porous soot layer, (2) solution impregnation of the porous layer, (3) drying of the porous layer, and (4) sintering of the layer and collapse of the preform. From reference [3].

strate tube with a rare earth containing sol, which subsequently gels and leaves a thin dopant layer (see Figure 2.10).

Rare earth and index-raising dopants may be combined. The coating sol is formed by hydrolysing a mixture of a soluble rare earth compound with $SiO(C_2H_5)_4$ (TEOS). The viscosity of the gel slowly increases with time as the hydrolysis polymerizes the reactants. Deposition of the film then proceeds by filling the inside of the MCVD support tube with the gel, followed by draining. The gel layer thickness is controlled by the viscosity of the gel, which in turn is determined by its age and the rate at which the gel liquid is drained. Film thickness of a fraction of a micrometer is typical, thereby allowing a well-confined dopant region. The coated tube is returned to the glass working lathe for subsequent collapse.

2.3.3 Rod and Tube Methods

The first optical fibers were made by drawing a preform assembly made of a core rod and cladding tube of the proper dimensions and indices.[48] Recent adaptations of this method have been demonstrated for making compound glass core compositions.[49] To retain the overall compatibility with communication grade doped silica fiber, a small compound glass rod is inserted into a thick-walled silica tube. The combination is then drawn at the high temperatures required by the silica tube. As a result, a few of the

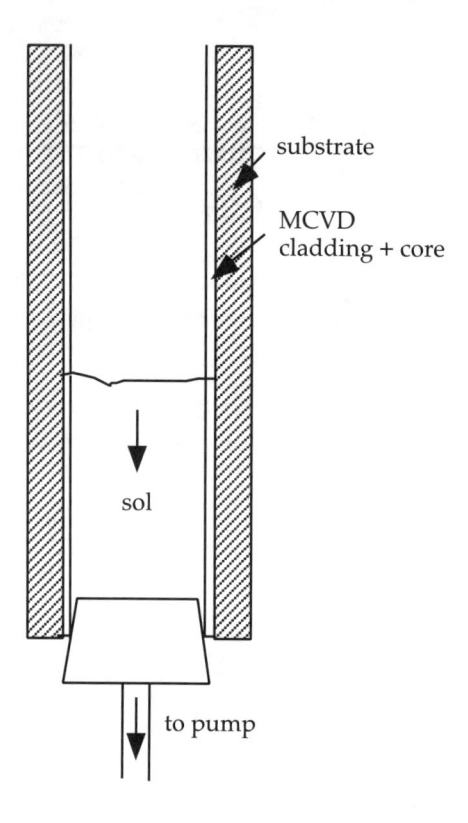

Figure 2.10: Sol-gel dipcoating process for MCVD. From reference [47].

less stable constituents of the compound glass are volatilized. In spite of this, lengths of fiber can be drawn that are long enough for practical use.

Given the interest in distributed erbium-doped fiber amplifiers, a need arises for a method to produce uniform and very low dopant levels. Solution-doping and outside process methods have been used for these low levels of dopants. In addition, a rod-and-tube-like technique was also devised to provide the low dopant levels. Here the rare earth is introduced into an MCVD preform as the core of a fiber with a 150 μm outside diameter and a 10 μm core diameter (see Figure 2.11).

Fibers fabricated by this method have demonstrated background losses as low as 0.35 dB/km at 1.62 μm. A ground state absorption at 1.53 μm of 1 dB/km (nearly 1 ppm Er) resulted from a 10 μm diameter seed fiber core with an erbium concentration of 1400 ppm.[50] This method is different than traditional rod-and-tube processes in that the rod is effectively dissolved in the host preform core. The dissolving seed is evident by the difference in fluorescence spectra of the seed composition and the resultant distributed amplifier fiber core composition.

2.4. PUMP-SIGNAL INTERACTION METHODS

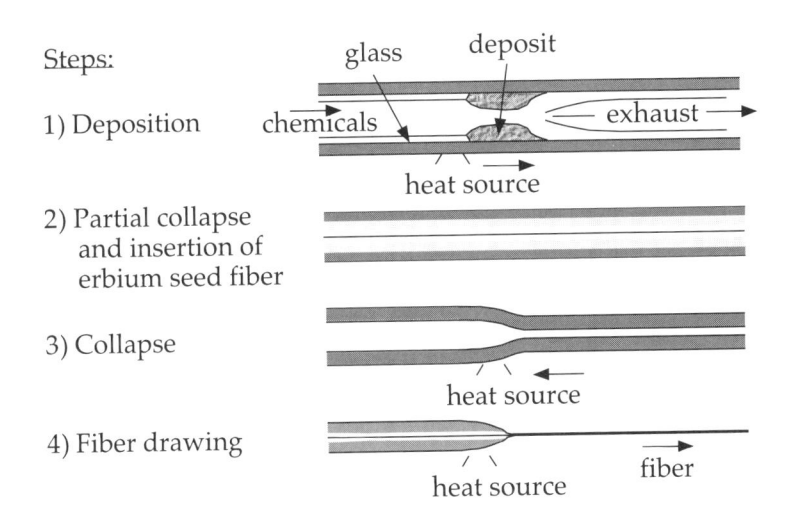

Figure 2.11: Seed fiber doping of the MCVD method. From reference [50] (©1991 IEEE).

2.4 PUMP-SIGNAL INTERACTION METHODS

There are a variety of configurations used to bring the pump, signal, and active media together. Although propagating the signal and pump together in a doped single-mode fiber is the configuration most widely used, Figure 2.12 shows that there are a number of alternative fiber and bulk optic device geometries of interest. The intersecting-beam approach, however simple, suffers from a short pump-signal interaction length. The zigzag approach increases the interaction length but does not result in a low threshold device, lacking the waveguide to confine the signal and pump. As alternatives to the now-traditional fiber-confined pump and signal, there are the evanescent field and double clad configurations that provide challenges to fabrication.

2.4.1 Evanescent Field

The wings or evanescent field of the optical signal guided by a single-mode fiber may be used to interact with an active material outside of the core region. This method suffers from the inherent disadvantage that the dopant resides where the optical field intensity is low. One approach to enhancing the pumping efficiency is to locally taper the guide, thereby causing the optical power to increase outside the glass material bounds of the fiber over a length of several millimeters. This tapering method has been used to demonstrate a 20 dB gain amplifier for a pump power below 1 W with a dye solution circulating around the tapered fiber region.[51] Both the signal at 750 nm and pump at 650 nm were copropagating in the core. Similarly, the active media can be incorporated in the cladding glass, as has been demonstrated for erbium or neodymium.[52, 53]. Reported gain for an erbium-doped cladding structure was 0.6 dB for a 1.55 μm signal with a 1.48 μm pump power of 50 mW.

Intersecting Beams

Active Media
Signal
Pump

Zig-Zag

Pump
Signal

Double-clad

Signal Pump

Evanescent Field

Pump + Signal
Dye Solution

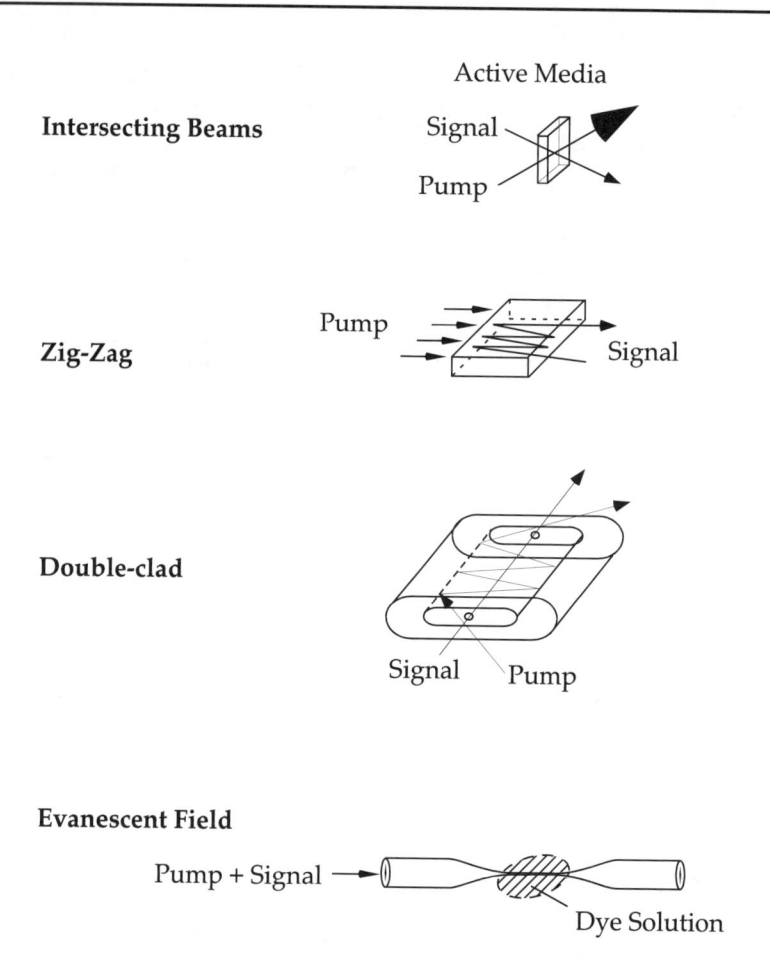

Figure 2.12: Pump-signal interaction configurations indicating alternate ways of combining pump and signal, compared to propagating both in a single mode fiber.

The evanescent field may also be accessed by polishing away a portion of the fiber cladding, thus creating a structure similar to a D-shaped fiber. Pulsed amplification of 22 dB for one such dye evanescent amplifier has been achieved.[54]

The pump power required for these devices to obtain a sizable gain far exceeds that needed for schemes in which active media is contained in the core. However, by using this evanescent interaction, active media, such as dyes that cannot be incorporated into a glass, can be explored.

2.4.2 Double Clad Fiber Design

Another approach to achieving interaction of guided pump light with an active medium uses a single-mode guide for the signal surrounded by a multimode pump guide. This design is called double clad given that there are two guiding structures, the core and the glass cladding surrounding the core. The glass cladding is surrounded by a low index polymer second cladding which allows it to become a guiding structure. Pump light is launched from the fiber end into the undoped cladding, propagating in a multimode fashion and interacting with the doped core as it travels along the fiber.[31, 55] Configurations with the core offset in a circular cladding and a core centered in an elliptical cladding have been demonstrated.[55] High brightness Nd^{3+} and Yb^{3+} fiber lasers based on the latter design have provided outputs of 5 W.[56] The effective absorption of the pump light in these asymmetric cladding structures with cladding numerical apertures of 0.5 has been studied previously.[57] Here, the effective numerical aperture of the rectangular fiber was observed to be less than that of circular designs. Also observed was an increase in both absorptive loss and Rayleigh scattering in the rectangular cladding design. The ability of the cladding configuration to accept light from the lower brightness multimode semiconductor pump-laser arrays allows pump redundancy and substantial pump power scaling not readily available to the all single-mode pump designs.

The most efficient conversion of pump to signal photons uses the design in which both the pump and signal are confined in the fiber core. This configuration has been made especially attractive by the availability of commercial, low insertion loss, low reflectivity fiber couplers, which can be chosen to combine a variety of pump and signal wavelengths onto a common output fiber.

2.5 COMPOSITIONS

The glass host composition impacts the solubility and environment of the rare earth dopant, which may in turn affect the fluorescence lifetime, absorption, emission, and excited-state absorption cross sections of the dopant transitions. Devices of general interest span rare earth concentrations of one to several thousand parts per million (ppm), resulting in devices of millimeters to kilometers long. For all designs, the rare earth should ideally be confined as a delta function in the center of the core for maximum gain per unit pump power. Practically speaking, there is a necessary trade-off between the confinement and the rare earth concentration. The more confined structures require a higher rare earth concentration for an equivalent length, eventually reaching the clustering limit for the particular host glass composition.[58] Clustering and high concentration effects are to be avoided in that they induce fluorescence quenching and reduces the perfomance of the device (see Chapter 4, sections 4.5.1 and 4.6, and Chapter 6, section 6.7). Commercially available bulk laser glasses are typically based on phosphate or multicomponent silicate host compositions, which have been developed to accommodate several weight percent concentrations of rare earth oxides without clustering. Host glasses compatible with this relatively high concentration of rare earth oxide, without giving rise to clustering, require the open chainlike structure of phosphate glasses. Alternatively, the addition of modifier ions (Ca, Na, K, Li, etc.) can be used to open the silicate structure and increase solubility, as shown in Figure 2.13.[59, 60]

The limitation due to clustering in a predominantly silica host without modifier ions has been well documented.[61] The maximum erbium concentration in silica for optimum amplifier performance has been suggested to be under 100 ppm.[62] However, a 14.4 dB gain, 900 ppm erbium-doped silica fiber amplifier has been reported,

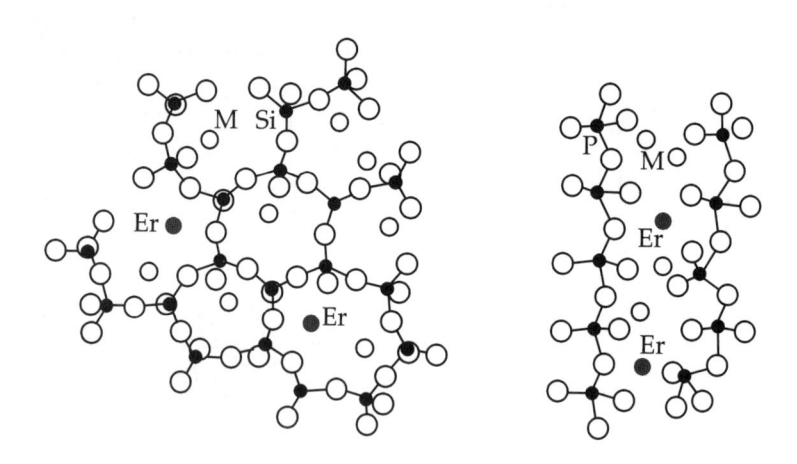

Figure 2.13: Placement of the rare earth ion Er^{3+} in a multicomponent glass structure. The modifier ions are indicated by the symbol M. The silica glass structure is on the left and the phosphate glass structure is on the right. From reference [59](reprinted with permission of Springer Verlag).

indicating that higher concentrations can produce useful devices.[63] When only index-raising dopants of germanium and phosphorus are used to modify the silica structure as in standard telecommunications fiber (typically 5.0 mole % GeO$_2$, 0.5 mole % P$_2$O$_5$), the limit to rare earth incorporation before the onset of fluorescence quenching is thought to be near 1000 ppm for Nd.[64] The addition of aluminum oxide, considered to be in part a network modifier, has also been used to improve the solubility of rare earth ions.[37, 43, 64, 65, 66, 67] Rare earth concentrations of 2% have been claimed without clustering using fiber host compositions of 8.5 mole % Al$_2$O$_3$.[43] A detailed study of the fluorescence quenching in Nd-doped SiO$_2$-Al$_2$O$_3$-P$_2$O$_5$ for Nd levels up to 15 wt% indicated that a phase separation of the rare earth oxide occurs at high concentrations.[64]

In addition to the solubility and fluorescence line shape, excited-state absorption can be substantially affected by the host composition.[43] This competing absorption phenomenon (see Chapter 6) can seriously diminish the efficiency of an active fiber device. A decrease in the excited-state absorption for erbium-doped fibers, when changing from a germano-silica host to an alumino-silica host, has been verified, demonstrating the importance of host selection for a given rare earth ion and/or laser transition.[43]

In addition to interactions between host and rare earth ions, it is necessary to consider background losses from impurity absorption and scattering mechanisms, which decrease the efficiency of the fiber device. The effect of internal loss is most dramatic in distributed amplifiers where pump light must travel long distances in the process of distributing gain. The magnitude of this effect has been calculated for the case of the distributed erbium-doped fiber amplifier by adding a pump and signal loss term to the rate equations.[50] These calculations indicate that for a 50 km distributed amplifier with near-optimum Er-dopant level for transparency, an increase in the background loss from 0.2 to 0.3 dB/km (10−15 dB per span) results in a fivefold increase in the required pump power. This emphasizes the need for state-of-the-art low background

2.6. PHYSICAL PROPERTIES

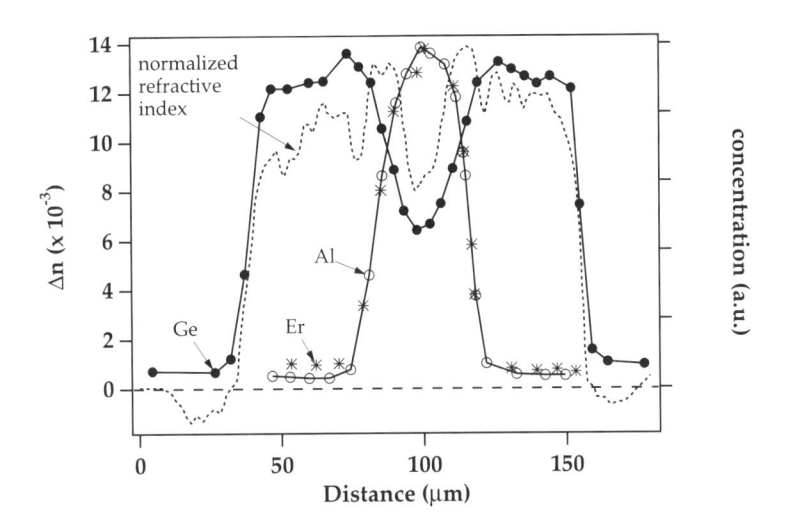

Figure 2.14: Distribution of dopant ions (Al, Ge, and Er) and refractive index profile for an erbium-doped fiber preform. From reference [21].

attenuation in fibers for distributed amplifiers. For devices a few meters long, the background loss may be kept near the fusion splice losses,—namely up to a few tenths of a dB,—with little reduction in performance.

2.6 PHYSICAL PROPERTIES

2.6.1 Fiber Refractive Index and Composition Profile

As we will discuss in Chapter 6, mode field diameter and confinement of the rare earth affect the performance of the fiber device. Both are controlled by the composition profile determined during fabrication. The typical index-raising dopants used in communication-compatible fiber are germanium, aluminum, and phosphorus. The incorporation of these dopants during deposition depends on a number of factors including partial pressure of dopant reactant, partial pressure of oxygen, and deposition temperature.[17, 68, 69] The dopants GeO_2 and P_2O_5 are unstable at high temperatures, causing a depletion and corresponding refractive index depression during the collapse stage of MCVD. A similar behavior has been reported for the incorporation of erbium in a GeO_2-SiO_2 host.[21] When Al_2O_3 or Al_2O_3 with P_2O_5 were added, however, no depletion of the erbium in the center was observed, as shown in Figure 2.14.

Several methods to eliminate the depression in refractive index from germanium depletion have been demonstrated for the MCVD process. One method uses a flow of $GeCl_4$ during substrate tube collapse, providing a concentration of germanium to compensate for the burn off of this dopant.[70] Alternatively, the central, germanium-depleted portion of the preform may be removed by etching with a flow of SiF_4 prior to the last collapse pass.[71] The high-temperature stability of Al_2O_3 results in a smooth refractive index profile when it is used as a dopant.[72, 73, 74] Concentrations of Al_2O_3 are limited to a few mole percent in binary compositions before crystallization occurs. For this reason GeO_2 is added, providing an additional rise in the refractive index. The ability of the MCVD process to control both the refractive index profile and the dopant placement within the core is shown by the experimental dopant distribution in Figure 2.14. This overlay of refractive index profile and corresponding dopant profiles indicates the complexity and control available in the fabrication process for doped fiber.

2.6.2 Strength and Reliability

The strength of high-quality optical fiber, a principal concern for reliability, is primarily determined by submicrometer flaws on the glass surface. Intrinsic strengths of silica fiber are near 800 kpsi. However, moisture and surface damage may easily bring this value down to tens of kpsi by static fatigue over a time scale of seconds to years. Commercial fibers are commonly proof-tested at 50 kpsi for terrestrial applications and 200 kpsi for undersea cables. Excluding the distributed amplifier, most of the laser or amplifier devices are typically packaged in coils a few centimeters in diameter with negligible strength reduction over decades. In the interest of creating small packaged amplifiers, however, hermetic-coated erbium-doped fibers have been fabricated for spools as small as 15 mm in inside diameter.[75]

Other factors that are important determinants of reliability include mechanisms that may reduce the transparency of the fiber with time, notably high-energy radiation and hydrogen in-diffusion. Radiation measurements of erbium-doped fiber amplifiers have been reported indicating a predicted gain reduction of less than 0.1 dB for a typical terrestrial exposure dose rate of 0.5 rad/year over 25 years. A wavelength dependence of the reduced gain was observed, with 1.536 μm signals suffering nearly twice the loss as 1.555 μm signals.[76] Measurements of radiation induced degradation in operational erbium-doped fiber amplifier indicate added losses of approximately 5×10^{-6} dB to 150×10^{-6} dB per km per rad.[77, 78, 79] These induced loss values are nearly 25 to 100 times higher than for standard telecomunications grade fiber. This is to be expected given the presence of Al in the erbium-doped fiber, a known source of radiation-induced color centers.

Hydrogen-induced losses in these doped fibers are generally higher than standard telecomunications grade fiber as well.[80] Although Al is thought to be the ion causing this higher sensitivity, an Al-La combination is reported to have higher immunity to hydrogen. An exposure of 0.1 atm of hydrogen at 150°C did not result in any degradation of a working amplifier pumped at 1480 nm.[81]

2.6.3 Alternate Glass Host Fabrication

In addition to silica based glasses, a number of other glass forming systems have been considered as hosts for rare earth dopants: fluorides, chlorides, sulfides, iodides, selenides, tellurites and germanates. There are several reasons for considering the use of glass hosts other than silica. The glass host, through the characteristic phonon energies of the lattice, governs the nonradiative transition rates, as discussed in Chapter 4, section 4.4.2. This influences the lifetimes of the levels and can make a key difference in the ability of an ion to possess an amplifying transition. For example, Pr^{3+} is an

2.6. PHYSICAL PROPERTIES

effective amplifier ion at 1.3 μm in a fluoride host, and not in silica, only because the lifetime of the $^1G_4$ state is long enough in fluorides but not in silica (see Chapter 10, section 10.2.2). The host material will also impact the transition rates for energy transfer between rare earth ions, in the event of co-doping. Another factor is that since the energy splitting of the Stark levels of each rare earth ion multiplet is dependent on the host, one expects a change in the spectral shape and strength of various transitions. A fluoride host, for example, gives rise to a flatter transition for the Er^{3+} transition at 1.5 μm as compared to a silica host (see Chapter 8). The absolute values of the transition cross section will also change. The excited state cross sections will also be modified, perhaps allowing for the reduction in strength of a competing ESA effect in an amplifier. We will discuss in this section fabrication methods for some glass hosts other than silica.

Glasses made from halides (F, Cl, Br, I, At) have been considered as a potential replacement for doped silica as a low attenuation material.[82, 83, 84, 85] It has been predicted that fluoride based glasses could have absorption losses substantially below silica based glasses, 10^{-2} to 10^{-3} dB/km at wavelengths near 3 μm. Impurity levels of 0.1 parts per billion would be necessary to achieve this low loss along with a low scattering loss component. A substantial effort toward realizing this occured during the late 1970s to mid 1980s without a practical substitute for silica. The logic behind using the halide materials was based on the expectation of both a lower Rayleigh scattering loss and longer wavelength infrared absorption edge.[86] This longer wavelength absorption edge is the result of a lower frequency resonant absorption due to the heavier atoms with low bond strengths between them. This is in contrast to Si or Ge bonded strongly with oxygen as is the case for traditional oxide glasses.

Fabrication and use of halide glass fiber is substantially more difficult than silica with issues of purification, fiber drawing, and fiber reliability. Silica is inherently more stable as a glass than halides, as it is less likely to crystallize, and it has a broad temperature range where its viscosity is suitable for drawing into fiber. For example ZBLAN, a suitable ZrF_4 based glass, may be drawn into fiber over a temperature range of only 30 degrees.[87] Fabrication of pure, silica based fiber is readily achieved by the gas phase reaction of pure sources of Si, Ge, P, etc., which are liquids at room temperature. Control of the partial pressure of oxygen during high temperature processing also diminishes the incorporation of the key optical absorbing species of transition metals (Cu^{2+}, Ni^{2+}, Fe^{2+}) and OH.

For low loss, fluorides also require very low levels of transition metals and hydrogen impurities as well as low levels of oxygen to diminish OF^{-3} and O^{-2} levels. Gas phase reaction and deposition of halide glasses has been performed with difficulties of stable source materials, corrosive nature of reactants such as HF, and efficiency of deposition.[88]

A typical ZBLAN fluoride composition for the core and cladding is (the numbers preceding each compound denote the mole percentages):

- $54.9ZrF_4$ - $17.7BaF_2$ - $3.9LaF_3$ - $3.7AlF_3$ - $14.7NaF$ - $0.2InF_3$ - $4.9PbF_3$ (core)

- $54.9ZrF_4$ - $22.6BaF_2$ - $3.9LaF_3$ - $3.7AlF_3$ - $14.7NaF$ - $0.2InF_3$ (cladding)

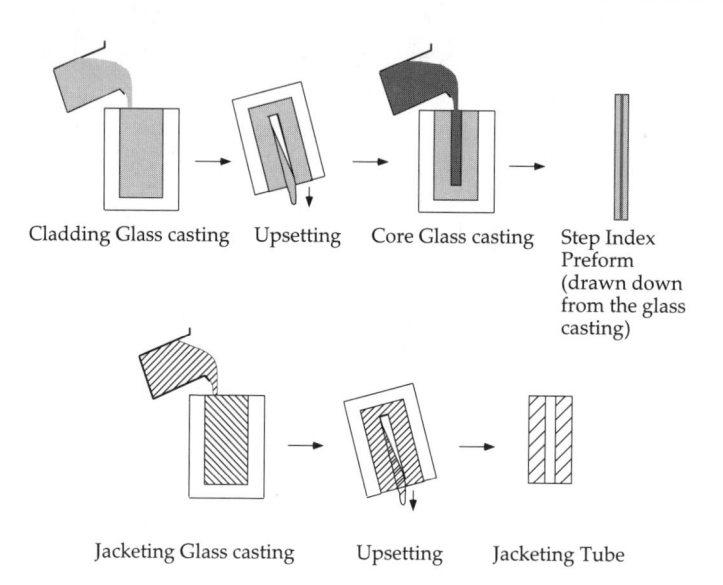

Figure 2.15: The casting steps used to fabricate the preform core and cladding (top) and jacketing tube (bottom) used in the subsequent rod-in-tube or jacketing method of fluoride preform fabrication. The upsetting step removes the central unsolidified portion of the cladding or jacketing glass. From reference [91] (©1984 IEEE).

Additions of PbF_2 and BiF_3 are made to the base glass to raise the refractive index of the core and additions of LiF, NaF and AlF_3 are used to lower the index for the cladding.[89]

In addition to the difficulties in fluoride glass making from the point of view of fundamental glass stability, rare earth doped fibers must be single-mode to provide for high pump and signal intensities for amplifier purposes. As an example, a high performance single-mode Pr^{3+} doped ZBLAN fiber may have a core diameter of 1.7 μm and numerical aperture of 0.39.[90] Accurate control of this small core diameter along the preform length presents a fabrication challenge. Methods of making halide glass multimode fiber by double crucible, vapor deposition, or rotational casting are not as readily used in fabricating the small single-mode fiber core required to produce a high pump and signal power intensities.[89] Single-mode halide glass fibers have been made using the built-in casting and jacketing technique, as well as the double crucible method shown in Figures 2.15, 2.16, 2.17, and 2.18.

Building up glass rods and tubes to form a preform which is then drawn into fiber can be accomplished in a number of ways. One method independently fabricates a preform and a jacket. The preform consists of a core surrounded by a cladding with a cladding to core ratio of at least 5:1 as shown in the upper portion of Figure 2.15. A jacketing tube which provides the bulk of the fiber is independently fabricated by casting the glass in a mold, upsetting the inner molten portion to create a tube, as shown in the lower portion of Figure 2.15. The inner diameter of the jacket tube must be made to closely fit the preform rod for subsequent assembly and fiber drawing (see Figure 2.16). This close fit may be achieved by ultrasonically boring and polishing,

2.6. PHYSICAL PROPERTIES 33

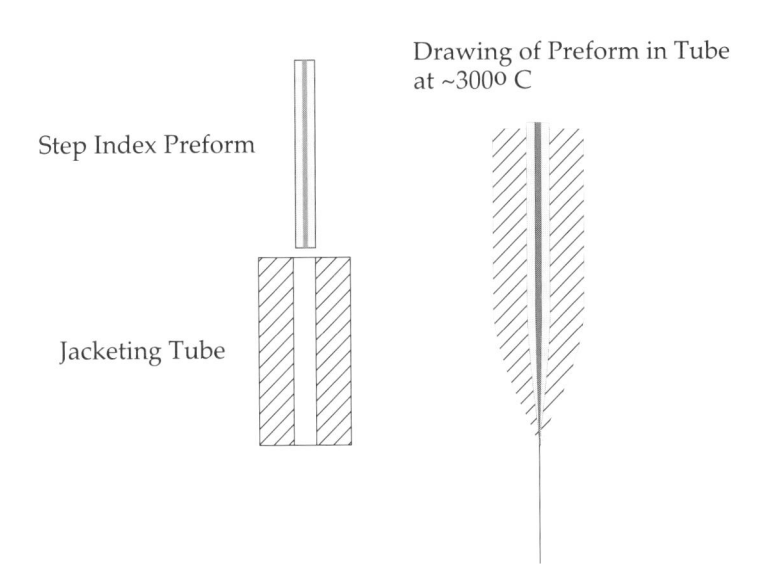

Figure 2.16: The jacketing method of fabricating a fluoride single-mode fiber. A step index core with a thin cladding is inserted into the jacketing tube followed by fiber drawing. From reference [91] (©1984 IEEE).

although a smooth and clean surface is difficult to realize. Smooth and clean surfaces are necessary to prevent crystallization and bubble formation during drawing.

An alternate way of creating a smooth inner surface jacketing tube is the rotational casting method as shown in Figure 2.17.[92] Rotational casting is performed by first pouring the fluoride glass melt into a cylindrical mold in a near vertical orientation. The mold is then sealed and positioned horizontally where it is then rotated at several thousand RPM untill the glass has solidified. A variation on this rotational casting method surrounds the mold with a reduced pressure chamber. Multimode fiber made using this technique demonstrated a minimum loss of 0.65 dB/km at a wavelength of 2.59 μm.[93]

Yet another method used to combine the core and cladding uses the contraction of the cooling cladding glass to draw a tube of the core into the axis of a cladding. This method is referred to as the suction casting method.[94] The fluoride mixtures are melted at 900°C, cast into molds preheated to 250°C, and drawn at 370°C.

Drawing of halide glass preforms into a fiber is performed at a rate of 10 m/min (compared to 20 m/s for silica). The jacketing tube is evacuated to 0.5 atm. to reduce the occurence of bubbles and incorporation of airborn impurities. A teflon-FEP tube surrounding the preform prior to drawing may be used to protect the fiber surface from chemical and mechanical degradation. This opportunity to apply a polymer coating material prior to drawing, possible for these low melting glasses, is not an option at the silica drawing temperatures of greater than 1000°C. Alternatively, a polymer coating may be applied following drawing, as typical of the silica drawing process.

The double crucible method produces a fiber by directly pulling the glass from coaxial reservoirs containg the core and cladding glass (see Figure 2.18).[95] Given

34   CHAPTER 2. OPTICAL FIBER FABRICATION

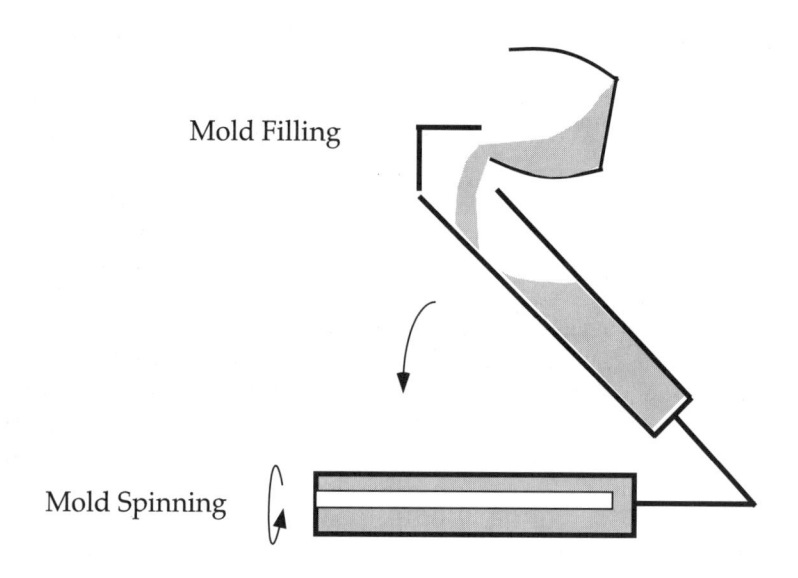

Figure 2.17: The rotational casting method for fabrication of a cladding tube. The molten glass is poured into a near vertical stationary mold and from there moved to a horizontal spinning position. This creates a cylindrical tube with a smooth inside surface. From reference [92].

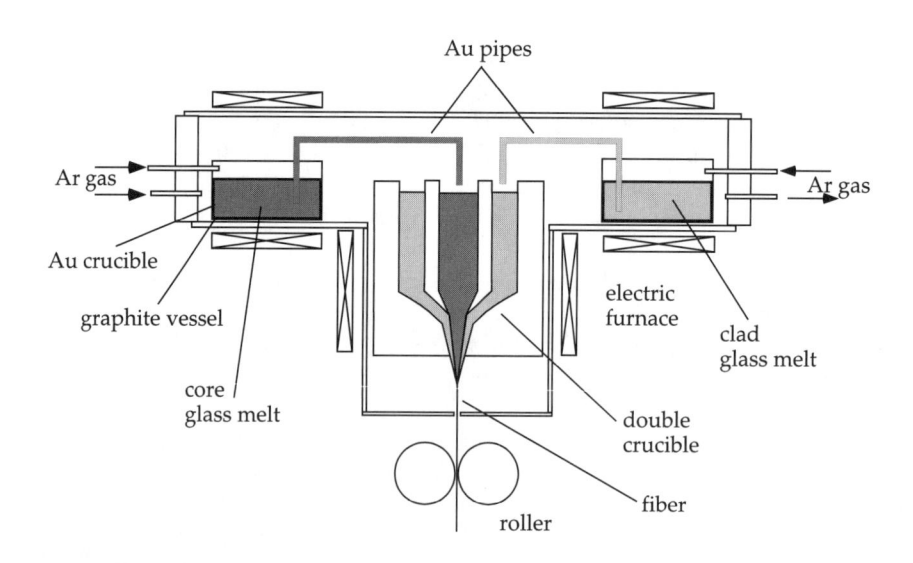

Figure 2.18: The double crucible method of fluoride fiber fabrication. From reference [95].

2.6. PHYSICAL PROPERTIES

the reactive nature of the fluoride glass, it is necessary to first melt the glass contained in Au crucibles surrounded by a dry and inert atmosphere. These melts are then transferred to the concentric reservoirs from which the fiber is drawn. As in the preform draw method, the glass is drawn at a speed of 15 m/min and a glass temperature of 320°C.[95]

The fabrication methods just described have allowed fiber amplifiers with novel properties to be realized. We discuss in chapters 4 and 8 applications where Er^{3+} has been doped in tellurite and fluoride glasses. This flattens the gain spectrum of the Er^{3+} ion near 1.5μm. Some of these amplifiers have been used in WDM transmission experiments. Chapter 10 discusses the 1.3μm amplifier properties of Pr^{3+} doped in fluoride glass. As amplifiers with desired spectroscopic properties are developed, alternate glass host fabrication methods will be instrumental in fabricating them.

Finally, we mention that special fabrication methods are employed in making rare earth doped planar waveguides. These guides have been fabricated using plasma-enhanced vapor deposition, ion exchange, implantation, RF-sputtering, chelate transport, and flame hydrolysis.[96, 97, 98, 99, 100, 101] The challenge for these waveguide fabrication methods is the ability to dope these structures with high concentrations of rare earths without the deleterious effects of ion-ion interactions. The hope is that some day these amplifiers can become part of photonic integrated circuits. As glass fabrication techniques progress, the development of amplifiers with desired properties, both spectroscopic and size, should further the goals of all optical networking.

Bibliography

[1] S. Poole, D. N. Payne, R. J. Mears, M. E. Fermann, and R. Laming, *J. Light. Tech.* **LT-4**, 870 (1986).

[2] P. Urquhart, *IEE Proc. J. Optoelectronics* **135**, 385 (1988).

[3] J. Simpson, "Fabrication of rare earth doped glass fibers," in *Fiber Laser Sources and Amplifiers*, M. J. F. Digonnet, Ed., *Proc. SPIE* **1171**, pp. 2–7 (1989).

[4] D. J. DiGiovanni, "Fabrication of rare earth doped optical fiber" in *Fiber Laser Sources and Amplifiers II*, Michel J. F. Digonnet, Ed., *Proc. SPIE* **1373**, pp. 2–8 (1990).

[5] B. J. Ainslie, *J. Light Tech.* **9**, 220 (1991).

[6] T. Izawa, S. Kobayashi, S. Sudo, and F. Hanawa, "Continuous fabrication of high silica fiber preform", in *1977 International Conference on Integrated Optics and Optical Fiber Communication*, Proceedings, Part 1, pp. 375–378.

[7] D. B. Keck and P. C. Schultz, U.S. Patent 3,737,292 (1973).

[8] J. B. MacChesney, P. B. O'Connor, F. V. DiMarcello, J. R. Simpson, and P. D. Lazay, "Preparation of low loss optical fibers using simultaneous vapor phase deposition and fusion," *10th International Congress on Glass* (Kyoto, Japan), Proceedings, Part 1, pp. 6-40–6-45 (1974).

[9] J. Koenings, D. Kuppers, H. Lydtin, and H. Wilson, "Deposition of SiO_2 with low impurity content by oxidation of $SiCl_4$ in nonisothermal plasma," *Proceedings of the Fifth International Conference on Chemical Vapor Deposition*, pp. 270–281 (1975).

[10] E. M. Dianov, K. M. Golant, V. I. Karpov, R. R. Khrapko, A. S. Kurkov, and V. N. Protopopov, "Efficient amplification in erbium-doped high-concentration fibers fabricated by reduced-pressure plasma CVD," in *Optical Fiber Communication Conference*, Vol. 8, OSA Technical Digest Series (Optical Society of America, Washington, D.C., 1995), pp. 174–175.

[11] L. Stensland and P. Gustafson, *Ericsson Rev.* **4**, 152 (1989).

[12] J. B. MacChesney and D. W. Johnson Jr., "Large silica bodies by sol-gel for production of optical fibers," in *Optical Fiber Communication Conference*, Vol. 6, 1997 OSA Technical Digest Series (Optical Society of America, Washington, D.C., 1997), p. 3.

[13] S. E. Miller and I. P. Kaminow, Eds., *Optical Fiber Telecommunications* **II** (Academic Press, New York, 1988).

[14] T. Li, Ed., *Optical Fiber Communications* I: *Fiber Fabrication* (Academic Press, New York, 1985).

[15] D. L. Wood, K. L. Walker, J. B. MacChesney, J. R. Simpson, and R. Csencsits, *J. of Light. Tech.* **LT-5**, 277 (1987).

[16] J. R. Simpson, and J. B. MacChesney, "Alternate dopants for silicate waveguides", in *Fifth Topical Meeting on Optical Fiber Communications* (Optical Society of America, Washington, D.C., 1982), p. 10.

[17] P. Kleinert, D. Schmidt, J. Kirchhof, and A. Funke, *Kristall und Technik* **15**, K85, (1980).

[18] E. Shimazaki and N. Kichizo, *Z. Anorg. Allg. Chem.* **314**, 21 (1962).

[19] F. H. Spedding and A. H. Daane, *The Rare Earths* (Wiley, New York, 1961), p. 98.

[20] J. E. Sicre, J. T. Dubois, K. J. Eisentraut, and R. E. Sievers, *J. Am. Chem. Soc.* **91**, 3476 (1969).

[21] B. J. Ainslie, J. R. Armitage, C. P. Craig, and B. Wakefield, "Fabrication and optimisation of the erbium distribution in silica based doped fibres," *Fourteenth European Conference on Optical Communications*, Proceedings Part 1, pp. 62−65 (1988).

[22] A. Wall, H. Posen, and R. Jaeger, "Radiation hardening of optical fibers using multidopants Sb/P/Ce," in *Advances in Ceramics Vol. 2, Physics of Fiber Optics*, Bendow and Mitra, Eds. pp. 393−397 (1980).

[23] S. B. Poole, D. N. Payne, and M. E. Fermann, *Elect. Lett.* **21**, 737 (1985).

[24] M. Watanabe, H. Yokota, and M. Hoshikawa, "Fabrication of Yb_2O_3-SiO_2 core fiber by a new process," in *Eleventh European Conference on Optical Communications*, Technical Digest, Volume 1, pp. 15−18 (1985).

[25] R. Laoulacine, T. F. Morse, P. Charilaou, and J. W. Cipolla, "Aerosol delivery of non-volatile dopants in the MCVD system," *Extended Abstracts of the AIChE Annual Meeting*, Washington DC, paper 62B (1988).

[26] T. F. Morse, L. Reinhart, A. Kilian, W. Risen, and J. W. Cipolla, "Aerosol doping technique for MCVD and OVD," in *Fiber Laser Sources and Amplifiers*, M. J. F. Digonnet, Ed., *Proc. SPIE* **1171**, pp. 72−79 (1989).

[27] T. F. Morse, A. Kilian, L. Reinhart, W. Risen Jr., and J. W. Cipolla, Jr., "Aerosol techniques for fiber core doping," in *Optical Fiber Communication Conference*, Vol. 4, 1991 OSA Technical Digest Series (Optical Society of America, Washington, D.C., 1991), p. 63.

[28] T. F. Morse, T. F., A. Kilian, L. Reinhart, W. Risen, J. W. Cipolla *J. Non-Cryst. Solids* **129**, 93 (1991).

[29] R. P. Tumminelli, B. C. McCollum, and E. Snitzer, *J. Light. Tech.* **LT-8**, 1680 (1990).

[30] K. Sanada, T. Shioda, T. Moriyama, K. Inada, S. Takahashi and, M. Kawachi, "PbO doped high silica fiber fabricated by modified VAD," *Sixth European Conference on Optical Communications*, Proceedings, pp. 14–17 (1980).

[31] P. L. Bocko, "Rare-earth doped optical fibers by the outside vapor deposition process," in *Optical Fiber Communication Conference*, Vol. 5, 1989 OSA Technical Digest Series (Optical Society of America, Washington, D.C., 1989), p. 20.

[32] D. A. Thompson, P. L. Bocko, J. R. Gannon, "New source compounds for fabrication of doped optical waveguide fibers," in *Fiber Optics in Adverse Environments II*, R. A. Greenwell, Ed., *Proc. SPIE* **506**, pp. 170–173 (1984).

[33] A. A. Abramov, M. M. Bubnov, E. M. Dianov, A. E. Voronkov, A. N. Guryanov, G. G. Devjatykh, S. V. Ignatjev, Y. B. Zverev, N. S. Karpychev, and S. M. Mazavin, "New method of production of fibers doped by rare earths," in *Conference on Lasers and Electro Optics*, Vol. 7, 1990 OSA Technical Digest Series (Optical Society of America, Washington, D.C., 1990), pp. 404–406.

[34] M. Shimizu, F. Hanawa, H. Suda, and M. Horiguchi, *J. Appl. Phys.* **28**, L476 (1989).

[35] W. H. Dumbaugh and P. C. Schultz, "Method of producing glass by flame hydrolysis," U.S. Patent 3,864,113 (1975).

[36] A. Wada, D. Tanaka, T. Sakai, T. Nozawa, K. Aikawa, and R. Tamauchi, "Highly Er-doped alumino-silicate optical fiber synthesized by high temperature plasmas and its application to compact EDF coils," in *Optical Amplifiers and Their Applications*, Vol. 17, 1992 OSA Technical Digest Series (Optical Society of America, Washington, D.C., 1992), pp. 222–225.

[37] K. Arai, H. Namikawa, K. Kumata, T. Honda, Y. Ishii, and T. Handa, *J. Appl. Phys.* **59**, 3430 (1986).

[38] H. Namikawa, K. Arai, K. Kumata, Y. Ishii and H. Tanaka, *Jpn. J. Appl. Phys.* **21**, L360 (1982).

[39] P. C. Schultz, *J. Am. Ceram. Soc.* **57**, 309 (1974).

[40] T. Gozen, Y. Kikukawa, M. Yoshida, H. Tanaka, and T. Shintani, "Development of high Nd^{3+} content VAD single-mode fiber by molecular stuffing technique," in *Optical Fiber Communication Conference*, Vol. 1, 1988 OSA Technical Digest Series (Optical Society of America, Washington, D.C., 1988), p. 98.

[41] M. A. Saifi, M. J. Andrejco, W. A. Way, A. Von Lehman, A. Y. Yan, C. Lin, F. Bilodeau, and K. O. Hill, "Er^{3+} doped GeO_2-CaO-Al_2O_3 silica core fiber amplifier pumped at 813 nm," in *Optical Fiber Communication Conference*, Vol.

4, 1991 OSA Technical Digest Series (Optical Society of America, Washington, D.C., 1991), p. 198.

[42] J. E. Townsend, S. B. Poole, and D. N. Payne, *Elect. Lett.* **23**, 329 (1987).

[43] S. B. Poole, "Fabrication of Al_2O_3 co-doped optical fibres by a solution-doping technique," in *Fourteenth European Conference on Optical Communications*, Proceedings, Part 1, pp. 433–436 (1988).

[44] C. C. Larsen, *Lucent Technologies, Specialty Fiber Products*, Denmark, private communication.

[45] L. Cognolato, B. Sordo, E. Modone, A. Gnazzo, and G. Cocito, "Aluminum/erbium active fibre manufactured by a non-aqueous solution doping method," in *Fiber Laser Sources and Amplifiers*, M. J. F. Digonnet, Ed., *Proc. SPIE* **1171**, pp. 202–208 (1989).

[46] A. L. G. Carter, S. B. Poole, and M. G. Sceats, *Elect. Lett.* **28**, 2009 (1992).

[47] D. J. DiGiovanni and J. B. MacChesney, "New optical fiber fabrication technique using sol-gel dipcoating," in *Optical Fiber Communication Conference*, Vol. 4, 1991 Technical Digest Series (Optical Society of America, Washington, D.C., 1991), p. 62.

[48] E. Snitzer, *J. Appl. Phys.* **32**, 36 (1961).

[49] E. Snitzer and R. Tumminelli, *Opt. Lett.* **14**, 757 (1989).

[50] J. R. Simpson, H. -T. Shang, L. F. Mollenauer, N. A. Olsson, P. C. Becker, K. S. Kranz, P. J. Lemaire, and M. J. Neubelt, *J. Light. Tech.* **LT-9**, 228 (1991).

[51] H. S. Mackenzie and F. P.Payne, *Elect. Lett.* **26**,130 (1990).

[52] I. Sankawa, H. Izumita, T. Higashi, and K. Ishihara, *IEEE Phot. Tech. Lett.* **2**, 41 (1990).

[53] A. V. Astakhov, M. M. Butusov, and S. L. Galkin, *Opt. Spektrosk.* **59**, 913 (1985).

[54] W. V. Sorin, K. P. Jackson, and H. J. Shaw, *Elect. Lett.* **19**, 820 (1983).

[55] H. Po, E. Snitzer, R. Tumminelli, L. Zenteno, F. Hakimi, N. M. Cho, and T. Haw, "Double clad high brightness Nd fiber laser pumped by GaAlAs phased array," in *Optical Fiber Communication Conference*, Vol. 5, 1989 OSA Technical Digest Series (Optical Society of America, Washington, D.C., 1989), pp. 220–223.

[56] H. Po, J. D. Cao, B. M. Laliberte, R. A. Minns, R. F. Robinson, B. H. Rockney, R. R. Tricca, and Y. H. Zhang, *Elect. Lett.* **17**, 1500 (1993).

[57] A. Liu and K. Ueda, *Opt. Eng.* **11**, 3130 (1996).

[58] E. Desurvire, J. L. Zyskind, and C. R. Giles, *J. Light. Tech.* **LT-8**, 1730 (1990).

[59] T. S. Izumitani, *Optical Glass* (American Institute of Physics, New York, 1986), pp. 162–172, originally published as *Kogaku Garasu* (Kyoritsu Shuppan, Ltd., 1984).

[60] T. Yamashita, S. Amano, I. Masuda, T. Izumitani, and A. Ikushima, in *Conference on Lasers and Electro Optics*, Vol. 7, 1988 OSA Technical Digest Series (Optical Society of America, Washington, D.C., 1988) p. 320.

[61] E. Snitzer, *Appl. Opt.* **5**, 1487 (1966).

[62] M. Shimizu, M. Yamada, M. Horiguchi, and E. Sugita, *IEEE Phot. Tech. Lett.* **2**, 43 (1990).

[63] M. Suyama, K. Nakamura, S. Kashiwa, and H. Kuwahara, "14.4-dB Gain of erbium-doped fiber amplifier pumped by 1.49-μm laser diode," in *Optical Fiber Communication Conference*, Vol. 5, 1989 OSA Technical Digest Series (Optical Society of America, Washington, D.C., 1989), pp. 216–219.

[64] B. J. Ainslie, S. P. Craig, S. T. Davey, D. J. Barber, J. R. Taylor, and A. S. L. Gomes, *J. Mater. Sci. Lett.* **6**, 1361 (1987).

[65] J. Stone and C. R. Burrus, *Appl. Phys. Lett.* **23**, 388 (1973).

[66] J. B. MacChesney and J. R. Simpson, "Optical waveguides with novel compositions," in *Optical Fiber Communication Conference*, 1985 OSA Technical Digest Series (Optical Society of America, Washington, D.C., 1985), p. 100.

[67] B. J. Ainslie, S. P. Craig, and S. T. Davey, *Mater. Lett.* **5**, 143 (1987).

[68] D. L. Wood, K. L. Walker, J. B. MacChesney, J. R. Simpson, and R. Csencsits, *J. Light. Tech.* **LT-5**, 277 (1987).

[69] D. J. Digiovanni, T. F. Morse, and J. W. Cipolla, *J. Light Tech.* **LT-7**, 1967 (1989).

[70] T. Akamatsu, K. Okamura, and Y. Ueda, *Appl. Phys. Lett.* **31**, 515 (1977).

[71] S. Hopland, *Elect. Lett.* **14**, 7574 (1978).

[72] Y. Ohmori, F. Honawa, and M. Nakahara, *Elect. Lett.* **10**, 410 (1974).

[73] J. R. Simpson and J. B. MacChesney, *Elect. Lett.* **19**, 261 (1983).

[74] C. J. Scott, "Optimization of composition for Al_2O_3/P_2O_5 doped optical fiber," in *Optical Fiber Communication Conference*, 1984 OSA Technical Digest Series (Optical Society of America, Washington, D.C., 1984), pp. 70–71.

[75] A. Oyobe, K. Hirabayashi, N. Kagi, and K. Nakamura, "Hermetic erbium-doped fiber coils for compact optical amplifier modules," in *Optical Fiber Communication Conference*, Vol. 4, 1991 OSA Technical Digest Series (Optical Society of America, Washington, D.C., 1991), p. 114.

BIBLIOGRAPHY

[76] A. Wada, T. Sakai, D. Tanaka, and R. Yamauchi, "Radiation sensitivity of erbium-doped fiber amplifiers," in *Optical Amplifiers and Their Applications*, 1990 OSA Technical Digest Series (Optical Society of America, Washington, D.C., 1990), pp. 294–297.

[77] J. R. Simpson, M. M. Broer, D. J. DiGiovanni, K. W. Quoi and S. G. Kosinski, "Ionizing and optical radiation-induced degradation of erbium-doped-fiber amplifiers," in *Optical Fiber Communication Conference*, Vol. 4, 1993 OSA Technical Digest Series (Optical Society of America, Washington, D.C., 1993), pp. 52–53.

[78] R. B. J. Lewis, E. S. R. Sikora, J. V. Wright, R. H. West, and S. Dowling, *Elect. Lett.* **28**, 1589 (1992).

[79] G. M. Williams, M. A. Putnam, C. G. Askins, M. E. Gingerich, and E. J. Freibele, *Elect. Lett.* **28**, 1816 (1992).

[80] P. J. Lemaire, H. A. Watson, D. J. DiGiovanni, and K. L. Walker, *IEEE Phot. Tech. Lett.* **5**, 214 (1993).

[81] C. C. Larsen and B. Palsdottir, *Elect. Lett.* **30**, 1414 (1994).

[82] R. M. Almeida, Ed., *Halide Glasses for Infrared Fiberoptics*, Martinus Nijhoff Publishers (1987).

[83] P. W. France, Ed., *Optical Fibre Lasers and Amplifiers* Blackie, Glasgow and London (1991).

[84] I. D. Aggarwal and G. Lu, Eds., *Fluoride Glass Fiber Optics*, Academic Press, Inc. (1991).

[85] J. Nishii, S. Morimoto, I. Inagawa, R. Iizuka, T. Yamashita, and T. Yamagishi, *J. Non-Cryst. Solids* **140**, 199 (1992).

[86] M. E. Lines *J. Appl. Phys.* **55**, 4052 (1984).

[87] P. W. France, S.F. Carter, M. W. Moore and C.R. Day, *Br. Telecom. Technol. J.* **5**, 28 (1987).

[88] A. Sarahangi, "Vapor Deposition of Fluoride Glasses," in *Halide Glasses for Infrared Fiberoptics*, R. M. Almeida, Ed., Martinus Nijhoff Publishers (1987), pp. 293–302.

[89] S. Takahashi and H. Iwasaki, "Preform and Fiber Fabrication," Chapter 5 of *Fluoride Glass Fiber Optics*, I. D. Aggarwal and G. Lu, Eds., Academic Press, Inc. (1991).

[90] V. Morin, E. Taufflieb and I. Clarke, "+20 dBm Praseodymium doped fiber amplifier single-pumped at 1030 nm," *OSA Trends in Optics and Photonics*, Vol. 16, Optical Amplifiers and their Applications, M. N. Zervas, A. E. Willner, and S. Sasaki, eds. (Optical Society of America, Washington D.C., 1997), pp. 76–79.

[91] Y. Ohishi, S. Mitachi, and S. Takahashi, *J. Light. Tech.* **LT-2**, 593 (1984).

[92] D. C. Tran, C. F. Fisher, and G.H. Sigel, *Elect. Lett.* **18**, 657 (1982).

[93] S. F. Carter, M. W. Moore, D. Szebesta, J. R. Williams, D. Ranson, and P. W. France, *Electron. Lett.* **26**, 2116 (1990).

[94] Y. Ohishi, S. Sakagashi, and S. Takahashi, *Elect. Lett.* **22**, 1034 (1986).

[95] H. Tokiwa, Y. Mimura, T. Nakai, and O. Shinbori, *Elect. Lett.* **21**, 1132 (1985).

[96] K. Shuto, K. Hattori, T. Kitagawa, Y. Ohmori, and M. Horiguchi, *Elect. Lett.* **29**, 139 (1993).

[97] R. V. Ramaswamy and R. Srivastava, *J. Light. Tech.* **6**, 984 (1988).

[98] A. Polman, D. C. Jacobson, D. J. Eaglesham, R. C. Kistler, and J. M. Poate, *J. Appl. Phys.* **70**, 3778 (1992).

[99] J. Schumulovich, A. Wong, Y. H. Wong, P. C. Becker, A. J. Bruce, and R. Adar, *Elect. Lett.* **28**, 1181 (1992).

[100] R. Tumminelli, F. Hakimi, and J. Haavisto, *Opt. Lett.* **16**, 1098 (1991).

[101] J. R. Bonar and J. S. Aitchison, *IEE Proc. Optoelectron.* **143**, 293 (1996).

Chapter 3
Components and Integration

3.1 INTRODUCTION

Erbium-doped fiber amplifiers are typically constructed by connecting fibers (erbium-doped as well as transmission fiber) with other components necessary for the amplifier's operation. These other components are either passive (e.g., isolators) or active (e.g., pump lasers). Such bulk components usually have fiber pigtails to make it easier to integrate them in a fiber-based system by fusion splicing the fibers together. A typical two-stage amplifier is shown in Figure 3.1. Many components of different type are clearly needed to obtain an amplifier with the desired performance characteristics.

In this chapter we will discuss the various key components that make up a real amplifier. We will first discuss fusion-splicing techniques and fiber connectors used to piece together different elements of an amplifier. We will then move on to more complex elements such as isolators, circulators, filters, and gratings. We will also touch on components that increase the amplified system performance, such as dispersion compensators and add/drop filters. Finally, we will describe the various pump lasers used to pump erbium-doped fiber amplifiers.

3.2 FIBER CONNECTORS

The loss and reflectivity of the component connections will critically influence the efficiency and stability of the amplifier. Polished fiber connectors, although convenient, have the disadvantage of higher reflectivity and higher insertion loss than fused fiber splices. Connectors are therefore typically limited to the input and output of an erbium-doped fiber amplifier with the internal, high-gain elements protected from reflections by isolators at the input and output ports. Reflections from connections within the span of amplified systems can also impact the system performance.[1] These effects are of particular concern in analog modulation systems. A number of connector designs, now standard in laboratory and system use, are shown in Figure 3.2.[3] The variety in these connectors arises from trade-offs between manufacturability and performance. They are connected by screw threads or snap-together mechanisms, and commonly use precision cylinder ferrules to align the fiber cores.[2]

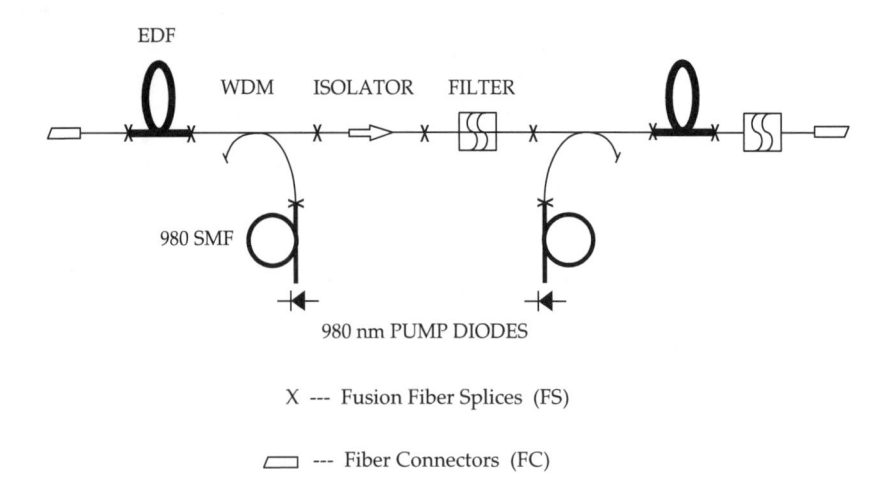

Figure 3.1: Typical two-stage erbium-doped fiber amplifier. The various components needed are pump lasers, isolators, wavelength division multiplexers (WDM), filters, connectors, and various types of transmission fiber.

The performances of some standard fiber connectors are shown in Table 3.1. These performances are largely determined by the polishing method and less by the connector type. As an example, the FC connector can be specified with polish designations of flat end (FL), physical contact (PC), super PC (SPC), ultra PC (UPC), and angled physical contact (APC), with the reflectivity decreasing from -14 dB to less than -60 dB, respectively. Variations in physical design can be seen in Figure 3.2. In addition to the designs shown, the CECC-LSH design offers a shutter that closes over the end of the fiber when disconnected. This shutter provides both eye safety and protection of the ferrule endface.

The insertion loss for connectors is determined by the alignment of the fiber cores and by the spatial match of the mode fields. Positioning of the cores is typically achieved by alignment of the opposing fiber outside glass surfaces with precision bore ferrules. An alternate, more accurate method machines or crimps the connector around the axis of the core. Ferrules position the core with an accuracy determined by the fiber core to cladding concentricity error, cladding noncircularity, and core eccentricity. Fiber manufacturers offer high concentricity fiber specifically for these applications with core to cladding concentricity error of less than 0.5 μm and cladding noncircularity of less than 1.0%. The induced losses for lateral and angular misalignment are given in equations 3.1 and 3.2. The lateral core offset induced loss for displaced fiber ends has been approximated by the overlap of two Gaussian beams, as in equation 3.1,

$$\text{Lateral Displacement Induced Loss (dB)} = 4.343 \left(\frac{d}{\omega}\right)^2 \quad (3.1)$$

where d is the lateral offset and ω is the mode field radius (assumed to be equal for both fibers).[5] Here it is assumed that the two fibers have an index-matching media between them. Increasing the mode field diameter or decreasing the lateral offset reduces this

3.2. FIBER CONNECTORS 45

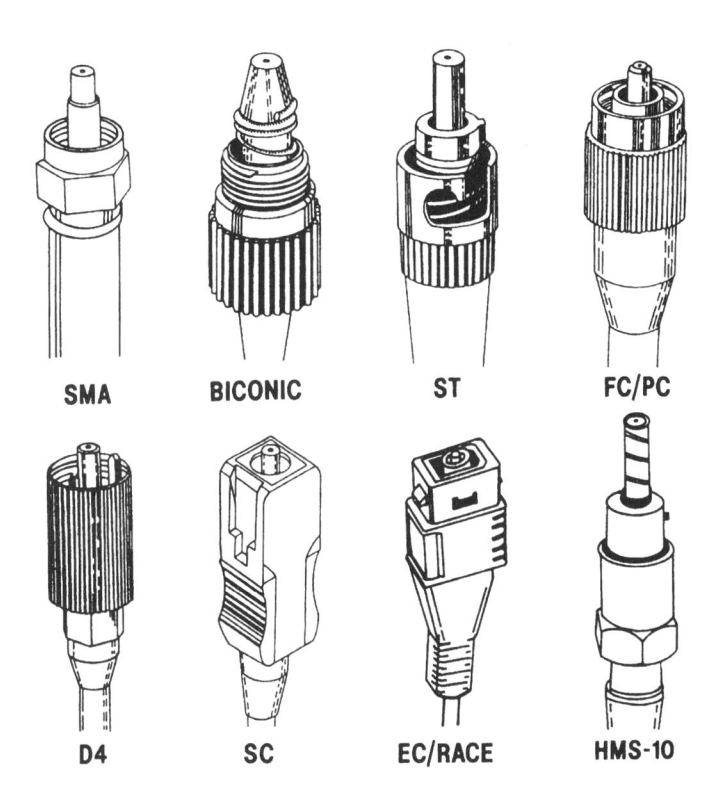

Figure 3.2: Single-mode fiber connectors. From reference [3].

loss component. In practice, the highest performance connectors position the cores to within 0.25 μm and compress the fiber ends together for a physical contact to diminish the induced loss and reflectivity.[4]

Increasing the mode field radius to reduce the axial misalignment induced loss unfortunately increases the angular misalignment component of loss. The angular offset (ϕ) core induced loss can be estimated using equation 3.2,

$$\text{Angular Misalignment Induced Loss (dB)} = 4.343 \left(\frac{\pi n_{\text{clad}} \omega \phi}{\lambda} \right)^2 \quad (3.2)$$

where n_{clad} is the refractive index of the cladding, ω is the mode field diameter (assumed to be equal for both fibers), λ is the wavelength of the light being transmitted, and ϕ is the angle between the two longitudinal fiber axes.[5]

It is common that fibers within the amplifier will have different mode field diameters. For example, low threshold optimized erbium-doped fiber common in preamplifier and remote amplifier designs has a smaller mode field diameter (e.g., 3 to 5 μm) than the transmission fiber (e.g., 6 to 11 μm). The loss induced by such a mismatch in

Connector Type	Developer	Typical Insertion Loss (dB)	Typical Reflectivity (dB)
SMA/PC	Amphenol	≤ 1.0	−45
BICONIC	Bell Labs	0.3	−14
ST/PC (physical contact)	AT&T	0.2	−50
FC/FL (flat end)	NTT	≤ 1.0	−14
FC/PC (physical contact)	NTT	≤ 0.5 0.25	≥ −27
FC/SPC (super physical contact)	NTT	≤ 0.5	≥ −40
FC/UPC (ultra physical contact)	NTT	≤ 0.5	≥ −50
FC/APC (angled physical contact)	NTT	0.17	−67
D4/PC	NEC	≤ 0.8	−40
SC/APC	NTT	≤ 0.5	≤ −27 to ≥ −60
HRL10 APC (angled physical contact)	Diamond SA	0.12	−66
CECC-LSH	European Committee	0.2	−50 to −70

Table 3.1: Characteristics of common connectors used in integrating amplifiers and fiber transmission systems. The data is derived from a variety of product catalogs. An important determinant of the reflectivity and the insertion loss are the fiber alignment and polishing methods.

mode field diameters can be estimated as,

$$\text{Mode Field Mismatch Induced Loss (dB)} = 10\log\left[\frac{(2\omega_{r1}\omega_{r2})}{(\omega_{r1}^2 + \omega_{r2}^2)}\right]^2 \quad (3.3)$$

where $\omega_{r1,r2}$ are the mode field diameters of the two fibers.[2] The dependence of connectivity loss on differences in mode field diameters between opposing cores can be seen in Figure 3.3, where the splice loss is plotted as a function of the mode field diameter ratio ω_{r1}/ω_{r2}. The range of mode field diameters for standard communications fiber is indicated in Table 3.2 The resulting induced loss between typical erbium-doped fiber and standard transmission fiber is thus 2 to 4 dB, when using connectors or poor splicing techniques. A detailed analysis of losses created in connecting single-mode fiber has been presented by Marcuse.[6]

Early versions of flat end connectors positioned the fiber ends close to each other with a small air gap between them. The reflectivity in this case is the glass to air reflectivity of near 4% (−14 dB). The physical contact method, in which the glass

3.2. FIBER CONNECTORS

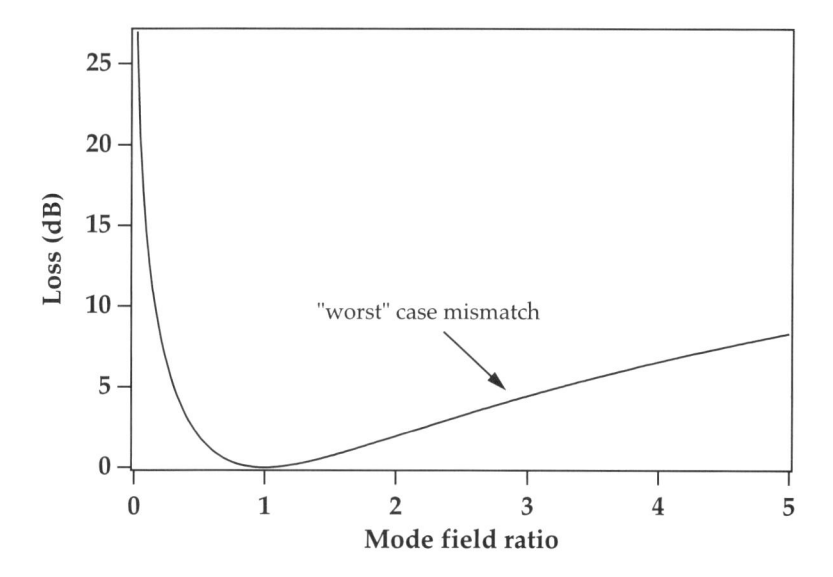

Figure 3.3: Splice loss versus mode field diameter ratio as expressed in equation 3.3. The "worst case" mismatch is highlighted, as an example, for the combination of a 3.6 μm mode field diameter erbium-doped fiber and a 10.5 μm mode field diameter transmission fiber.

cores touch, reduces the reflectivity to near 0.01% (−40 dB).[7] Further reduction in reflectivity is achieved using 8° angled polished fiber connections (APC), resulting in better than −60 dB reflectivities. These connectors have the additional advantage of maintaining better than −60 dB reflection even when unmated, unlike the flat end physical contact type which drop to −15 dB. An alternative to the physical contact method is a controlled air gap (CAG) design. Here, a 12° angled end polish along with a precise end separation results in reflectivities as low as −80 dB.[8] It is claimed as well that this controlled air gap method has the advantage of being less susceptible to scratches or dust, which degrade the connector performance.[8]

Falling somewhere between connectors and splicing is the concept of expanding the mode field of the fiber to diminish the alignment sensitivity. Matching of mode fields to diminish the loss may be accomplished by a single-mode fiber adiabatic taper. With a taper, the waveguide mode size is changed along a portion of the guide at a gradual rate.[9, 10]

The expansion of the mode field is typically accomplished by either expanding or tapering the fiber's outside dimension or by thermal diffusion of the core dopants. It is common practice in silica-based fiber to accomplish an adiabatic taper with a fusion splice. When fluoride fiber is interconnected with silica-based fiber, these tapers are typically inserted and are often referred to as thermally expanded core (TEC) sections. Heating a section of fiber for tens of minutes expands the core mode field diameter from near 10 μm to as high as 50 μm. Fiber ends with such mode field expansions

Fiber Type	Manufacturer	Mode Field Diameter (μm)
Dispersion normal depressed clad	Lucent Technologies	10.0 (1550 nm) 5.0 (980 nm)
Dispersion normal matched clad SMF-28™	Corning, Inc.	10.5 (1550 nm)
SMBD0980B	SpecTran	5.0 ± 1.0 (980 nm)
CS-980	Corning, Inc.	4.2 (980 nm)
Dispersion normal silica core Z-fiber™	Sumitomo	9.5 (1550 nm)
Dispersion shifted SMF/DS	Corning, Inc.	8.1 (1550 nm)
Non-zero dispersion shifted TrueWave™	Lucent Technologies	8.4 (1550 nm)
Dispersion compensating	Lycom DK-SM	5.0 (1550 nm)
980 nm single-mode	Lucent Technologies	9.7 (1550 nm)
1060 nm single-mode Flexcore-1060™	Corning Inc.	5.9 (1550 nm)

Table 3.2: Mode field diameters for representative transmission fibers encountered in systems involving erbium-doped fiber amplifiers. The wavelength corresponding to the listed mode field diameter is indicated in parentheses.

have been used in fiber-pigtailed components where filters or other optical elements may be placed in the region of the expanded beam with moderate attenuation.

3.3 FUSION SPLICING

State-of-the-art fusion splicing provides low-loss and high-strength fiber joining between like and unlike fibers. One has access today to commercial equipment that automatically fuses two fibers with very little operator assistance other than preparation and placement of the fiber ends in the machine. Commercial cleaving devices can achieve perpendicular and nearly flat fiber ends. Splicing, in general, relies on the radiant heating of the two fiber ends by an arc, filament, or flame source while pushing the two fiber ends toward one another. Analytical models of this process have been developed that include details of the heating, viscous flow, and resulting stress.[11] Although the cores are aligned prior to fusion splicing, the surface tension forces during the viscous sintering of the two fiber ends may misalign the cores. It is therefore important to provide fiber with low concentricity error to achieve the lowest possible losses.

Important qualities of the fusion-splicing process in fiber amplifier assembly are low loss, low reflectivity, and high strength. Arc fusion splices between identical fibers may yield losses as low as 0.01 dB, reflectivities that are too low to be measured, and tensile strengths as high as 5.5 GPa.[12, 13] We show the evolution of the splice loss during fusion splicing for three durations of heating in Figure 3.4. By monitoring the splice loss during splice heating, the optimum heating duration may be determined.

3.3. FUSION SPLICING

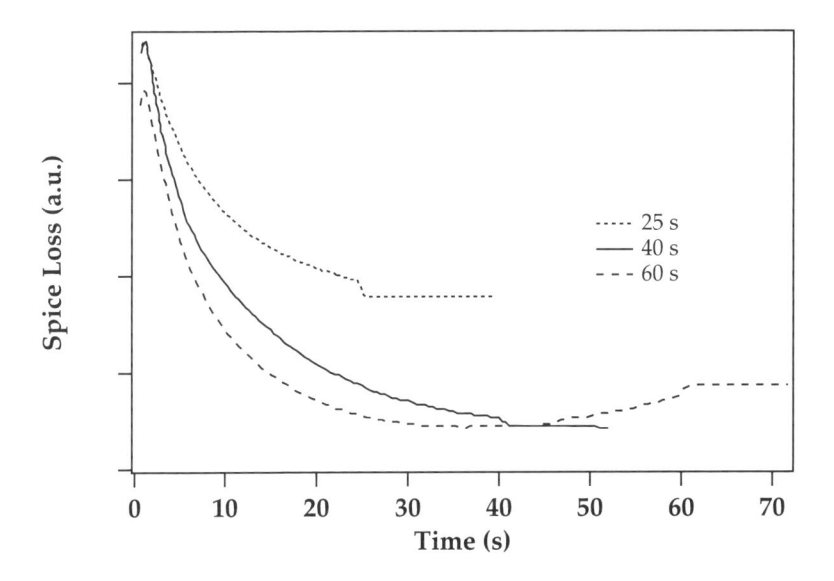

Figure 3.4: Experimental splice loss for two single-mode fibers as a function of time, for 25, 40, and 60 second heating times (courtesy of J. Krause, Lucent Technologies, Murray Hill, NJ).

Figure 3.4 shows the cases of a heating duration that is too short (25 seconds), too long (60 seconds), and nearly optimal (40 seconds). The optimal heating durations will be affected by heating temperature, fiber types, and other splice conditions.

The ability to achieve low-loss and high-strength splice properties when connecting substantially different mode field diameter fiber types is possible by making use of the different diffusion rates of index-altering dopants (e.g., Ge, P, Al, and F). As the fibers are heated during fusion splicing, the ions diffuse thereby expanding the mode field radius. When ions within the fiber of smaller mode field diameter diffuse faster than ions in the larger mode field diameter, there will be a time when the mode field diameters will be equal. Realistically, the mode field diameters may never become equal. However a small difference will be adequate to result in a small splice loss. Incorporating the longer time, high-temperature method of diffusing the core to match mode fields in commercial arc-fusion splicing has resulted in splice losses near 0.1 dB and splice strengths of 400 kpsi for erbium-doped fiber (3.4 μm mode field radius) to standard fiber (9.0 μm mode field radius).[20, 21] Note that the diffusion of core and cladding dopants was first studied to explain the influence of fusion time and temperature on the splice loss of standard telecomunications fiber.[14, 15, 16]

Mode fields expanded by tapering the fiber near the splice have also been proposed as a method to reduce splice loss between dissimilar fibers. This method would tend, however, to concentrate any stress on the fiber in the splice region, through the reduced cross-sectional area.[17] Using a propagating beam method of analysis, it has been predicted that a taper transition length of near 1000 wavelengths for a spot size 3.4 times larger does not add significant loss.[18] It has also been suggested that a stepwise al-

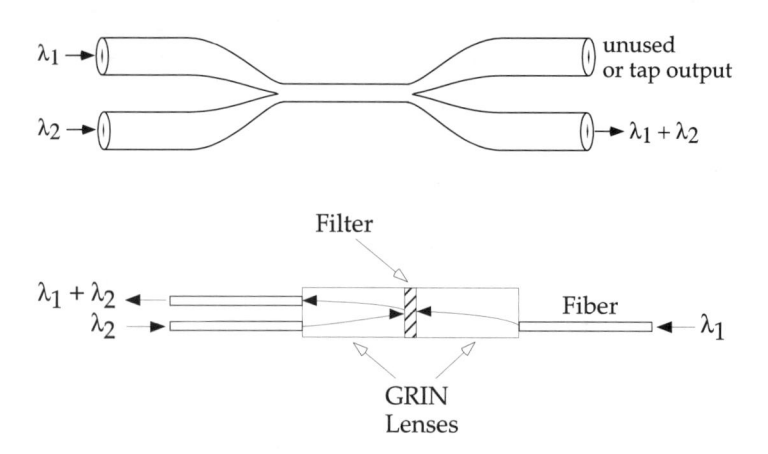

Figure 3.5: WDM designs to combine signal and pump (upper figure: fused fiber type; lower figure: miniaturized filter type).

ternative using one or more nontapered intermediate fiber splices can also substantially diminish the overall splice loss.[19]

The length of the tapered region when using a commercial arc splicer is determined by the heating profile of the arc and may present a limitation to the overall change in mode field size possible with low loss. These methods have been applied to commercial splicing equipment along with the real-time monitoring of the mode field match to control the fusion time to produce the lowest splice loss.[22] Here the fiber radial-dependent thermal luminesence (hot-fiber image) of the two fibers are monitored during the splice operation. These images are then translated into relative refractive index differences. This real-time monitoring then ends the splice heating when the core refractive index differences are nearly the same for each fiber.[22]

3.4 PUMP AND SIGNAL COMBINERS

Combining the signal and pump paths to allowing copropagation or counterpropagation in the doped fiber of amplifiers, is provided by a wavelength division multiplexer (WDM).The most significant properties of these wavelength-dependent couplers for signal and pump are the loss in both paths and the splitting ratio or isolation (i.e., how completely the two channels are separated). Significant as well is the polarization sensitivity of these properties, given that both channel polarizations will likely wander during operation. These devices are typically made as either fused fiber couplers or miniaturized interference filter reflectors, as shown in Figure 3.5.

A comparison of typical device performance for 1480 nm & 1550 nm and 980 nm & 1550 nm WDM's are shown in Table 3.3 and Table 3.4. In general, the fused fiber devices have an advantage in insertion loss, reliability, and cost, while the interference-filter-based devices show an advantage in wider passband, lower polarization-dependent loss while maintaining higher isolation.[23] The isolation is defined as a measure of the

3.4. PUMP AND SIGNAL COMBINERS

	Fused Fiber	Interference Filter
Isolation (dB)	≥ 10	12 (1480) 30 (1550)
Max. insertion Loss (dB)	0.5	0.4
Back-reflection (dB)	−55	−68
Polarization-dependent loss (dB)	≤ 0.1	0.015
Thermal stability (dB/°C)	≤0.002	≤ 0.005
Passband (nm)	10	33 (1480) 35 (1550)

Table 3.3: Comparison of fused fiber and interference filter WDMs for combining 1480 nm and 1550 nm signals. The interference filter data is from reference [25] (courtesy E-TEK Dynamics, Inc., San Jose, CA).

	Fused Fiber	Interference Filter
Isolation (dB)	20	60 pump
Max. insertion Loss (dB)	0.4	0.4
Back-reflection (dB)	−55	−68
Polarization-dependent loss (dB)	≤ 0.1	0.015
Thermal stability (dB/°C)	≤ 0.002	≤ 0.005
Passband (nm)	20	30 (980) 60 (1550)

Table 3.4: Comparison of fused fiber and interference filter WDMs for combining 980 nm and 1550 nm signals. The interference filter data is from reference [25] (courtesy E-TEK Dynamics, Inc., San Jose, CA).

amount of light at an undesired wavelength at any given port, relative to the input power of that light. The trade-off in performance is highly dependent on the application.

When combining 980 nm pump and 1550 nm signals, special consideration must be given to the fiber used in the pump path. Normal telecom fiber supports more than one mode at 980 nm for cutoff wavelengths between 1.20 and 1.45 μm. For stability, it is desirable to use a pump input fiber that is single mode at 980 nm. Examples of this fiber type are shown in Table 3.2. It is possible to make use of fiber with a cutoff wavelength slightly higher than 980 nm such as Flexcore-1060™. In particular, a fused fiber coupler may be fabricated with such fiber if it is designed with a surrounding glass with an index lower than the cladding.[26, 27]

The challenge of producing a fused fiber coupler with low polarization sensitivity while maintaining low insertion loss and low crosstalk has been addressed by twisting the fibers within the coupling region and potting the coupler in a low-index polymer. For a narrow coupler channel spacing such as the 1480 nm/1550 nm pump coupler, it is predicted that a polarization sensitivity of ≤ 0.1 dB and insertion loss of ≤ 1.2 dB is achievable for a twist rate of 1.5 T/B (T = number of full turn twists, B = number of beat maxima observed while pulling the coupler during fabrication).[28] Polarization sensitivity of ≤ 0.2 dB has been reported for a similar 1480 nm/1550 nm pump coupler potted in a silicone elastomer.[29]

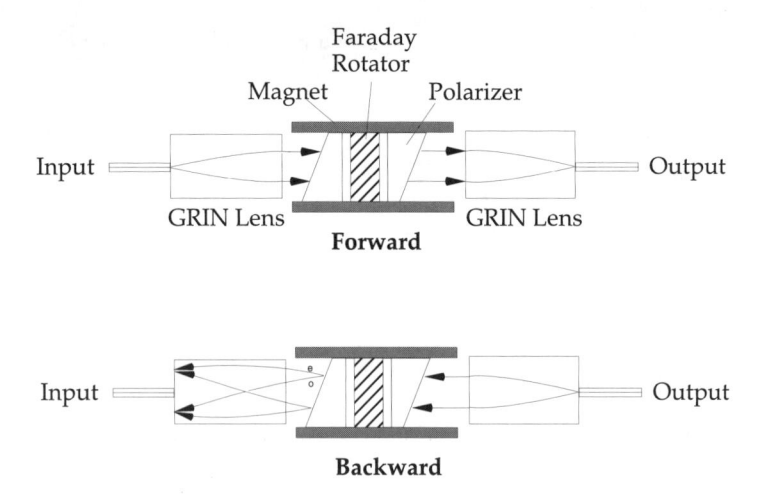

Figure 3.6: The forward- and backward-propagating optical paths for a typical isolator design. Notice that the backward paths for the ordinary and extraordinary light do not couple into the input fiber. Figure courtesy E-TEK Dynamics, Inc., San Jose, CA.

3.5 ISOLATORS

Reflections at the input and output ports of an erbium-doped fiber amplifier or laser may have a profound effect on their performance. These reflections may result from Fresnel reflections from connectors, Rayleigh scattering from lengths of fiber connected to either end, or reflections from within the amplifier itself.[30] Early demonstrations of erbium-doped fiber amplifiers used oil-immersed fiber ends and angle-polished fiber ends to reduce the Fresnel reflections.[31] For example, reflections at the input end degrades the noise figure by reflecting the backward ASE, which in turn lowers the input population inversion.[32]

The structure for a typical polarization-independent isolator is shown in Figure 3.6. In the forward direction, the fiber output is first collimated using a GRIN lens followed by a birefringent rutile (TiO_2) wedge. The outputs of the wedge are then a pair of ordinary and extraordinary rays that proceed through a Faraday rotator consisting of a length of Yittrium Iron Garnet (YIG) crystal $Y_3Fe_5O_{12}$ surrounded by a permanent magnet, which results in a 45° rotation of the two polarization axes. These rotated optical axes then proceed through a second birefringent wedge which recombines them for launching into the exit fiber. A reverse signal will experience first the separation into ordinary and extraordinary beams, rotated by the YIG to an angle now 90° from the input polarization, and sent on divergent paths by the second wedge. These divergent paths do not focus onto the input fiber and are therefore excluded from coupling back through the input fiber. This "walk-off" extinction effect takes the place of conventional isolators, which use polarizers to provide the extinction of counterpropagating signals.[33] Key parameters that describe the performance of this device are insertion loss, isolation, and return loss. The spectral and temperature dependence of this device are related to the Faraday rotating material response whereas the insertion and return

3.6. CIRCULATORS

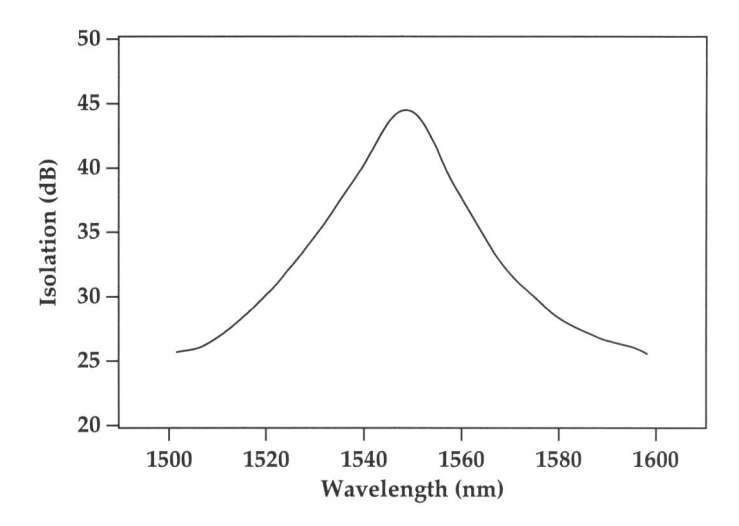

Figure 3.7: Isolation dependence on wavelength and temperature for a commercial isolator. From reference [34] (courtesy of E-TEK Dynamics, Inc., San Jose, CA).

losses are more a function of the antireflection coatings and alignment precision of the elements. The high end performance of these devices provide a near wavelength independent isolation of 75 dB over the range 1460 nm to 1570 nm, an insertion loss of 0.5 dB, and return loss of 65 dB. Representative isolation dependence on wavelength for a typical commercial isolator is shown in Figure 3.7.

Demonstrations of fiber-integrated isolators have shown promise in terms of reducing the size of these devices by designing an isolator chip that may be positioned between a linear array of thermally expanded core fiber ends.[35]

3.6 CIRCULATORS

The circulator component is a passive multifiber junction where an incoming signal is routed to another fiber. This component is functionally described by Figure 3.8, where signals entering the circulator are routed to the next fiber encountered in a clockwise direction. Within the circulator are a number of optical components similar to an isolator. One example of a four port polarization-independent circulator structure is shown in Figure 3.9. The incoming signal is first split into two orthogonal polarization states by a birefringent plate ($CaCO_3$ crystal). The two resulting paths are then reflected and transmitted to separate paths through a Faraday rotator (GBIG crystal with a Sm-Co magnet) and an optically active rotator (SiO_2 crystal) in series. Finally, the paths are recombined through a polarization beam splitter and a birefringent plate. A typical circulator, as commercially offered, performs with a loss between ports of 0.9 dB, isolation between ports greater than 50 dB, polarization sensitivity of 0.05 dB, and polarization dispersion of 0.07 ps.[37]

Amplifier designs that have used the circulator include a midstage dispersion compensating element in a reflective amplifier design, a gain equalizing amplifier, an am-

54   CHAPTER 3. COMPONENTS AND INTEGRATION

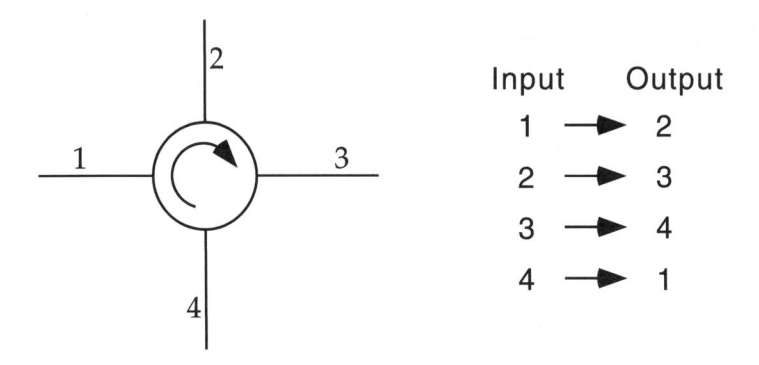

Figure 3.8: A four-port circulator flow diagram showing the redirection of input fiber ports 1 through 4 to the corresponding outputs.

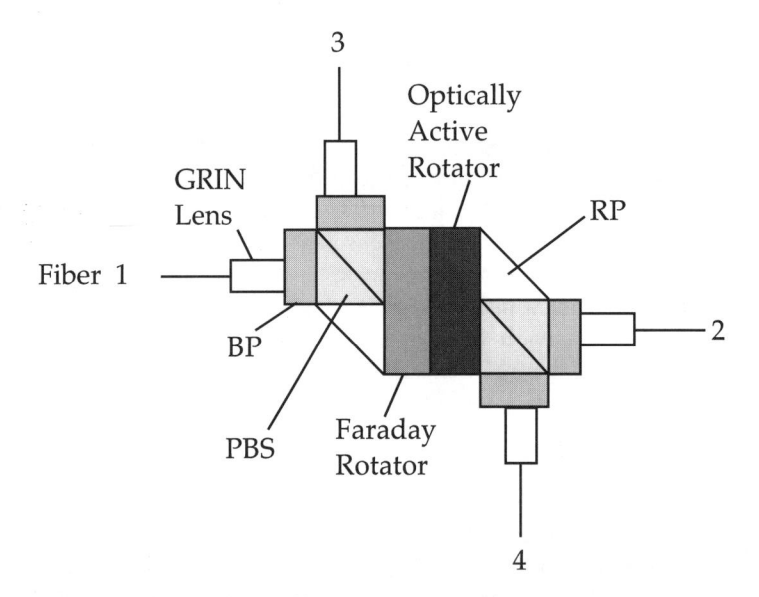

Figure 3.9: Circulator component diagram showing the placement of four fiber/GRIN lens ports, four birefringent plates (BP), two polarization beam splitters (PBS), two reflective prisms (RP), one Faraday rotator, and one optically active rotator. These collectively function as a circulator. From reference [36] (©1991 IEEE).

plifier with signal filters, and an amplifier that monitors noise figure by detecting backward-amplified spontaneous emission.[38, 39, 40, 41] System applications include its use in repeaterless transmission, add/drop methods (see Figure 3.13), grating dispersion compensation and OTDR transparent systems.[42, 43, 44, 45]

3.7 FILTERS

Erbium-doped fiber amplifiers may use optical filters to enhance performance by suppressing amplified spontaneous emission either as it builds within the amplifier or at the end of a system span before detection. Filters may also be used to alter the gain spectral profile or to enhance the properties of a component within the amplifier.

Inserting a filter inside the amplifier, as shown in Figure 3.1, can be beneficial in realizing a lower noise figure. Amplifiers employing these filters within a two-stage amplifier design have been suggested for preamplifiers and inline applications.[46, 47] The filters chosen for this application are typically interference filters fabricated by multiple dielectric coatings on a glass substrate and typically have a passband of 10 nm. These filters are then made compatible with fiber connections by placing them between GRIN lens/fiber pigtail connections, such as the straight-through path of the interference filter design shown in Figure 3.5.

Narrowband Fabry-Perot filters are typically used following an erbium-doped fiber preamplifier and preceding the receiver detector. These limit the amplified spontaneous noise of the preamplifier arriving at the detector (see Chapter 7, section 7.3.3).[48, 49, 50] The impact of the bandwidth of the filter on the signal to noise ratio is discussed in Chapter 7, section 7.3.3. System implementation with a Fabry-Perot filter requires an automatic tracking function to keep the filter centered on the signal wavelength.[51]

Filters added to an amplifier to alter the gain profile are often inserted between gain stages to reduce the impact of added loss on the amplifier noise figure. For this application Fourier filters, twin core fiber, and long period gratings have been used.[52, 53, 54, 55] The Fourier filter is created by inserting a 154 μm thick glass plate into an expanded beam between two fiber pigtails. The spectral-dependent absorption is tuned by changing the insertion distance and angle of the glass plate in the beam. Filters based on twin core fiber are in effect Mach-Zehnder interferometers within a fiber. Filters based on fiber gratings will be discussed in the following section.

3.8 FIBER GRATINGS

3.8.1 Introduction

A fiber grating may be manufactured by exposing a length of the core to a nearly sinusoidal varying intensity of UV light. This effect was first observed as the result of a standing wave pattern created in a germanium doped fiber by 488 nm light from an argon ion laser.[56] The standing wave pattern arises from the reflection of the laser light from the two ends of the fiber. This sinusoidal variation is nowadays created by side illumination of the fiber. The fiber is exposed to the interference pattern of two laser beams or to that created by a laser beam traversing a phase mask.[57, 58, 59] This UV light intensity exposure then imposes a periodic index along the length of the fiber core by creating a corresponding periodic concentration of glass defects. These defects are usually associated with the germanium dopant in the silica host.

The amplitude of the induced refractive index variation controls the magnitude of the grating effect. More intense UV exposures create higher concentrations of defects. It has been found that the diffusion of hydrogen into the fiber prior to UV exposure

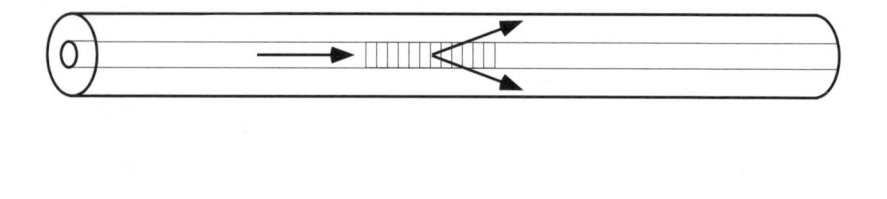

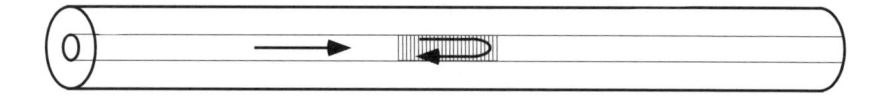

Figure 3.10: Long period grating fiber filter (upper sketch) and Bragg grating fiber filter (lower sketch).

substantially increases the refractive index change.[60] The maximum achievable amplitude of this index variation is about $\Delta n = 10^{-2}$.[61]

Fiber gratings created in this fashion are divided into two categories: Bragg gratings and long period gratings. Bragg gratings generally refer to devices with refractive index maximum spacings on the order of 1/2 to 1 times the wavelength of the guided mode (0.5 μm to 1.0 μm for single-mode fibers at 1.55 μm). These gratings couple the forward-propagating core modes to backward-propagating guided modes, as shown in Figure 3.10 (lower sketch). The grating reflects light of wavelength λ_{Bragg}, such that

$$\lambda_{Bragg} = 2n_{core}^{eff} \Lambda \qquad (3.4)$$

where n_{core}^{eff} is the mode effective index of the guided LP$_{01}$ core mode and Λ is the periodicity of the index grating (in units of length).[62] The mode effective index of refraction is the weighted average of the core and cladding indices, where the weighting is derived from the proportions of the mode that are in the core and cladding regions. Bragg gratings are available with bandwidths from 0.1 to 10 nm and reflectivities from 1 to nearly 100. Reflection and transmission spectra for a representative commercial Bragg grating are shown in Figure 3.11

Long period gratings are created by refractive index periodicities Λ on the order of 100 times the operating wavelength (i.e., 100 to 500 μm). This periodicity causes coupling between the forward-guided LP$_{01}$ mode and forward-guided cladding modes. Coupling between the modes occurs when the following condition is met:

$$n_{core}^{eff} - n_{clad}^{eff} = \lambda/\Lambda \qquad (3.5)$$

where n_{core}^{eff} and n_{clad}^{eff} are the mode effective indices of the core and cladding modes, respectively, and λ is the wavelength of the light propagating in the fiber. There are a number of modes that can be supported by the cladding. As a result, the above equation may be satisfied for a number of wavelengths. There are then a number of notch filter responses that result from a grating with one periodicity, as shown in Figure 3.12.

Both Bragg and long period grating devices are especially attractive because they are readily fusion spliced onto other fiber devices.

3.8. FIBER GRATINGS

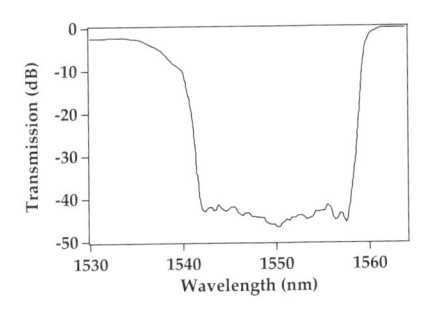

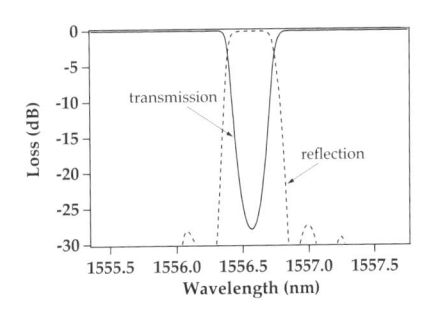

Figure 3.11: Wideband chirped grating (left) and narrowband (for 50 GHz channel spaced DWDM applications) linear grating (right) fiber Bragg gratings. Courtesy of JDS Fitel, Inc., Nepean, Canada.

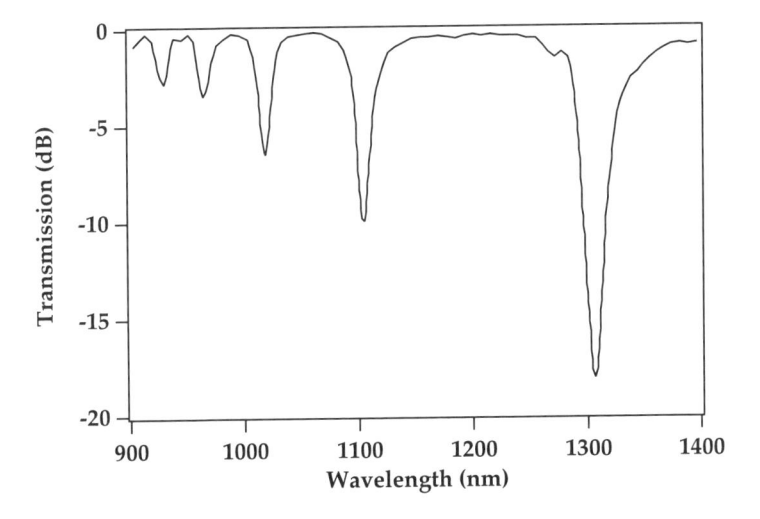

Figure 3.12: Long period grating fiber filter spectral response. From reference [63] (©1996 IEEE). The multiple notches result from the coupling between the fundamental mode of the core and various cladding modes.

3.8.2 Applications of Bragg Gratings

Reflective filters based on fiber Bragg gratings include the following applications related to erbium-doped fiber amplifiers or systems containing them:

- stabilization of fiber amplifier pump lasers
- add/drop devices
- pump power reflection at amplifier ends
- dispersion compensation

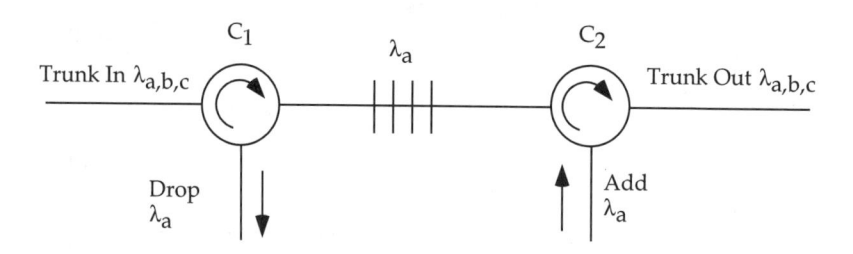

Figure 3.13: Add/drop unit for a wavelength λ_a, constructed with circulators C_1 and C_2 and one fiber Bragg grating. From reference [43].

The stability of 980 nm pump lasers is critical to the stable operation of erbium-doped amplifiers. Given the relatively narrow spectral pump band at 980 nm, spectral change in the output of the laser can cause significant variations in the gain and noise figure of the amplifier. These spectral changes can come about from aging of the laser, mutual interaction of several pump sources, and reflections from passive components, which change with time and temperature. Weakly reflecting fiber Bragg gratings, installed in the 980 nm laser pigtails, have improved the stability of the lasers. These gratings do not form an external cavity but provide enough feedback to lock the laser oscillation wavelength to the grating reflection wavelength.[64]

The narrow, spectral-dependent reflectivity of the Bragg gratings has been used as an add/drop component as well. The add/drop function is created by combining fiber Bragg gratings at multiple wavelengths with either circulators or 3 dB couplers.[43] As an example, Figure 3.13 shows an add/drop unit for a wavelength λ_a. The operation of the add/drop unit is as follows. Suppose that a number of wavelengths, including λ_a, are incident on the add/drop unit. Wavelength λ_a is reflected by the grating, returns into circulator C_1, and is then routed to the next port, which is the drop port. When adding wavelength λ_a, one injects it into the add port of circulator C_2, it is routed back into the unit toward the Bragg grating, reflected back into the circulator C_2, and from there routed to the trunk output. The circulators may be replaced with 3 dB couplers to provide the same add/drop function with higher losses.

Erbium-doped fiber amplifiers typically use pump power somewhat inefficiently, allowing unused pump power to escape through one of the erbium fiber ends. By reflecting the pump back into the amplifier, a more efficient operation can be obtained. This concept has been demonstrated in remote pumping of erbium-doped fiber amplifiers. In one particular remote pumping experiment, a broadband reflective grating (near 100% reflection for a 2 nm FWHM) allowed the amplifier pump power to be reduced by 33% for equivalent gain and noise figure performance.[65]

Dispersion compensation is accomplished by using a very long chirped Bragg grating where the periodicity of the grating decreases continuously along the length of the grating. This provides a spectrally dependent delay, the priciple of which is depicted in Figure 3.14. The long wavelengths are reflected from the front of the grating, where the spacing matches the long wavelengths. The short wavelengths penetrate farther into the grating before being reflected. This provides the wavelength-dependent time delay for reflection from the grating. The maximum delay time that the grating can

3.8. FIBER GRATINGS

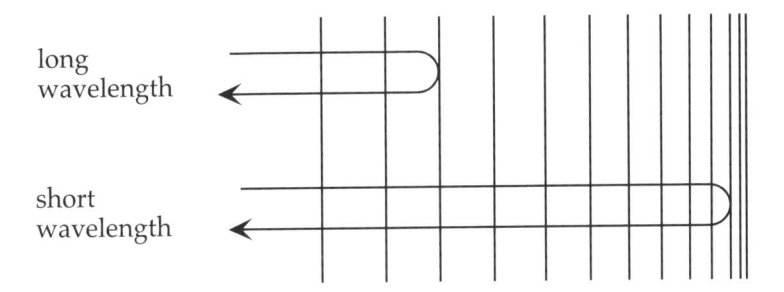

Figure 3.14: Dispersion compensation with a chirped Bragg grating.

offer is twice the transit time through the grating. The use of such gratings in systems requires the use of an accompanying circulator, since the gratings are reflective and not transmissive.

Gratings have been demonstrated that offer up to 1800 ps/nm delay compensation over a bandwidth of 5.2 nm with a grating 1 m in length.[66] Compensation of dispersion using inline fiber gratings has allowed the propagation of high bit rate (10–40 Gb/s) signals at 1.5 μm over long lengths (50–100 km) of standard fiber with a power penalty of only a few dB. Using two 40 cm long gratings with delay compensations on the order of 850 ps/nm over a 4 nm bandwidth, the transmission of 40 Gb/s over 109 km of dispersion normal fiber was demonstrated with a power penalty of 6 dB.[66] The gratings compensated for a broadening of the pulses from their original value of 6.4 ps to about 350 ps. Without the grating compensators, the maximum transmission distance would have been only 4 km at the same bit rate.[66]

3.8.3 Long Period Gratings

Notch filters based on long period gratings induced in the core of optical fiber have been demonstrated as useful components within erbium-doped fiber amplifiers.[55, 63, 67, 68, 69] A typical spectral filter shape resulting from one grating period is shown in Figure 3.15.

The amplitude of the index variation, controlled by the UV light exposure and the fundamental defect level achievable, determine the depth of the notch filter. Strong gratings can be written with maximum loss values of 32 dB.[63] Additionally, the number of periods in the grating controls the spectral width of the notch filter. Properties that make these devices attractive are the low insertion loss of $\leq$ 0.2 dB, low back-reflection of $\leq$ −80 dB, low polarization mode dispersion of $\leq$ 0.01 ps, and low polarization-dependent loss of $\leq$ 0.02 dB. The main challenge in using these devices is their packaging. A successful package must address performance sensitivities to fiber bending, fiber coating refractive index, strain, and temperature.

Filters with complex spectral shape may also be created by cascading a number of long period gratings. The resulting filter may be used to flatten the spectral gain of an erbium-doped amplifier for WDM applications.[70, 71] As an example, the shape of the long period grating filter used to flatten the gain of an erbium-doped fiber amplifier over 40 nm, centered at 1550 nm, is shown in Figure 3.16.[71] Gain-flattened amplifiers

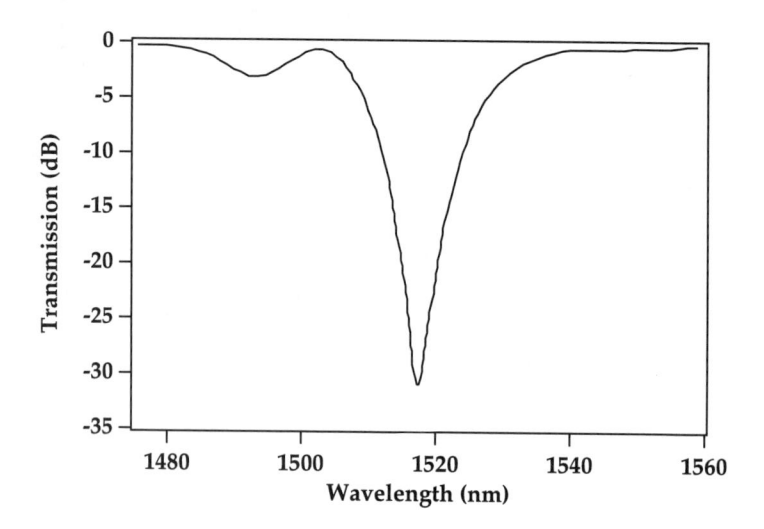

Figure 3.15: Spectral transmission characteristic of a long period grating fiber filter. From reference [63] (©1996 IEEE).

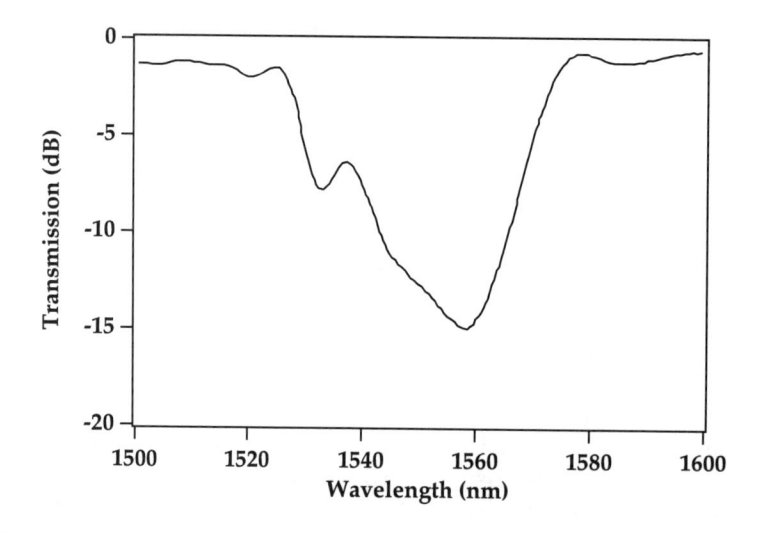

Figure 3.16: Spectral transmission characteristic of a long period grating fiber filter used to flatten the gain (to within 1 dB) over 40 nm centered at 1550 nm, of an erbium-doped fiber amplifier. From reference [71].

are discussed in Chapter 9.

Tunability of long period gratings can be achieved by bending the fiber in the region of the grating. An acoustically generated long period grating has also been demonstrated with the ability to be tuned by changing the acoustic frequency used to excite the grating.[72]

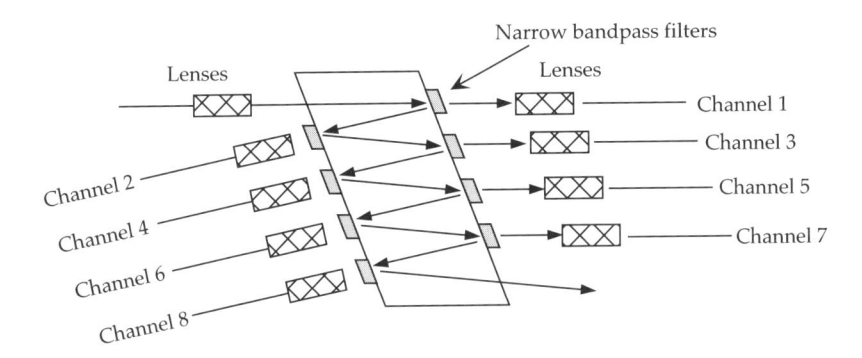

Figure 3.17: Signal multiplexer/demultiplexer based on multiple interference filters. From reference [73] (courtesy Corning OCA Corporation).

3.9 SIGNAL MULTIPLEXERS AND DEMULTIPLEXERS

To make the most use of the spectral gain bandwidth of optical amplifiers or to design all optical networks, it may be necessary to combine many wavelength signal channels onto one fiber. The implementation of such WDM systems will be discussed in Chapter 9. In general, such systems need to combine N input wavelengths onto one output fiber or equally divide a signal channel between N output fibers. Key to these systems are multiplexer components that exhibit low insertion loss to each channel and low crosstalk between channels. This multiplexing function may be accomplished using a variety of methods including fiber splitters, miniature bulk optic filters, arrayed waveguide gratings, and grating/fiber coupler combinations. The 1 x N splitter approach suffers from high signal loss given that the power at the input port is divided evenly between all N outputs. This loss may be offset by amplifiers for each output port at substantial cost and complexity.

A lower-loss method employs a series of interference filters arranged in a zigzag geometry (see Figure 3.17). Narrow channels may be added to or taken from the optical path through GRIN-lensed fiber ports. Channel spacings of 0.8 nm (100 GHz) for this design have been demonstrated with up to 32 channels total. Channels spaced by 1.6 nm have been multiplexed in this manner with an adjacent channel isolation of 30 dB and maximum insertion loss per channel of less than 4 dB. Thermal stability of these devices is quoted as better than 0.004 nm/°C and even better for filters especially engineered for stability.[73] Efficient multiplexing may also be accomplished using a waveguide grating router (WGR), also referred to as an arrayed waveguide grating (AWG).[74, 75, 76, 77, 78] Here, two star couplers are connected by an array of waveguides. This structure is formed lithographically in a planar configuration using a substrate of Si or InP, as shown in Figure 3.18. The length of each guide connecting the two couplers provides the necessary phase shift for the wavelength spacing desired. Optical power entering from the left port is first divided between the collection of curved waveguides. These waveguides, of different lengths, when combined with the star splitter created by the output free space region, act as a grating. This effective grating then maps the wavelength bands onto the four output waveguides as shown.

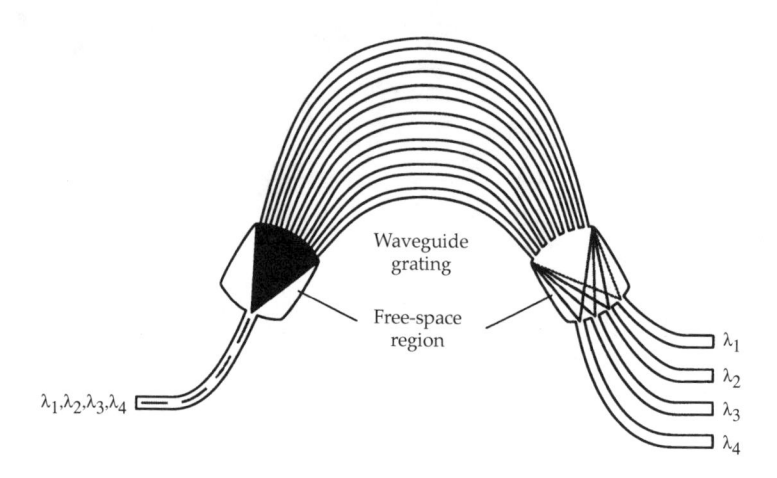

Figure 3.18: An arrayed waveguide router typically fashioned as a planar waveguide (courtesy of M. Zirngibl, Lucent Technologies, Crawford Hill, NJ).

Arrayed waveguide gratings may also be fashioned as an N X N star coupler (N input guides with the power in each input guide divided between the N output guides) or an add-drop multiplexer. This approach is especially attractive given that the loss associated with the splitting and combining is independent of the splitting number (unlike fused fiber star couplers). In general, these devices have a higher temperature sensitivity (on the order of 0.01 nm/°C and therefore generally require active temperature control. Multiplexers of this type with up to 144 x 144 (input channels x output channels) have been fabricated. This component has also been used to create a multiwavelength laser with accurately spaced wavelength channels suitable for WDM applications.[79]

3.10 SIGNAL ADD/DROP COMPONENTS

The flexibility of amplified systems may be increased by allowing a signal wavelength to be added and removed at a point in the fiber network. This add/drop function is typically accomplished using a combination of filters that transmit or reflect channel wavelengths along with couplers or circulators. Ideal devices would allow one of many channels, closely spaced in wavelength, to be added to or removed from a number of channels with sufficient isolation between adjacent channels, low sensitivity to the input state of polarization, and immunity to environmental effects. Presently, channel spacings of 0.81 nm (100 GHz) are standards recognized by the International Telecommunications Union for the 1550 nm wavelength range.

One version of an add/drop component using a combination of Bragg gratings and circulators was discussed in Section 3.8.2.[43] Other approaches use Bragg gratings within the two arms of an all-fiber Mach-Zehnder interferometer or in both arms separating two polarization beam splitter couplers (see Figure 3.19).[80, 81, 82, 83] Here, the narrow band add/drop function may be controlled by the Bragg reflective gratings. Any imbalance in the interferometer caused by nonidentical gratings is adjusted by a

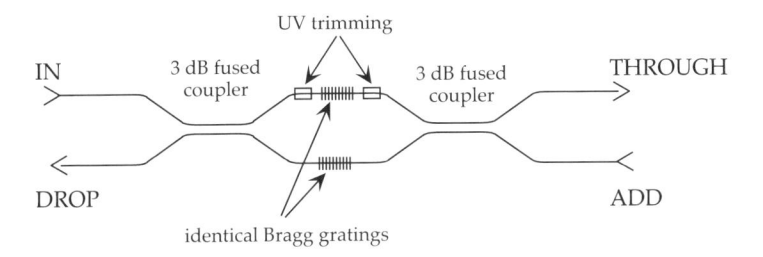

```
                    UV trimming
IN     3 dB fused   ↘ ↙    3 dB fused        THROUGH
       coupler      ▭▦▦▭   coupler
───┐  ┌──────────┐       ┌──────────┐  ┌───▶
   │  │          │       │          │  │
   └──┘          └───┬───┘          └──┘
                     │
───┐  ┌──────────┐  ▦▦▦  ┌──────────┐  ┌───
◀──┘  └──────────┘   │   └──────────┘  └───
DROP                 ↗↖                    ADD
            identical Bragg gratings
```

Figure 3.19: Add/drop component based on a Mach-Zehnder interferometer, consisting of two 3 dB couplers and two identical Bragg gratings at the drop wavelength. From reference [80]. The drop and add wavelength channels are reflected from the device to the appropriate port.

UV exposure in one or both arms to trim the performance. The UV exposure changes the refractive index of the glass, allowing for adjustment of the optical path length. This design has been demonstrated in both fiber and planar waveguide structures.

The add/drop function may also be performed by an interference filter arranged much the same as the pump and signal combiner previously shown in Figure 3.5. A commercial example of this device exhibits an insertion loss of less than 2 dB for add and drop with an adjacent channel isolation of greater than 30 dB. The wavelength temperature sensitivity is less than 0.0005 nm/°C.[85]

In addition to the single channel add/drop components mentioned above, there are potential applications for multiple channel add/drop components. A three wavelength add/drop structure using a combination of a Bragg reflector and four fiber biconic taper filters has been demonstrated.[84] A reconfigurable add/drop has also been demonstrated with arrayed-waveguide-grating multiplexers selecting a wavelength channel to be added or dropped with a thermo-optic switch.[86]

3.11 DISPERSION COMPENSATION COMPONENTS

The use of optical amplifiers has extended the span lengths to a point where the dispersion of the span may limit the system performance. Components which compensate for the span dispersion are then necessary to make the most use of the amplified systems. For dispersion normal fiber with a dispersion minimum near 1300 nm, dispersion is accumulated at 1550 nm at a rate of near 17 ps/nm · km. With chirp free sources, this chromatic dispersion limits the transmission distance of dispersion normal fiber to about 900 km at 2.5 Gbps and about 200 km at 5.0 Gbps. Additionally, for wavelength multiplexed systems which may use a 40 nm wide gain band and a span of 100 km there would be a delay difference of 68 nanoseconds for the extreme ends of the gain band. Equalizing this large dispersion spread is an additional challenge for compensation components in WDM amplified systems. In addition to linear dispersion, compensation is necessary in long haul systems to control optical nonlinearity induced impairments to system performance. The combination of self-phase modulation and chromatic dispersion may also generate degradation in digital systems as well as intermodulation distortion in analog modulated systems.[87]

64　　CHAPTER 3. COMPONENTS AND INTEGRATION

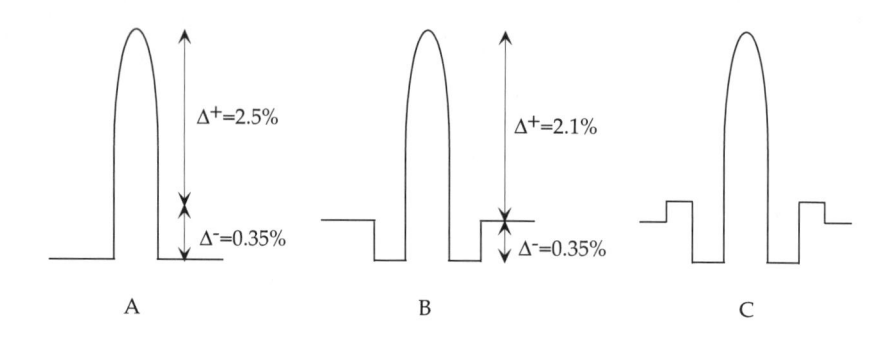

Figure 3.20: Refractive index profiles for three commercial dispersion compensating fiber designs. Designs A and B are from reference [91], design C is from reference [92]. The refractive index differences (Δs) are referenced to undoped silica.

Dispersion compensation is employed at the receiver, or the transmitter, or periodically along the span or at midspan locations. Components which provide the required negative dispersion are based upon dispersion compensating fiber, fiber Bragg gratings, higher-order spatial mode compensators, a bulk optic phased array, or by spectral inversion techniques.[88] With the exception of the spectral inversion technique, all methods provide inline fiber component blocks which cancel the dispersion by providing an opposite dispersion to that experienced in the fiber span. Dispersion compensation using chirped Bragg gratings has already been discussed in section 3.8.2.

Dispersion compensating fiber appears at present to be the most practical method, especially when compensation over a broad spectral range is required. Fiber designs which yield the desired negative dispersion require a core refractive index difference near 2%, higher than conventional transmission fiber designs, as well as a smaller core diameter and longer cutoff wavelength. As shown in Figure 3.20 these fiber designs may assume a step index, depressed clad or multiple clad structure.

The combination of high index and high germanium concentration as well as a higher order mode cutoff at a longer wavelength than standard fiber all contribute to a higher attenuation. In addition, these designs are more prone to bending induced losses and higher splice loss to standard fiber.[94] For this reason, a figure of merit (FOM) for the compensating fiber has been adopted with units of absolute chromatic dispersion per loss of the process (ps/nm · dB). Values of 418 ps/nm · dB and a dispersion of −295 ps/nm · km have been demonstrated.[89] More typical values for commercial fiber designs as shown in Figure 3.20 are a FOM of 250 and a dispersion of about −90 ps/nm · km.[91] In addition to this FOM there is an additional concern for the system impairment due to fiber nonlinearity which may arise from the small effective area ($\simeq 20\ \mu m^2$) of this fiber. When DCF is used in long distance systems with amplifier and DCF combined, an alternate FOM has been proposed which includes the influence of nonlinearity.[93]

Polarization mode dispersion for these fibers are on the order of 0.1 ps/$\sqrt{km}$. The attenuation spectrum and corresponding dispersion for a a commercial product are shown in Figure 3.21.[95] In addition to these fibers used to compensate for dispersion at 1550 nm in dispersion normal fiber, there is interest in fiber which would flatten

3.11. DISPERSION COMPENSATION COMPONENTS

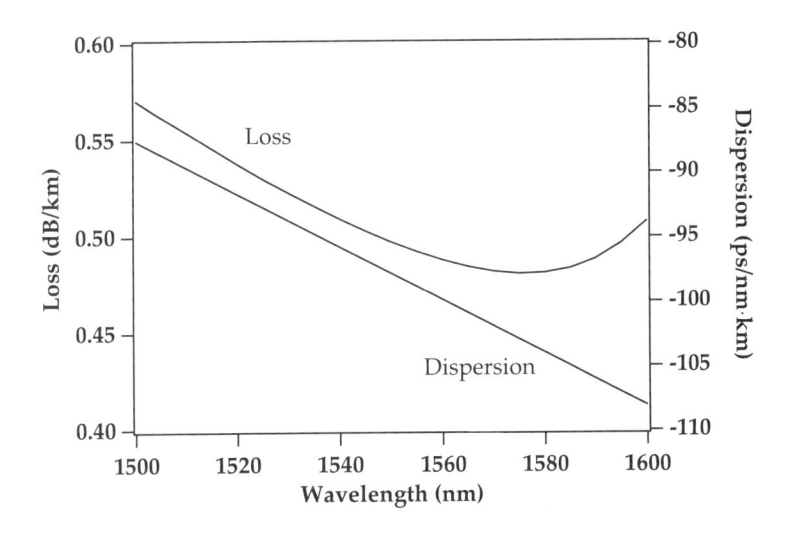

Figure 3.21: Loss and dispersion spectra for a commercially available dispersion compensating fiber. From reference [95] (courtesy L. Gruner-Nielsen, Specialty Fiber Devices Group, Lucent Technologies, Brondby, Denmark).

the wavelength dependent dispersion in dispersion shifted fiber.[96] These fibers would compensate for non-zero dispersion values at wavelengths near 1550 nm thereby flattening the dispersion slope, extending the useful wavelength range for WDM systems.

A bulk optic dispersion compensator has also been described which can provide a chromatic dispersion of -2000 ps/nm to $+2000$ ps/nm over a spectral range of 50 nm.[97, 98] Using this device, collimated light from a fiber is sent through a semi-cylindrical lens to a 1 mm thick angled glass plate. A single channel, 1800 ps/nm compensated, 10 Gb/s system demonstration has shown the potential of this device.[98] Due to the periodic transmission attenuation characteristics of this device, alignment of channel wavelength spacings with the peak device transmission is required.

The large negative waveguide dispersion of the LP_{11} higher-order mode near cutoff in a two mode fiber may also be used as a compensating fiber mechanism.[99] A fiber index difference of greater than 2% is necessary here as well. Theory suggests that a dispersion of 70 times that of conventional fiber ($\simeq 1000$ ps/nm · km) is possible with this technique. Low loss conversion of the LP_{01} transmission fiber mode to the LP_{11} mode and back is also a requirement for this method.

The mid-span spectral inversion technique creates an optical phase conjugate of the signal pulse at a position exactly half way in the transmission fiber length. The fiber length following the mid-span reverses the dispersion by propagating the reversed spectral components of the pulse through an equal length.[90]

The practicality of all dispersion compensating methods is primarily determined by the attenuation caused by the compensation method. Secondary issues are the ease of accomodating multiple wavelength channels, influence of nonlinear optical effects and

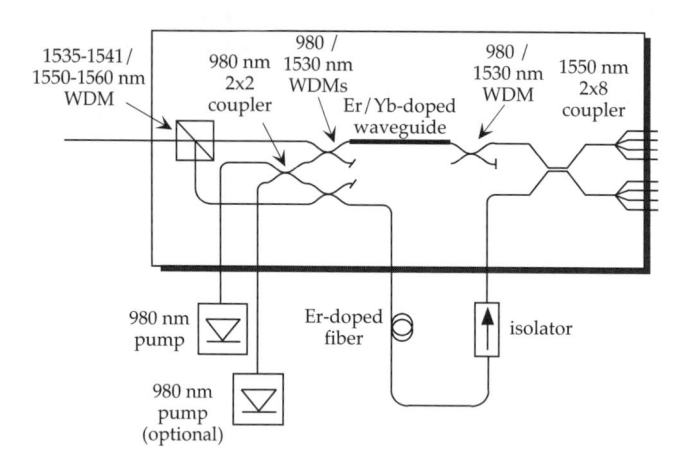

Figure 3.22: Hybrid two-window optical branching amplifier. From reference [101].

polarization mode dispersion. At this time, the dispersion compensating fiber appears to be the most practical solution.

3.12 INTEGRATED COMPONENTS

For amplifiers to become widely used, both the cost and size must be reduced. Ideally, as in the electronics industry, all components would be integrated on a common platform such as a planar integrated optic waveguide to reduce the cost of manufacture. Integration of some amplifier components has been demonstrated. For example, an isolator, signal monitor, and wavelength division multiplexer have been integrated in a 14-pin butterfly package.[100] Hybrid assemblies which include combinations of planar waveguide couplers, WDMs, and Er-Yb waveguides along with erbium-doped fiber and fiber isolators have been proposed for broadband interactive services (Figure 3.22).[101, 102] Ideally, all components as well as the doped waveguide could be fabricated as an integrated optic amplifier (see section 8.3.8).

3.13 PUMP LASERS

The success of erbium-doped fiber amplifiers has been predicated on the commercial availability of reliable diode laser pumps with power sufficient to stimulate gain from the device. The first experiments with diode lasers were reported in 1989, and commercial lasers made a widespread appearance a few years later.

There are several types of laser diode based pump lasers which have been demonstrated as pumps for erbium-doped fiber amplifiers:

- 1480 nm diode pump lasers
- 980 nm diode pump lasers
- 800 nm diode pump lasers

3.13. PUMP LASERS

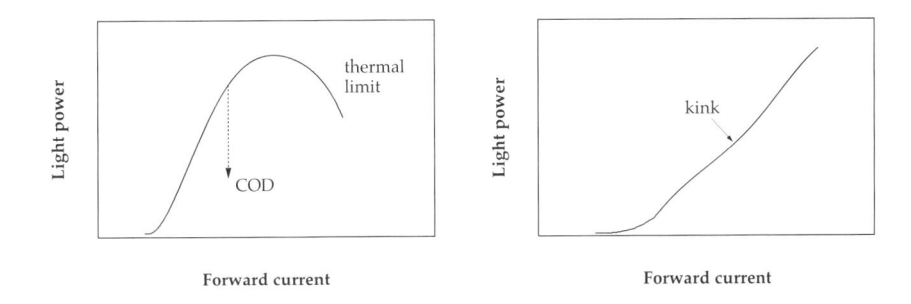

Figure 3.23: Typical L-I curves for diode lasers. The figure on the left shows the thermal limit and a possible COD (catastrophic optical damage) effect. The curve on the right shows a nonlinear kink. From reference [104] (reprinted with permission of Springer Verlag).

- 670 nm diode pump lasers
- high power solid state lasers pumped by laser diode arrays
- high power fiber lasers pumped by laser diode arrays
- MOPA (Master Oscillator Power Amplifier) lasers

Of these, the most commonly used today are the 1480 nm and 980 nm diode lasers. 1480 nm lasers were the first diode lasers to demonstrate acceptable reliability for use in telecommunications systems. They have been used in the first optically amplified undersea cable systems such as TAT-12,13. 980 nm pump lasers, which provide a lower noise figure than 1480 nm, have taken a longer time to prove adequate reliability for integration into field deployable systems. 670 nm diode lasers have been demonstrated in the laboratory.[103] The high power solid state and fiber lasers are used for high output power booster amplifiers. We briefly describe below some of the characteristics and typical operating parameters of these pump lasers.

Diode lasers are often characterized by their L-I curves (light output power vs forward current across the diode). Some general comments can be made about these L-I curves. The typical shapes of L-I curves are shown in Figure 3.23. The light power increases with current until the heating of the semiconductor by the electrical power causes the output power to roll over. This phenomenon is typically reversible. COD (catastrophic optical damage) causes a permanent damage in the laser. It occurs at the facet of the laser where there can be localized heating due to laser light absorption. Excess carriers created at the surface recombine nonradiatively thus heating the volume affected. An avalanche effect can occur leading to increased absorption and finally meltdown at a point in the laser facet. The laser is damaged and the output power suddenly drops to a very low value. For systems with low carrier surface recombination velocity, such as is the case for 1480 nm lasers, this effect is negligible. 980 nm lasers containing Al based layers can, however, suffer significantly from this effect.

Kinks, as shown in the right hand sketch of Figure 3.23, are caused by a lateral mode change with a change in current. Such kinks are usually accompanied by a

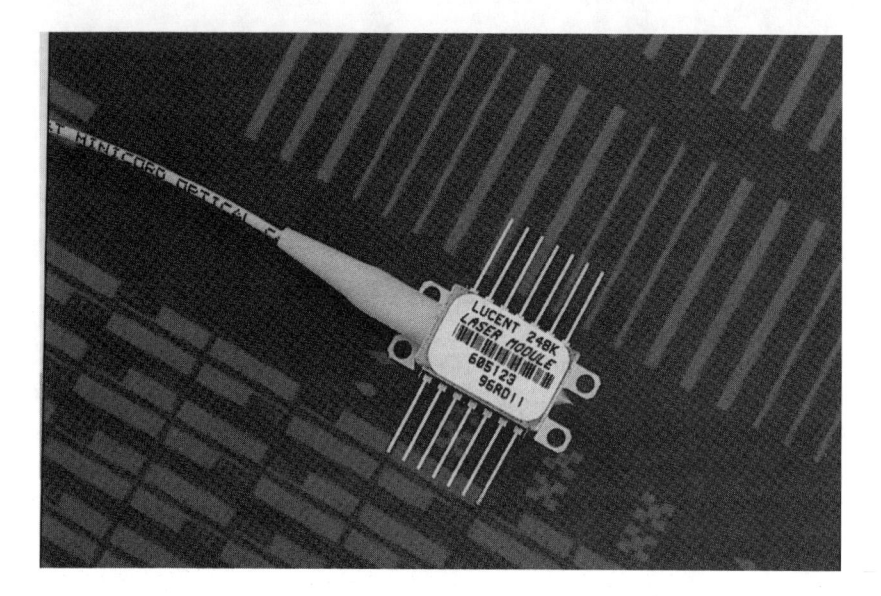

Figure 3.24: Packaged 1480 nm laser with single mode fiber pigtail, in a 14 pin butterfly package. From reference [105].

sudden change in the coupling efficiency to a single mode fiber pigtail. Kinks can be avoided by designing a laser to operate only in the fundamental mode. This can be accomplished by designing an active region in the shape of a stripe approximately 1 to 2 μm wide, or by creating a situation whereby the loss is very high for higher order modes.[104]

Diode laser pumps at 1480 nm and 980 nm are typically packaged in a 14 pin butterfly package, as shown in Figure 3.24. The laser is usually sold with an integrated single mode fiber pigtail. This fiber pigtail can be made from polarization maintaining fiber if the laser is to be polarization multiplexed with another one. The hermetically sealed package provides environmental and mechanical protection for the laser diode and the coupling region to the fiber pigtail. The pins deliver current to the laser and to a thermoelectric cooler if it is integrated with the laser. A rear facet monitor photodiode is usually included to provide a measure of the laser output power.

1480 nm lasers have a structure similar to that of 1.55 μm telecommunications lasers. They are typically multi-quantum well buried heterostructure Fabry-Perot lasers, where the multi-quantum well active region has a higher index of refraction and is surrounded on all sides by several layers of lower index semiconductor compounds. Whereas older laser structures were based on an InGaAsp/InP structure, nowadays 1480 nm lasers are based on a quarternary/quarternary structure, InGaAsP/InGaAsP. Such strongly index guided structures provide good mode confinement and thus the laser has a well defined single mode output. Figure 3.25 shows the structure and mounting arrangement of a commercial 1480 nm pump laser of the CMBH (capped mesa buried heterostructure) type. The active region is made up of InGaAsP/InGaAsP quantum wells. The laser chip has been flipped over before bonding such that the InP

3.13. PUMP LASERS 69

1.48μm FIBER PUMP LASER
(CMBH-FP LASER DIODE)

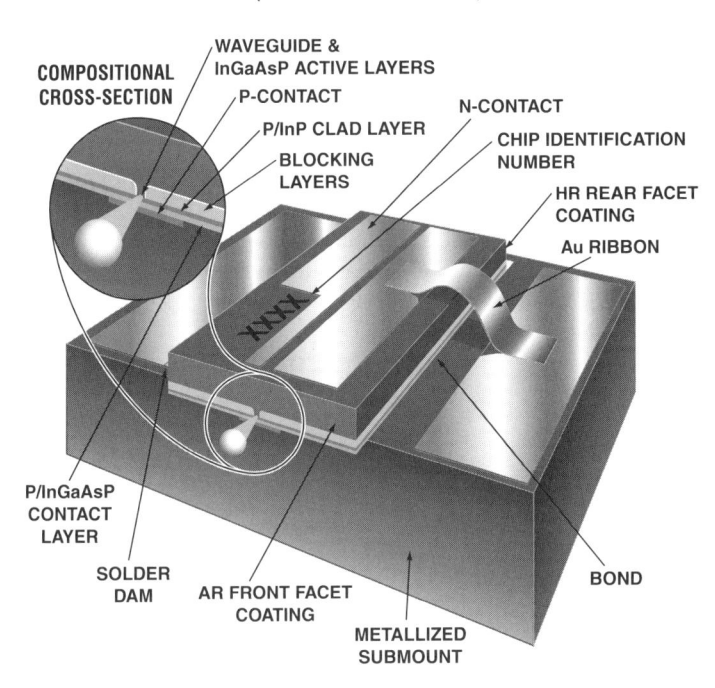

Figure 3.25: Structure and mounting arrangement of a commercial 1480 nm pump laser (courtesy D. Wilt, Microelectronics Group, Lucent Technologies, Breinigsville, PA).

substrate appears at the top. Because manufacturing and reliability of the buried heterostructure laser is relatively well understood, from its use as a transmitter laser, it provides a natural technology platform for 1480 nm pump lasers. The reliability of commercial 1480 nm lasers available (1998 data) is characterized by an approximate MTTF (mean time to failure) of 10^6 hours. Some characteristic operating parameters of a commercial 1480 nm laser diode are shown in Table 3.5. Such commercial lasers are usually rated to operate between -40 °C and +70 °C with forward currents allowed up to 750 mA. Temperature control is achieved with an integrated thermoelectric cooler. The L-I curve and spectrum for a commercial 1480 nm pump laser are shown in Figure 3.26. High power at 1480 nm from laser diode pumps can be obtained by wavelength multiplexing 1480 nm laser operating at different wavelengths (e.g., 1463 nm, 1478 nm, 1493 nm).[107] This can yield total pump powers approaching 1 W when polarization multiplexing is also used. The absorption in the 1480 nm region (see Figure 6.1) is a guide to the available range for pump lasers operating near 1480 nm.

Parameter	Condition	Min.	Typ.	Max.	Unit
Threshold current	CW operation	-	20	50	mA
Optical output power	$I_f = 600$mA	140	-	-	mW
Forward voltage	$I_f = 600$mA	-	2	2.5	V
Center wavelength	$I_f = 600$mA	1465	-	1495	nm
3 dB Spectral width	$I_f = 600$mA	-	-	10	nm

Table 3.5: Selected operating characteristics of a commercial 1480 nm InGaAsP/InP multi-quantum well laser diode pump under CW operation (I_f is the forward current across the diode; optical power is at the output of a single mode fiber pigtail attached to the laser). From reference [106] (courtesy Sumitomo Electric Lightwave Corporation, Research Triangle Park, NC).

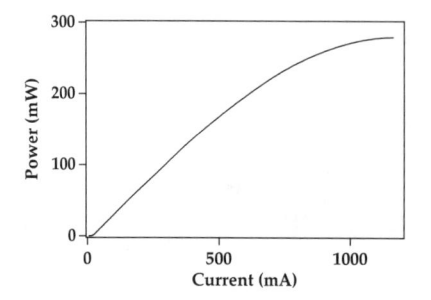

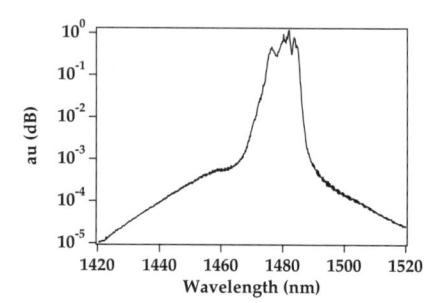

Figure 3.26: Left: L-I (light output power vs forward current) curve of a commercial 1480 nm pump laser, measured directly from the chip. The coupling efficiency to a single fiber pigtail packaged with the laser is typically between 60 % and 85 %. Right: output spectrum (on a relative dB scale) of the same 1480 nm laser diode (measurement resolution 0.2 nm). Data courtesy K. Wang, Microelectronics Group, Lucent Technologies, Breinigsville, PA.

Diode laser pumps operating at 980 nm provide a lower noise figure for erbium-doped fiber amplifiers, as discussed in Chapter 7 (section 7.3.3), and are thus more desirable for a system where low noise figure is of paramount concern. These lasers are also electrically more efficient than 1480 nm lasers, as can be seen from a comparison of the L-I curves for the two types of lasers. 980 nm lasers are made from the GaAs/GaAlAs materials system. Because the materials growth required for making buried heterostructure lasers is very difficult in this material system, 980 nm lasers are usually of the ridge type that does not require regrowth steps. The result is an inferior mode confinement as compared to buried heterostructure lasers. This causes the lasers to be more susceptible to kinks and to possess an asymmetric far field pattern. This results in more difficult coupling into single mode fibers and renders the laser packaging more complex. The typical structure of a 980 nm laser is shown in Figure 3.27. The laser shown in Figure 3.27 has an active region consisting of a single 7 nm GaAs quantum well sandwiched between two AlGaAs graded-index regions. The ridge width is 3 μm.

3.13. PUMP LASERS 71

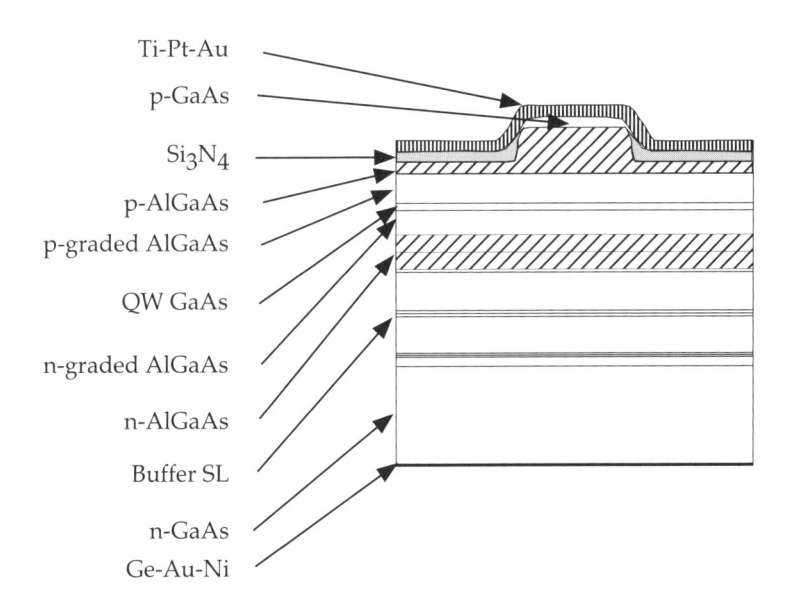

- Ti-Pt-Au
- p-GaAs
- Si$_3$N$_4$
- p-AlGaAs
- p-graded AlGaAs
- QW GaAs
- n-graded AlGaAs
- n-AlGaAs
- Buffer SL
- n-GaAs
- Ge-Au-Ni

Figure 3.27: Schematic of the composition and band structure of a strained SQW 980 nm pump laser. From reference [108]. Reprinted with permission from F. R. Gfeller and D. J. Webb, *J. App. Phys.* Vol. 68, p. 14 (1990). Copyright 1990 American Institue of Physics.

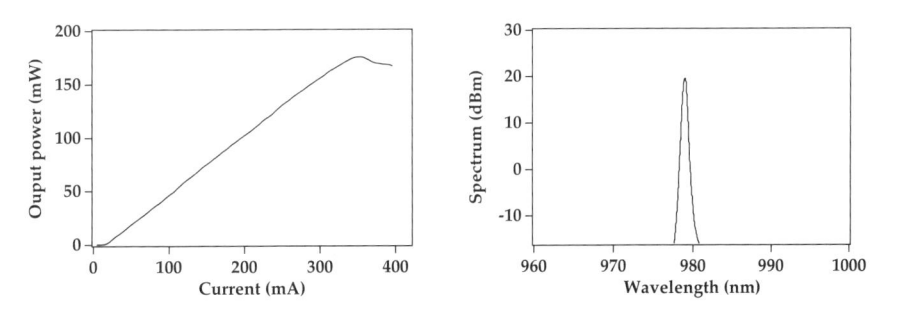

Figure 3.28: L-I curve (left) and spectrum (right) of a commercial 980 nm pump laser, spectrum stabilized by a fiber Bragg grating, measured at the output of the single mode fiber pigtail attached to the laser. From reference [111] (courtesy S. Pontelet, SDL, Inc., San Jose, CA).

The L-I curve and spectrum for a commercial cooled 980 nm pump laser are shown in Figure 3.28. This 980 nm laser is wavelength stabilized by a fiber Bragg grating. It was shown early on that 980 nm lasers were subject to hopping in the central wavelength of operation. This causes changes in the gain and noise figure of the amplifier pumped since the 980 nm region absorption is relatively narrow. A small amount of reflection from a fiber Bragg grating locks the operating wavelength of the laser and

Parameter	Condition	Min.	Max.	Unit
Operating current	CW operation	-	350	mA
Threshold current	CW operation	-	25	mA
Optical output power	$P_{kink-free}$ $T_{laser} = 25°C$	145	-	mW
Forward voltage	I_{op}	-	2.5	V
Spectral cutoff (low)	I_{op}	974	-	nm
Spectral cutoff (high)	I_{op}	-	985	nm
Spectrum stability	I_{op}, t = 60 seconds T_{case} = constant	-	0.1	nm

Table 3.6: Selected operating characteristics of a fiber grating stabilized commercial 980 nm laser diode pump under CW operation (I_{op} is the operating current across the diode; $P_{kink-free}$ is the output power in the kink-free region of the output power vs current curve). From reference [112] (courtesy SDL, Inc., San Jose, CA).

prevents the hopping.[109] Some characteristic operating parameters of a commercial 980 nm laser diode are shown in Table 3.6.

980 lasers were beset by reliability problems in the early 1990's. Degradation has been documented at both the laser and packaging level. Recently, due to advances in fabrication and processing, commercial 980 nm lasers have achieved the reliablity levels necessary for commercial deployment in telecom systems. Some research groups have suggested that one can mitigate the reliability problem of 980 nm lasers by using an Al free structure, such as InGaAs/InGaAsP/InGaP, as opposed to InGaAs/GaAs/AlGaAs structures.[113, 114] This reduces the rate of formation of defects which might lead to sudden failure, in particular on the facets of the lasers.[113, 114, 115] Proper packaging has been shown to be key in achieving 980 nm laser reliability.[116] A major advance has been in laser chip processing methods, which significantly reduces facet corrosion and catastrophic optical mirror damage.[117] This has lead to high reliability for 980 nm pump lasers.[118]

The reliability of laser diodes is often quoted in FITS. One FIT corresponds to 1 Failure in a Time of 10^9 device hours. Commercial pump lasers today have FITS of about 1,000, much improved from the values of 10,000 which were quoted only a few years ago. Pump lasers for undersea cable applications are quote with a FIT of about 50 as they are operated at low output powers. Measurements of the reliability and degradation mechanisms of pump lasers are conducted by accelerated aging tests. The lasers are run in a high temperature environment which accelerates the degradation. The degradation times are then extrapolated back to operating temperature. The failures are characterized either as gradual degradation or sudden failures. Gradual degradation is often defined to have occured when the current has to be increased by 50% to maintain a constant optical output power. Sudden failures are the COD failures described previously.

MOPA (Master Oscillator Power Amplifier) lasers have been used in a research environment for both high power erbium-doped fiber amplifiers, and also for 1.3 μm amplifiers using Pr^{3+} doped fluoride fibers.[122, 123] The principle of operation of a

3.13. PUMP LASERS

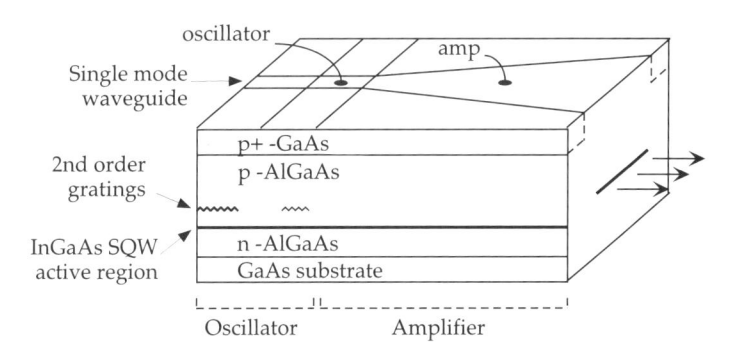

Figure 3.29: Structure of a MOPA laser. From reference [110] (©1993 IEEE). The two oscillator sections are each 750 μm in length, while the amplifier section is 2 mm in length and the width of the laser output stripe is 200 μm. The oscillator and amplifier regions have separate electrical contacts.

MOPA is based on the following. To dramatically increase the output power from a semiconductor laser, a semiconductor amplifier in placed directly after the laser, and integrated with it on the same substrate. This is the so-called MOPA configuration, and is shown in Figure 3.29.[110] The amplifier section is tapered with a progressively wider opening so as to allow the laser beam to freely diffract as it propagates along the power amplifier length.

Diode laser pumped solid state laser sources have, to date, been used for pumping Er^{3+}-Yb^{3+} and Pr^{3+}-doped fiber amplifiers.[119, 120] Diode arrays are used to pump a a Nd^{3+} doped crystal (typically YLF, YAG, or YVO_4) which lases in the vicinity of 1.05 to 1.06 μm. High output powers are available by using multiple diode array pumps, arranged in a suitable fashion around the rare earth doped crystal. The laser beam, which is defined by the laser resonator the crystal is placed in, is typically a TEM_{00} mode beam. This makes for easier coupling into a single mode fiber. One drawback is that thermal lens effects at high powers can sometimes render TEM_{00} operation difficult to achieve. Output powers from commercial versions of these lasers is of the order of several W, in CW operation. Demonstrations of output powers in excess of 10 W have also been reported.[121]

High power fiber lasers pumped by diode laser arrays are finding increased applications as pumps for high power fiber amplifiers.The fiber lasers can be based either on rare earth doped cladding pumped lasers, or cascaded Raman resonator lasers. A sketch of a cladding pumped laser is shown in Figure 3.30. The double clad fiber is constituted by a doped single mode fiber core surrounded by a multimode core.[124] The pump light from the diode array can be readily coupled into the multimode fiber core. A rectangular shape for the multimode core is preferred for efficient coupling of radiation to the inner single mode core, and matches well the geometric aspect ratio of diode laser pumps.[124] The light overlaps the single mode fiber core as it illuminates the multimode core, and inverts the dopant ions. The laser cavity reflectors include flat microsheets places against the fiber end as well as fiber gratings.[125] The output of these devices is constrained to be single mode by the single mode fiber core, hence

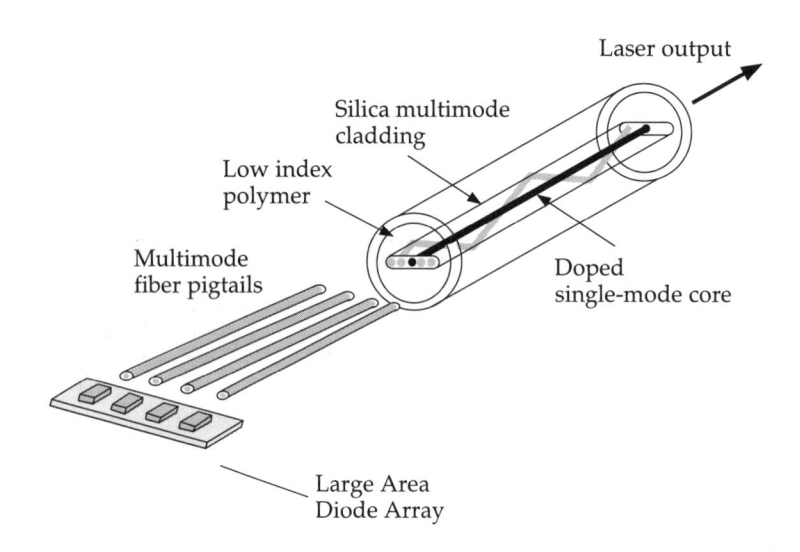

Figure 3.30: Structure of a double clad fiber laser. The rare earth doped single mode core is surrounded by a cladding in which the pump light propagates. Only one pump ray is shown for illustrative purposes. In practice, there are multiple rays propagating simultaneously within the cladding. Reflectors are placed at each end to form the laser cavity.

the name brightness converters (from large area multimode to single mode) often given to these devices. This facilitates coupling to fiber amplifiers with similar single mode cores. These lasers have become ever more popular as the advantages of diode laser arrays have become apparent. The advantages include the large scaleability of diode laser array pump sources as well as their continuing decline in cost.

An early demonstration showed that a double-clad Er^{3+}-Yb^{3+}-doped fiber amplifier could be pumped by a diode array operating at 960 nm. The pump light was absorbed by the Yb^{3+} ions, and then transfered to the Er^{3+} ions. This yielded an output power of 17 dBm at 1535 nm, with a pump power of 620 mW.[126] Since then, substantial increases in the power output from double clad fiber lasers has been recorded. For the purpose of pumping erbium-doped fiber amplifiers, double clad fiber lasers have been fabricated with both Nd^{3+} and Yb^{3+} dopants. The Nd^{3+} laser emission at 1.06 μm can be used to pump a co-doped Yb^{3+}-Er^{3+} amplifier.[127] The Yb^{3+} fiber laser, on the other hand, can be used to pump a cascaded Raman resonator laser which can then be used to pump an erbium-doped fiber amplifier. Using Nd^{3+} as an active medium, an output power of 5 W at 1064 nm was demonstrated, with a pump power of 13.5 W from a diode laser array at 807 nm.[128]

Very high output powers have been achieved with Yb^{3+} cladding pumped fiber lasers. An output power of 20.4 W at 1101 nm was obtained with a Yb^{3+} doped cladding pumped fiber laser, using a commercial 915 nm diode laser bar.[129] Using

3.13. PUMP LASERS

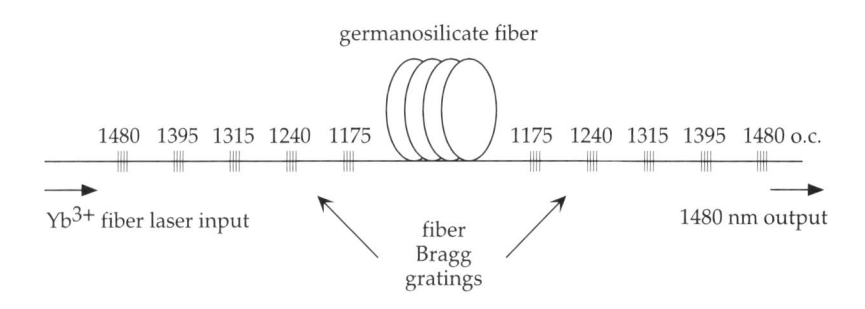

Figure 3.31: Structure of a cascaded Raman resonator fiber laser for output at 1480 nm when pumping at 1100 nm (o.c.: output coupler). The pairs of Bragg gratings form laser resonators, where each frequency pumps the next order Stokes emission. From reference [131].

this laser in turn to pump a a cascaded Raman resonator laser, an output power of 8.5 W at 1472 nm was obtained.[129] High power commercial fiber lasers are now becoming available. Fiber lasers with an output power of 9 W (pumped by a 20 W fiber coupled diode bar) at 1.1 μm can be obtained.[130]

Cascaded Raman resonator fiber lasers are fabricated by defining Raman resonators with UV induced Bragg gratings, where each Raman laser pumps the subsequent Stokes emission.[132, 133] This is shown in Figure 3.31 for the case of a laser pumped at 1100 nm with an output at 1480 nm.[131] This particular laser achieved an output power of 1.5 W at 1484 nm with a slope efficiency of 46% when pumped with an Yb^{3+} laser at 1117 nm. The spectral width is about 2 nm, with a high suppression ratio between the 1484 nm line and the intermediate Stokes emission lines. These lasers have shown great promise as high power 1480 nm pump sources, and can be used for remote pumping of erbium-doped fiber amplifiers or for power amplifier applications.

Bibliography

[1] J. L. Gimlett, M. Z. Iqbal, N. K. Cheung, A. Righetti, F. Fontana, and G. Grasso, *IEEE Phot. Tech. Lett.* **2**, 211 (1990).

[2] C. M. Miller, S. C. Mettler, and I. A. White, *Optical Fiber Splices and Connectors: Theory and Methods* (Marcel Dekker, NY), p. 152.

[3] R. G. Ajemian, "A selection guide for fiber optic connectors," *Optics and Photonics News*, pp. 32–36 (June 1995).

[4] N. Suzuki, M. Saruwatari, and M. Okuyama, *Elect. Lett.* **22**, 110 (1986).

[5] H. R. D. Sunak and S. P. Bastien, *IEEE Phot. Tech. Lett.* **1**, 146 (1989).

[6] D. Marcuse, *Bell Syst. Tech. J.* **56**, 703 (1977).

[7] M. Takahashi, "Novel stepped ferrule for angled convex polished optical-fiber connector," in *Conference on Optical Fiber Communication*, Vol. 8, 1995 OSA Technical Digest Series, (Optical Society of America, Washington, D.C., 1995), pp. 184–185.

[8] Product sheet, Ultra high return loss connectors, RIFOCS Corp., Camarillo, CA, 1992.

[9] K. Furuya, T. C. Chong, and Y. Suematsu, *Trans. IEICE Japan*, **E61**, 957 (1978).

[10] N. Amitay, H. M. Presby, F. V. DiMarcello, and K. T. Nelson, *J. Light. Tech.* **LT-5**, 70 (1987).

[11] W. Frost, P. Ruffin and W. Long, "Computational model of fiber optic, arc fusion splicing; analysis," in *Fiber Optics Reliability: Benign and Adverse Environments II, (Proc. SPIE)* **992**, pp. 296–311 (1988).

[12] J. T. Krause and C. R. Kurkjian, *Elect. Lett.* **21** 533 (1985).

[13] J. T. Krause, S. N. Kher, and D. Stroumbakis, "Arc fusion splices with near pristine strengths and improved optical loss," in *22nd European Conference on Optical Communication*, Proceedings Vol. 2, pp. 2.237–2.240 (1996).

[14] J. T Krause, W. A. Reed, K. L Walker, *J. Light. Tech.* **LT-4**, 837 (1986).

[15] J. S. Harper, C. P. Botham and S. Hornung, *Elect. Lett.* **24**, 245 (1988).

[16] C. P. Botham, *Elect. Lett.* **24**, 243 (1988).

[17] D. B. Mortimore and J. W. Wright, *Elect. Lett.* **22**, 318 (1986).

[18] K. Shiraishi, Y. Aizawa, and S. Kawakami, *J. Light Tech.* **8**, 1151 (1990).

[19] M. J. Holmes, F. P. Payne and D. M. Spirit, *Elect. Lett.* **26**, 2103 (1990).

[20] H. Y. Tam, *Elect. Lett.* **27**, 1597 (1991).

[21] S. G. Kosinski, K. W. Quoi, and J. T. Krause, "Low-loss (0.15 dB) arc fusion splicing of erbium-doped fibers with high strength (400 ksi) for highly reliable optical amplifier systems," in *Conference on Optical Fiber Communication*, Vol. 5, 1992 OSA Technical Digest Series (Optical Society of America, Washington, D.C., 1992), p. 231.

[22] W. Zheng, O. Hulten, and R. Rylander, *J. Light. Tech.* **12**, 430 (1994).

[23] A. Lord, I. J. Wilkinson, A. Ellis, D. Cleland, R. A. Garnham, and W. A. Stallard, *Elect. Lett.* **26**, 900 (1990).

[24] This reference number has not been used.

[25] Product catalog, E-TEK Dynamics, Inc., San Jose, CA, 1998.

[26] D. W. Hall, C. M. Truesdale, D. L. Weidman, M. E. Vance and L. J. Button, "Wavelength division multiplexers for 980 nm pumping of erbium-doped fiber optical amplifier," in *Optical Amplifiers and Their Applications*, Vol. 13, OSA Technical Digest Series (Optical Society of America, Washington, D.C., 1990), pp. 222–225.

[27] D. L. Weidman and D. W. Hall, "Low-loss dissimilar-fiber wavelength division multiplexers for 980nm pumped fiber-optic amplifiers," in *Conference on Optical Fiber Communications*, Vol. 5, 1992 OSA Technical Digest Series (Optical Society of America, Washington, D.C., 1992), p. 232.

[28] H. Bulow and R. H. Robberg, "Performance of twisted fused fiber EDFA pump and WDM-couplers," in *17th European Conference on Optical Communication*, Proceedings Part 1, pp. 73–76 (1991).

[29] J. D. Minelly and M. Suyama, *Elect. Lett.* **26**, 523 (1990).

[30] S .L. Hansen, K. Dybdal, and C. C. Larsen, "Upper gain limit in Er-doped fiber amplifiers due to internal Rayleigh backscattering," in *Conference on Optical Fiber Communications*, Vol. 5, 1992 OSA Technical Digest Series (Optical Society of America, Washington, D.C., 1992), p. 68.

[31] E. Desurvire, J. R. Simpson, and P. C. Becker, *Opt. Lett.* **12**, 888 (1987).

[32] R. G. Mckay, R. S. Vodhanel, R. E. Wagner and R. I. Laming, "Influence of forward and backward travelling reflections on the gain and ASE spectrum of EDFA's," in *Conference on Optical Fiber Communication*, Vol. 5, 1992 OSA Technical Digest Series (Optical Society of America, Washington, D.C., 1992), pp. 176–177.

[33] K. Chang, S. Schmidt, W. Sorin, J. Yarnell, H. Chou and S. Newton, "A high-performance optical isolator for lightwave systems," *Hewlett-Packard Journal*, pp. 45–50 (1991).

[34] Polarization insensitive fiber isolator, Classic series, 1998 Product catalog, E-TEK Dynamics, Inc.

[35] T. Irie, K. Shiraishi, T. Sato, R. Kashara and S. Kawakami, "Fiber-integrated isolators with high performance," in *Conference on Optical Fiber Communication*, Vol. 2, 1996 OSA Technical Digest Series (Optical Society of America, Washington, D.C., 1996), pp. 53–54.

[36] Y. Fujii, *J. Light. Tech.* **9**, 456 (1991).

[37] CR1500 series optical circulators, Product Review 1995/96 Performance Highlights, JDS Fitel Inc., Nepean, Ontario, Canada.

[38] J.-M. P. Delavaux, J. A. Nagel, K. Ogawa, and D. DiGiovanni, *Opt. Fiber Tech.* **1**, 162 (1995).

[39] M. Koga, J. Minowa, and T. Matsumoto, *Trans. IEICE Japan* **E72**, 1086 (1989).

[40] D. R. Huber, "Erbium-doped fiber amplifier with a 21 GHz optical filter based on a in-fiber Bragg grating," *18th European Conference on Optical Communications*, Proceedings Volume 1, pp. 473–476 (1992).

[41] Y. Sato, Y. Yamabayashi, and K. Aida, "Monitoring the noise figure of EDFAs with circulators via backward amplified spontaneous emission," in *Optical Amplifiers and Their Applications*, Vol. 14, 1993 OSA Technical Digest Series, (Optical Society of America, Washington, D.C., 1993), pp. 202–205.

[42] J.-M. P. Delavaux and T. Strasser, *Opt. Fiber Tech.* **1**, 318 (1995).

[43] C. R. Giles and V. Mizrahi, "Low-loss ADD/DROP multiplexers for WDM lightwave networks," *Tenth International Conference on Integrated Optics and Optical Fibre Communication, IOOC-95*, Technical Digest Vol. 3, pp. 66–67 (1995).

[44] M. J. Cole, H. Geiger, R. I. Laming, S. Y. Set, M. N. Zervas, W. H. Loh, and V. Gusmeroli, "Continuously chirped, broadband dispersion-compensating fibre gratings in a 10 Gb/s 110 km standard fibre link," *22nd European Conference on Optical Communication*, Proceedings Vol. 5, pp. 5.19–5.22 (1996).

[45] Y. Sato, and K. Aoyama, *IEEE Phot. Tech. Lett.* **3**, 1001 (1991).

[46] J. H. Povlsen, A. Bjarklev, O. Lumholt, H. Vendeltorp-Pommer, K. Rottwitt, and T. Rasmussen, "Optimizing gain and noise performance of EDFA's with insertion of a filter or an isolator," *Fiber Laser Sources and Amplifiers* III, M. J. F. Digonnet and E. Snitzer, Ed., *Proc. SPIE* **1581**, pp. 107–113 (1991).

[47] J. L. Zyskind, R. G. Smart and D. DiGiovanni, "Two-stage EDFA's with counterpumped first stage suitable for long-haul soliton systems," in *Conference on Optical Fiber Communication*, Vol. 4, 1994 OSA Technical Digest Series (Optical Society of America, Washington, D.C., 1994), pp. 131–132.

[48] Y. K. Park and S. W. Granlund, *Opt. Fiber Tech.* **1,** 59 (1994).

[49] S. R. Chinn, *Elect. Lett.* **31**, 756 (1995).

[50] S. R. Chinn, D. M. Boroson, and J. C. Livas, *J. Light. Tech.* **14**, 370 (1996).

[51] C. M. Miller, "A field-worthy high-performance tunable fiber Fabry-Perot filter," in *16th European Conference on Optical Communication*, Proceedings Volume 1, pp. 605–608 (1990).

[52] R. A. Betts, S. J. Frisken, and D. Wong "Split-beam Fourier filter and its application in a gain-flattened EDFA," in *Conference on Optical Fiber Communication*, Vol. 8, 1995 OSA Technical Digest Series (Optical Society of America, Washington, D.C., 1995), pp. 80–81.

[53] G. Grasso, F. Fontana, A. Righetti, P. Scrivener, P. Turner and P. Maton, "980-nm diode pumped Er-doped fiber optical amplifiers with high gain-bandwidth product," in *Conference on Optical Fiber Communication*, Vol. 4, 1991 OSA Technical Digest Series (Optical Society of America, Washington, D.C., 1991), p. 195.

[54] C. V. Poulsen, O. G. Graydon, R. I. Laming, M. N. Zervas, and L. Dong, *Elect. Lett.* **32**, 2166 (1996).

[55] A. M. Vengsarkar, P. J. Lemaire, J. B. Judkins, J. E. Sipe, and T. Erdogan, "Long-period Fiber Gratings as Band-Rejection Filters," in *Conference on Optical Fiber Communication*, Vol. 8, 1995 OSA Technical Digest Series (Optical Society of America, Washington, D.C., 1995), pp. 339–342.

[56] K. O. Hill, Y. Fujii, D. C. Johnson, and B. S. Kawasaki, *Appl. Phys. Lett.* **32**, 647 (1978).

[57] G. Meltz, W. W. Morey, and W. H. Glenn, *Opt. Lett.* **14**, 823 (1989).

[58] D. Z. Anderson, V. Mizrahi, T. Erdogan, and A. E. White, "Phase-mask method for volume manufacturing of fiber gratings," in *Conference on Optical Fiber Communication*, Vol. 4, 1993 OSA Technical Digest Series (Optical Society of America, Washington, D.C., 1993), pp. 335–337.

[59] K. O. Hill, F. Bilodeau, B. Malo, J. Albert, and D. C. Johnson, "Application of phase masks to the photolithographic fabrication of Bragg gratings in conventional fiber/planar waveguides with enhanced photosensitivity," in *Conference on Optical Fiber Communication*, Vol. 4, 1993 OSA Technical Digest Series (Optical Society of America, Washington, D.C., 1993), pp. 331–334.

[60] P. J. Lemaire, R. M. Atkins, V. Mizrahi, and W. A. Reed, *Elect. Lett.* **29**, 1191 (1993).

[61] V. Mizrahi, P. J. Lemaire, T. Erdogan, W. A. Reed, D. J. DiGiovanni, and R. M. Atkins, *Appl. Phys. Lett.* **63**, 1727 (1993).

[62] V. Mizrahi and J. E. Sipe, *J. Light. Tech.* **11**, 1513 (1993).

[63] A. M. Vengsarkar, P. J. Lemaire, J. B. Judkins, V. Bhatia, T. Erdogan, and J. E. Sipe, *J. Light. Tech.* **14**, 58 (1996).

[64] C. R. Giles, T. Erdogan, and V. Mizrahi, "Simultaneous wavelength-stabilization of 980-nm pump lasers," in *Optical Amplifiers and Their Applications*, Vol. 14, 1993 OSA Technical Digest Series, (Optical Society of America, Washington, D.C., 1993), pp. 380–383.

[65] C. E. Soccolich, V. Mizrahi, T. Erdogan, P. J. Lemaire, and P. Wysocki, "Gain enhancement in EDFas by using fiber-grating pump reflectors," in *Conference on Optical Fiber Communication*, Vol. 4, 1994 OSA Technical Digest Series (Optical Society of America, Washington, D.C., 1994), pp. 277–278.

[66] L. Dong, M. J. Cole, A. D. Ellis, M. Durkin, M. Ibsen, V. Gusmeroli, and R. I. Laming, "40 Gbit/s 1.55 um transmission over 109 km of non-dispersion shifted fibre with long continuously chirped fibre gratings," in *Conference on Optical Fiber Communication*, Vol. 6, 1997 OSA Technical Digest Series (Optical Society of America, Washington, D.C., 1997), pp. 391–394.

[67] A. M. Vengsarkar, P. J. Lemaire, G. Jacobovitz-Veselka, V. Bhatia, and J. B. Judkins, "Long-period fiber gratings as gain-flattening and laser stabilizing devices," *Tenth International Conference on Integrated Optics and Optical Fibre Communication, IOOC-95*, Technical Digest Vol. 5, pp. 3–4 (1995).

[68] E. M. Dianov, V. I. Karpov, A. S. Kurkov, O. I. Medvedkov, A. M. Prokhorov, V. N. Protopopov, and S. A. Vasil'ev, "Gain spectrum flattening of erbium-doped fiber amplifier using long period fiber grating," in *Photosensitivity and Quadratic Nonlinearity in Glass Waveguides: Fundamentals and Applications*, Vol. 22, 1995 OSA Technical Digest Series (Optical Society of America, Washington, D.C., 1995), pp. 14–17.

[69] J. B. Judkins, J. R. Pedrazzani, D. J. DiGiovanni, and A. M. Vengsarkar, "Temperature-insensitive long-period fiber gratings," in *Conference on Optical Fiber Communication*, Vol. 2, 1996 OSA Technical Digest Series (Optical Society of America, Washington, D.C., 1996), pp. 331–334.

[70] A. M. Vensarkar, J. R. Pedrazzani, J. B. Judkins, P. J. Lemaire, N. S. Bergano, and C. R. Davidson, *Opt. Lett.* **21**, 336 (1996).

[71] P. F. Wysocki, J. Judkins, R. Espondola, M. Andrejco, A. Vengsarkar, and K. Walker, "Erbium-doped fiber amplifier flattened beyond 40 nm using long-period gratings," in *Conference on Optical Fiber Communication*, Vol. 6, 1997

OSA Technical Digest Series (Optical Society of America, Washington, D.C., 1997), pp. 375–378.

[72] H. S. Kim, S. H. Yun, I. K. Hwang, and B. Y. Kim, "Single-mode-fiber acousto-optic tunable notch filter with variable spectral profile," in *Conference on Optical Fiber Communication*, Vol. 6, 1997 OSA Technical Digest Series (Optical Society of America, Washington, D.C., 1997), pp. 395–399.

[73] 8 channel dense WDM, 1995 product sheet, Optical Corporation of America, Marlborough, MA.

[74] M. K. Smit, *Elect. Lett.* **24**, 385 (1988).

[75] A. R. Vallekoop and M. K. Smit, *J. of Light. Tech.* **9**, 310 (1991).

[76] H. Takahashi et. al., *Elect. Lett.* **26**, 87 (1990).

[77] C. Dragone, *J. Opt. Soc. Amer.* **A7**, 2081 (1990).

[78] C. Dragone, *IEEE Phot. Tech. Lett.* **3**, 9, 812 (1991).

[79] M. Zirngibl, B. Glance, L. W. Stulz, C. H. Joyner, G. Raybon, and I. P. Kaminow, *IEEE Phot. Tech. Lett.* **6**, 1082 (1994).

[80] D. C. Johnson, K. O. Hill, F. Bilodeau, and S. Faucher, *Elect. Lett.* **23**, 668 (1987).

[81] F. Bilodeau, B. Malo, D. C. Johnson, J. Albert, S. Theriault, and K. O. Hill, "High Performance wavelength-division-multiplexing/demultiplexing device using an all-fiber Mach-Zehnder interferometer and photoinduced Bragg grating," in *Conference on Optical Fiber Communication*, Vol. 8, 1995 OSA Technical Digest Series (Optical Society of America, Washington, D.C., 1995), pp. 130–132.

[82] D. Bilodeau, C. Johnson, S. Theriault, B. Malo, J. Albert, and K. O. Hill, *IEEE Phot. Tech. Lett.* **7**, 388 (1995).

[83] S. Y. Kim, S. B. Lee, J. Chung, S. Y. Kim, J. C. Jeong, and S. S. Choi, "Highly stable optical add/drop multiplexer using polarization beam splitters and fiber Bragg gratings." in *Conference on Optical Fiber Communication*, Vol. 6, 1997 OSA Technical Digest Series (Optical Society of America, Washington, D.C., 1997), pp. 282–283.

[84] G. Nykolak, M. R. X. de Barros, S. Celaschi, J. T. Jesus, D. S. Shenk and T. A. Strasser, "An all-fiber multiwavelength add-drop multiplexer," in *Conference on Optical Fiber Communication*, Vol. 6, 1997 OSA Technical Digest Series (Optical Society of America, Washington, D.C., 1997), pp. 281–282.

[85] 4 port channel add/drop filter (200 GHz), 1996 preliminary data sheet, Optical Corporation of America, Marlborough, MA.

[86] K. Okamoto and Y. Inoue, "Silica-based planar lightwave circuits for WDM systems," in *Conference on Optical Fiber Communication*, Vol. 8, 1995 OSA Technical Digest Series (Optical Society of America, Washington, D.C., 1995), pp. 224–225.

[87] M. R. Phillips, T. E. Darcie, D. Marcuse, G. E. Bodeep and N. J. Frigo, *IEEE Phot. Tech. Lett.* **3**, 481 (1991).

[88] M. Artiglia, "Upgrading Installed Systems to Multigigabit bit-rates by means of dispersion compensation," in *22nd European Conference on Optical Communications*, Proceedings Volume 1, pp. 1.75–1.82 (1996).

[89] D. W. Hawtof, G. E. Berkey, and A. J. Antos, "High Figure of Merit Dispersion Compensating Fiber," in *Conference on Optical Fiber Communication*, Vol. 2, 1996 OSA Technical Digest Series (Optical Society of America, Washington, D.C., 1996), pp. 350–355.

[90] A. Yariv, D. Fekete, and D. M. Pepper, *Opt. Lett.* **4**, 52 (1979).

[91] Y. Koyano, M. Onishi, K. Tamano, and M. Nishimura, "Compactly-packaged high performance fiber-based dispersion compensation modules," in *22nd European Conference on Optical Communications*, Proceedings Volume 3, pp. 3.221–3.224 (1996).

[92] A. J. Antos, D. W. Hall, and D. K. Smith, "Dispersion-compensating fiber for upgrading existing 1310-nm-optimized systems to 1550 nm operation" in *Conference on Optical Fiber Communication*, Vol. 4, 1993 OSA Technical Digest Series (Optical Society of America, Washington, D.C., 1993), pp. 204–205.

[93] F. Forghieri, R. W. Tkach and A. R. Chraplyvy, *IEEE Phot. Tech. Lett* **9**, 970 (1997).

[94] B. Edvold and L. Gruner-Nielsen, "New Technique for Reducing the Splice Loss to Dispersion Compensating Fiber," in *22nd European Conference on Optical Communications*, Proceedings Volume 2, pp. 2.245–2.248 (1996).

[95] Technical specification of DK-SM Dispersion Compensating Fiber, Lucent Technologies, Specialty Fiber Group, Franklin Township, NJ, 1998.

[96] Y. Akasaka, R. Sugizaki, S. Arai, Y. Suzuki and T. Kamiya, "Dispersion Flat Compensation Fiber for Dispersion Shifted Fiber," in *22nd European Conference on Optical Communications*, Proceedings Volume 2, pp. 2.221–2.224 (1996).

[97] M. Shirasaki, *Opt. Lett.* **21**, 366 (1996).

[98] M. Shirasaki, "Chromatic Dispersion Compensation Using Virtually Imaged Phased Array," in *OSA Trends in Optics and Photonics*, Vol. 16, Optical Amplifiers and Their Applications, M. N. Zervas, A. E. Willner, and S. Sasaki, Eds. (Optical Society of America, Washington, DC, 1997), pp. 274–277.

[99] C.D. Poole, J.M. Wiesenfeld, A.R. McCormick and K.T. Nelson, *Opt. Lett.* **17**, 985 (1992).

[100] J. J. Pan, M. Shih, P. Jiang, J. Chen, J. Y. Xu, and S. Cao, "Miniature integrated isolator with WDM module and super-compact fiber amplifiers," *Tenth International Conference on Integrated Optics and Optical Fibre Communication, IOOC-95*, Technical Digest Vol. 2, pp. 56–57 (1995).

[101] A. M. J. Koonen, F. W. Willems, J. C van der Plaats, and W. Muys, "HDWDM Upgrade of CATV Fibre-coax Networks for Broadband Interactive Services," in *22nd European Conference on Optical Communications*, Proceedings Volume 3, pp. 3.19–3.25 (1996).

[102] D. Barbier, M. Rattay, S. Saint Andre, A. Kevorkian, J.M.P. Delavaux and E. Murphy, "Amplifying Four Wavelengths Combiner Based on Erbium-Ytterbium doped Planar Integrated Optical Modules," in *22nd European Conference on Optical Communications*, Proceedings Volume 3, pp. 3.161–3.164 (1996).

[103] M. Horiguchi, K. Yoshino, M. Shimizu, and M. Yamada, *Elect. Lett.* **29**, 593 (1993).

[104] H. Kressel, M. Ettenberg, J.P. Wittke, and I. Ladany, in *Semiconductor Devices for Optical Communications* (*Topics in Applied Physics*, Vol. 39), H. Kressel, Ed. (Springer-Verlag, Berlin, 1982), pp. 9–62.

[105] Optoelectronics Components Handbook, Microelectronics Group, Lucent Technologies, Breinigsville, PA, 1997.

[106] Technical specification of 1.48 μm laser diode module, SLA5600 series, Sumitomo Electric Lightwave Corporation, Research Triangle Park, NC, 1998.

[107] Y. Tashiro, S. Koyanagi, K. Aiso, and S. Namiki, "1.5 W erbium-doped fiber amplifier pumped by the wavelength division-multiplexed 1480 nm laser diodes with fiber Bragg grating," in *OSA Trends in Optics and Photonics*, Vol. 25, Optical Amplifiers and Their Applications, D. M. Baney, K. Emura, and J. M. Wiesenfeld, Eds. (Optical Society of America, Washington, DC, 1998), pp. 18–20.

[108] F. R. Gfeller and D. J. Webb, *J. App. Phys.* **68**, 14 (1990).

[109] C. R. Giles, T. Erdogan, and V. Mizrahi, *IEEE Phot. Tech. Lett.* **6**, 903 (1994).

[110] R. Parke, D. F. Welch, A. Hardy, R. Lang, D. Mehuys, S. O'Brien, K. Dzurko, and D. Scifres, *IEEE Phot. Tech. Lett.* **5**, 297 (1993).

[111] 980 nm laser module test data, SDL, Inc., San Jose, CA (1998).

[112] SDLO-2500 series data sheet, SDLO-2500-145 laser, SDL, Inc., San Jose, CA (1998).

[113] H. Asonen, J. Nappi, A. Ovtchinnikov, P. Savolainen, G. Zhang, R. Ries, and M. Pessa, *IEEE Phot. Tech. Lett.* **6**, 589 (1993).

[114] P. Savolainen, M. Toivonen, H. Asonen, M. Pessa, and R. Murison, *IEEE Phot. Tech. Lett.* **8**, 986 (1996).

[115] M. Fukuda, M. Okayasu, J. Temmyo, and J. Nakano, *IEEE J. Quant. Elect.* **30**, 471 (1994).

[116] J. A. Sharps, P. A. Jakobson, and D. W. Hall, "Effects of packaging atmospheric and organic contamination on 980 nm laser diode reliability," in *Optical Amplifiers and Their Applications*, Vol. 14, OSA Technical Digest Series (Optical Society of America, Washington, D.C., 1994), pp. 46–48.

[117] H. P. Meier, C. Harder, and A. Oosenburg, "A belief has changed: highly reliable 980-nm pump lasers for EDFA applications," in *Conference on Optical Fiber Communication*, Vol. 2, 1996 OSA Technical Digest Series (Optical Society of America, Washington, D.C., 1996), p. 229.

[118] C. S. Harder, L. Brovelli, and H. P. Meier, "High reliability 980-nm pump lasers for Er amplifiers," in *Conference on Optical Fiber Communication*, Vol. 6, 1996 OSA Technical Digest Series (Optical Society of America, Washington, D.C., 1997), p. 350.

[119] S. G. Grubb, W. H. Humer, R. S. Cannon, S. W. Vendetta, K. L. Sweeney, P. A. Leilabady, M. R. Keur, J. G. Kwasegroch, T. C. Munks, and D. W. Anthon, *Elect. Lett.* **28**, 1275 (1992).

[120] T. Whitley, R. Wyatt, D. Szebesta, S. Davey, and J. R. Williams, *IEEE Phot. Tech. Lett.* **4**, 399 (1993).

[121] J. Zhang, M. Quade, K. M. Du, Y. Liao, S. Falter, M. Baumann, P. Loosen, and R. Poprawe, *Elect. Lett.* **33**, 775 (1997).

[122] S. Sanders, F. Shum, R. J. Lang, J. D. Ralston, D. G. Mehuys, R. G. Waarts, and D. F. Welch, "Fiber-coupled M-MOPA laser diode pumping of a high-power erbium-doped fiber amplifier," in *Conference on Optical Fiber Communication*, Vol. 2, 1996 OSA Technical Digest Series (Optical Society of America, Washington, D.C., 1996), pp. 31–32.

[123] S. Sanders, K. Dzurko, R. Parke, S. O'Brien, D. F. Welch, S. G. Grubb, G. Nykolak, and P. C. Becker, *Elect. Lett.* **32**, 343 (1996).

[124] H. Po, E. Snitzer, R. Tumminelli, L. Zenteno, F. Hakimi, N. M. Cho, and T. Haw, "Double clad high brightness Nd fiber laser pumped by GaAlAs phased array" in *Conference on Optical Fiber Communication*, Vol. 5, 1989 OSA Technical Digest Series (Optical Society of America, Washington, D.C., 1989), pp. 395–398.

[125] C. Headley, S. Grubb, T. Strasser, R. Pedrazzani, B. H. Rockney, and M. H. Muendel, "Reduction of the noise power in rare-earth-doped fiber lasers by broadening the laser linewith," in *Conference on Optical Fiber Communication*, Vol. 6, 1997 OSA Technical Digest Series (Optical Society of America, Washington, D.C., 1997), p. 170.

[126] J. D. Minelly, W. L. Barnes, R. I. Laming, P. R. Morkel, J. E. Townsend, S. G. Grubb, and D. N. Payne, *IEEE Phot. Tech. Lett.* **5**, 301 (1993).

[127] S. G. Grubb, D. J. DiGiovanni, J. R. Simpson, W. Y. Cheung, S. Sanders, D. F. Welch, and B. Rockney, "Ultrahigh power diode-pumped 1.5-μm fiber amplifiers," in *Conference on Optical Fiber Communication*, Vol. 2, 1996 OSA Technical Digest Series (Optical Society of America, Washington, D.C., 1996), pp. 30–31.

[128] H. Po, J. D. Cao, B. M. Laliberte, R. A. Minns, R. F. Robinson, B. H. Rockney, R. R. Tricca, and Y. H. Zhang, *Elect. Lett.* **29**, 1500 (1993).

[129] D. Inniss, D. J. DiGiovanni, T. A. Strasser, A. Hale, C. Headley, A. J. Stentz, R. Pedrazzani, D. Tipton, S. G. Kosinski, D. L. Brownlow, K. W. Quoi, K. S. Kranz, R. G. Huff, R. Espindola, J. D. LeGrange, and G. Jacobovitz-Veselka, "Ultrahigh-power single-mode fiber lasers from 1.065 to 1.472 μm using Yb-doped cladding-pumped and cascaded Raman lasers," in *Conference on Lasers and Electro-Optics*, Vol. 11, 1997 OSA Technical Digest Series ((Optical Society of America, Washington, D.C., 1997), pp. 651–652.

[130] SDL-FL10 Fiber Laser, Product data sheet, SDL, Inc., San Jose, CA (1998).

[131] S. G. Grubb, T. Strasser, W. Y. Cheung, W. A. Reed, V. Mizrahi, T. Erdogan, P. J. Lemaire, A. M. Vengsarkar, D. J. DiGiovanni, D. W. Peckham, and B. H. Rockney, "High-power 1.48 μm cascaded Raman laser in germanosilicate fibers," in *Optical Amplifiers and Their Applications*, Vol. 18, OSA Technical Digest Series (Optical Society of America, Washington, D.C., 1995), pp. 197–199.

[132] S. G. Grubb, T. Erdogan, V. Mizrahi, T. Strasser, W. Y. Cheung, W. A. Reed, P. J. Lemaire, A. E. Miller, S. G. Kosinski, G. Nykolak, P. C. Becker, and D. W. Peckham, "1.3 μm cascaded Raman amplifier in germanosilicate fibers," in *Optical Amplifiers and Their Applications*, Vol. 14, OSA Technical Digest Series (Optical Society of America, Washington, D.C., 1994), pp. 187–190.

[133] E. M. Dianov, A. A. Abramov, M. M. Bubnov, A. V. Shipulin, S. L. Semjonov, A. G. Schebunjaev, A. N. Guryanov, and V. F. Khopin, "Raman amplifier for 1.3 μm on the base of low-loss high-germanium-doped silica fibers," in *Optical Amplifiers and Their Applications*, Vol. 18, OSA Technical Digest Series (Optical Society of America, Washington, D.C., 1995), pp. 189–192.

Chapter 4
Rare Earth Ions – Introductory Survey

4.1 INTRODUCTION

In this chapter we will review the fundamental atomic properties of trivalent rare earth ions, their behavior in a glass matrix, and the optical properties that result from these. The optical properties of the rare earth ions underlie all the bulk and fiber lasers and amplifiers made with these ions. In Section 4.2 we will examine the peculiar and unique atomic characteristics of the 4f electrons of rare earth ions and the models that have been developed to understand and parameterize the optical spectra of rare earth ions in crystals and glasses. This section is a somewhat detailed discussion of the atomic physics of rare earth ions and is provided as background for the reader; it is not essential for an understanding of rare earth amplifiers. In Section 4.3 we will discuss the Judd-Ofelt model of the intensities of the one-photon transitions within the 4f shell. A superficial overview of this section is sufficient for future reference. In Section 4.4 we will discuss the spectroscopic fundamentals, lifetimes, and cross sections of the transitions, which will be needed in subsequent chapters to model the behavior of fiber amplifiers. We will also introduce the McCumber theory for cross section determination. In Section 4.5 we will discuss the specific case of erbium with data for lifetimes, spectra, cross sections, and linewidth broadening. Finally, in Section 4.6, we will discuss ion-ion interaction effects that enter into play at high erbium concentrations.

4.2 ATOMIC PHYSICS OF THE RARE EARTHS

4.2.1 Introduction — The 4f Electron Shell

Rare earth atoms are divided in two groups: the lanthanides (the Greek *lanthanos* means hidden) with atomic number 57 through 71, and the actinides with atomic number 89 through 103. The rare earths are not so rare. Cerium, the second lanthanide element, is present at a level of 46 ppm in the earth's crust. The vast majority of rare earth doped fibers have been doped with lanthanide elements—for example, erbium (atomic number 68); there are many more known lasers with lanthanides as the active element than with actinides. The key to the optical behavior of lanthanide elements is contained in their very particular atomic structure, which we review first.

The optical spectra of rare earths were first observed in the 1900s by J. Becquerel, who observed sharp absorption lines in the spectrum of rare earth salts cooled to low temperatures (less than 100 K).[1, 2, 3] The theoretical explanation for this came about first from the work of M. Mayer in 1941, who calculated the atomic structure of the lanthanides from first principles.[4]

The classic picture of atoms is that of a nucleus surrounded by shells of electrons, which are gradually filled as one moves along the periodic table. In general, the succesive shells have monotonically increasing radii. However, at the atomic number $Z = 57$, an abrupt contraction takes place. The situation is as follows. The 5s and 5p shells ($5s^2 5p^6$) are full and one adds next a 4f shell in which electrons are progressively inserted. The 4f shell, instead of having a larger radius than the 5s and 5p shells, actually contracts and becomes bounded by these shells. This was very simply explained by Mayer by considering the effective radial potential of the electrons:[4]

$$V(r) = -\frac{e^2}{r}[1 + (Z-1)\varphi(\frac{r}{\mu})] + \frac{h^2}{8\pi^2 m} \cdot \frac{\ell(\ell+1)}{r^2} \quad (4.1)$$

which is the Coulomb potential energy plus the centrifugal potential energy, where e is the electronic charge, r is the distance from the nucleus, h is Planck's constant, m is the mass of the electron, ℓ the angular momentum quantum number of the electron, μ is defined as in reference [4], and $\varphi(\frac{r}{\mu})$ is the so-called Thomas-Fermi function. This expression is derived from the Thomas-Fermi model and predicts the behavior of atomic electrons by solving the quantum mechanical Schroedinger equation with V(r) as the effective potential energy. It allows the separation of a multi-electron problem into a separable and much more tractable form. For f electrons, ℓ is equal to 3. The minima of V(r) vary in a very interesting way with Z, the atomic number. V(r) is sketched in Figure 4.1. For low values of Z, V(r) has only one minimum, situated at roughly r = 6Å. At higher values of Z a second minimum develops, at a smaller value of r, to which the 4f wavefunction collapes when the potential well becomes deep enough. From Mayer's model, this is predicted to occur abruptly at Z = 60. In fact, this happens first with lanthanum (Z = 57), the first lanthanide element. This illustrates how the important properties of a complex atomic system with sixty interacting electrons can be understood from a simple model incorporating first principles, without complex computer simulations.

As one progresses along the lanthanide series, the average radius of the 4f shell slowly decreases.[5] This so-called lanthanide contraction is about 10% from the beginning to the end of the lanthanide series. The average radius of the 4f shell is about 0.7 times the Bohr radius. The shielding of the 4f electron shell from its environment by the outermost 5s and 5p electrons is responsible for the rare earths' rich optical spectrum and for the many laser transitions that have been observed.

The most common form of rare earth elements is the ionic form, in particular the trivalent state $(Ln)^{3+}$. Neutral lanthanide elements have the atomic form $(Xe)4f^{N'}6s^2$ or $(Xe)4f^{N'-1}5d6s^2$, where (Xe) represents a Xenon core. The ionization of the lanthanides involves the removal of the two loosely bound 6s electrons, and then of either a 4f or a 5d electron. Thus, the trivalent rare earth ions, on which most active rare earth

4.2. ATOMIC PHYSICS OF THE RARE EARTHS

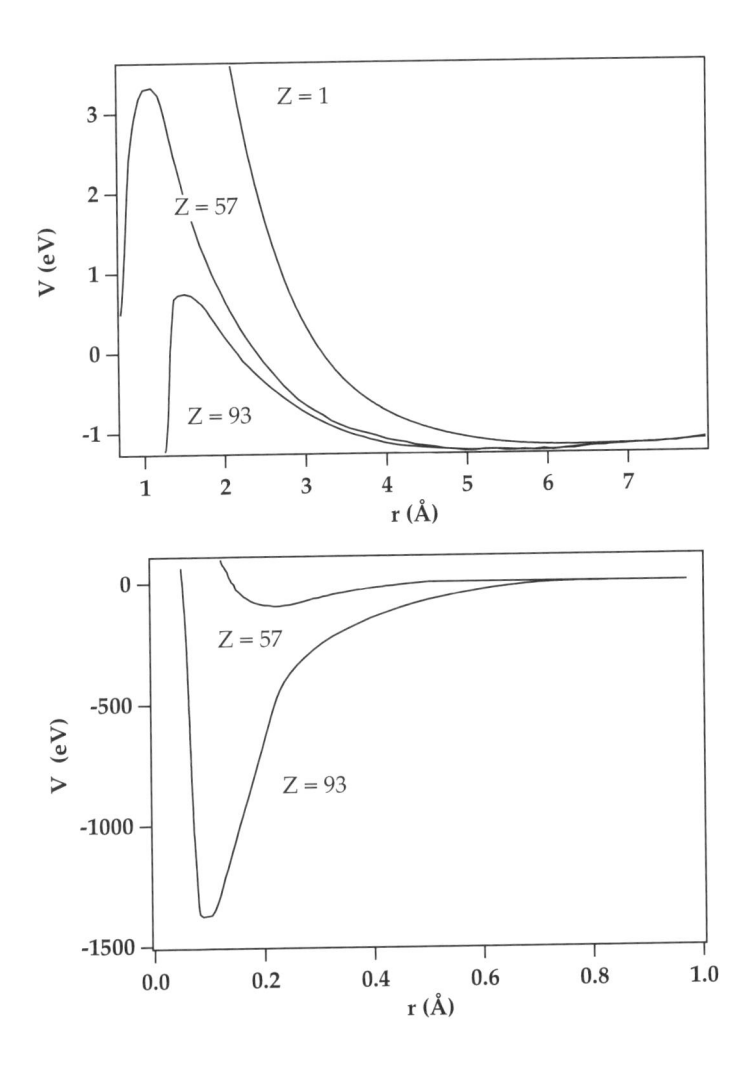

Figure 4.1: Effective potential V(r) for 4 f electrons as a function of atomic number Z (top: large values of r; bottom: small values of r). From reference [4].

doped devices are based, have a Xenon core to which N 4f electrons have been added. The electrostatic shielding of the 4f electrons by the $5s^2 5p^6$ shells, which constitute something akin to a metal sphere, is responsible for the atomiclike properties of the lanthanides when present in a solid environment such as a crystal (ordered solid) or glass (disordered solid).

Normally, atoms in a semiconductor or metal give up their electrons to the solid as a whole since their wavefunctions are very delocalized. This gives rise to the very wide energy spectrum of such materials. In contrast, for a rare earth doped insulator, the energy spectrum is composed of a series of narrow lines. Figure 4.2 shows the radial distribution function (the absolute square of the radial wavefunctions) of the 4f,

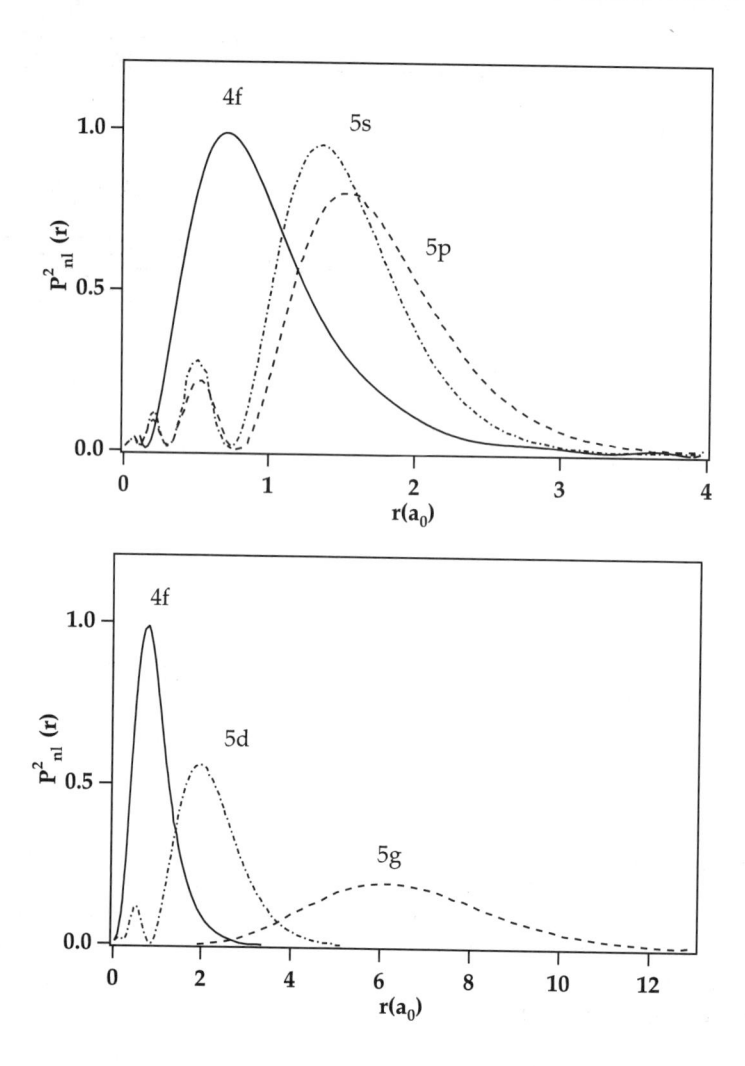

Figure 4.2: Radial distribution functions of the 4f, 5s, 5p, 5d, and 5g orbitals for the Pr^{3+} free ion, from Hartree-Fock calculations, from reference [6]. Top: 4f, 5s, and 5p orbitals of the ground configuration 4f^{2}5s^{2}5p^6, showing that the 4f orbitals are within the 5s and 5p orbitals. Bottom: 4f, 5d, and 5g orbitals of the ground configuration 4f^{2}5s^{2}5p^6, and of the excited configurations 4f5d5s^{2}5p^6 and 4f5g5s^{2}5p^6, respectively, showing that when electrons are excited from a 4f to a 5d or 5g orbital its orbital radius is larger that the 5s or 5p electrons.

5s, 5p, 5d, and 5g orbitals of the Pr^{3+} ion as obtained from computer simulations that include all the electrons of the atom and their mutual interaction as well as that with the nucleus.[6] Clearly the 4f orbitals are closer to the nucleus than the 5s and 5p orbitals, by roughly one Bohr radius. The 5d and 5g orbitals, to which a 4f electron can be excited, are much farther removed and in a solid will overlap neighboring ions.

4.2. ATOMIC PHYSICS OF THE RARE EARTHS

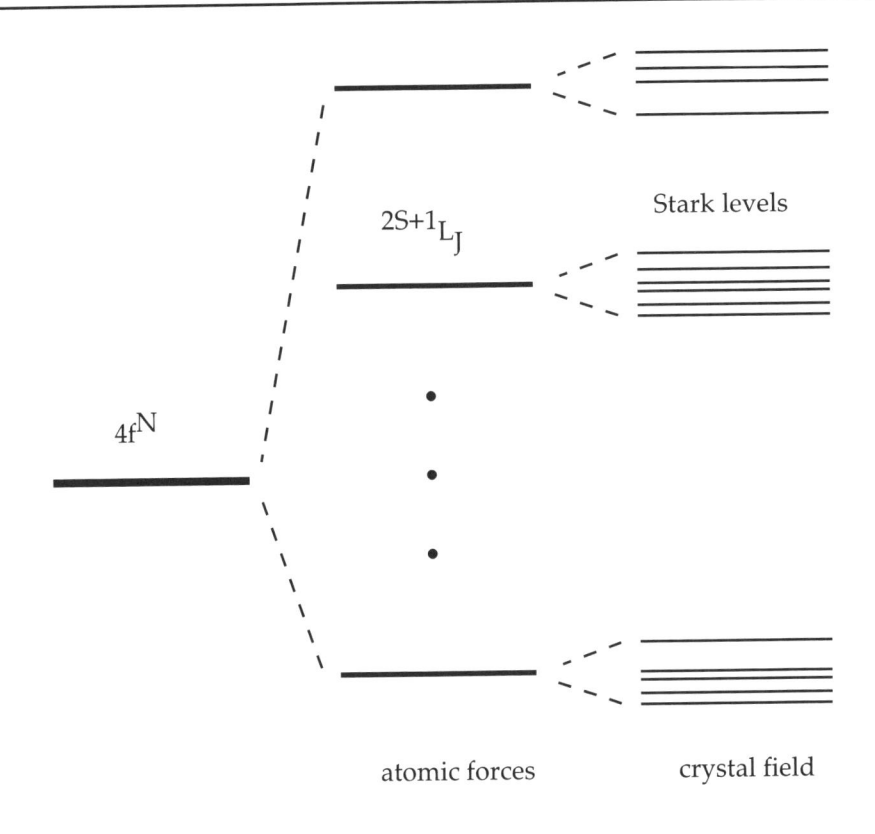

Figure 4.3: Schematic representation of the splitting of the $4f^N$ ground configuration under the effect of progressively weaker perturbations, the atomic and crystal field Hamiltonians. First, the atomic forces split the original $4f^N$ configuration of one electron orbital into $^{2S+1}L_J$ levels, then the (roughly) 100 times weaker electrostatic crystal-field Hamiltonian splits each free-ion atomic level into a collection of Stark levels. The nomenclature is described in the text.

4.2.2 The "Puzzle" of 4f Electron Optical Spectra

The $4f^N$ configuration is composed of a number of states, and before a 4f electron is excited to a higher lying orbital, such as 4d or 5g, it can be excited within the $4f^N$ set of states. The spread in the $4f^N$ energy levels arises from various atomic interactions between the electrons. A further splitting of the energy levels comes about when the ion is placed in a crystalline host. The environment provided by crystalline hosts destroys the spherically symmetric environment that rare earth ions enjoy in the vapor phase. Thus the degeneracy of the 4f atomic states will be lifted to some degree. This splitting is also referred to as Stark splitting, and the resulting states are called Stark components (of the parent manifold). This splitting is depicted in Figure 4.3, where the energy level nomenclature described in more detail below is used.

The crystal lattice, or more appropriately the crystal field produced by the ions at the lattice points, provides a certain symmetry at the sites occupied by the lanthanide ion. This site symmetry can be classified by one of the 32 crystal point groups. This was recognized by Bethe, who in 1929 showed how the crystal-field splitting of the free ion levels could be characterized group theoretically.[7] This paved the way for a fuller understanding of the multitude of absorption lines that appear in the visible spectra of lanthanide doped crystals. These transitions occur between the crystal field states of the unfilled 4f shell and are, as a consequence, not allowed in the electric dipole approximation, since the parity must change in a dipole transition (the Laporte rule). Thus, these transitions are "forbidden electric dipole" in nature, with small oscillator strengths. Their presence was explained in 1937 by Van Vleck, who referred to this problem as "The Puzzle of Rare Earth Spectra in Solids."[8] The principal mechanism which allows intra $4f^N$ transitions is the admixture into the $4f^N$ configuration of a small amount of excited opposite parity configurations.

The necessary conceptual foundation for the study of the visible spectra of rare earth ions was thus established by 1941. Qualitatively, the spectra were reasonably well understood, if one took into account the fact that both the atomic physics of the 4f shell and the group theoretical symmetry of the host crystal were needed to explain the observed pattern of visible absorption and emission lines. Quantitative studies were made possible by the atomic and crystal field models, which use the mathematical techniques developed by G. Racah in the 1940s.[5, 9, 10] The state of the spectroscopic investigations on rare earth ions in crystals was summarized up to 1968 by Dieke.[11] Hüfner's work is a more recent exposition.[5]

4.2.3 Semiempirical Atomic and Crystal—Field Hamiltonians

The configurations of the atomic electrons of the lanthanide ions are specified by a set of one-electron orbitals. Since the closed shells are assumed to be inert with respect to optical excitations, they will not be referred to and only the ground and excited configurations of the 4f electrons will be used to specify the states. The Hamiltonian operator H that determines the wavefunctions of the 4f electrons contains terms that describe the atomic interactions of the free ion (which might be modified by the presence of the surrounding ligands) and terms that describe the interaction of the electrons with the crystal field produced by the surrounding charges. For such high atomic number systems as the rare earths, it is extremely unwieldy to calculate H and the resulting 4f energy levels from first principles. Rather, H is written in a parameterized form and the parameters are then found by adjusting them to obtain the best fit of calculated eneregy levels to experimental energy levels.

The Hamiltonian operator for the 4f electrons is written as follows:[5, 9, 12]

$$H = H_{atomic} + H_{cf} \tag{4.2}$$

where

$$H_{atomic} = -\frac{\hbar^2}{2m}\sum_{i=1}^{N}\nabla_i^2 - \sum_{i=1}^{N}\frac{Z^*e^2}{r_i} + \sum_{i<j}^{N}\frac{e^2}{r_{ij}} + \sum_{i=1}^{N}\zeta(r_i)\vec{s}_i \cdot \vec{\ell}_i + \text{smaller terms} \tag{4.3}$$

and

$$H_{cf} = \sum_{k,q,i} B_q^k C_q^{(k)}(i) \tag{4.4}$$

4.2. ATOMIC PHYSICS OF THE RARE EARTHS

with k, q even. H_{atomic} is the free-ion Hamiltonian in the central field approximation. N is the number of 4f electrons, Z^*e is the effective nuclear charge (which takes into account the closed shells that lie between the nucleus and the 4f electrons), and $\zeta(r_i)$ the spin-orbit coupling function. The first two terms of H_{atomic} represent the kinetic energy and the Coulomb interaction of the nucleus with the 4f electrons. They are spherically symmetric and do not lift the degeneracy of the $4f^N$ configuration. However, the third and fourth terms, which represent the mutual Coulomb repulsion and spin-orbit interactions of the 4f electrons, are responsible for the spread of the $4f^N$ free-ion levels over tens of thousands of wavenumbers. H_{cf} is the crystal-field Hamiltonian where the summation over i is over all the 4f electrons of the ion. The B_q^k are constants, and the $C_q^{(k)}$ are proportional to the spherical harmonics:

$$C_q^{(k)}(\theta, \phi) = \left(\frac{4\pi}{2k+1}\right)^{\frac{1}{2}} Y_{kq}(\theta, \phi) \tag{4.5}$$

There are several schemes for describing the states of a many-electron system. For rare earth ions, it is customary to use Russell-Saunders, or LS, coupled states. To write Russell-Saunders states, one first couples the orbital angular momenta $\vec{l}_i$ of the electrons to form a resultant total orbital angular momentum $\vec{L}$, and similarly for the spins $\vec{s}_i$ to form a total spin $\vec{S}$. $\vec{L}$ and $\vec{S}$ are then coupled to form $\vec{J}$, the total angular momentum. Thus a state of the $4f^N$ configuration will be described by the ket $|4f^N \alpha SLJJ_z\rangle$, where α refers to the other quantum numbers needed to label the state. These quantum kets form a complete set. One calculates the matrix elements of H between these states and diagonalizes H to obtain the eigenvalues and eigenvectors. When the mutual electrostatic repulsion of the electrons is much stronger than the spin-orbit coupling, the eigenfunctions are nearly pure LS-coupled states. The electrostatic interaction splits the configuration into terms ^{2S+1}L, and the spin-orbit interaction lifts the degeneracy with respect to J and splits the terms into levels $^{2S+1}L_J$.

In the case of lanthanide ions, it is not quite correct to consider the electrostatic and spin-orbit interactions as successively smaller perturbations. The actual eigenstates of the system are usually linear combinations of $^{2S+1}L_J$ states, since these form a complete set of functions for functions of the angular variable. The only caveat to this statement is that for some configurations an additional quantum number is needed in addition to the Russell-Saunders label to uniquely specify a state. In general, when one refers to a $^{2S+1}L_J$ level of a lanthanide ion, it constitutes the leading term in the expansion of the state in Russell-Saunders coupled states. For example, the $^4I_{15/2}$ ground multiplet of Er^{3+} contains a small admixture of $^2K_{15/2}$ states. Similarly, it is not a good approximation to use a spin-orbit interaction of the form $\lambda \vec{L} \cdot \vec{S}$ since this operator has no off-diagonal elements in L and S, whereas $\sum_i \vec{l}_i \cdot \vec{s}_i$ does. For rare earth ions there is substantial LS mixing, an effect that grows stronger as the atomic number increases. This is due to the fact that the spin-orbit interaction increases faster than the mutual electrostatic repulsion between the 4f electrons, as one moves up the lanthanide series.[13, p. 238]

There still remains a degeneracy with respect to J_z since spherical symmetry has not been broken by H_{atomic}. The crystal field removes that symmetry by means of the non spherically symmetric distribution of charge surrounding the free ion. For 4f electrons, the crystal-field Hamiltonian is a small perturbation compared to the electrostatic and spin-orbit interactions since it is roughly a 100 times smaller in magnitude. In contrast, the crystal-field splittings of the excited configurations are quite large, — on the order of 20,000 cm^{-1}, — as the excited orbitals (e.g., 5d or 5g) are not shielded as the 4f orbitals are.

Figure 4.4, a chart originally drawn by Dieke, is the energy level diagram of the low-lying f^N states of the trivalent lanthanide ions doped into the crystal LaCl$_3$, with the multiplets labeled by the leading $^{2S+1}L_J$ component.[11] The widths of the multiplets show the extent of the crystal-field splitting. Since the crystal field of the host lattice is usually such a small perturbation, the Dieke chart is an accurate representation of the f^N electronic energy level structure for trivalent rare earth ions doped into most ionic insulating crystals and glasses.

A number of other terms are added into the Hamiltonian in an attempt to model some of the more complex effects that can occur in many-electron systems. A discussion of these atomic calculations is beyond the scope of this chapter; however, a modern review can be found in an article by Judd.[14]

4.2.4 Energy Level Fitting

The atomic Hamiltonian that was introduced in equation 4.3 is usually calculated using a semiempirical model, by replacing each term by ones that contain "effective" operators. For example, the Coulomb interaction term can be simplified to the sum $\sum_{k=0,2,4,6} F^k(nf, nf) f_k$. The $F^k(nf, nf)$ are proportional to radial integrals and are usually treated as adjustable parameters; the f_k are matrix elements of products of spherical harmonics and are calculated using standard tensor operator techniques. The f_k are an example of "effective" operators. These operators were added in a piecemeal fashion as the art of fitting complex spectra advanced throughout the 1960s. The reason that this parameter-based method has remained so prevalent is heuristic in nature, in the sense that with a relatively small number of parameters one can reasonably accurately reproduce observed spectra.

In practice, one performs a least-squares fit of the calculated energy levels to the observed energy levels, or more appropriately to the observed transitions. The accuracy of the fit is characterized by the reduced root-mean-square energy deviation between the observed and calculated energy levels. The prototypical example of the use of this type of Hamiltonian in systematizing the 4f^N spectra of the trivalent lanthanides is the work of Carnall, Crosswhite, and Crosswhite, using the host crystals LaF$_3$ and LaCL$_3$.[15]

Typically, 50 to 150 energy levels up to 50,000 cm^{-1} are assigned and can be fit with 10 to 20 parameters, yielding rms energy deviations in the range of 10 to 30 cm^{-1}. This is quite good if one considers that the 4f^N configuration can go up to 200,000 cm^{-1} in energy so that the rms energy deviation is on the order of 0.1 to 0.01 of the spread in energy of the configuration. Experimentally, however, these energies can be found to within 0.1 cm^{-1} accuracy, so there is still quite a lot of room for the theory to improve. An energy level fit for Er^{3+} in the crystal LuPO$_4$ has been reported by Hayhurst.[16]

4.3. OPTICAL SPECTRA OF RARE EARTH IONS 95

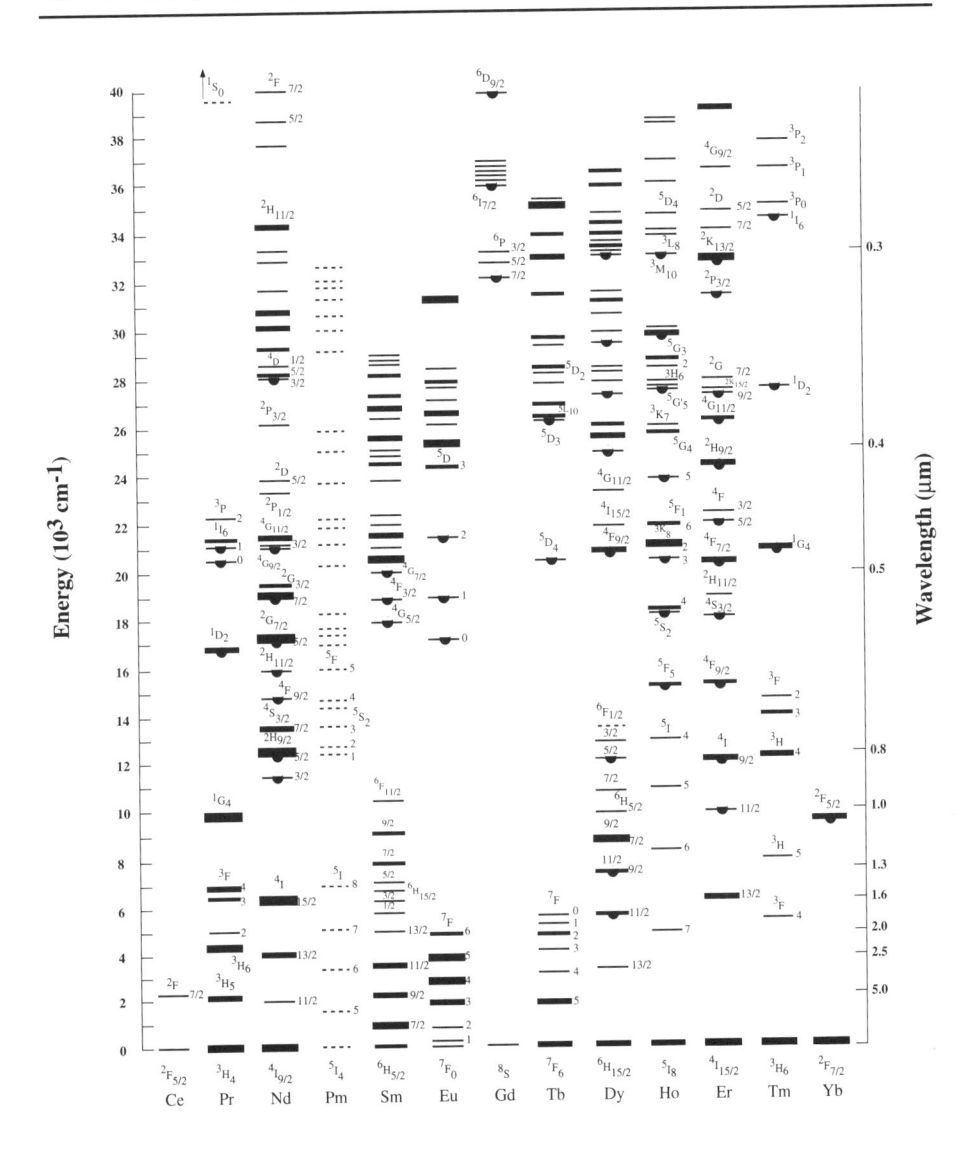

Figure 4.4: Energy level diagram of the low-lying $4f^N$ states of trivalent ions doped in LaCl$_3$. From references [12] and [5]. The pendant semicircles indicate fluorescing levels.

4.3 OPTICAL SPECTRA OF RARE EARTH IONS

4.3.1 *The Character of* $4f^N - 4f^N$ *Optical Transitions*

Several mechanisms come into play to give the forbidden electric dipole transitions small but significant amplitude. A brief description is given of these mechanisms. Peacock has written a review that covers work through 1975.[17]

- In cases where the site symmetry is not a center of inversion, the crystal-field Hamiltonian can contain odd-parity terms. These terms introduce into the 4f wavefunctions small components of opposite-parity and higher-energy configurations, such as $4f^{N-1}5d$. The intensity of a particular transition is determined by the amount of opposite-parity configurations that have been mixed into the initial and final states.

- Crystal vibrations of odd symmetry can admix odd-parity $4f^{N-1}nl$ configurations into the $4f^N$ manifold. For a centrosymmetric crystal, this is the only source of "forbidden electric dipole transitions."

- Magnetic dipole transitions are allowed since no change in parity is involved for this process. The intensities of magnetic dipole transitions are comparable to those of the "forbidden electric dipole transitions" (for non-centrosymmetric crystals). There are fewer magnetic dipole transitions since they must satisfy the selection rules $\Delta J = 0, \pm 1$, $\Delta L = 0$, $\Delta S = 0$, and $J = 0 \rightarrow J = 0$ transitions forbidden.

Most of the intensity calculations have centered only on the rare earth ion, using its wavefunctions in the free-ion sense. More recent work has taken into consideration the surrounding ligands as active participants in the interaction between the radiation field and the solid.[18, 19, 20] In this picture the f electrons of the rare earth ion are considered to polarize the charge clouds of the surrounding ions, which then interact with the radiation field via a nonzero total dipole moment (if the rare earth ion is not situated at a center of inversion). These effects are usually called ligand polarization effects.

4.3.2 Intensities of One-Photon Transitions — Judd-Ofelt Theory

The bulk of experimental evidence supports Van Vleck's conjecture, namely, that the mechanism responsible for the sharp-line optical absorption and emission transitions in rare earth compounds is forbidden electric dipole in nature. The oscillator strengths of these transitions, being on the order of 10^{-6}, are too weak to be the allowed 4f – 5d transitions. Electric quadrupole transitions are predicted to be even weaker and have not yet been observed. Magnetic dipole transitions are observed only when the selection rules for such transitions are satisfied. Thus the major part of the 4f – 4f one-photon lanthanide intensities have their origin in static or dynamic admixtures of opposite-parity configurations. The static admixture is caused by odd terms in the crystal-field potential (for a non-centrosymmetric rare earth site), and the dynamic part is caused by odd-parity vibrations. An order of magnitude estimate for the amount of admixture of the opposite-parity configurations can be obtained from perturbation theory by the ratio

$$\frac{B_q^k(k_{\text{odd}})}{E(4f^{N-1}nl) - E(4f^N)} \tag{4.6}$$

where B_q^k is on the order of a few hundred cm^{-1} and the configuration energy difference $E(4f^{N-1}nl) - E(4f^N)$ is on the order of 100,000 cm^{-1} for the next highest configuration, $4f^{N-1}5d$. As a consequence, we expect that the $4f^{N-1}5d$ configuration

4.3. OPTICAL SPECTRA OF RARE EARTH IONS

will contribute only 0.1 % to the composition of the ground configuration. For energy level calculations, this effect can be neglected and the ground configuration assumed to be pure $4f^N$. Insofar as intensity calculations are concerned, this tiny admixture becomes extremely important. An order of magnitude estimate of the oscillator strength of the 4f → 4f transitions is obtained by squaring equation 4.6, which yields the value 10^{-6}.

The theory of the intensities of rare earth, one-photon $4f^N$ transitions was cast into a mathematical framework by Judd and Ofelt and has become known as the Judd-Ofelt theory.[21, 22] The intensities of the one-photon transitions are due mostly to the admixture of the excited $4f^{N-1}5d$ configuration to the ground configuration. Other excited configurations, however, have been demonstrated to contribute to higher order light processes. The study of two-photon processes in rare earth ions by Becker and Judd, via electronic Raman scattering, has shown that excited g orbital based configurations can play a significant role in the interaction of rare earth ions with light fields.[23]

The wavefunction for a particular state is written using first-order perturbation theory:

$$|i\rangle = \sum_{S,L,J,J_z} \left[|4f^N \alpha SLJJ_z\rangle + \sum_{\psi'} \frac{|\psi'\rangle\langle\psi'|V_{odd}|4f^N \alpha SLJJ_z\rangle}{E(\psi') - E(4f^N \alpha SLJJ_z)} \right] \quad (4.7)$$

where V_{odd} represents the odd-parity part of the crystal field and $|\psi'\rangle$ represents the possible intermediate states of the system. The matrix element of the electric dipole operator $\vec{D}$ between the states $|i\rangle$ and $|f\rangle$ of the $4f^N$ configuration will thus involve a sum over intermediate states of the form

$$\sum_{\psi'} \frac{\langle f|\vec{D}|\psi'\rangle\langle\psi'|V_{odd}|i\rangle}{E(\psi') - E(4f^N \alpha SLJJ_z)} \quad (4.8)$$

where the states $|i\rangle$ and $|f\rangle$ are written in zeroth order. Barring some drastic simplification, this infinite sum is mathematically intractable. The key idea, originally due to Griffith, is to approximate the energy denominator by an average denominator, which then allows use of the closure relation $\sum_{\psi'} |\psi'\rangle\langle\psi'|$ to dispense with the sum over intermediate states. The details of the calculation, which makes extensive use of tensor operator techniques, can be found in the work published by Judd.[21] The excited configurations that enter equation 4.7 as intermediate states are $4f^{N-1}nd$, $4f^{N-1}ng$, $4f^{N+1}nd^{-1}$, and the continuum d and g configurations.

Several points can be made about the derivation and its underlying assumptions.[17]

- The closure is performed over each individual excited configuration. As such it is not a full quantum mechanical closure over a complete set of eigenfunctions of the Schrodinger equation. Rather, it is a closure over the angular variables only. The assumption here is that the spread in energy of the excited configuration is small compared to the average difference in energy between that configuration and the ground $4f^N$ configuration. This is not a very good approximation for rare earth ions for which the $4f^{N-1}5d$ configuration does not lie very much higher in energy than the $4f^N$ ground configuration.

- The Judd-Ofelt theory involves three parameters, usually labeled $\Omega_2, \Omega_4, \Omega_6$. They parameterize the oscillator strengths for the absorption and fluorescence transitions between various multiplets of the $4f^N$ configuration. The Ω_k are usually calculated to provide the best fit to the experimental data. The scheme has met with relatively good success, considering the approximations involved. Except for the case of the so-called hypersensitive transitions, it is a good phenomenological model, although errors on the order of 50% are not uncommon in predicting radiative transition rates from Judd-Ofelt parameters.[17, 24, 25]

The intensity of a transition between two states is often characterized by its oscillator strength. The radiative transition rates and the radiative lifetimes can be obtained from the oscillator strengths. The oscillator strength for the forced electric dipole transition is usually defined in terms of the three Judd-Ofelt parameters:[21]

$$f_{ij}^{ED} = \frac{8\pi^2 m\nu}{3h(2J+1)} \cdot \frac{(n^2+2)^2}{9n} \sum_{\lambda=2,4,6} \Omega_\lambda \left|\left\langle f^N J||U_\lambda||f^N J'\right\rangle\right|^2 \qquad (4.9)$$

where i and j are the initial and final states, respectively, ν is the transition frequency between the two states, n is the index of refraction of the host medium, m is the electron mass, and the reduced matrix elements $\langle f^N J||U_\lambda||f^N J'\rangle$ have been tabulated by Nielson and Koster.[26] Equation 4.9 contains a correction, expressed in terms of the refractive index n, for the local electric field in the host medium. The Judd-Ofelt parameters for Er^{3+} in various hosts are listed in Table 4.1. The error bars on the Judd-Ofelt parameters can be rather large. Those reported for fluoride glass, for example, are $\Omega_2 = 1.54 \pm 0.25$, $\Omega_4 = 1.13 \pm 0.40$, and $\Omega_6 = 1.19 \pm 0.20$.[30] For glasses, since there is a much greater variation in the kind of sites that the Er^{3+} ion occupies and the environment it experiences, the calculated Judd-Ofelt parameters are average values of the parameters for each given site. It has been noted that the parameter Ω_2 is the most strongly dependent on the site environment and the amount of covalent bonding, is higher in oxide glasses than in fluoride glasses, and is higher in glasses than in crystals.[27]

The oscillator strength for magnetic dipole transitions is given by the expression

$$f_{ij}^{MD} = \frac{8\pi^2 m\nu}{3he^2(2J+1)} n^3 |\mu_B|^2 \left|\left\langle f^N J||\vec{L}+2\vec{S}||f^N J'\right\rangle\right|^2 \qquad (4.10)$$

where μ_B is the Bohr magneton, equal to $e\hbar/2m$, and $\vec{L}+2\vec{S}$ is the magnetic dipole operator. In cases where the electric dipole matrix element is small, the magnetic dipole component of an absorption or fluorescence transition can be significant. This is the case for the Er^{3+} transition between $^4I_{13/2}$ and $^4I_{15/2}$, where, for example, the spontaneous emission probabilities were computed to be 73.5 s^{-1} for the electric dipole component, and 34.7 s^{-1} for the magnetic dipole component, yielding an overall spontaneous emission lifetime of 9.2 ms (in ZBLA glass).[29]

Finally, it is important to note that the Judd-Ofelt theory yields the total oscillator strength for a transition between two $^{2S+1}L_J$ multiplets, the summation over all the

Host Matrix	Ω_2	Ω_4	Ω_6	Ref.
Phosphate	9.92	3.74	7.36	[27]
Borate	11.36	3.66	2.24	[27]
Germanate	6.40	0.75	0.34	[27]
Tellurite	7.84	1.37	1.14	[27]
ZBLA	3.26	1.85	1.14	[28]
ZBLA	2.54	1.39	0.97	[29]
Fluoride glass	1.54	1.13	1.19	[30]
ZBLAN	2.3	0.9	1.7	[31]
LaF$_3$	1.1	0.3	0.6	[32]
Y$_2$SiO$_5$	2.84	1.42	0.82	[33]

Table 4.1: Judd-Ofelt parameters Ω_2, Ω_4, and Ω_6 for Er^{3+} in various glass and crystalline hosts, in units of 10^{-20}cm^2.

Stark components having been performed in the sum over J$_z$. For example, the Judd-Ofelt theory estimates the integrated electric dipole oscillator strength for the ^{4}I$_{13/2}$ to ^{4}I$_{15/2}$ transition of Er^{3+}, and not the strengths of the individual transitions between the various states of each multiplet.

4.4 FUNDAMENTAL PROPERTIES

4.4.1 Transition Cross Sections

Introduction

Cross sections quantify the ability of an ion to absorb and emit light. They are related to the Einstein A and B coefficients. Stated simply, the cross section of a particular transition between two states of an ion represents the probablity for that transition to occur with the concurrent emission or absorption of light.[34] Given two states 1 and 2, with respective energies E$_1$ and E$_2$ (E$_1$ less than E$_2$), the transition probability for the absorption of a photon of energy (E$_2$ − E$_1$) is proportional to the cross section σ_{12}, and that for the emission of a photon is proportional to the emission cross section σ_{21}. The dimension of the cross section is that of an area. The amount of light power P$_{abs}$ absorbed by an ion with light incident upon it at a frequency ω is given by

$$P_{abs} = \sigma_{12} I \quad (4.11)$$

where I is the intensity of the light incident upon the ion. Dividing by the photon energy, $\hbar\omega$, we obtain the rate of absorption in number of photons,

$$N_{abs} = \sigma_{12} \frac{I}{\hbar\omega} = \sigma_{12} \Phi(\omega) \quad (4.12)$$

where $\Phi(\omega)$ is the photon flux in units of number of photons per unit area per unit time. Similarly, the amount of stimulated light power emitted by an ion with light of intensity I incident upon it is given by

$$P_{em} = \sigma_{21} I \quad (4.13)$$

Intuitively, the absorption cross section can be thought of as a sort of target area that can intercept a light flux by "catching" the photons that flow through it. The emission probability has an analogous interpretation. For a collection of identical ions, with a population N_1 in the lower state and a population N_2 in the upper state, the total change in power of a light of intensity I traversing the collection of ions is given by

$$\Delta P = P_{em} - P_{abs} = (N_2\sigma_{21} - N_1\sigma_{12})I \tag{4.14}$$

and the amount of light power absorbed by the medium is the negative of this quantity. It should be noted that the emission and absorption probabilities are proportional to the light intensity, not its power. The smaller the area over which the light is concentrated, the higher the probability for emission or absorption of light.

For two nondegenerate states 1 and 2, the emission and absorption cross sections will be equal: $\sigma_{12} = \sigma_{21}$. Most textbooks treat the simple case of two individual levels and present relationships linking the cross sections, Einstein A and B coefficients, and radiative lifetimes. Complications, which have produced a substantial amount of discussion in the literature, arise in the case of rare earth ions in solids. One often sees the statement that the emission and absorption cross sections are different.[35, 36, 37] The reason for this comes from the fact that, in the case of rare earth ions, the two states referred to,—e.g., the $^4I_{15/2}$ and $^4I_{13/2}$ levels of Er^{3+},—are really a comb of sublevels that are populated to various extents depending on the thermal distribution. The cross section is then meant to specify the cross section at a particular frequency within the spectral bandwidth of the transition.

The spectral shape of the emission and absorption cross sections for transitions between two manifolds will be, in general, different since they depend on the thermal populations in the various sublevels. The emission and absorption cross sections, at a particular frequency, are equal only if the various sublevels are equally populated, or if the transition strengths between any of the individual sublevels are all equal.[34, 38] The cross sections that are derived from experiment yield the effective spectral absorption and emission strength for a given ion's transition, at a given temperature, which is what is usually needed to accurately model and predict the performance of an amplifier. The key issue is how to correctly extract the emission and absorption cross sections from the measured absorption and fluorescence spectra. We discuss here the theory in more detail, reviewing the standard Einstein theory and the more general McCumber treatment.

Einstein Theory and Ladenburg-Fuchtbauer Relation

Absorption and emission of light in a degenerate two-level system, where level 1 has degeneracy g_1 and level 2 has degeneracy g_2, can be treated with basic quantum mechanical principles.[34, Appendix 7.A] The population changes in levels 1 and 2 are governed by the transition strengths between the individual sublevels that make up each level. Unless certain conditions are met, the general equations cannot be simplified. Two assumptions that allow for reduction of the equations to a more simple form are that the sublevels are all equally populated, or that the transition strengths between the sublevels are all equal. If one of these conditions is met, then the Einstein A and B

4.4. FUNDAMENTAL PROPERTIES

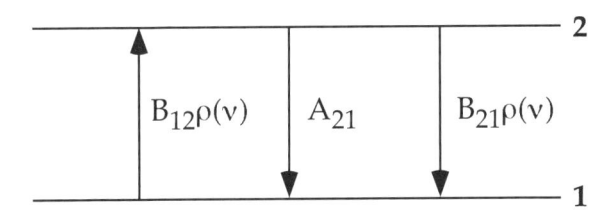

Figure 4.5: Transitions corresponding to the absorption and emission of light transitions in a two-level system. A_{21} is the spontaneous emission rate, $B_{21}\rho(\nu)$ the stimulated emission rate, and $B_{12}\rho(\nu)$ the absorption rate, where $\rho(\nu)$ is the photon flux density.

coefficients can be used in the form

$$\left(\frac{dN_2}{dt}\right)_{abs} = B_{12}\rho(\nu)N_1 \qquad (4.15)$$

$$\left(\frac{dN_2}{dt}\right)_{emis} = -(A_{21} + B_{21}\rho(\nu))N_2 \qquad (4.16)$$

where N_1 and N_2 are the populations of the lower and upper states, respectively, and $\rho(\nu)$ is the photon flux density in units of number of photons per unit frequency bandwidth per unit volume, as the ion is assumed to be interacting with broadband light.[34] The transitions corresponding to the Einstein coefficients are depicted pictorially in Figure 4.5, for a simple two-level system.

In the general case where the levels are degenerate, with level 1 having g_1 sublevels m_1 and level 2 having g_2 sublevels m_2, then $(dN_2/dt)_{abs}$ and $(dN_2/dt)_{emis}$ will contain sums of terms corresponding to transitions between sublevels. Following the derivation of reference [34], we write the stimulated transition rates between sublevels m_1 and m_2 as $R(m_1, m_2)$, and we have

$$\left(\frac{dN_2}{dt}\right)_{abs} = \sum_{m_1,m_2} R(m_1, m_2)N_{m_1} \qquad (4.17)$$

The corresponding term for the emission is

$$\left(\frac{dN_2}{dt}\right)_{emis} = -\sum_{m_1,m_2} (A(m_1, m_2) + R(m_1, m_2))N_{m_2} \qquad (4.18)$$

where $A(m_1, m_2)$ is the spontaneous transition rate between sublevels m_1 and m_2. The transition rates are proportional to the matrix elements of the light operator (e.g., electric dipole and magnetic dipole) between the two states. One can then show that if all the sublevels are equally populated—so that for each sublevel m_1 of level 1 we have $N_{m_1} = N_1/g_1$ and similarly for level 2 $N_{m_2} = N_2/g_2$, where N_1 and N_2 are the total populations of levels 1 and 2—then

$$B_{21}\rho(\nu) = \frac{1}{g_2}\sum_{m_1,m_2} R(m_1, m_2) \qquad (4.19)$$

and

$$B_{12}\rho(\nu) = \frac{1}{g_1} \sum_{m_1,m_2} R(m_1, m_2) \qquad (4.20)$$

Thus, in the context of level degeneracy, the B_{12} coefficient characterizes the transition strength from an "average" sublevel of level 1, to all the possible sublevels of level 2, with a similar interpretation for B_{21}. It follows immediately from equations 4.19 and 4.20 that

$$g_1 B_{12} = g_2 B_{21} \qquad (4.21)$$

and one can also show that

$$A_{21} = \frac{8\pi h \nu^3 n^3}{c^3} B_{21} = \frac{8\pi h}{\lambda^3} B_{21} \qquad (4.22)$$

where ν and λ are the frequency and wavelength of the transition, respectively. The key assumption in the previous derivation was that the level population was equally distributed among the sublevels.

We can extend the previous relationships to the case where the transition lines have a finite linewidth and are characterized by a lineshape function. The absorption cross section at a frequency ν is defined for a broadened line, by comparing the expression for light absorption 4.11 to that for an equivalent expression involving the B coefficient, as

$$\sigma_{12}(\nu) = \frac{h\nu n}{c} B_{12} g_{12}(\nu) = \frac{h}{\lambda} B_{12} g_{12}(\nu) \qquad (4.23)$$

and similarly for the emission cross section:

$$\sigma_{21}(\nu) = \frac{h\nu n}{c} B_{21} g_{21}(\nu) = \frac{h}{\lambda} B_{21} g_{21}(\nu) \qquad (4.24)$$

where $g_{21}(\nu)$ and $g_{12}(\nu)$ are the normalized emission and absorption lineshapes, respectively, and n is the index of refraction of the medium.[34] Maintaining the assumption that the sublevels are equally populated leads to the identity $g_{12}(\nu) = g_{21}(\nu)$. However, equations 4.23 and 4.24 have been also used in situations where the emission lineshape $g_{21}(\nu)$ is different from the absorption lineshape $g_{12}(\nu)$, as has been observed experimentally.[38] Since in this case the populations are not, in general, distributed equally among sublevels, the relationships between the B coefficients (such as equation 4.21), as well as those that flow from them, will no longer be accurate.[38] We will see later how to deal with this situation by means of the McCumber theory.

By integrating equations 4.23 and 4.24, and by taking λ to be the center wavelength of the transition, we can derive the following relationship:

$$g_1 \int \sigma_{12}(\nu) d\nu = g_2 \int \sigma_{21}(\nu) d\nu \qquad (4.25)$$

The upper state 2 will have a radiative lifetime τ_{21}, considering only the decay pathway to the lower state 1 via the spontaneous emission of a photon. The emission and absorption cross sections are directly related to the inverse of this lifetime, since all three

4.4. FUNDAMENTAL PROPERTIES

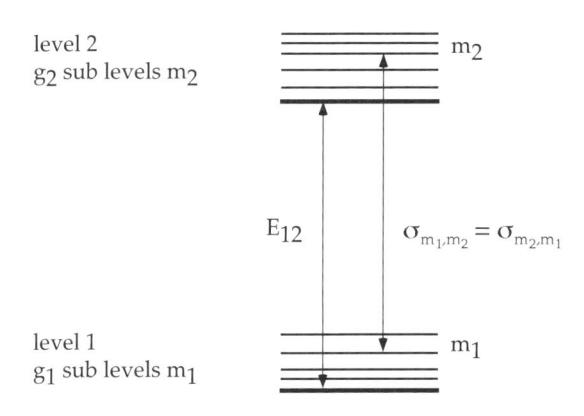

[Figure showing energy level diagram with level 2 (g₂ sub levels m₂) and level 1 (g₁ sub levels m₁), separated by energy E₁₂, with $\sigma_{m_1,m_2} = \sigma_{m_2,m_1}$]

Figure 4.6: Energy level structure for two multiplets 1 and 2, where sublevel degeneracy has been lifted by an external field. The emission and absorption cross section between two sublevels are equal by definition. E_{12} is the energy separation between the lowest-lying states of each multiplet.

parameters are governed by the same quantum mechanical matrix elements. We can write the radiative lifetime as

$$\frac{1}{\tau_{21}} = A_{21} = \frac{8\pi}{\lambda^2} \int \sigma_{21}(\nu) d\nu = \frac{8\pi}{\lambda^2} \frac{g_1}{g_2} \int \sigma_{12}(\nu) d\nu \qquad (4.26)$$

where λ, the wavelength of the transition in the medium, is equal to λ_0/n, with λ_0 the wavelength in vacuum and n the refractive index. The relations presented here also go by the name of the Ladenburg-Fuchtbauer relations. In general, one measures the absorption cross section exactly, by spectral attenuation measurements in a known length of material. The fluorescence spectral profile is then measured and its absolute amplitude is then obtained by integrating the absorption cross section and using the relationships developed above (e.g., one can obtain B_{21} by integrating the absorption cross section and then use it in the definition of the emission cross section).

The Ladenburg-Fuchtbauer approach was used by Sandoe et al, to calculate the emission cross sections of Er^{3+}.[39] This was also the case in the first studies of the cross sections for erbium-doped silica fibers.[40, 41, 42] It was later observed that the emission cross sections calculated in this fashion did not agree with experiment.[38, 43] This problem had already been noted in the 1960s in the context of transition metal ions, and a more adequate treatment was given by McCumber.[37]

McCumber Theory of Emission Cross Sections

To extend the previous discussion to the case where levels 1 and 2 have their degeneracy lifted and the populations in the sublevels are not all equal, we will consider that the populations are determined by the Boltzmann distribution. A simple derivation can be made of the original McCumber result, following Payne et al.[44] Consider the energy level structure of Figure 4.6, which generalizes the picture of Figure 4.5. We can derive an exact expression for the overall cross section of the level 1 to level 2 transition as the weighted sum of the intersublevel transition cross sections, where the weight is the population in the sub level where the transition originates. We write the emission and absorption cross sections for the 1 ↔ 2 transitions as

$$\sigma_{em}(\nu) = \sum_{m_1,m_2} \left(\frac{e^{-E_{m_2}/kT}}{Z_2}\right) \sigma_{m_2,m_1}(\nu) \tag{4.27}$$

and

$$\sigma_{abs}(\nu) = \sum_{m_1,m_2} \left(\frac{e^{-E_{m_1}/kT}}{Z_1}\right) \sigma_{m_1,m_2}(\nu) \tag{4.28}$$

The Z_i are the partition functions, $Z_i = \sum_{m_1,m_2} e^{-E_{m_i}/kT}$. The intersublevel cross sections $\sigma_{m_2,m_1}(\nu)$ contain all the lineshape information.[44]

First, note that one can recover as a limit the Ladenburg-Fuchtbauer situation, by assuming the high temperature limit $kT \to \infty$. In this case all the sublevels are equally populated and expressions 4.27 and 4.28 reduce to

$$\sigma_{21}(\nu) = \frac{1}{g_2} \sum_{m_2,m_1} \sigma_{m_2,m_1}(\nu) \tag{4.29}$$

and

$$\sigma_{12}(\nu) = \frac{1}{g_1} \sum_{m_1,m_2} \sigma_{m_2,m_1}(\nu) = \frac{g_2}{g_1} \sigma_{21}(\nu) \tag{4.30}$$

We obtain this equality also from equations 4.23 and 4.24 since $g_{12}(\nu) = g_{21}(\nu)$ when the sublevels are equally populated.

When the spread in energy of the sublevels is larger than the thermal energy kT, then departure from the Ladenburg-Fuchtbauer limit occurs. Note that in the case of the lower levels of the erbium ion (e.g., $^4I_{15/2}$, $^4I_{13/2}$) the total Stark splittings, which are on the order of 300–400 cm^{-1}, are quite a bit larger than the average thermal energy, kT, which is approximately 200 cm^{-1}. We thus expect significant differences among the thermal populations of the different sublevels. The assumption that the Stark levels are equally populated will then be violated.

By dividing equation 4.27 by equation 4.28, and using the fact that $E_{m_2} = E_{m_1} + h\nu - E_{12}$, we obtain

$$\begin{aligned}\frac{\sigma_{em}(\nu)}{\sigma_{abs}(\nu)} &= \frac{Z_1}{Z_2} \frac{\sum_{m_1,m_2} e^{-E_{m_2}/kT} \sigma_{m_2,m_1}(\nu)}{\sum_{m_1,m_2} e^{-E_{m_1}/kT} \sigma_{m_1,m_2}(\nu)} \\ &= \frac{Z_1}{Z_2} e^{(E_{12}-h\nu)/kT} \frac{\sum_{m_1,m_2} e^{-E_{m_1}/kT} \sigma_{m_1,m_2}(\nu)}{\sum_{m_1,m_2} e^{-E_{m_1}/kT} \sigma_{m_1,m_2}(\nu)} \\ &= \frac{Z_1}{Z_2} e^{(E_{12}-h\nu)/kT} \end{aligned} \tag{4.31}$$

4.4. FUNDAMENTAL PROPERTIES

This relationship between the emission and absorption cross sections replaces that obtained from the Einstein coefficients. Originally derived by McCumber, equation 4.31 is often seen with the quantity $Z_1/Z_2 e^{E_{12}/kT}$ replaced by the expression $e^{\epsilon/kT}$.[37]

The McCumber theory was reintroduced, in the context of Er^{3+}-doped glasses, by Miniscalco and Quimby.[45] The McCumber derivation dispenses with the assumptions needed to reach the Ladenburg-Fuchtbauer relations in the context of multistate levels, namely that the states be equally populated or have equal transition probabilities. The key McCumber assumption is that thermal equilibrium be reached within a multistate manifold in a time short compared to the overall lifetime of the manifold. This case is satisfied in most cases, especially that of the Er^{3+} $^4I_{15/2} \leftrightarrow {}^4I_{13/2}$ transition. The McCumber relationship between the emission cross section σ_{21} and the absorption cross section σ_{12} is thus written as

$$\sigma_{21}(\nu) = \sigma_{12}(\nu) e^{(\varepsilon - h\nu)/kT} \qquad (4.32)$$

where ε is the "mean" transition energy between the two manifolds, as seen from the derivation above involving the partition functions.[37]

Equation 4.32 implies that the emission and absorption cross sections are equal only at one frequency and that for frequencies higher than the crossing point, the absorption cross section is larger than the emission cross section and vice versa for frequencies below the crossing point. This point is verified experimentally, in particular for Er^{3+}. An additional relationship is provided by the McCumber analysis, linking the radiative lifetime and the emission cross section:

$$\frac{1}{\tau_{21}} = \frac{8\pi n^2}{c^2} \int \nu^2 \sigma_{21}(\nu) d\nu \qquad (4.33)$$

The additional information provided by equation 4.33 allows the determination of the absolute values of the emission cross section, after the absorption cross sections and τ_{21} lifetime have been measured. A summary discussion of both the Einstein coefficient relationships and the cross section relationships for multiplets with thermally distributed populations has been given by McCumber, based on the principles of detailed balance.[46]

The McCumber treatment has led to excellent agreement with experiment. Figure 4.7 shows the measured and calculated cross sections for an erbium-doped fluorophosphate glass, from Miniscalco's and Quimby's work.[45] The cross sections derived from the various models can be used in modeling the behavior of a doped fiber amplifier made from such a glass, as discussed in Chapter 6.

4.4.2 Lifetimes

The lifetime of a level is inversely proportional to the probability per unit time of the exit of a ion from that excited level. The decay of the population in a given level, for an ensemble of excited ions, then drops exponentially with a time constant equal to the lifetime. When there are several pathways for the population to decay, the total probability is equal to the sum of the individual probabilities for each pathway. Individual lifetimes can also be assigned to these decay paths. Usually, the lifetime for a given

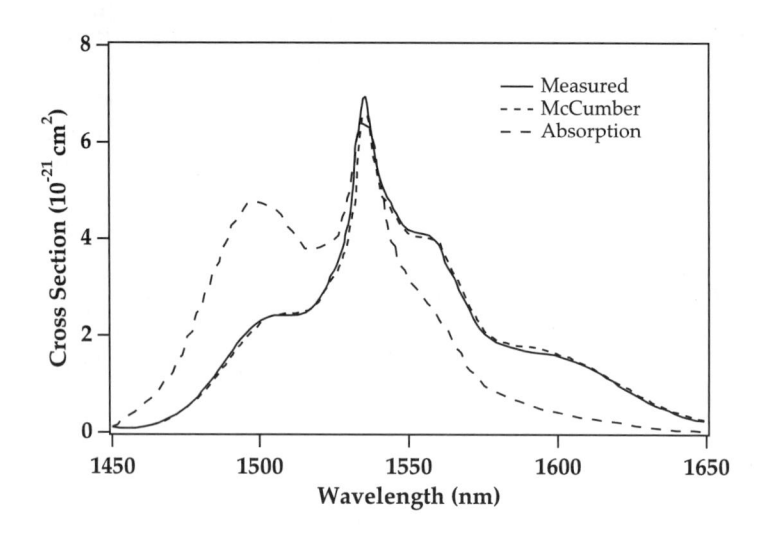

Figure 4.7: Comparison of the shape of the measured stimulated emission cross section with that calculated from the absorption cross section using the McCumber theory, for the Er^{3+} $^4I_{15/2} \leftrightarrow\, ^4I_{13/2}$ transition. From reference [45].

rare earth level follows from consideration of the two main paths for decay, radiative and nonradiative:

$$\frac{1}{\tau} = \frac{1}{\tau_r} + \frac{1}{\tau_{nr}} \tag{4.34}$$

where τ is the total lifetime, τ_r is the radiative lifetime, and τ_{nr} is the non-radiative lifetime. The radiative lifetime arises from the fluorescence from the excited level to all the levels below it. It can be calculated from the Judd-Ofelt analysis. Since the radiative transitions are forbidden to first order, radiative lifetimes tend to be long, on the order of microseconds to milliseconds.

Nonradiative lifetimes depend largely on the nature of the glass or crystal host and the coupling between the vibrations of the lattice ions and the states of the rare earth ions. In the nonradiative process, the deactivation process from the excited rare earth level is accompanied by the emission of one or several phonons (i.e., elementary vibrations of the host). Non-radiative processes in rare earth doped systems have been well studied. The rule of thumb is to consider the level directly below the excited level and measure the energy difference in units of the highest energy phonon of the host, which will be an optical phonon. The higher the number of phonons needed to bridge the gap, the less likely the probability of the transition. The nonradiative transition probability drops exponentially with the number of phonons required to bridge the energy gap to the next lowest level. Given that the nonradiative rate will increase with temperature, since the phonon population increases with temperature according to Bose statistics, for the nonradiative transition rate at temperature T one obtains

$$(1/\tau_{nr})_{n,T} = (1/\tau_{nr})_{n,0}\left[1 - \exp(-\hbar\omega/kT)\right]^{-n} \tag{4.35}$$

4.4. FUNDAMENTAL PROPERTIES

Host	B(s^{-1})	α (cm)	$\hbar\omega$ (cm^{-1})	Ref.
Tellurite	$6.3 \cdot 10^{10}$	$4.7 \cdot 10^{-3}$	700	[27]
Phosphate	$5.4 \cdot 10^{12}$	$4.7 \cdot 10^{-3}$	1200	[27]
Borate	$2.9 \cdot 10^{12}$	$3.8 \cdot 10^{-3}$	1400	[27]
Silicate	$1.4 \cdot 10^{12}$	$4.7 \cdot 10^{-3}$	1100	[49]
Germanate	$3.4 \cdot 10^{10}$	$4.9 \cdot 10^{-3}$	900	[27]
ZBLA	$1.59 \cdot 10^{10}$	$5.19 \cdot 10^{-3}$	500	[27]
ZBLA	$1.88 \cdot 10^{10}$	$5.77 \cdot 10^{-3}$	460-500	[29]
Fluoroberyllate	$9 \cdot 10^{11}$	$6.3 \cdot 10^{-3}$	500	[49]

Table 4.2: Nonradiative phenomenological transition parameters for equation 4.36, for different glasses.

where $\hbar\omega$ is the energy of the phonon, $n = \Delta E / h\nu_m$ is the number of phonons needed to bridge the gap (ΔE is the energy gap and $h\nu_m$ is the maximum phonon energy of the phonons that can couple to the ion), and $(1/\tau_{nr})_{n,0}$ is the transition rate at T = 0.[47, 48] The nonradiative transition rate at T = 0 is also exponentially dependent on the number of phonons required to bridge the energy gap. Equation 4.35 can thus be rewritten using the phenomenological parameters B and α:

$$(1/\tau_{nr})_{n,T} = B \exp(-\alpha \Delta E) \left[1 - \exp(-\hbar\omega/kT)\right]^{-n} \quad (4.36)$$

In a given multiplet, the higher-lying states tend to relax very fast to the lowest-lying state of the multiplet. The lifetime of the lowest-lying state is then mostly governed by the transition rate to the next-lower multiplet. This model accounts reasonably well for the nonradiative transitions in rare earth ions in condensed matter. Values for the parameters B and α, along with the energy of the dominant phonon involved in the nonradiative process, are tabulated in Table 4.2. The nonradiative transition rates for Er^{3+} in several glass hosts are plotted in Figure 4.8.[50, 51] Phonon-aided optical transitions in rare earth ions, mediated by the same electron-phonon coupling as in the previously discussed nonradiative relaxation processes, can manifest themselves both in absorption and emission. Absorption can be observed even when the incident radiation is not coincident with the rare earth ion level, as emission or absorption of one or more phonons can help bridge the gap.[52] In fact, multiphonon sideband absorption has been proposed as a contributor to background loss in Er^{3+}-doped fluoride fibers.[53, 54]

In some cases, a higher nonradiative transition rate is desired. In Yb co-doped erbium-doped fiber, for example, it is often desired to absorb the pump radiation with the Yb^{3+} ion and transfer this energy to the Er^{3+} $^4I_{11/2}$ level, without any back-transfer to the Yb^{3+} ion. One way to accomplish this is to increase the speed of the decay from the $^4I_{11/2}$ level to the $^4I_{13/2}$ level of Er^{3+}, so that there is no time for the back-transfer process to develop, by choosing a host with a larger phonon energy than silica. Certain phosphoaluminosilicate glasses, with a maximum phonon energy of 1330 cm^{-1}, are thus good candidates for Yb-Er amplifiers.[55] Section 4.5.1 discusses observed lifetimes in the case of Er^{3+}.

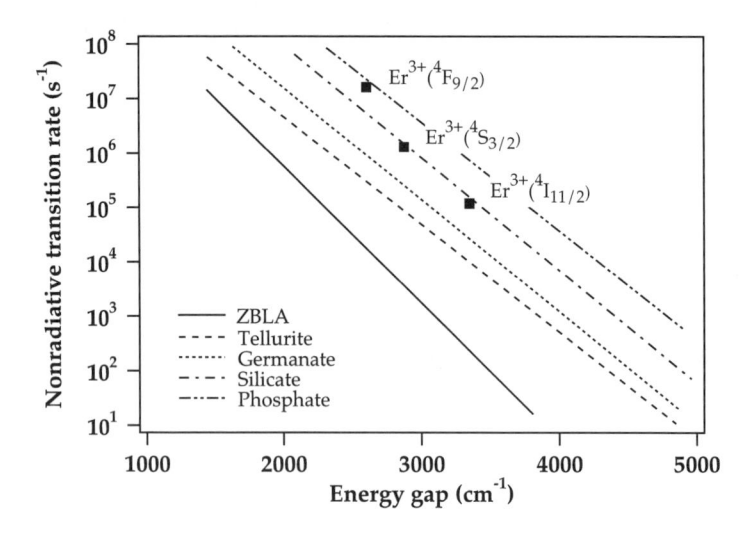

Figure 4.8: Nonradiative transition rates for Er^{3+} in various glass hosts, as a function of the energy gap to the next-lowest level. From references [50] and [51]. The squares are measured nonradiative transition rates for energy levels of Er^{3+} in silicate glass from reference [51]. Reprinted with permission from D. C. Yeh, W. A. Sibley, M. Suscavage, and M. G. Drexhage, *J. Appl. Phys.* Vol. 62, p. 266 (1987). Copyright 1987 American Institute of Physics.

4.4.3 Linewidths and Broadening

The linewidth of a transition contains contributions from various effects. For a transition between two given eigenstates of a rare earth ion, the linewidth, or breadth, of a transition, contains both a homogeneous and inhomogeneous contribution. The homogeneous, or natural, broadening arises from the lifetime and dephasing time of the state and depends on the both radiative and nonradiative processes.[34] The faster the lifetime or dephasing time, the broader the state. The inhomogeneous broadening is a measure of the various different sites in which an ensemble of ions can be situated. With variations in the local environment of an ion, there will be shifts in the energy levels of the ion and the fluorescence or absorption observed from this collection of ions will be smeared by the inhomogenous broadening. An inhomogeneous line is thus a superposition of a set of homogeneous lines. Such homogeneous and inhomogeneous lineshapes are depicted pictorially in Figure 4.9.

In the presence of a strong signal that saturates the transition, the absorption or emission lineshape will be affected in a different way, depending on whether the line is homogeneously or inhomogeneously broadened. In short, one can say that in the case of homogeneous broadening, the line will saturate uniformly as the population inversion is reduced (for example, under the effect of a strong signal). On the other hand, in the case of inhomogeneous broadening, the population inversion can be "locally" affected in a subset of the entire energy space of the considered transition. The gain spectrum then changes non-uniformly across the transition, with a "hole" in the

4.4. FUNDAMENTAL PROPERTIES

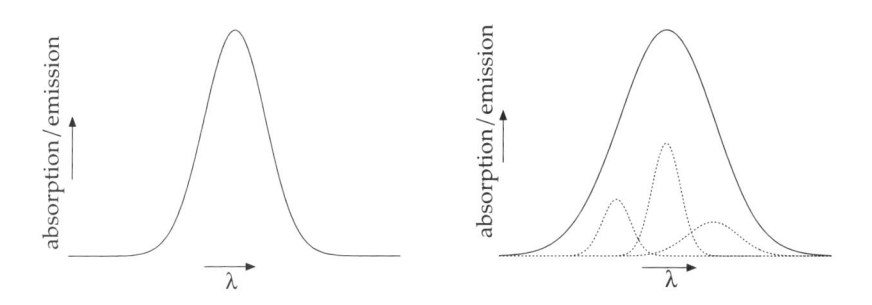

Figure 4.9: Left: a homogeneously broadened line for a collection of ions with identical transition frequencies and lifetimes. Right: inhomogeneously broadened line made up of a collection of homogeneously broadened lines with different center frequencies and linewidths.

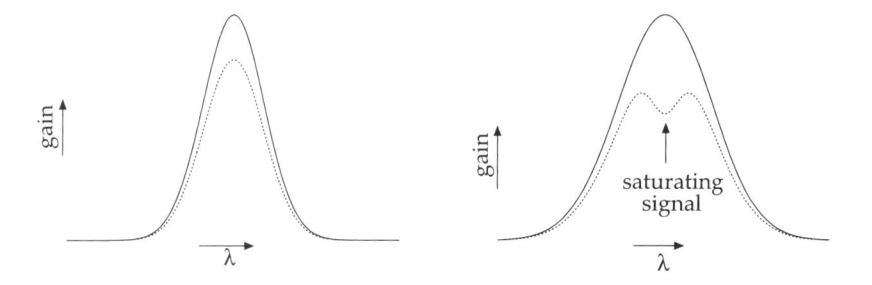

Figure 4.10: Gain saturation for a broadened line (solid line: unsaturated gain; dotted line: saturated gain in the presence of a strong signal). Left: gain saturation for a homogeneously broadened line. Right: gain saturation for an inhomogeneously broadened line (the spectral position of the narrow band strong signal is indicated by the arrow).

vicinity of the energy level where the population inversion was depleted. These two different fashions of gain saturation are depicted in Figure 4.10.

Another broadening mechanism can be considered to arise from the Stark splitting of the two multiplets. When these levels are broadened and the center of the transitions between the different Stark levels are close in energy, the transition lines overlap and appear to form one large transition with a clearly visible substructure. In general, the larger the number of Stark components, the broader will be the total splitting of the manifold and the larger the breadth of the transition between the manifolds. If the population redistribution between the Stark levels is fast enough (i.e., faster than the time scale over which the light signal of interest is interacting with the ion), then the entire transition will take on a homogeneously broadened character. That is, the saturation of a transition between Stark levels saturates the transitions between the other Stark levels of the multiplets.

In glasses, both the homogeneous and inhomogeneous broadening can be quite large, as compared to crystals. The homogeneous broadening of a given level is determined mostly by the nonradiative transitions, when there are levels located close in energy and below the level in question. In silica fibers where the phonon energy is rather strong and the electron-phonon coupling strength is significant, the homogeneous broadening is quite large. The inhomogeneous broadening is also quite large due to the multiplicity of sites and environments available to the ion given how it intercalates itself among the vitreous rings and chains.[56, 57] Section 4.5.2 discusses linewidth broadening studies of the Er^{3+} $^4I_{15/2} \leftrightarrow {}^4I_{13/2}$ transition.

The exact character of the line broadening of the $^4I_{13/2} \leftrightarrow {}^4I_{15/2}$ transition is important in terms of its impact on the gain process, especially as regards gain saturation behavior and WDM amplification. In the presence of homogeneous broadening, a strong enough signal can extract all the energy stored in the amplifier, while for an inhomogeneously broadened amplifier only the energy stored in the subset of ions interacting with the incident radiation can be extracted. Thus, homogeneously broadened amplifiers are more efficient in providing energy to a signal than inhomogeneously broadened amplifiers. On the other hand, a homogeneously broadened amplifier used in a WDM system is very susceptible to adding and dropping of wavelength channels, or their power, in that this will affect the gain and energy extraction of other channels. All the wavelength channels are coupled as they interact with the same homogeneously broadened line. In the case of inhomogeneously broadened amplifiers, however, sufficiently separated wavelength channels can propagate quasi-independently. Any differences in gain will cumulate with the number of amplifiers and can give rise to different system characteristics for long distance propagation.

4.5 SPECTROSCOPY OF THE ER^{3+} ION

Er^{3+} is the ion of choice for lasing and amplification in the 1.5 μm region, due to its $^4I_{13/2} \leftrightarrow {}^4I_{15/2}$ transition. We will discuss in this section the linear absorption and emission spectroscopy of Er^{3+} in glass hosts. The next section will treat ion-ion interaction effects which can appear at high ion concentrations. The discussion of pump and signal excited state absorption will be deferred to Chapter 6.

There is a rather large body of work in the literature on which we can draw to understand the properties of Er^{3+} in various hosts. The energy level scheme of Er^{3+}, up to the blue-green region of the spectrum, is shown in Figure 4.11. The transitions between the higher-lying energy levels and the ground level can be observed by absorption measurements. Figure 4.12 shows the absorption measured in an erbium-doped fiber at room temperature. The erbium is doped in a core of germano-alumino-silica glass. The measurement was made by the cutback technique using a white light source and an optical spectrum analyzer. The various peaks correspond to transitions between the $^4I_{15/2}$ ground state and the higher-lying states. The two main pump regions at 1480 nm and 980 nm are seen to provide significant absorption.

The key to the success of erbium is that the upper level of the amplifying transition, $^4I_{13/2}$, is separated by a large energy gap from the next-lowest level, so that its lifetime is very long and mostly radiative. The value of the lifetime is around 10 ms and varies depending on the host and erbium concentration. This long lifetime permits the inver-

4.5. SPECTROSCOPY OF THE ER^{3+} ION   111

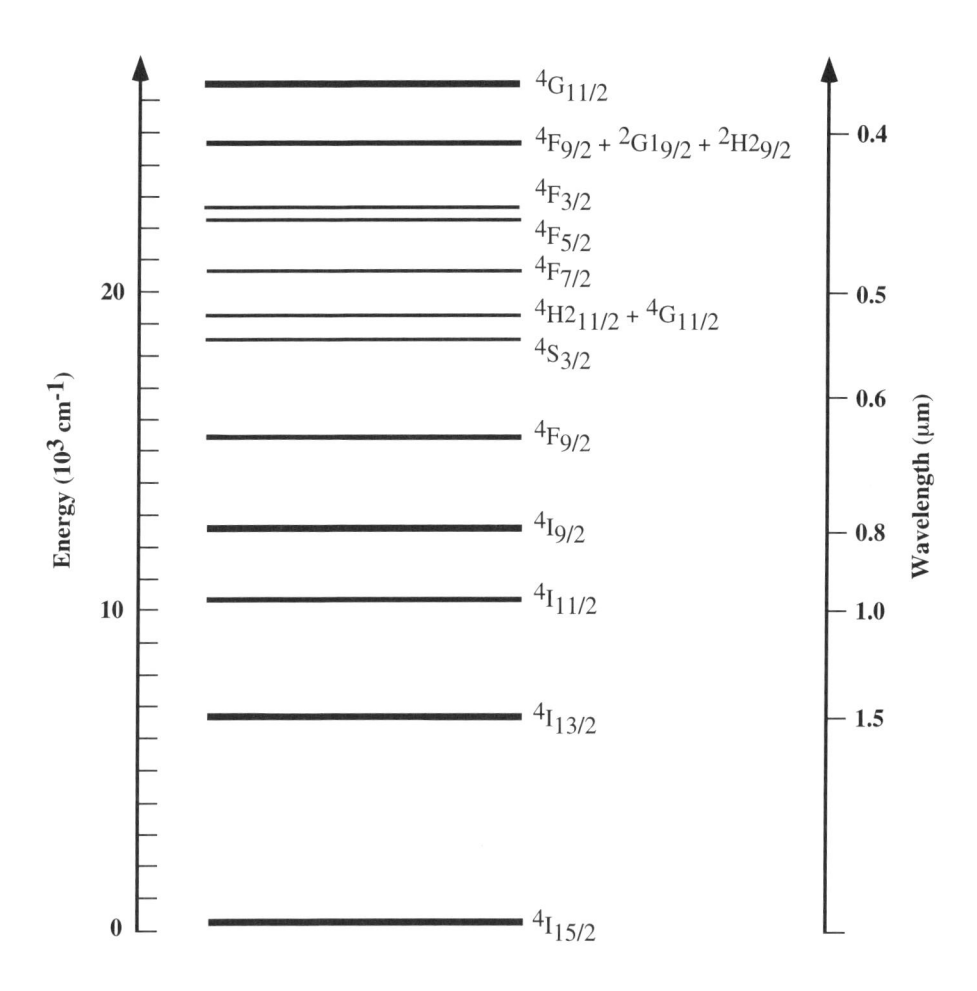

Figure 4.11: Energy level structure of Er^{3+}. The wavelength scale corresponds to the wavelength of the transition from a given energy level to the ground state.

sion of the population between the $^4I_{13/2}$ and $^4I_{15/2}$ levels with an a priori weak, and thus practical, pump source.

4.5.1 Lifetimes

The glass composition has a major effect on the lifetimes, radiative and nonradiative, of the excited states of Er^{3+}.[58, 59] The lifetime of the Er^{3+} $^4I_{13/2}$ level, in various hosts, is given in Table 4.3, as collected from various reports. The phosphate glasses have higher indices of refraction than the silica-based glasses. This contributes to a larger radiative cross section, from the local field effect, and thus reduces the lifetime for the $^4I_{13/2}$ level. The transition rates from the higher-lying states, $^4I_{11/2}$ and above, are significantly faster due to their high nonradiative transition rates. The transition rate out of the $^4I_{11/2}$ level, for example, is about 10^5 s^{-1} for silicate hosts (corresponding

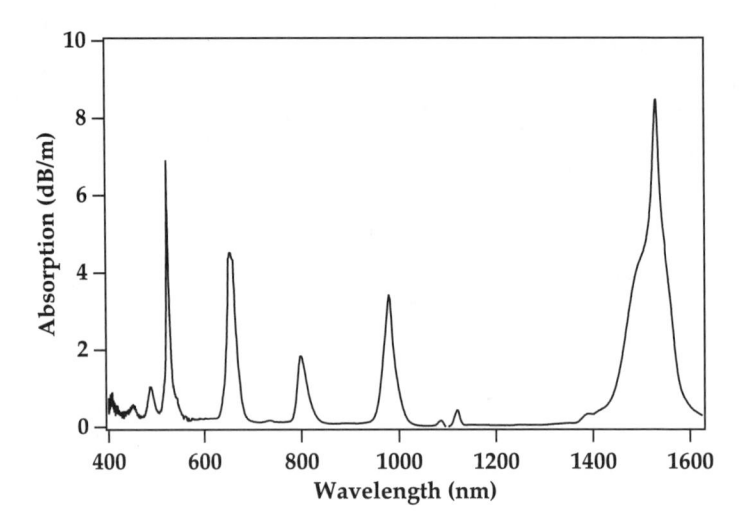

Figure 4.12: Experimentally measured absorption spectrum of an Er^{3+}-doped germano-alumino-silica fiber. The absorption in the 400–600 nm region has been divided by a factor of 10. The small oscillatory structure near 1100 nm corresponds to the cutoff of the second-order mode of the fiber.

Host Glass	Lifetime (ms)	Ref.
Na-K-Ba-silicate	14	[60]
ED-2 (silicate)	12	[60]
Silicate	14.7	[63]
Silicate L-22	14.5	[58]
Al-P silica	10.8	[58]
Al-Ge silica	9.5–10.0	[64]
Na-Mg-phosphate	8.2	[60]
LGS-E (phosphate)	7.7	[60]
LGS-E7 (phosphate)	7.9	[60]
Phosphate	10.7	[65]
Phosphate	8.5	[63]
Fluorophosphate	8.0	[60]
Fluorophosphate L11	8.25	[58]
Fluorophosphate L14	9.5	[58]
Ba-La-borate	8.0	[60]
Na-K-Ba-Al-germanate	6.5	[60]
Fluoride	10.3	[63]
Fluorozirconate F88	9.4	[58]
Tellurite	4	[66]

Table 4.3: Lifetime of the $^4I_{13/2}$ level of Er^{3+} in various glass hosts.

4.5. SPECTROSCOPY OF THE ER³⁺ ION

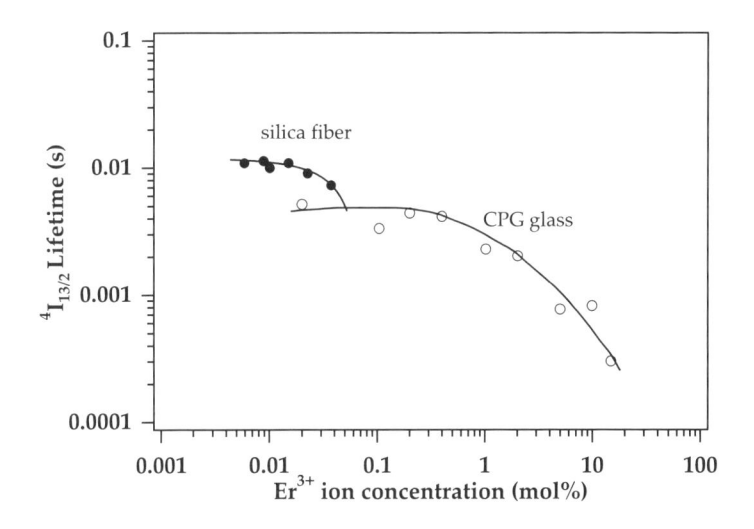

Figure 4.13: Room-temperature fluorescence lifetime measured at 1.55 μm for Er^{3+} in a silica glass fiber and a bulk sample of CPG (calcium metaphosphate) glass. From reference [62]. Reprinted with permission from Y. Mita, T. Yoshida, T. Yagami, and S. Shionoya, *J. Appl. Phys.* Vol. 71, p. 938 (1992). Copyright 1992 American Institute of Physics.

to a 10 μs lifetime) and increases to about 10^6 s^{-1} for phosphate hosts.[60] A value of 7 μs has been reported for the $^4I_{11/2}$ lifetime in a silica fiber.[61] In silica, the $^4S_{3/2}$ level, responsible for the green fluorescence sometimes observed from erbium-doped fibers, has a lifetime on the order of only 1 μs.[55, 62]

At high concentrations, an effect known as *concentration quenching* can occur. This is the shortening of the lifetimes of excited states, below the value measured in low concentration samples.[58] Figure 4.13 shows this effect for the lifetime of the $^4I_{13/2}$ state of Er^{3+}, as a function of concentration, for Er^{3+} in silica glass fiber and a bulk sample of CPG (calcium metaphosphate) glass.[62] CPG is a better host matrix for erbium ions in the sense that there is less tendency for the ions to cluster, as compared to silica. This translates into concentration quenching only settling in at higher concentrations. The drop in lifetime with concentration has also been seen in high-concentration thin films doped with erbium.[67, 68] The concentration quenching can arise from upconversion effects, as discussed in Section 4.6. It can also arise from the fact that, in high concentration samples, the excitation can migrate from one ion to another ion and has thus a greater probability of encountering a quenching center where a local phonon or deformation can deactivate the excitation.[69] Detailed studies of concentration quenching for Nd^{3+} in glasses have been performed, showing that at concentrations of about 3×10^{20} to 6×10^{20} ions per cm^3, the lifetime of the Nd^{3+} $^4F_{3/2}$ level is reduced by one half.[70]

Host Glass	Wavelength (nm)	σ_{abs}	σ_{em}	Ref.
Al-P silica	1531	6.60	5.70	[58]
Silicate L-22	1536	5.80	7.27	[58]
Fluorophosphate L11	1533	6.99	7.16	[58]
Fluorophosphate L14	1532	5.76	5.79	[58]
Fluorozirconate F88	1531	4.98	4.95	[58]
GeO$_2$-SiO$_2$	1530	7.9 ± 0.3	6.7 ± 0.3	[38]
Al$_2$O$_3$-SiO$_2$	1530	5.1 ± 0.6	4.4 ± 0.6	[38]
GeO$_2$-Al$_2$O$_3$-SiO$_2$	1530	4.7 ± 1.0	4.4 ± 1.0	[38]
Tellurite	1532		6.6	[66]

Table 4.4: Absorption (σ_{abs}) and emission (σ_{em}) cross sections for the $^4I_{13/2} \leftrightarrow {}^4I_{15/2}$ transition in Er^{3+} at the indicated wavelengths in units of 10^{-21} cm^2 for various glass hosts.

4.5.2 Er^{3+} Spectra, Cross Sections, and Linewidths

The spectrum of Er^{3+} in the 1.5 μm region depends on the host glass. The peaks and valleys in the spectra have different shapes based on the precise location of the Stark levels, the intensities of the transitions between the Stark levels, and on the amount of inhomogeneous and homogeneous broadening of these levels. As an example, the absorption and emission spectra of Er^{3+} near 1.5 μm, representing the $^4I_{13/2} \leftrightarrow {}^4I_{15/2}$ transition, are shown in Figure 4.14 for tellurite, fluoride (ZBLYAN), and Al-Ge silica glass hosts. Spectra of Er^{3+} in other glasses,—for example, phosphates and borates,—have also been reported.[39, 58]

The absorption and emission cross sections for the $^4I_{13/2} \leftrightarrow {}^4I_{15/2}$ transition for various hosts are shown in Table 4.4. A full spectrum of the absorption and emission cross sections for a particular erbium-doped Al-Ge silica fiber is shown in Chapter 6. Table 4.5 details the absorption coefficients for the pump band $^4I_{15/2} \rightarrow {}^4I_{11/2}$.

Studies have been made of the line-broadening mechanisms for Er^{3+} in crystals and glasses. Experimental investigations have sought to determine the relative importance of homogeneous and inhomogeneous broadening in Er^{3+}-doped fibers, and the exact value of the homogeneous and inhomogeneous linewidths of the $^4I_{13/2} \leftrightarrow {}^4I_{15/2}$ transition. There have been reports of a difference in this regard between germano-silicate doped fibers, in which inhomogeneous broadening is reported to be more pronounced than homogeneous broadening as compared to alumino-silicate doped fibers, which lean more toward homogeneous broadening of the $^4I_{13/2} \leftrightarrow {}^4I_{15/2}$ transition.[71, 72] A summary of reported homogeneous and inhomogeneous linewidths of the transition is presented in Table 4.6. The values included in Table 4.6 were obtained from room temperature measurements and do not include those extrapolated from low-temperature measurements.

The homogeneous linewidth of a transition is temperature dependent since the non-radiative transition rate, as governed by multiphonon emission, is temperature dependent. At low temperatures the phonon population is very small and the nonradiative transition rates are commensurately lower. Thus the overall line broadening at low temperature arises mostly from inhomogeneous or Stark splitting of the line. As the

4.5. SPECTROSCOPY OF THE ER^{3+} ION

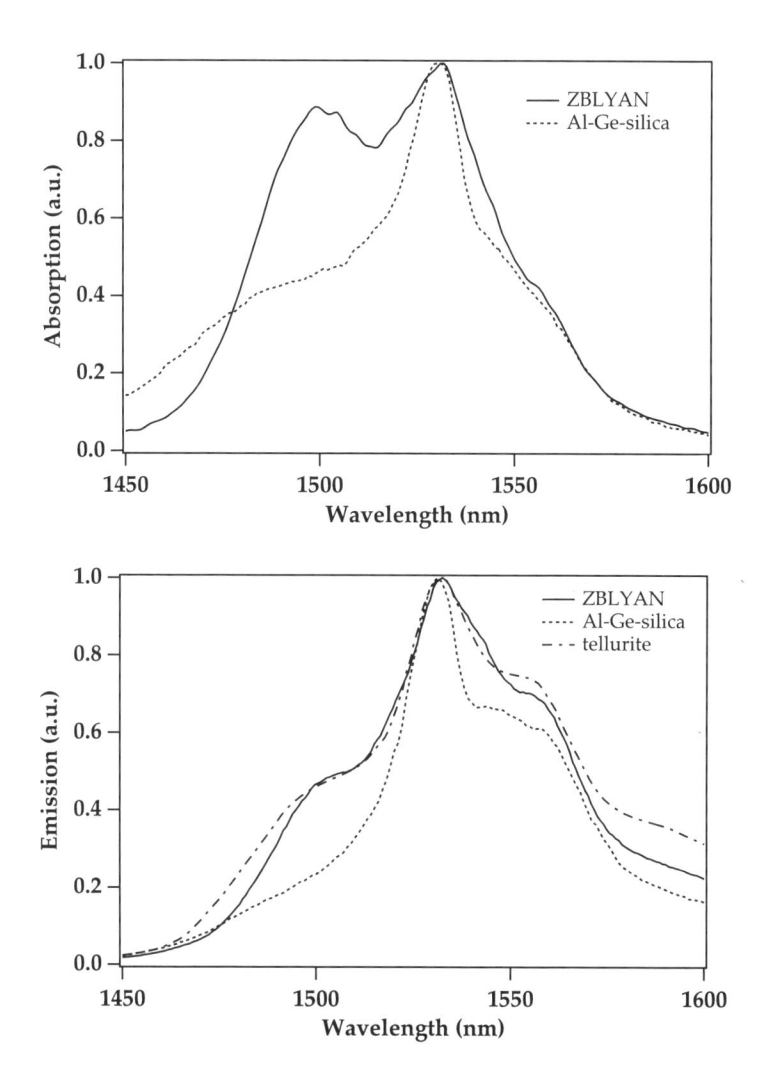

Figure 4.14: Absorption and emission profiles of Er^{3+} near 1.5 μm in ZBLYAN, tellurite, and in Al-Ge silica glasses. The tellurite spectrum is from reference [66], the fluoride spectrum is courtesy R. Tench, Lucent Technologies, Breinigsville, PA, and the silica spectrum is courtesy D. DiGiovanni, Lucent Technologies, Murray Hill, NJ.

temperature is increased, the linewidth increases along with its homogeneous component.

Two ways to measure the homogeneous linewidth are by the related techniques of fluorescence line narrowing and spectral gain hole burning. In fluorescence line narrowing (FLN), a narrow linewidth pump source excites a small subset only of the ions in the sample, i.e., those that are resonant with the excitation frequency.[56] Such a measurement needs to be done at low temperature, otherwise the homogeneous broadening of the line will be large enough that essentially all ions are excited (for the typical

Host Glass	Wavelength (nm)	σ_{abs}	Ref.
GeO$_2$-SiO$_2$	980	2.52 ± 0.03	[38]
Al$_2$O$_3$-SiO$_2$	980	1.9 ± 0.3	[38]
GeO$_2$-Al$_2$O$_3$-SiO$_2$	980	1.7 ± 0.3	[38]
Al-P silica	979	3.12	[58]
Silicate L-22	981	3.12	[58]
Phosphate L12	975	2.01	[58]
Phosphate L28	979	2.47	[58]
Fluorophosphate L11	974	2.46	[58]
Fluorophosphate L14	974	2.15	[58]
Fluorophosphate L11	974	2.46	[58]
Fluorophosphate L14	974	2.15	[58]
Fluorozirconate F88	974	2.15	[58]

Table 4.5: Absorption cross sections (σ_{abs}) for the $^4I_{15/2} \rightarrow {}^4I_{11/2}$ transition in Er^{3+}, in units of 10^{-21} cm^2 for various glass hosts.

Host Glass	Wavelength (nm)	Homogeneous Linewidth (nm)	Ref.
Al-Ge Silica	1545	2	[73]
Al-Ge Silica	1551	4	[73]
Al-Ge Silica	1562	5	[73]
Al-P Silica	1522	1.8	[74]
Ge Silica	1535	4	[75]
Fluorophosphate (low fluorine)	1522	1.6–2.2	[74]
Fluorophosphate (high fluorine)	1522	4.5–7.9	[74]
Fluorozirconate	1522	4.5–7.9	[74]

Table 4.6: Homogeneous linewidths of the Er^{3+} $^4I_{13/2} \leftrightarrow {}^4I_{15/2}$ transition at a specified center wavelength, in various glass hosts, reported from room-temperature measurements. In the FLN or hole-burning experiments, the homogeneous linewidth is taken to be one-half the width of the resonance observed.

case of a rare earth ion in a glass matrix). As the excitation source is swept in frequency, the FLN spectra then show the fluorescence spectra corresponding to the different sites that the ions can occupy, and an estimate can be made of both inhomogeneous and homogeneous broadening of the levels. It has been shown that the width of the resonant FLN signal is equal to twice the homogeneous linewidth, when spectral diffusion is negligible and the excitation is weak.[76] The first FLN measurement of Er^{3+} doped glass was reported by Zemon et al.[77] It was performed at low (T = 4.2K) temperatures to reduce the homogeneous linewidth and allow a study of the inhomogeneous broadening and Stark splitting. One general result is that site-to-site variation in the position of the energy levels is of the same order as the Stark splitting between adjacent levels, about 20 cm^{-1} to 60 cm^{-1}. The total Stark width of the $^4I_{15/2}$ and $^4I_{13/2}$ multiplets were extrapolated to be on the order of 330 cm^{-1} to 400 cm^{-1}. At room temperature the spectra were found to be representative of a mostly homogeneously broadened line.[77] Other FLN experiments have found that site distribution is wider in Al-doped silica fibers as compared to Ge-doped silica fibers, 42 cm^{-1} vs 10 cm^{-1}, respectively.[78] This is believed to originate from the fact that addition of Al disrupts the silica network, thereby creating a greater variety of sites for the Er^{3+} ion, as compared to silica without addition of Al.

In the spectral hole-burning technique, the ASE spectra (which are representative of the small signal gain spectrum) are recorded under the influence of a strong and narrow in-wavelength signal. The width of the hole surrounding the signal, in wavelength, is indicative of the homogeneous linewidth and is equal to twice the homogeneous linewidth under conditions of low power excitation.[76] Early experiments, done at low temperatures, extrapolated observed hole widths from 77 K to room temperature. Measurements done in this way claimed a homogeneous linewidth of 11.5 nm and an inhomogeneous linewidth of 12.5 nm for alumino-germano-silicate fibers, while for germano-silicate fibers the corresponding values were reported to be 4 nm and 8 nm, respectively, at wavelengths near 1530 nm.[71, 72]

Given the multitude of Stark levels and the large homogeneous broadening at room temperature, it is no surprise that at room temperature the 1.5 μm transition behaves essentially as a homogeneously broadened line, since the site-to-site variation in the centers of the Stark levels tend to be smeared out by the spectral overlap between the broadened adjacent Stark levels. Additionally, thermal redistribution between the Stark levels of a given manifold is extremely fast at room temperature, on the order of a few ps. Thus, on the time scale of typical pump pulses (cw, mostly), the population distribution within the Er^{3+} manifolds remains the same, and the transition spectra are not significantly changed by the pump or signal radiation. Since at room temperature the homogeneous linewidth is quite large, and especially so for alumino-germano-silicate fibers, any hole burning is quite hard to observe at room temperature.[80]

The appearance of inhomogeneous broadening and spectral hole-burning effects at room temperature have recently been observed with high-precision gain measurements.[81, 73] The results of a hole-burning study performed at room temperature in an alumino-germano-silicate fiber are shown in Figure 4.15. A hole is clearly observed at the wavelengths where the saturating signal is present. The hole depth, as well as its width, also increases as the signal strength, as measured by the overall gain compression, increases. Note that the homogeneous linewidth is observed to vary with wavelength. This can arise from the difference in nonradiative lifetimes among different Stark sublevels of a multiplet. For the purpose of determining the room-temperature homogeneous linewidths, the room-temperature spectral gain hole measurements yield the best estimates, as compared to extrapolations from low-temperature measurements.

Spectral hole burning can also be observed in a long haul system constituted of a chain of cascaded amplifiers. Bergano et al measured the difference in output spectra when the signal is present as opposed to absent, in a 6300 km chain where filters were used to equalize the gain in a WDM transmission experiment.[82] A spectral hole with a linewidth of 7 nm was observed (at a center wavelength of 1551 nm),

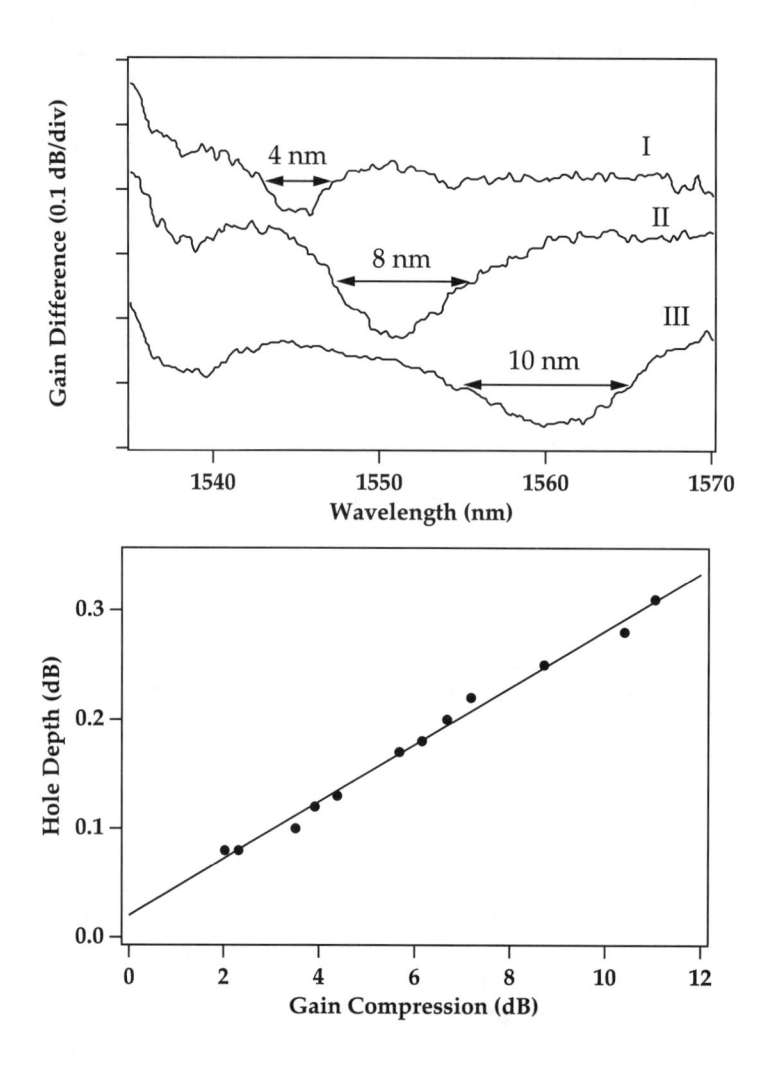

Figure 4.15: Spectral hole burning in the 1.5 μm region for Er^{3+} in a alumino-germano-silica glass fiber. From reference [73]. Top: wavelength dependence of the hole burning for a strong signal λ_{sat} with varying wavelength (curve I: $\lambda_{sat} = 1545$ nm, curve II: $\lambda_{sat} = 1551$ nm, curve III: $\lambda_{sat} = 1562$ nm). Bottom: depth of the spectral hole at 1551 nm for $\lambda_{sat} = 1551$ nm as a function of gain compression from the saturating signal.

consistent with the measurements of Figure 4.15. This hole burning will have an impact in WDM experiments as the strength of a given signal channel will impact the gain spectrum in nearby adjoining channels, if close enough, but not channels far removed in wavelength. This departure from homogeneous behavior of the gain profile will have to be mitigated in long haul dense WDM transmission systems by careful balancing of

4.5. SPECTROSCOPY OF THE Er^{3+} ION

Multiplet	Al-Ge Silica	Y_2SiO_5
$^4I_{15/2}$	0	0
	26–34	47
	51–59	100
	125–133	140.5
	201–268	250.5
		321.5
		428.5
		478.5
$^4I_{13/2}$	(6540)	(6471.5)
	0–8	0
		38.5
	96–104	79.5
	163–171	126.5
	222–230	256.5
		303.5
		382.5

Table 4.7: Positions of the Stark levels of the Er^{3+} $^4I_{13/2}$ and $^4I_{15/2}$ multiplets in Al-Ge silica and in a Y_2SiO_5 powder (cm^{-1}). From references [33] and [85]. Ranges are given for the glass due to greater experimental uncertainty. The energy position of the lowest level of the $^4I_{13/2}$ manifold is indicated in parentheses, and the energies of the other Stark levels of the manifold are listed with respect to the lowest-lying level.

the input signal powers, operating conditions of the amplifiers, and characteristics of the gain-flattening elements.

Another experimental demonstration of the inhomogeneous character of the Er^{3+} transitions in a glass fiber host is provided by a study of changes in the gain spectrum with pump wavelength variation. It has been found that changes of a few nm in the center wavelength of a narrow pump near 980 nm gives rise to non-uniform changes in the shape of the spectral gain curve.[83] The changes do not come about from variations in the average inversion of the fiber, as the pump power was corrected to maintain constant the gain at one signal channel. The effect comes about from the variation in excitation of different subsets of erbium ions in the glass, as the pump wavelength is changed. The non-uniformity is most pronounced in the 1530 nm region, where gain variations of up to 3 dB (relative to an average gain of about 30 dB) were observed for nm changes in the pump wavelength between 975 nm and 981 nm. In the 1550 nm region, the gain variations are only 0.5 dB (relative to an average gain of 30 dB).

Low-temperature (or room-temperature crystal host) measurements also give some idea of the position of the individual Stark levels of each multiplet. The $^4I_{15/2}$ multiplet has eight Stark components, while the $^4I_{13/2}$ multiplet has seven. Table 4.7 shows the experimentally determined positions of the Stark levels of the $^4I_{15/2}$ and $^4I_{13/2}$ manifolds, for Er^{3+} doped in an Al-Ge silica glass, and, for comparison, in a Y_2SiO_5 crystal (comparable to a phosphate glass in terms of the Er^{3+} environment).[33, 85] Given the larger inhomogeneous broadening in the glass, it is more difficult to unequivocally measure and assign the individual Stark levels.

4.6 ER^{3+}-ER^{3+} INTERACTION EFFECTS

The vast majority of erbium-doped fiber amplifiers use fiber lengths that are in the 1 m to 100 m range. It has been found that at short lengths, where the concentration of erbium is high, undesirable effects occur that reduce the amplifier performance.[86] These effects have been tied to Er^{3+}-Er^{3+} ion-ion interaction effects. Such effects can also occur at longer lengths in fibers where the erbium ions are confined to the center of the core region. This is because the erbium ion density will be high there, and it is the local concentration of Er^{3+} ions that is important in this context.[79] It is known that very low concentration erbium-doped fibers are more efficient in terms of overall amplifier performance.[87] As one might expect, the host glass composition is of paramount importance in determining the strength and nature of the interaction effects. These interactions become crucially important for the development of erbium-doped waveguide amplifiers where the amplifier length is reduced to a few cm.

The ion-ion interaction effects relate to the issue of *energy transfer* between rare earth ions. When the local concentration of rare earth ions becomes high enough, it is no longer valid to assume that each ion is an isolated ion that acts independently of its neighbors. One must consider the possibility of energy transfer between ions, and its consequences.[88, 89] This can have a negative impact on amplifiers when the upper state of an amplifying transition is being depleted by energy transfer. Energy transfer can occur via phonon (or virtual photon) mediated interactions. Energy transfer can sometimes be used advantageously, as for example in the case of infrared pumped visible lasers.[90]

The terminology used in this field can vary from author to author, so it is useful to first define and list some of the various effects that can occur. We follow Auzel's work for our definitions.[88] The word *upconversion* will return often and basically signifies that energy is being given to an excited ion thereby promoting it to an even higher energy state. There are many mechanisms that give rise to upconversion.

The simplest case of energy transfer involves the transfer of a given amount of energy from an excited ion to a nearby ion in its ground state, without gain or loss of energy on the part of the two-ion system as a whole. This is known as *resonant energy transfer* and is illustrated in Figure 4.16a. Nonresonant transfer can also occur where the difference in energies between the initial and final ionic states is made up with the assistance of a photon or a phonon. Energy transfer can also occur between ions in their excited states. This is illustrated in Figure 4.16b where ions 1 and 2—previously excited by some means—transfer energy, resulting in the promotion of ion 2 to a higher energy level. This is known as *stepwise upconversion*, or simply upconversion. *Cooperative upconversion* is a more subtle process that involves the behavior of two (or more) ions that are so closely coupled that they act as a single molecule.

A heuristic definition of cooperative luminescence processes, in the case of two ions, is that they involve the emission (or absorption) of one photon by two ions acting simultaneously. For example, we show in Figure 4.16c cooperative luminescence where two coupled excited ions in the same excited state simultaneously decay to their respective ground states and emit a single photon with twice the energy of the single-ion excited state.[93] Figure 4.16d shows a slightly more complicated process, that of cooperative energy transfer with simultaneous absorption of a photon. The erbium-doped

4.6. ER^{3+}-ER^{3+} INTERACTION EFFECTS

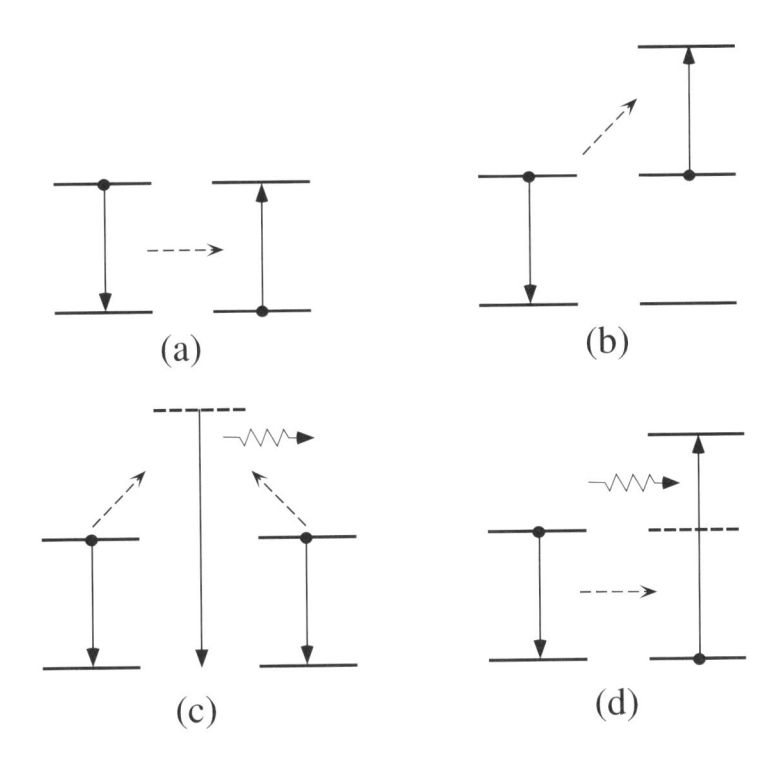

Figure 4.16: Ion-ion energy transfer mechanisms. (a): Resonant energy transfer; (b): stepwise upconversion; (c): cooperative luminescence; (d): cooperative energy transfer and simultaneous photon absorption.

fiber literature will often refer to process (a) of Figure 4.16 as cooperative upconversion, which can be confusing as, strictly speaking, it is not a cooperative process.

In the case of Er^{3+}, the upconversion mechanism of Figure 4.16b is the most prevalent one, where the excited state involved is the $^4I_{13/2}$ state with a radiative lifetime of 10 ms, which gives closely spaced neighbors ample time to interact. The upconversion process, since it deactivates ion 1 back to the ground state (where it now becomes an absorber rather than an emitter) is another pathway for population loss from the $^4I_{13/2}$ state and thus acts to reduce the overall lifetime of this state. The upconversion of 1.5 μm radiation by pairs of Er^{3+} ions excited to the $^4I_{13/2}$ levels, where one ion drops back down to the ground state and the other is promoted to a higher-lying state, was first observed by Brown et al in 1969 in an SrF$_2$ crystal.[91] Johnson et al, in 1972, used this effect to produce visible radiation from infrared radiation in an Er^{3+} doped BaYF$_5$ crystal.[92] Auzel demonstrated that this resulted from stepwise upconversion, rather than cooperative upconversion.[94] Upconversion has also been observed in erbium-doped silica fibers, from the visible radiation emitted by the fiber. With high power 1480 nm pumps (100–1000 mW), for example, emission at 550 nm, 660 nm, and 807 nm can be observed.[95]

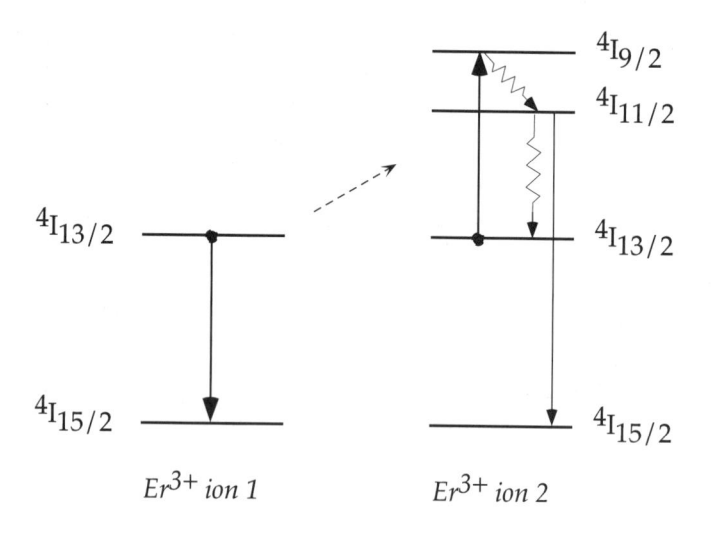

Figure 4.17: Upconversion process between two erbium ions excited to their respective $^4I_{13/2}$ levels.

We show the result of the upconversion process for the case of interest to us for erbium-doped amplifiers in Figure 4.17. Ion 2, which is excited to the $^4I_{9/2}$ level as a result of the upconversion, quickly cascades back down to the metastable $^4I_{13/2}$ level, with the $^4I_{11/2}$ level as a possible intermediate step. Radiative transitions from the $^4I_{11/2}$ level to the ground state—with emission of a 980 nm photon—can also occur, although the probability for this is low. This effect depends on pump power, since the effect grows with the probability of two closely spaced ions being both excited to the $^4I_{13/2}$ state. This is illustrated in Figure 4.18, which shows the lifetime of the $^4I_{13/2}$ state (as measured from the 1/e point of the 1.5 μm luminescence decay) in fibers and glass waveguides with varying concentrations of Er^{3+}, as a function of pump power at 980 nm.[64] This upconversion will of course decrease the amplifier efficiency, since the pump threshold and signal gain depend critically on the $^4I_{13/2}$ level lifetime.

An added complication is that the local structure can vary for a given bulk glass or fiber, such that clusters of ions can form in one region whereas another region might have isolated ions.[96, 97] One will then observe several components in the lifetime measured from the 1.5 μm luminescence, where the faster components arise from the clustered ions. Some host glasses, such as silica, are known to have more of tendency to incorporate Er^{3+} clusters. The clustering effect has been shown to lead to cooperative luminescence phenomena and has been evidenced even at quite low levels of rare earth dopant ($\sim$ 85 ppm).[98] It has been found, however, that the addition of Al reduces the tendency to cluster.[58] A detailed study of clustering by Nd^{3+} ions in silica glass, and its mitigation by Al co-doping, was done by Arai and coworkers.[99] Arai reported microclustering by Nd^{3+} in an SiO_2 host, due to the fact that the SiO_2 matrix is rigid and does not accept well the Nd^{3+} ions. The Nd^{3+} ions do not bond well with nonbridging oxygen ions of the host and tend to cluster together. The presence of Al is

4.6. ER^{3+}-ER^{3+} INTERACTION EFFECTS

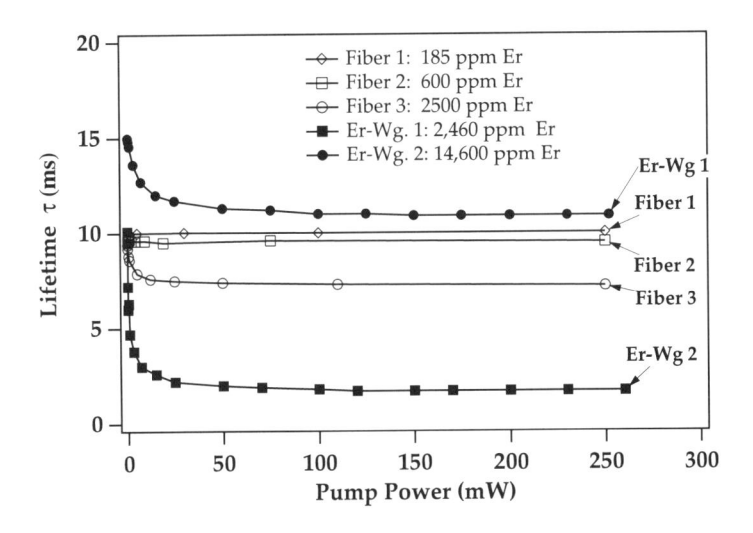

Figure 4.18: Plot of the spontaneous emission lifetime of the $^4I_{13/2}$ level (defined as the 1/e point of the fluorescence decay), versus pump power, for several different erbium-doped Ge-Al-P silica fibers and erbium-doped glass waveguides, pumped at 980 nm. From reference [64] (©1993 IEEE).

believed to produce solvation shells into which the Nd^{3+} ions insert themselves, thus mitigating the clustering effect.[99]

In terms of actual amplifier performance, it has been reported that for optimum performance, for germano-silicate fibers the Er^{3+} concentration should be held to less than about 100 ppm, while for alumino-germano-silicate fibers the Er^{3+} concentration should be less than about 1000 ppm.[101] Studies of pair-induced quenching, where the clusters of erbium ions are pairs of ions, indicate that the maximum concentration of Er^{3+} in alumino-germano-silicate fibers should be 400 mole ppm, to obtain near-optimum gain performance.[100] Nevertheless, Al doped silica fiber amplifiers with an Er^{3+} concentration of 8900 ppm have been reported, with reasonable gains (20–30 dB) in lengths of 0.5 m to 1 m, when pumped at 980 nm or 1480 nm.[101]

Bibliography

[1] J. Becquerel, *C. R. Acad. Sci.* **142**, 775 (1906).

[2] J. Becquerel, *Le Radium* **4**, 328 (1907).

[3] J. Becquerel and H. Kammerlingh Onnes, *Proc. Acad. Amsterdam* **10**, 592 (1908).

[4] M. Mayer, *Phys. Rev.* **60**, 184 (1941).

[5] S. Hüfner, *Optical Spectra of Rare Earth Compounds* (Academic Press, New York, 1978).

[6] P. C. Becker, Ph.D thesis, University of California, Berkeley, 1986.

[7] H. Bethe, *Ann. Physik* **3**, 133 (1929).

[8] J. H. Van Vleck, *J. Phys. Chem.* **41**, 67 (1937).

[9] B.G. Wybourne, *Spectroscopic Properties of Rare Earths* (Interscience, New York, 1965).

[10] B. R. Judd, *Operator Techniques in Atomic Spectroscopy* (McGraw-Hill, New York, 1963).

[11] G. H. Dieke, *Spectra and Energy Levels of Rare Earths Ions in Crystals*, H. M. Crosswhite and H. Crosswhite, Eds. (Interscience, New York, 1968).

[12] B. Henderson and G. F. Imbusch, *Optical Spectroscopy of Inorganic Solids* (Clarendon Press, Oxford, 1989).

[13] R. D. Cowan, *The Theory of Atomic Structure and Spectra* (University of California Press, Berkeley, 1981).

[14] B. R. Judd, *Rep. Prog. Phys.* **48**, 907 (1985).

[15] W. T. Carnall, H. Crosswhite, H. M. Crosswhite, J. P. Hessler, N. Edelstein, J. G. Conway, and G. V. Shalimoff, *J. Chem. Phys.* **72**, 5089 (1980).

[16] T. Hayhurst, G. Shalimoff, N. Edelstein, L. A. Boatner, and M. M. Abraham, *J. Chem. Phys.* **74**, 5449 (1981).

[17] R. D. Peacock, *Structure and Bonding* **22**, 83 (1975).

[18] B. R. Judd, *J. Chem. Phys.* **70**, 4830 (1979).

[19] M. F. Reid and F. S. Richardson, *J. Chem. Phys.* **79**, 5735 (1983).

[20] M. F. Reid and F. S. Richardson, *J. Chem. Phys.* **79**, 5743 (1983).

[21] B. R. Judd, *Phys. Rev.* **127**, 750 (1962).

[22] G. S. Ofelt, *J. Chem. Phys.* **37**, 511 (1962).

[23] P. C. Becker, N. Edelstein, B. R. Judd, R. C. Leavitt, and G. M. S. Lister, *J. Phys. C* **18**, L1063 (1985).

[24] C. K. Jørgensen and B. R. Judd, *Mol. Phys.* **8**, 281 (1964).

[25] S. F. Mason, R. D. Peacock, and B. Stewart, *Mol. Phys.* **30**, 1829 (1975).

[26] C. W. Nielson and G. F. Koster, *Spectroscopic Coefficients for the p^n, d^n, and f^n Configurations* (The M.I.T. Press, Cambridge, MA, 1963).

[27] R. Reisfeld, in *Spectroscopy of Solid-State Laser-Type Materials*, B. Di Bartolo, Ed. (Plenum Press, New York, 1987) pp. 343–396.

[28] R. Reisfeld, G. Katz, C. Jacoboni, R. de Pape, M. G. Drexhage, R. N. Brown, and C. K. Jørgensen, *J. Sol. St. Chem.* **48**, 323 (1983).

[29] M. D. Shinn, W. A. Sibley, M. G. Drexhage, and R. N. Brown, *Phys. Rev. B* **27**, 6635 (1983).

[30] R. Reisfeld, G. Katz, N. Spector, C. K. Jørgensen, C. Jacoboni, and R. de Pape, *J. Sol. St. Chem.* **41**, 253 (1982).

[31] F. Auzel, *Rivista della Statz. Sper. Vetro* **5**, 49 (1990).

[32] M. J. Weber, *Phys. Rev.* **157**, 262 (1967).

[33] C. Li, C. Wyon, and R. Moncorge, *IEEE J. Quant. Elect.* **28**, 1209 (1992).

[34] P. W. Milonni and J. H. Eberly, *Lasers* (Wiley, New York, 1988).

[35] W. Beall Fowler and D. L. Dexter, *Phys. Rev.* **128**, 2154 (1962).

[36] W. Beall Fowler and D. L. Dexter, *J. Chem. Phys.* **43**, 1768 (1965).

[37] D. E. McCumber, *Phys. Rev.* **134**, A299 (1964).

[38] W. L. Barnes, R. I. Laming, E. J. Tarbox, and P. R. Morkel, *IEEE J. Quant. Elect.* **27**, 1004 (1991).

[39] J. N. Sandoe, P. H. Sarkies, and S. Parke, *J. Phys. D* **5**, 1788 (1972).

[40] W. J. Miniscalco, L. J. Andrews, B. A. Thompson, T. Wei, and B. T. Hall, "$^4I_{13/2}-^4I_{15/2}$ emission and absorption cross sections for Er^{3+}-doped glasses," in *Tunable Solid State Lasers*, Vol. 5 of OSA Proceedings Series, M. L. Shand and H. P. Jenssen, Eds. (Optical Society of America, Washington, D.C., 1989), p. 354-357.

[41] W. J. Miniscalco, L. J. Andrews, B. A. Thompson, T. Wei, and B. T. Hall, "The effect of glass composition on the performance of Er^{3+} fiber amplifiers," in *Fiber Laser Sources and Amplifiers*, M. J. F. Digonnet, Ed., *Proc. SPIE* **1171**, pp. 93–102 (1990).

[42] E. Desurvire and J. R. Simpson, *J. Light. Tech.* **7**, 835 (1989).

[43] K. Dybdal, N. Bjerre, J. Engholm Pedersen, and C. C. Larsen, "Spectroscopic properties of Er-doped silica fibers and preforms," in *Fiber Laser Sources and Amplifiers*, M. J. F. Digonnet, Ed., *Proc. SPIE* **1171**, pp. 209–218 (1990).

[44] S. A. Payne, L. L. Chase, L. K. Smith, W. L. Kway, and W. F. Krupke, *IEEE J. Quant. Elect.* **28**, 2619 (1992).

[45] W. J. Miniscalco and R. S. Quimby, *Opt. Lett.* **16**, 258 (1991).

[46] D. E. McCumber, *Phys. Rev.* **136**, A954 (1964).

[47] L. A. Riseberg and H. W. Moos, *Phys. Rev.* **174**, 429 (1968).

[48] M. J. Weber, *Phys. Rev. B* **8**, 54 (1973).

[49] C. B. Layne and M. J. Weber, *Phys. Rev. B* **16**, 3259 (1977).

[50] D. C. Yeh, W. A. Sibley, M. Suscavage, and M. G. Drexhage, *J. Appl. Phys.* **62**, 266 (1987).

[51] C. B. Layne, W. H. Lowdermilk, and M. J. Weber, *Phys. Rev. B* **16**, 10 (1977).

[52] F. Auzel, *Phys. Rev. B* **13**, 2809 (1976).

[53] F. Auzel, *Elect. Lett.* **29**, 337 (1993).

[54] F. Auzel, *J. Lum.* **60** & **61**, 101 (1994).

[55] J. R. Lincoln, W. L. Barnes, W. S. Brocklesby, and J. E. Townsend, *J. Lum.* **60** & **61**, 204 (1994).

[56] M. J. Weber, in *Laser Spectroscopy of Solids*, Vol. 49 of *Topics in Applied Physics*, W. M. Yen and P. M. Selzer, Eds. (Springer-Verlag, New York, 1981), pp. 189–239.

[57] S. Todoriki, K. Hirao, and N. Soga, *J. Appl. Phys.* **72**, 5853 (1992).

[58] W. J. Miniscalco, *J. Light. Tech.* **9**, 234 (1991).

[59] W. Ryba-Romanowski, B. Jezowska-Trzebiatowska, J. Sarzynski, J. Nowak, and A. Nowak, *Act. Phys. Pol.* **A55**, 841 (1979).

[60] V. P. Gapontsev, S. M. Matitsin, A. A. Isinev, and V. B. Kravchenko, *Opt. Laser Tech.* **14**, 189 (1982).

[61] P. F. Wysocki, J. L. Wagener, M. J. F. Digonnet, and H. J. Shaw, "Evidence and modeling of paired ions and other loss mechanisms in erbium-doped silica fibers," in *Fiber Laser Sources and Amplifiers IV*, M. J. F. Digonnet and E. Snitzer, Eds., *Proc. SPIE* **1789**, pp. 66–79 (1993).

[62] Y. Mita, T. Yoshida, T. Yagami, and S. Shionoya, *J. Appl. Phys.* **71**, 938 (1992).

[63] F. Auzel, *Ann. Telecomm.* **24**, 199 (1969).

[64] G. Nykolak, P. C. Becker, J. Shmulovich, Y. H. Wong, D. J. DiGiovanni, and A. Bruce, *IEEE Phot. Tech. Lett.* **5**, 1014 (1993).

[65] T. Nishi, K. Nakagawa, Y. Ohishi, and S. Takahashi, *Jpn. J. Appl. Phys.* **31**, L177 (1992).

[66] A. Mori, Y. Ohishi, M. Yamada, H. Ono, Y. Nishida, K. Oikawa, and S. Sudo, "1.5 μm broadband amplification by tellurite-based EDFAs," in *Optical Fiber Communication Conference*, Vol. 6, 1997 OSA Technical Digest Series (Optical Society of America, Washington, D.C., 1997), pp. 371–374.

[67] A. Polman, D. C. Jacobson, D. J. Eaglesham, R. C. Kistler, and J. M. Poate, *J. Appl. Phys.* **70**, 3778 (1991).

[68] A. Lidgard, A. Polman, D. C. Jacobson, G. E. Blonder, R. Kistler, J. M. Poate, and P. C. Becker, *Elect. Lett.* **27**, 995 (1991).

[69] B. I. Denker, V. V. Osiko, A. M. Prokhorov, and I. A. Shcherbakov, *Sov. J. Quant. Elect.* **8**, 485 (1978).

[70] S. E. Stokowski, in *Handbook of Laser Science and Technology*, M. J. Weber, Ed., Vol. 1 (CRC Press, Boca Raton, FL, 1982), pp. 215–264.

[71] E. Desurvire, J. L Zyskind, and J. R. Simpson, *IEEE Phot. Tech. Lett.* **2**, 246 (1990).

[72] J. L. Zyskind, E. Desurvire, J. W. Sulhoff, and D. J. DiGiovanni, *IEEE Phot. Tech. Lett.* **2**, 869 (1990).

[73] A. K. Srivasta, J. L. Zyskind, J. W. Sulhoff, J. D. Evankow, Jr., and M. A. Mills, "Room temperature spectral hole-burning in erbium-doped fiber amplifiers," in *Optical Fiber Communication Conference*, Vol. 2, 1996 Technical Digest Series (Optical Society of America, Washington D.C., 1996) pp. 33–34.

[74] S. Zemon, G. Lambert, W. J. Miniscalco, and B. A. Thompson, "Homogeneous linewidths in Er^{3+}-doped glasses measured by resonance fluorescence line narrowing," in *Fiber Laser Sources and Amplifiers III*, M. J. F. Digonnet and E. Snitzer, Eds., *Proc. SPIE* **1581**, pp. 91–100 (1992).

[75] R. I. Laming, L. Reekie, P. R. Morkel, and D. N. Payne, *Elect. Lett.* **25**, 455 (1989).

[76] T. Kushida and E. Takushi, *Phys. Rev. B* **12**, 824 (1975).

[77] S. Zemon, G. Lambert, W. J. Miniscalco, L. J. Andrews, and B. T. Hall, "Characterization of Er^{3+}-doped glasses measured by fluorescence line narrowing," in *Fiber Laser Sources and Amplifiers*, M. J. F. Digonnet, Ed., *Proc. SPIE* **1171**, pp. 219–236 (1990).

[78] A. M. Jurdyc, B. Jacquier, J. C. Gacon, J. F. Bayon, and E. Delevaque, *J. Lum.* **58**, 316 (1994).

[79] D. J. DiGiovanni, P. F. Wysocki, and J. R. Simpson, in *Laser Focus World*, pp. 95–106, September 1993.

[80] M. Tachibana, R. I. Laming, P. R. Morkel, and D. N. Payne, *Opt. Lett.* **16**, 1499 (1991).

[81] V. J. Mazurczyk, "Spectral response of a single EDFA measured to an accuracy of 0.01 dB," in *Optical Fiber Communication Conference*, Vol. 4, 1994 Technical Digest Series (Optical Society of America, Washington D.C., 1994), pp. 271–272.

[82] N. S. Bergano, C. R. Davidson, M. A. Mills, P. C. Corbett, S. G. Evangelides, B. Pedersen, R. Menges, J. L. Zyskind, J. W. Sulhoff, A. K. Srivasta, C. Wolf, and J. Judkins, "Long-haul WDM transmission using optimum channel modulation:a 160 Gb/s (32 × 5 Gb/s) 9,300 km demonstration," in *Optical Fiber Communication Conference*, Vol. 6, 1997 OSA Technical Digest Series (Optical Society of America, Washington, D.C., 1997), pp. 432–435.

[83] K. W. Bennett, F. Davis, P. A. Jakobson, N. Jolley, R. Keys, M. A. Newhouse, S. Sheih, and M. J. Yadlowsky, "980 nm based pump wavelength tuning of the gain spectrum of EDFAs," OSA Trends in Optics and Photonics Vol. 16, Optical Amplifiers and their Applications, M. N. Zervas, A. E. Willner, and S. Sasaki, Eds. (Optical Society of America, Washington, D.C., 1997), pp. 152–155.

[84] P. C. Becker, M. M. Broer, V. G. Lambrecht, A. J. Bruce, and G. Nykolak, "Pr^{3+}:La-Ga-S glass: a promising material for 1.3 μm amplification," in *Optical Amplifiers and Their Applications Technical Digest*, Vol. 17 (Optical Society of America, Washington D.C., 1992), pp. 251–254.

[85] E. Desurvire and J. R. Simpson, *Opt. Lett.* **15**, 547 (1990).

[86] M. Shimizu, M. Yamada, M. Horiguchi, and E. Sugita, *IEEE Phot. Tech. Lett.* **2**, 43 (1990).

[87] N. Kagi, A. Oyobe, and K. Nakamura, *IEEE Phot. Tech. Lett.* **2**, 559 (1990).

[88] F. Auzel, *Proc. IEEE* **61**, 758 (1973).

[89] J. C. Wright, "Up-conversion and excited state energy transfer in rare-earth doped materials," in *Radiationless Processes in Molecules and Condensed Phases*, Vol. 15 of *Topics in Applied Physics*, F. K. Fong, Ed. (Springer-Verlag, New York, 1976), pp. 239–295.

[90] L. F. Johnson and H. J. Guggenheim, *Appl. Phys. Lett.* **19**, 44 (1971).

[91] M. R. Brown, H. Thomas, J. M. Williams, R. J. Woodward, and W. A. Shand, *J. Chem. Phys.* **51**, 3321 (1969).

[92] L. F. Johnson, H. J. Guggenheim, T. C. Rich, and F. W. Ostermayer, *J. Appl. Phys.* **43**, 1125 (1972).

[93] E. Nakazawa and S. Shionoya, *Phys. Rev. Lett.* **25**, 1710 (1970).

[94] F. Auzel, *J. Lum.* **31** & **32**, 759 (1984).

[95] M. Fukushima, Y. Tashiro, and H. Ogoshi, "Visible luminescence in 1480 nm high power pumped erbium-doped fiber amplifier," OSA Trends in Optics and Photonics Vol. 16, Optical Amplifiers and their Applications, M. N. Zervas, A. E. Willner, and S. Sasaki, Eds. (Optical Society of America, Washington, D.C., 1997), pp. 173–176.

[96] B. J. Ainslie, S. P. Craig-Ryan, S. T. Davey, J. R. Armitage, C. G. Atkins, J. F. Massicott, and R. Wyatt, *IEE Proceedings* **137**, Pt. J, 205 (1990).

[97] B.J. Ainslie, *J. Light. Tech.* **9**, 220 (1991).

[98] S. Magne, Y. Ouerdane, M. Druetta, J. P. Goure, P. Ferdinand, and G. Monnom, *Opt. Comm.* **111**, 310 (1994).

[99] K. Arai, H. Namikawa, K. Kumata, T. Honda, Y. Ishii, and T. Handa, *J. Appl. Phys.* **59**, 3430 (1986).

[100] J. L. Wagener, P. F. Wysocki, M. J. F. Digonnet, and H. J. Shaw, *Opt. Lett.* **19**, 347 (1994).

[101] Y. Kimura and M. Nakazawa, *Elect. Lett.* **28**, 1420 (1992).

Chapter 5

Erbium-Doped Fiber Amplifiers – Amplifier Basics

5.1 INTRODUCTION

In this chapter we will develop and review the fundamentals needed to model gain in erbium-doped fiber amplifiers. In Chapter 6 we will build on this foundation, to perform gain and noise modeling of the amplifiers. The underpinning of the gain process consists of coupled atomic population and light flux propagation equations. We will treat the three-level system appropriate for erbium-doped fiber amplifiers at 1.5 μm. We will discuss calculations of the gain in both the small signal and saturation regimes to reach an intuitive understanding of the gain process. Then we will show how the three-level system can be reduced, with certain assumptions, to an equivalent two-level system. The importance of the absorption and emission cross sections, and the difference between the two at a given transition wavelength, will be highlighted. We will cover the concept of the overlap parameter, representing the geometric overlap between the transverse erbium ion distribution and the transverse profile of the light intensity. We will then outline the importance of amplified stimulated emission and the fundamental mechanism by which it is intertwined into all aspects of the amplification process. Finally, we will discuss analytical models of the erbium-doped fiber amplifier that do not necessitate complex numerical procedures to solve for the gain.

5.2 AMPLIFICATION IN THREE-LEVEL SYSTEMS – BASICS

5.2.1 Three-Level Rate Equations

The most simple treatment of the erbium-doped fiber amplifier starts out by considering a pure three-level atomic system.[1] Most of the important characteristics of the amplifier can be obtained from this simple model and its underlying assumptions. An added complication–possible stimulated emission at the pump wavelength–will be treated in

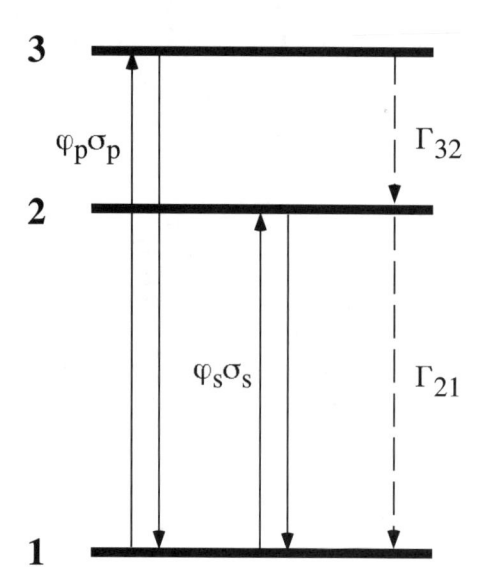

Figure 5.1: The three-level system used for the amplifier model. The transition rates between levels 1 and 3 are proportional to the populations in those levels and to the product of the pump flux ϕ_p and pump cross section σ_p. The transition rates between levels 1 and 2 are proportional to the populations in those levels and to the product of the signal flux ϕ_s and signal cross section σ_s. The spontaneous transition rates of the ion (including radiative and nonradiative contributions) are given by Γ_{32} and Γ_{21}.

Section 5.3. The rate equations can also be made more complex by considering such effects as excited state absorption and the three-dimensional character of the problem. These effects will be discussed in the next chapter.

Setting Up the Three-Level Rate Equations System

We consider a three-level system as depicted in Figure 5.1, with a ground state denoted by 1, an intermediate state labeled 3 (into which energy is pumped), and state 2. Since state 2 often has a long lifetime in the case of a good amplifier, it is sometimes referred to as the metastable level. State 2 is the upper level of the amplifying transition and state 1 is the lower level. The populations of the levels are labeled N_1, N_2, and N_3. This three-level system is intended to represent that part of the energy level structure of Er^{3+} that is relevant to the amplification process. To obtain amplification, we need a population inversion between states 1 and 2, and since state 1 is also the ground state, at least half of the total population of erbium ions needs to be excited to level 2 to have population inversion. This raises the threshold pump power needed for amplification and is a known drawback of three-level laser and amplifier systems.

One can take particular advantage, in the case of the erbium-doped fiber amplifier, of the fact that the light fields are confined in a core of very small dimensions. The light intensities reached are thus very high, over long distances, and population inversion is achieved with relatively small pump powers. We will initially consider the problem to

5.2. AMPLIFICATION IN THREE-LEVEL SYSTEMS – BASICS

be one-dimensional. That is, we assume that the pump and signal intensities as well as the erbium ion distribution are constant in the transverse dimensions, over an effective cross-sectional area of the fiber. We will consider in the next chapter the effects of the transverse variation in the light-field intensities and erbium ion distribution on the performance of the amplifier.

The incident light intensity flux at the frequency corresponding to the 1 to 3 transition (in number of photons per unit time per unit area) is denoted by ϕ_p and corresponds to the pump. The incident flux at the frequency corresponding to the 1 to 2 transition (in photons per unit time per unit area) is denoted by ϕ_s and corresponds to the signal field. The change in population for each level arises from absorption of photons from the incident light field, from spontaneous and stimulated emission, and from other pathways for the energy to escape a particular level. In particular, we write as Γ_{32} the transition probability from level 3 to level 2. This is the sum of the nonradiative and radiative transition probabilities, and in practice, for the most typical cases, is mostly nonradiative. Γ_{21} is the transition probability from level 2 to level 1. In the case of the Er^{3+} $^4I_{13/2}$ (level 2) to $^4I_{15/2}$ (level 1) transition, Γ_{21} is mostly due to radiative transitions. This is due to the fact that there are, for Er^{3+}, no intermediate states between levels 1 and 2 to which ions excited to level 2 can relax. We define $\Gamma_{21} = 1/\tau_2$. where τ_2 is the lifetime of level 2. Chapter 4 discusses the typical values of the transition rates between levels for Er^{3+} doped in glasses.

We denote the absorption cross section for the 1 to 3 transition by σ_p, and the emission cross section for the 2 to 1 transition by σ_s. We will assume for the time being that the absorption and emission cross sections we consider are those for transitions between individual nondegenerate states and are thus equal. We will consider in Section 5.3 the more practical case of erbium levels that consist of a set of states, and where the absorption and emission cross sections are different, as they incorporate information on the thermal population distribution.

The rate equations for the population changes are written as

$$\frac{dN_3}{dt} = -\Gamma_{32}N_3 + (N_1 - N_3)\phi_p\sigma_p \tag{5.1}$$

$$\frac{dN_2}{dt} = -\Gamma_{21}N_2 + \Gamma_{32}N_3 - (N_2 - N_1)\phi_s\sigma_s \tag{5.2}$$

$$\frac{dN_1}{dt} = \Gamma_{21}N_2 - (N_1 - N_3)\phi_p\sigma_p + (N_2 - N_1)\phi_s\sigma_s \tag{5.3}$$

In a steady-state situation, the time derivatives will all be zero,

$$\frac{dN_1}{dt} = \frac{dN_2}{dt} = \frac{dN_3}{dt} = 0 \tag{5.4}$$

and the total population N is given by

$$N = N_1 + N_2 + N_3 \tag{5.5}$$

Using equation 5.1, we can write the population of level 3 as

$$N_3 = \frac{1}{1 + \Gamma_{32}/\phi_p\sigma_p} N_1 \tag{5.6}$$

When Γ_{32} is large (fast decay from level 3 to level 2) compared to the effective pump rate into level 3, $\phi_p\sigma_p$, N_3 is very close to zero, so that the population is mostly in levels 1 and 2. Using equation 5.6 to substitute for N_3 in equation 5.2 we obtain

$$N_2 = \frac{(\phi_p\sigma_p/\Gamma_{32}) + \phi_s\sigma_s}{\Gamma_{21} + \phi_s\sigma_s} N_1 \qquad (5.7)$$

We then make use of equation 5.5 to derive the populations N_1 and N_2 and the population inversion $N_2 - N_1$:

$$N_2 - N_1 = \frac{\phi_p\sigma_p - \Gamma_{21}}{\Gamma_{21} + 2\phi_s\sigma_s + \phi_p\sigma_p} N \qquad (5.8)$$

The condition for population inversion, and thus for gain on the 2 to 1 transition (assuming no background loss), is that $N_2 \geq N_1$. The threshold corresponds to $N_1 = N_2$ and results in the following expression for the pump flux required:

$$\phi_{th} = \frac{\Gamma_{21}}{\sigma_p} = \frac{1}{\tau_2\sigma_p} \qquad (5.9)$$

In a situation where the signal intensity is very small, and the decay rate Γ_{32} is large compared to the transition rate induced by the pump field, $\phi_p\sigma_p$, we can thus write the population inversion as:

$$\frac{N_2 - N_1}{N} = \frac{\phi'_p - 1}{\phi'_p + 1} \qquad (5.10)$$

where

$$\phi'_p = \frac{\phi_p}{\phi_{th}} \qquad (5.11)$$

We plot the fractional population inversion, as given by equation 5.10, in Figure 5.2. Below the pump threshold the inversion is negative; above the pump threshold it is positive. When the inversion is negative, there are more absorptive transitions than emissive transitions at the signal wavelength, and the signal sees negative gain, i.e., attenuation. Conversely, when the inversion is positive, the signal experiences positive gain as it traverses the excited medium (assuming no background attenuation).

The pump intensity, in units of energy per unit area per unit time, is expressed as $I_p = h\nu_p\phi_p$. The threshold pump intensity is then given very simply by the expression:

$$I_{th} = \frac{h\nu_p\Gamma_{21}}{\sigma_p} = \frac{h\nu_p}{\sigma_p\tau_2} \qquad (5.12)$$

This equation is intuitively easy to understand. The higher σ_p is, the higher the probability that a pump photon is absorbed, which lowers the number of pump photons necessary to guarantee that enough are absorbed to reach threshold. In addition, the longer τ_2 is, the longer the energy stays in the reservoir formed by level 2, and, as a result, less pump photons are needed per unit time to keep energy in level 2. The conditions for low pump threshold are thus easily summarized as:

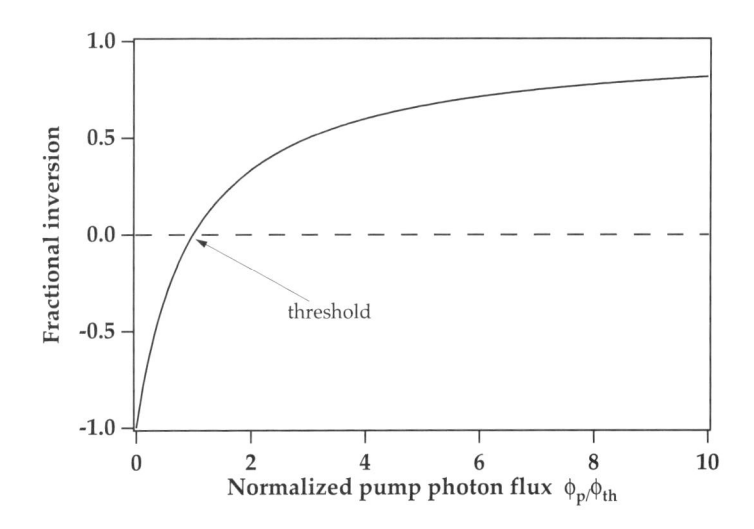

5.2. AMPLIFICATION IN THREE-LEVEL SYSTEMS — BASICS

Figure 5.2: Fractional population inversion $(N_2 - N_1)/N$ in a three-level system. The threshold corresponds to $\phi_p = \phi_{th}$ as defined in equation 5.9.

- high absorption cross section

- long lifetime of the metastable level

For erbium, the situation is particularly propitious from the point of view of τ_2, as the lifetime has the very large value of approximately 10 ms in silica glass.

We can estimate I_{th} by using some typical values for the erbium ion constants. We consider a pumping wavelength of 980 nm, $\sigma_p = 2 \times 10^{-21}$ cm^2, and a τ_2 lifetime of 10 ms to obtain $I_{th} \sim 10\,\text{kW/cm}^2$. Assuming that this pump intensity is distributed uniformly over an effective area (A_{eff}) of 5 μm^2 (the core area for a small-core, erbium-doped single-mode fiber), this corresponds to a power threshold $P_{th} = I_{th}\,A_{eff} \simeq 0.5\,\text{mW}$. Note that this corresponds to rendering transparent only an infinitesimally short length of erbium-doped medium. This ultra low threshold dramatically illustrates one of the main advantages of erbium-doped single-mode fiber amplifiers: a low pump threshold for gain, easily obtained with electrically pumped diode lasers.

Small Signal Gain

In this subsection we will calculate the gain or loss of pump and signal light propagating through a medium constituted by the ions characterized by the three-level system considered in the previous subsection.

We now consider that N, N_1, N_2, and N_3 are densities of populations, in units of number of ions per unit volume. Two light fields travel through the medium, interacting with the ions, and have intensities I_s (the signal field) and I_p (the pump field). The photon fluxes are given by:

$$\phi_s = \frac{I_s}{h\nu_s} \qquad (5.13)$$

and
$$\phi_p = \frac{I_p}{h\nu_p} \qquad (5.14)$$

We will treat the propagation of the signal along a single direction z (the axis of the fiber) as a one-dimensional problem, which is a simplification of the three-dimensional character of the erbium ion distribution in the fiber core and of the light modes. In Chapter 6 we will extend the discussion to the three-dimensional aspect of the field propagation, which involves the variation of the signal and pump intensities in the plane transverse to the axis of the fiber.

In the one-dimensional case, the light field intensities are derived from the light field powers by the following simplified relationship:

$$I(z) = \frac{P(z)\Gamma}{A_{\text{eff}}} \qquad (5.15)$$

where Γ is the overlap factor, representing the overlap between the erbium ions and the mode of the light field, and A_{eff} is the effective cross-sectional area of the distribution of erbium ions. The overlap factor and effective cross section are discussed in more detail in Section 5.2.2. Expression 5.15 essentially states that the intensity of the field at a point z will be taken to its cross-sectional average, computed as the amount of power traveling through the erbium-doped region of the fiber, divided by its cross-sectional area.

We will also assume in the following discussion that both pump and signal beams are propagating in the same direction, i.e., a copropagating configuration as opposed to a counterpropagating configuration.

The fields will be attenuated or amplified after an infinitesimal length dz by the combined effects of absorption arising from ions in their ground state (N_1) and stimulated emission from ions in the excited state (N_2 and N_3).

$$\frac{d\phi_s}{dz} = (N_2 - N_1)\sigma_s\phi_s \qquad (5.16)$$

$$\frac{d\phi_p}{dz} = (N_3 - N_1)\sigma_p\phi_p \qquad (5.17)$$

This leads, after some calculations, to the following equation for the signal intensity growth (or decay, as the case may be):

$$\frac{dI_s}{dz} = \frac{\frac{\sigma_p I_p}{h\nu_p} - \Gamma_{21}}{\Gamma_{21} + 2\frac{\sigma_s I_s}{h\nu_s} + \frac{\sigma_p I_p}{h\nu_p}} \sigma_s I_s N \qquad (5.18)$$

We can write an equation for the attenuation of the pump intensity as

$$\frac{dI_p}{dz} = -\frac{\Gamma_{21} + \frac{\sigma_s I_s}{h\nu_s}}{\Gamma_{21} + 2\frac{\sigma_s I_s}{h\nu_s} + \frac{\sigma_p I_p}{h\nu_p}} \sigma_p I_p N \qquad (5.19)$$

From equation 5.18, it is clear that the condition for gain for the signal field is that

$$I_p \geq I_{th} = \frac{h\nu_p}{\sigma_p \tau_2} \qquad (5.20)$$

5.2. AMPLIFICATION IN THREE-LEVEL SYSTEMS – BASICS

where we again used $\Gamma_{21} = 1/\tau_2$ and I_{th} is the pump threshold intensity for gain at the signal wavelength. This is equivalent to the condition derived above for population inversion.

We can write the equations in a somewhat simpler fashion by defining the intensities in units of the pump threshold. These "normalized" intensities are given by

$$I'_p = \frac{I_p}{I_{th}} \tag{5.21}$$

$$I'_s = \frac{I_s}{I_{th}} \tag{5.22}$$

We further define the quantity η as

$$\eta = \frac{h\nu_p}{h\nu_s} \frac{\sigma_s}{\sigma_p} \tag{5.23}$$

and the saturation intensity $I_{sat}(z)$ as

$$I_{sat}(z) = \frac{1 + I'_p(z)}{2\eta} \tag{5.24}$$

We can then write the propagation equations for the normalized intensities as

$$\frac{dI'_s(z)}{dz} = \frac{1}{1 + I'_s(z)/I_{sat}(z)} \left(\frac{I'_p(z) - 1}{I'_p(z) + 1} \right) \sigma_s I'_s(z) N \tag{5.25}$$

and for the pump

$$\frac{dI'_p(z)}{dz} = - \frac{1 + \eta I'_s(z)}{1 + 2\eta I'_s(z) + I'_p(z)} \sigma_p I'_p(z) N \tag{5.26}$$

Equations 5.25 and 5.26 determine the behavior of erbium-doped fiber amplifiers, at the simplest level. As we shall discuss in the following sections and chapters, modifications will need to be made to this system of equations to accurately model real erbium-doped fibers. Nevertheless, the basic characteristics of three-level fiber amplifiers can be understood from the above equations. We now explore equations 5.25 and 5.26 in more detail.

The signal propagation equation will lead to gain only if $I_p \geq I_{th}$. This is the expected threshold condition. When the pump intensity is less than the threshold, the signal is attenuated; when it is larger, the signal is amplified. Under the conditions of small signal gain, where $I'_s \ll I_{sat}$ (this condition is satisfied when the signal is weak and the pump is strong), and assuming for simplicity that the pump is constant as a function of z (the fiber is uniformly inverted), the signal propagation equation is easily integrated to yield the signal as a function of position along the fiber:

$$I'_s(z) = I'_s(0) \exp(\alpha_p z) \tag{5.27}$$

where we defined the gain coefficient α_p as

$$\alpha_p = \frac{I'_p - 1}{I'_p + 1}\sigma_s N \tag{5.28}$$

The signal grows exponentially, with a coefficient proportional to the signal emission cross section and the degree of population inversion. The latter is determined by the pump intensity relative to threshold. When the pump intensity is very strong and several times the threshold, such that the erbium ions are all inverted, the gain coefficient becomes approximately

$$\alpha_p = \sigma_s N \tag{5.29}$$

The small signal gain per unit length of fiber for a strong pump is determined very simply by the amount of erbium and the signal emission cross section. This implies that in the case of Er^{3+} doped in an Al-Ge silica fiber, since the emission cross section is equal to the absorption cross section, near 1535 nm, the 1535 nm small signal gain is roughly equal to the 1535 nm small signal attenuation under strong pumping conditions. In practice this does not happen, as we discuss in Section 5.4, due to the presence of amplified spontaneous emission, which robs the gain at the expense of the signal.

Saturation Regime

Equation 5.28 loses its validity when the signal grows to a large enough value and enters what is known as the saturation regime. This occurs when I'_s becomes comparable in value to I_{sat}. The signal growth is then damped by the saturation factor $1/(1+I'_s/I_{sat})$. In fact, when I'_s becomes very large and its ratio to I_{sat} becomes large compared to unity, the growth of the signal is determined by the approximate equation

$$\frac{dI'_s}{dz} = I_{sat}\left(\frac{I'_p - 1}{I'_p + 1}\right)\sigma_s N \tag{5.30}$$

so that now the signal growth is linear. The two regimes of signal growth are clear in the graph of Figure 5.3, which plots the gain in dB of a weak signal as a function of pump power. The gain, in dB, of the signal after a length L of fiber is defined as $G = 10\log(I_s(z = L)/I_s(z = 0))$. Figure 5.3 is derived using some typical values for an alumino-germano-silica erbium-doped fiber of length 15 m, with a signal at 1550 nm with a launched power of −40 dBm, and a pump of wavelength of 980 nm. Also shown is the gain obtained when modeling the fiber with the added effect of amplified spontaneous emission (ASE), discussed in Section 5.4. The effect of ASE is to reduce the available gain for the signal field. In the case shown, the ASE begins significantly impairing the gain process above signal levels of approximately 20 dB.

An interesting phenomenon is that the saturation power I_{sat} is not a constant, but instead increases linearly with pump power. This is expected to occur for a three-level laser system.[7] It arises from the following fact. In a three-level laser, ions driven down from the excited level 2 by stimulated emission by the signal are immediately available for pump absorption and can be returned to the excited level almost "instantaneously," given a high enough pumping rate. Maintaining a high level of inversion in the presence of a high signal power yields a high saturation value for the signal.

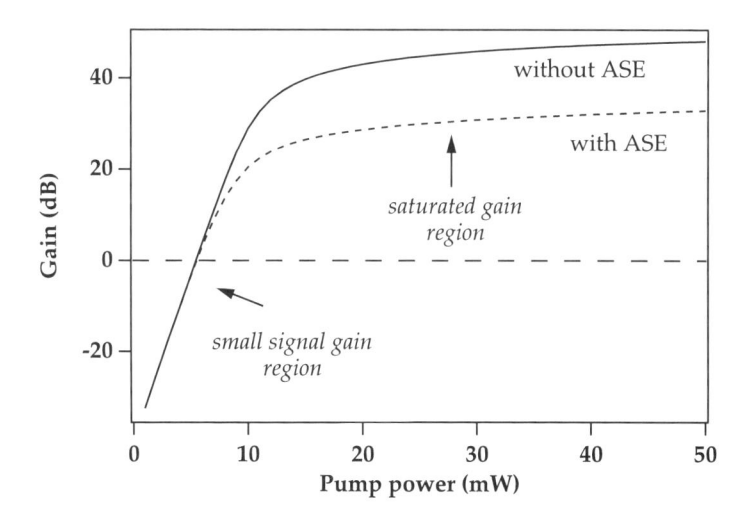

5.2. AMPLIFICATION IN THREE-LEVEL SYSTEMS – BASICS

Figure 5.3: Signal gain (in dB) at 1550 nm as a function of pump power at 980 nm for a typical erbium-doped fiber (15 m length of fiber A of Chapter 6), from numerical integration of the pump and signal propagation equations. The dashed curve shows the computed gain when amplified spontaneous emission is included in the simulation of the gain of the fiber amplifier; the solid curve includes only the pump and launched signal of power −40 dBm. At low pump powers the amplifier is in the small signal gain regime; at higher powers when it is strongly inverted the gain is saturated.

The experimentally determined saturation output power is defined as the signal output power at which the gain has been reduced (compressed) by 3 dB. The saturation output power is higher at 1.55 μm than at 1.53 μm because the emission cross section is lower for the latter, and the saturation output power is proportional to the inverse of the emission cross section.

Optimal Length for a Fiber Amplifier

For a given pump power, to obtain the maximum amount of gain for a given erbium concentration in the fiber core, the fiber length should be increased to the point at which the pump power becomes equal to the intrinsic pump threshold (for the model without ASE considered up to the present point). For axial points z, along the length of the fiber, prior to that point the gain is positive, after that point the gain is negative, and so the fiber should be terminated at the point where the pump power has decreased to the threshold level. This determines the optimal length of the fiber. This length is optimal only in the sense that the small signal gain of the amplifier is maximized. When ASE is included the optimum length for gain is determined by maximizing the signal gain in the presence of forward and backward ASE, which is also a function of the length.

When we need to optimize another parameter, say the noise figure or output saturation power, the optimal length determination will proceed differently. For example, it can be helpful to use an amplifier length such that one is operating in the saturation regime where fluctuations in the pump source power do not have a large effect on the

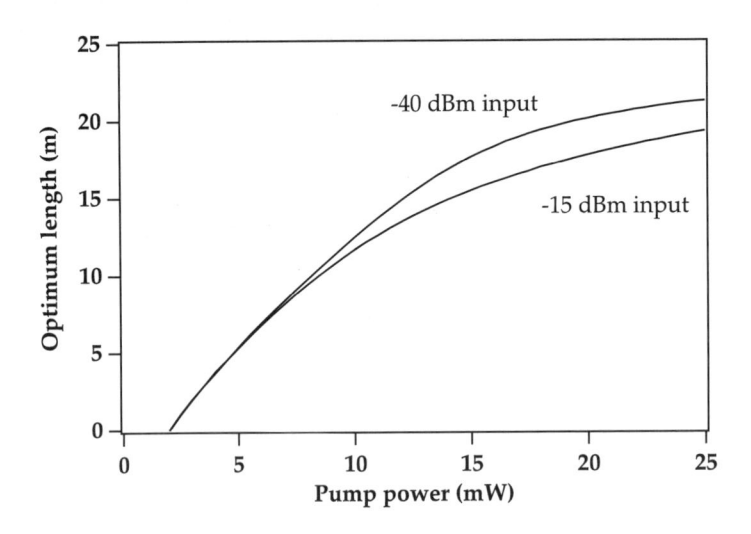

Figure 5.4: Optimal fiber length (in m) for gain at 1530 nm as a function of pump power, for signal inputs of −40 and −15 dBm, from numerical simulations of an erbium-doped fiber (fiber A of Chapter 6) pumped at 980 nm.

gain, as would be the case in the small signal gain regime. We plot in Figure 5.4 the optimal length for small signal gain for two different small signal inputs. The optimum length is 0 up until the point at which the pump power is equal to the pump power needed to render transparent an infinitesimal length of fiber, i.e., P_{th}.

5.2.2 The Overlap Factor

It is important to realize that even in a simple one-dimensional model of the fiber amplifier, the transverse shape of the optical mode and its overlap with the transverse erbium ion distribution profile are important.

In general, this can be parameterized by a factor known as the overlap factor.[2] Simply put, only that portion of the optical mode that overlaps with the erbium ion distribution will stimulate absorption or emission from the Er^{3+} transitions. The entire mode, however, will experience gain or attenuation as a result of this interaction. This situation is schematized in Figure 5.5 for the case of a constant erbium ion density, over a given area, in a fiber with cylindrical geometry. For example, for a step index fiber, if the erbium is doped only in the fiber core, since part of the optical mode extends into the cladding, only that part of the optical mode that is in the core will experience the effects of the Er^{3+} presence. This can be used advantageously. By doping the erbium only in the very center of the fiber core, the erbium ions will only "see" the very high intensity portion of the optical mode and the pump will more easily invert the maximum number of Er^{3+} ions.[3] In general, the erbium density is not necessarily constant and can vary significantly across the fiber core and cladding regions.

We consider a given mode of the fiber and write the light intensity $I(r, \phi)$ for either pump, signal, or ASE signal, as a function of the transverse position (r, ϕ), where r

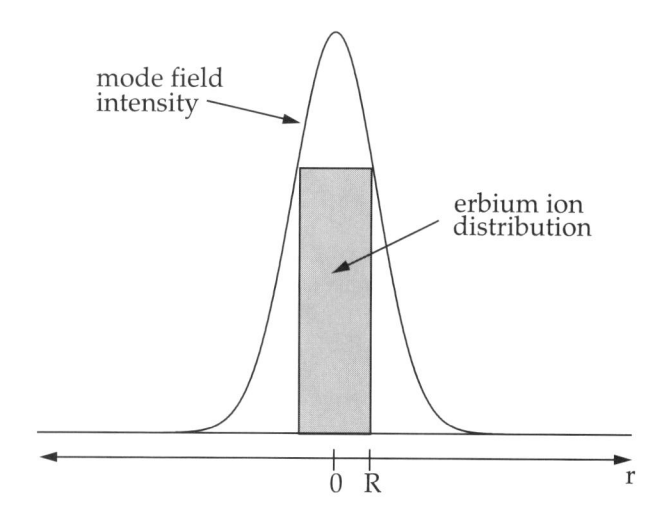

5.2. AMPLIFICATION IN THREE-LEVEL SYSTEMS — BASICS

Figure 5.5: Transverse optical mode profile overlapping with a uniform erbium ion distribution, where r is the distance from the fiber axis. In this example, the erbium distribution is uniform within the core of radius R and the mode field extends beyond the core radius.

is the distance from the fiber axis and ϕ is the azimuthal angle (the angle relative to a chosen axis perpendicular to the fiber axis), as

$$I(r, \phi) = P\, I^{(n)}(r, \phi) \tag{5.31}$$

where $I^{(n)}(r, \phi)$ is the normalized transverse mode intensity profile such that

$$\int_0^{2\pi} \int_0^\infty I^{(n)}(r, \phi)\, r\, dr\, d\phi = 1 \tag{5.32}$$

and P is the total power in the mode. $I^{(n)}(r, \phi)$ has the dimensions of inverse area. At each point (r, ϕ), the density of erbium ions will be written as $n(r, \phi)$, in ions per unit volume. When the intensity $P\, I^{(n)}(r, \phi)$ is incident on the erbium ions at (r, ϕ), the small signal absorption (assuming the ions are all in the ground state) is given by

$$\sigma^{(a)}\, P\, I^{(n)}(r, \phi)\, n(r, \phi) \tag{5.33}$$

where $\sigma^{(a)}$ is the absorption cross section. The total small signal absorption coefficient $\alpha^{(a)}$, in units of inverse length, is obtained by integrating the absorption over the transverse dimensions:

$$\alpha^{(a)} = \sigma^{(a)} \int_0^{2\pi} \int_0^\infty I^{(n)}(r, \phi)\, n(r, \phi)\, r\, dr\, d\phi \tag{5.34}$$

It is helpful at this point to construct a "flat top" erbium distribution, which is equivalent to the actual transverse distribution of erbium. For this purpose we assume

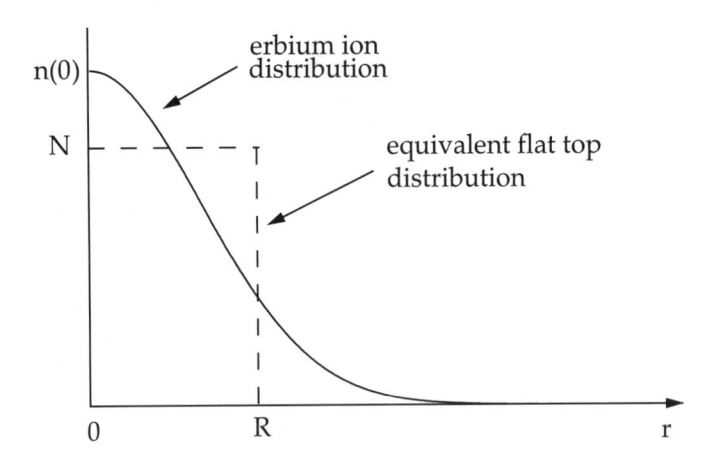

Figure 5.6: Example of a radial distribution of the erbium ion density in a single-mode fiber and the equivalent "flat top" distribution, which has a constant ion density N stretching from $r = 0$ to $r = R$.

that the erbium distribution is independent of ϕ. The situation is depicted in Figure 5.6. The radius R of the flat top distribution is determined by the geometric extent of the actual erbium distribution. We compute R by performing an area integral where the annular area increments are weighted by the relative proportions of erbium ions that they contain:

$$\pi R^2 = 2\pi \int_0^\infty \left(\frac{n(r)}{n(0)}\right) r\, dr \tag{5.35}$$

where n(0) is the erbium density at r = 0. Thus we have

$$R = \left(2 \int_0^\infty \left(\frac{n(r)}{n(0)}\right) r\, dr\right)^{\frac{1}{2}} \tag{5.36}$$

πR^2 is the effective cross-sectional area that was previously used in equation 5.15. The average density N is determined by conservation of the total quantity of erbium ions:

$$\pi R^2 N = 2\pi \int_0^\infty n(r)\, r\, dr \tag{5.37}$$

so that

$$N = \frac{2 \left(\int_0^\infty n(r)\, r\, dr\right)}{R^2} \tag{5.38}$$

Considering the average erbium ion density of N just determined, we can write the absorption coefficient in a more intuitive fashion as

$$\alpha^{(a)} = N\sigma^{(a)} 2\pi \int_0^\infty I^{(n)}(r) \left(\frac{n(r)}{N}\right) r\, dr = N\sigma^{(a)} \Gamma \tag{5.39}$$

5.2. AMPLIFICATION IN THREE-LEVEL SYSTEMS — BASICS

where Γ is the overlap factor, defined as the dimensionless integral overlap between the normalized optical intensity distribution and the "normalized" erbium ion distribution $n(r)/N$:

$$\Gamma = 2\pi \int_0^\infty I^{(n)}(r) \frac{n(r)}{N} r \, dr \tag{5.40}$$

The overlap factor is then a purely geometric correction factor, which accounts for the proportion of the optical power that propagates through the erbium-doped region. An expression similar to equation 5.39 can be written for the gain coefficient for a fully inverted medium.

When modeling the gain of the amplifier, the signal and pump modes interact with an erbium population in both the upper and lower states. The amount of population inversion depends on (r, ϕ), since the intensities governing the populations depend on (r, ϕ). We label the upper- and lower-state populations as $n_2(r, \phi)$ and $n_1(r, \phi)$. The gain in power of a signal field with power P_s, due to stimulated emission, will be proportional to the expression

$$P_s N \sigma^{(e)} \int_0^{2\pi} \int_0^\infty I^{(n)}(r, \phi) \left(\frac{n_2(r, \phi)}{N} \right) r \, dr \, d\phi \tag{5.41}$$

where $\sigma^{(e)}$ is the emission cross section corresponding to the wavelength of the signal considered. Similarly, the absorption experienced by the same signal is proportional to the expression

$$P_s N \sigma^{(a)} \int_0^{2\pi} \int_0^\infty I^{(n)}(r, \phi) \left(\frac{n_1(r, \phi)}{N} \right) r \, dr \, d\phi \tag{5.42}$$

where $\sigma^{(a)}$ is the absorption cross section corresponding to the wavelength of the signal considered.

Expressions 5.41 and 5.42 state that the stimulated emission and absorption experienced by the signal depend on the particular shape of the intensity profile in the direction transverse to the axis of the fiber, as well as that of the populations (which are governed by the transverse intensity profile of all light fields impinging upon the ions). It has been shown, however, that 5.41 can be simplified to $P_s N \sigma^{(e)} \Gamma$ and 5.42 to $P_s N \sigma^{(a)} \Gamma$, with good accuracy, when the effective radius of the erbium-doped region is less than or equal to that of the core.[2] This was established by explicitly performing the integrals in 5.41 and 5.42 under several different conditions typical in erbium-doped fiber amplifiers. The net effect is that the gain or absorption transverse mode issue can be simplified by simply using the overlap factor Γ as a multiplicative factor. Thus, our approach to perform the one dimensional modeling will be the following. The field propagation equations will involve the field powers, with the overlap factor accounting for the radial variation in the mode intensity profile relative to the erbium ion distribution. The population equation for the upper-state population will involve the intensities, as determined using the equivalent flat top distribution with the parameter R as computed above. Thus, we calculate an effective average intensity throughout the erbium-doped region

$$I(z) = \frac{P(z) \Gamma}{\pi R^2} = \frac{P(z) \Gamma}{A} \tag{5.43}$$

where I(z) is the intensity of the field at point z, P(z) is the power of the field at point z, Γ is the overlap factor between the field considered and the erbium ion distribution, and $A = \pi R^2$ is the transverse area occupied by the erbium ions. This can also be justified by integrating the population rate equations over the transverse dimensions.[2]

A standard case is to consider that the actual erbium ion distribution is constant from r = 0 to r = R. We then have

$$\Gamma = 2\pi \int_0^R I^{(n)}(r)\, r\, dr \tag{5.44}$$

with

$$2\pi \int_0^\infty I^{(n)}(r)\, r\, dr = 1 \tag{5.45}$$

The lowest-order mode for a step index fiber can be well represented by a Gaussian approximation:

$$I^{(01)}(r) = \left(\frac{1}{\pi \omega^2}\right) e^{-r^2/\omega^2} \tag{5.46}$$

where ω is the spot size, which can be calculated as a function of the V parameter of the fiber:[4]

$$\omega = \frac{1}{\sqrt{2}} \left(0.65 + \frac{1.619}{V^{1.5}} + \frac{2.879}{V^6} \right) \tag{5.47}$$

We assume a step index fiber for simplicity, and then obtain

$$\Gamma = \left(1 - e^{-R^2/\omega^2}\right) \tag{5.48}$$

The overlap factor Γ will depend on the frequency of the mode considered, as the spot size ω will vary with frequency. Note that definitions of the spot size other than equation 5.47 have been used to obtain better agreement with experimentally observed gain factors.[5]

5.3 REDUCTION OF THE THREE-LEVEL SYSTEM TO THE TWO-LEVEL SYSTEM

5.3.1 Validity of the Two-Level Approach

As described in Chapter 4, the energy levels of rare earth ions are composed of relatively well-separated multiplets, each of which is made up of a certain number of broadened individual levels.

We assume that the pumping level 3 belongs to a multiplet different than that of level 2, and we assume that there is rapid relaxation from level 3 to level 2. For all practical purposes, the population in level 3 is then effectively zero and the rate equations involve only the two levels 1 and 2, with level 3 being involved only via the value of the pump absorption cross section from level 1 to level 3. Examples of pump wavelengths in this case are 0.98 μm, 0.80 μm, 0.65 μm, and 0.54 μm.

In certain pumping configurations, level 3 can be identical to level 2 in the sense that the upper pump level and the upper amplifier level belong to the same multiplet.

5.3. REDUCTION OF THE THREE-LEVEL SYSTEM TO THE TWO-LEVEL SYSTEM

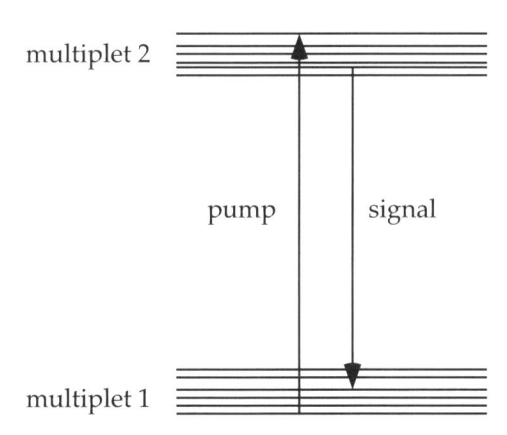

Figure 5.7: Energy levels of a two-multiplet system, where the pumping state 3 is a higher-lying state of multiplet 2, and the signal is resonant with a lower energy transition between the two multiplets, as compared to the pump transition.

This particular case is that of a 1.48 μm pumping wavelength and is depicted in Figure 5.7. In this case, the population of level 3 will not necessarily be equal to zero. At all times, there will be thermal equilibrium established within a given multiplet, assuming that the thermal equilibration time is very short compared to the overall multiplet lifetime (which is governed by decay to lower-lying multiplets).[6] Thus, the pumping level 3, depending on its energy separation from the bottom of the multiplet, will have some finite thermal population. There will then be stimulated emission at the pump frequency as well as at the signal frequency, the amount of which depends on the thermal population of the various states involved as well as the strength of their interaction with a light field.

The behavior of the entire system can be represented through the absorption and emission cross sections introduced in Chapter 4. The difference in spectral shape between the absorption and emission spectra is due to the thermal distribution of energy within the multiplets. The absorption and emission cross sections, as discussed in Chapter 4, contain the thermal population distribution information. For example, if level 3 is high above the bottom of the multiplet and its thermal occupation low, then its emission cross section for emission to the bottom of the ground multiplet will be relatively small, indicating the low probability of stimulated emission at that frequency. The absorption cross section at that frequency will be significantly higher, since most of the population of the ground multiplet will be near the bottom of the multiplet and, as a consequence, available for transitions to level 3.

Based on the previous discussion, we will refer only to the total populations of levels 1 and 2, N_1 and N_2, and use the cross sections to model the system's interaction with the pump and signal fields. In general, the emission and absorption cross sections will be related by the McCumber relationship discussed in Chapter 4. The case where level 3 belongs to a higher-lying multiplet can be reduced to the two-level picture just described by simply setting the pump emission cross section to zero, which effectively accounts for the fact that the population of level 3 is in this case equal to zero.

5.3.2 Generalized Rate Equations

Having reduced the three-level system to an effective two-level system, we can write the rate equations so as to involve only the total population densities of multiplets 1 and 2.

$$\frac{dN_2}{dt} = -\Gamma_{21}N_2 + (N_1\sigma_s^{(a)} - N_2\sigma_s^{(e)})\phi_s - (N_2\sigma_p^{(e)} - N_1\sigma_p^{(a)})\phi_p$$

$$\frac{dN_1}{dt} = \Gamma_{21}N_2 + (N_2\sigma_s^{(e)} - N_1\sigma_s^{(a)})\phi_s - (N_1\sigma_p^{(a)} - N_2\sigma_p^{(e)})\phi_p \quad (5.49)$$

where $\sigma_s^{(a)}$, $\sigma_s^{(e)}$, $\sigma_p^{(a)}$, and $\sigma_p^{(a)}$ represent the signal and pump absorption and emission cross sections, respectively.[9] Since the total population density N is given by

$$N = N_1 + N_2 \quad (5.50)$$

we have

$$\frac{dN_1}{dt} = -\frac{dN_2}{dt} \quad (5.51)$$

and only one of the equations from system 5.49 is an independent equation. We can calculate N_2, for example, in terms of the signal and pump intensities. N_1 is then simply given by $N - N_2$. We find from equations 5.49, for the case of one pump field and one signal field, that the population density $N_2(z)$, as a function of position z along the fiber, is given by

$$N_2(z) = \frac{\frac{\tau\sigma_s^{(a)}}{h\nu_s}I_s(z) + \frac{\tau\sigma_p^{(a)}}{h\nu_p}I_p(z)}{\frac{\tau\left(\sigma_s^{(a)} + \sigma_s^{(e)}\right)}{h\nu_s}I_s(z) + \frac{\tau\left(\sigma_p^{(a)} + \sigma_p^{(e)}\right)}{h\nu_p}I_p(z) + 1} N \quad (5.52)$$

In general, we will assume that N is independent of z. The pump and signal propagation equations are then written, in a very similar fashion, as

$$\frac{dI_p(z)}{dz} = \left(N_2\sigma_p^{(e)} - N_1\sigma_p^{(a)}\right)I_p(z)$$

$$\frac{dI_s(z)}{dz} = \left(N_2\sigma_s^{(e)} - N_1\sigma_s^{(a)}\right)I_s(z) \quad (5.53)$$

Stimulated emission from level 2 contributes to field growth, absorption from level 1 contributes to field attenuation. The equations needed to simulate the amplifying properties of the fiber are thus the population equation and the propagation equations, one for each field. The condition for population inversion, $N_2 - N_1 > 0$, in the presence of a small signal field, corresponds to the pump being greater than the threshold value:

$$I_{th} = \frac{h\nu_p}{\left(\sigma_p^{(a)} - \sigma_p^{(e)}\right)\tau_2} \quad (5.54)$$

5.4. AMPLIFIED SPONTANEOUS EMISSION

The pump threshold that corresponds to signal gain at the signal wavelength ($dI_s/dz > 0$) is slightly different and is equal to

$$I_{th} = \frac{h\nu_p}{\tau_2} \frac{1}{\sigma_p^{(a)}\left(\frac{\sigma_s^{(e)}}{\sigma_s^{(a)}}\right) - \sigma_p^{(e)}} \quad (5.55)$$

The equations above can be easily generalized to the case of multiple signals and multiple pumps.[2, 10] For the case of several signals s_i and several pumps p_i, the population equation becomes

$$N_2(z) = \frac{\sum_{s_i} \frac{\tau \sigma_{s_i}^{(a)}}{h\nu_{s_i}} I_{s_i}(z) + \sum_{p_i} \frac{\tau \sigma_{p_i}^{(a)}}{h\nu_{p_i}} I_{p_i}(z)}{\sum_{s_i} \frac{\tau\left(\sigma_{s_i}^{(a)}+\sigma_{s_i}^{(e)}\right)}{h\nu_{s_i}} I_{s_i}(z) + \sum_{s_i} \frac{\tau\left(\sigma_{p_i}^{(a)}+\sigma_{p_i}^{(e)}\right)}{h\nu_{p_i}} I_{p_i}(z) + 1} N \quad (5.56)$$

The field propagation equations are identical to those of equations 5.53, with the appropriate cross sections. Such multisignal systems of equations will be used when computing, for example, the spectral distribution of the ASE or the amplification of multiple-signal channels in a WDM system.

5.4 AMPLIFIED SPONTANEOUS EMISSION

The treatment outlined in the previous sections neglects an important factor present in all optical amplifiers, that of spontaneous emission. All the excited ions can spontaneously relax from the upper state to the ground state by emitting a photon that is uncorrelated with the signal photons. This spontaneously emitted photon can be amplified as it travels down the fiber and stimulates the emission of more photons from excited ions, photons that belong to the same mode of the electromagnetic field as the original spontaneous photon. This parasitic process, which can occur at any frequency within the fluorescence spectrum of the amplifier transitions, obviously reduces the gain from the amplifier. It robs photons that would otherwise participate in stimulated emission with the signal photons. It is usually referred to as ASE (amplified spontaneous emission). Ultimately, it limits the total amount of gain available from the amplifier. This is clearly evident in Figure 5.3.

To compute the ASE at the output of the fiber, we need to first calculate the spontaneous emission power at a given point in the fiber. This power is sometimes referred to as an equivalent noise power. For a single transverse mode fiber with two independent polarizations for a given mode at frequency ν, the noise power in a bandwidth $\Delta\nu$, corresponding to spontaneous emission, is equal to

$$P_{ASE}^0 = 2h\nu\Delta\nu \quad (5.57)$$

We now derive this expression from simple considerations.

Consider an ion excited to a higher-lying energy state that can relax down to the ground state by either spontaneous or stimulated emission of a photon. One of the principles of quantum mechanics is that the spontaneous emission rate into a given mode is the same as the stimulated emission rate into that mode with one photon already present in that mode.[11, 12] We consider this one photon and calculate its power. The photon is assumed to occupy a volume of length L (which may or may not correspond to the fiber length). The photon, of energy hν, has a velocity c so that it traverses length L in the time L/c. The noise power of the photon in a given mode is thus hνc/L. We then need to count the number of modes in the bandwidth $\Delta\nu$. The mode density in one dimension, in frequency space, for a medium of length L, can be easily calculated to be $2L\Delta\nu/c$. Since two polarizations are allowed in a single mode fiber, the mode density is actually twice this quantity, or $4L\Delta\nu/c$. The total noise power in bandwidth $\Delta\nu$ is the noise power of a one photon per mode multiplied by the total number of modes in $\Delta\nu$. The length L drops out as it should.[13] Since each mode is actually made up of a wave traveling in the forward direction and a wave traveling in the backward direction, the noise power traveling in one direction is half the total noise power. As a result, the noise power propagating in a given direction, spontaneously emitted at any given point along the fiber, is given by Expression 5.57. The derivation for the noise power is similar to that for Nyquist noise and has also been carried out in the case of ASE in Raman amplifiers, as well as in other laser amplifiers.[13, 14, 15]

The total ASE power at a point z along the fiber is the sum of the ASE power from the previous sections of the fiber and the added local noise power P^0_{ASE}. This local noise power will stimulate the emission of photons from excited erbium ions, proportionally to the product $\sigma^{(e)}(\nu)N_2$, where $\sigma^{(e)}(\nu)$ is the stimulated emission cross section at frequency ν. The propagation equation for the ASE power propagating in a given direction is thus

$$\frac{dP_{ASE}(\nu)}{dz} = \left(N_2\sigma^{(e)}(\nu) - N_1\sigma^{(a)}(\nu)\right) P_{ASE}(\nu) + P^0_{ASE}(\nu)N_2\sigma^{(e)}(\nu) \quad (5.58)$$

The population equation also needs to be modified in that signal terms corresponding to the ASE signal need to be added in as well. The generalized Expression 5.56 can be used for this purpose.

The modeling of the ASE can be done in a simple way, by treating it as one extra signal with a bandwidth corresponding to an effective bandwidth for the entire transition. A more complex treatment, detailed in Chapter 6, divides the ASE into small frequency segments of width $\Delta\nu$ much smaller than the transition bandwidth. The power within each frequency strip can then be propagated as an independent signal and the spectral shape of the output ASE can be computed. An added complication is that the ASE can actually propagate in both directions along the fiber, both copropagating and counterpropagating with the pump. As an example, we show the ASE power as a function of position along an erbium-doped fiber, in Figure 5.8. The backward ASE output at z = 0 is more intense than the forward ASE output at z = L, since the beginning section of the fiber, with a forward pump, is more inverted than the end section. The interplay between signal amplification and ASE will be explored in Chapter 6, and the noise properties of the amplifier resulting from the ASE will be discussed in Chapter 7.

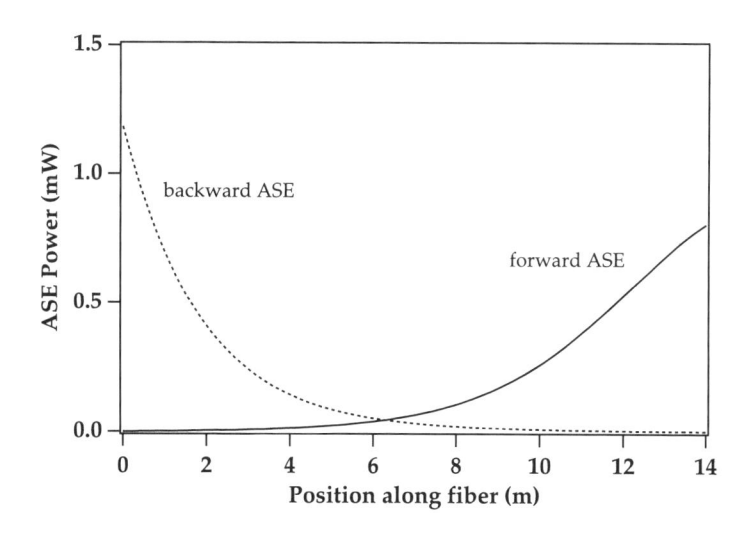

Figure 5.8: Total forward- and backward-propagating ASE power as a function of position along a 14 m length of erbium-doped fiber (fiber A of Chapter 6) pumped at 980 nm with 20 mW of power.

5.5 ANALYTICAL SOLUTIONS TO THE TWO-LEVEL SYSTEM

With certain approximations, the rate equation for the three-level system reduced to a two-level system can be solved analytically.

This approach was investigated by Saleh et al.[16] The key initial assumption is to ignore saturation effects due to the presence of ASE, which is equivalent to neglecting the spontaneous emission power in the fiber. An extension of this approach has included the effect of ASE.[19] We follow the derivation presented in reference [16] and apply it to a situation with one signal field and one co or counterpropagating pump field. The full model actually allows for an arbitrary number of co and counterpropagating signals and pumps.[16]

The key is to write the excited-state population in terms of derivatives of the field with respect to axial position, so that the propagation equations for the fields can be integrated along the entire fiber length, resulting in a transcendental equation for the pump and signal powers at input and output. The propagation equations for the pump and signal field powers are

$$\frac{\partial P_p(z,t)}{\partial z} = \rho u \Gamma_p \left[(\sigma_p^e + \sigma_p^a) N_2(z,t) - \sigma_p^a \right] P_p(z,t) \quad (5.59)$$

$$\frac{\partial P_s(z,t)}{\partial z} = \rho \Gamma_s \left[(\sigma_s^e + \sigma_s^a) N_2(z,t) - \sigma_s^a \right] P_s(z,t) \quad (5.60)$$

where $u = +1$ when the pump is propagating from $z = 0$ to $z = L$ (copropagating pump and signal) and $u = -1$ when the pump is propagating from $z = L$ to $z = 0$ (counterpropagating pump and signal), L is the length of the fiber, Γ_p and Γ_s are the pump and signal mode overlap factors with the erbium ion distribution, and ρ is the number of ions per unit volume. The signal is assumed to propagate from $z = 0$ to $z = L$. $P_p(z, t)$ and $P_s(z, t)$ are the pump and signal powers in units of number of photons per unit time. The expressions can be simply derived by adding the stimulated emission photons to the fields, and subtracting the absorbed photons, where the identity $N_1 + N_2 = 1$ is used (the populations are considered on a per ion basis).

The last equation to be added to the model is the rate equation for the population,

$$\frac{\partial N_2(z, t)}{\partial t} = -\frac{N_2(z, t)}{\tau} - \frac{1}{\rho A}\left(\frac{\partial P_s(z, t)}{\partial z} + u\frac{\partial P_p(z, t)}{\partial z}\right) \quad (5.61)$$

where A is the cross-sectional area of the erbium ion distribution in the fiber. The first term on the right-hand side of equation 5.61 is the decay in the upper-state population originating from spontaneous emission, while the second term combines both absorption and emission due to the fields. The second term on the right-hand side of equation 5.61 can be understood as follows: The change in the field power $\partial P_p(z, t)/\partial z$ (in units of photons) is the net number of photons deposited or acquired in the length dz per unit time at time t. Dividing this quantity by the cross-sectional area A traversed by the beam yields the number of photons deposited per unit volume, in the volume Adz. In turn, dividing this quantity by ρ yields the number of photons absorbed or emitted per ion. The fractional population of the upper state will change accordingly.

With steady-state conditions, the populations and the fields do not depend on time and equation 5.61 can be set equal to zero, which yields an expression for $N_2(z)$. Substituting this in equations 5.59 and 5.60, we obtain for the pump:

$$u\frac{dP_p(z)}{P_p(z)} = \rho\left[-\left(\sigma_p^e + \sigma_p^a\right)\frac{\tau\Gamma_p}{\rho A}\left(\frac{dP_s(z, t)}{dz} + u\frac{dP_p(z, t)}{dz}\right) - \Gamma_p\sigma_p^a\right]dz \quad (5.62)$$

and for the signal:

$$\frac{dP_s(z)}{P_s(z)} = \rho\left[-\left(\sigma_s^e + \sigma_s^a\right)\frac{\tau\Gamma_p}{\rho A}\left(\frac{dP_s(z, t)}{dz} + u\frac{dP_s(z, t)}{dz}\right) - \Gamma_s\sigma_s^a\right]dz \quad (5.63)$$

We define the attenuation constants α_p^a and α_s^a, for the pump and signal, respectively, and the saturation powers for the pump and signal, P_p^{sat} and P_s^{sat}, respectively, where

$$\alpha_{p,s} = \rho\Gamma_{p,s}\sigma_{p,s}^a \quad (5.64)$$

$$P_{p,s}^{sat} = \frac{A}{\left(\sigma_{p,s}^e + \sigma_{p,s}^a\right)\tau\Gamma_{p,s}} \quad (5.65)$$

and the pump and signal propagation equations can then be written as

$$u\frac{dP_p(z)}{P_p(z)} = -\left[\alpha_p + \frac{1}{P_p^{sat}}\left(\frac{dP_s(z, t)}{dz} + u\frac{dP_p(z, t)}{dz}\right)\right]dz \quad (5.66)$$

$$\frac{dP_s(z)}{P_s(z)} = -\left[\alpha_s + \frac{1}{P_s^{sat}}\left(\frac{dP_s(z, t)}{dz} + u\frac{dP_p(z, t)}{dz}\right)\right]dz \quad (5.67)$$

These equations can be integrated, from $z = 0$ to $z = L$ for the variable z, and from $P_{p,s}^{in}$ to $P_{p,s}^{out}$ for the pump powers. We define the input and output signal powers, P_s^{in} and

5.5. ANALYTICAL SOLUTIONS TO THE TWO-LEVEL SYSTEM

P_s^{out}, as those at z = 0 and z = L, respectively. We define P_p^{in} and P_p^{out} similarly for u = +1; if the pump is counterpropagating (u = −1), P_p^{in} and P_p^{out} are defined at z = L and z = 0, respectively. Integrating both sides of equations 5.66 and 5.67, we obtain

$$P_{p,s}^{out} = P_{p,s}^{in} \, e^{-\alpha_{p,s} L} e^{(P^{in} - P^{out})/P_{p,s}^{sat}} \tag{5.68}$$

where P^{in} and P^{out} are the total input and output powers, defined as the sums of the signal and pump input and output powers, respectively. We can sum independently equation 5.68 applied to both pump and signal to obtain an equation for the total output power, P^{out}, as a function of P^{in}, assuming that the attenuation coefficients and saturation powers are known quantities. The resulting equation is transcendental in P^{out}:

$$P^{out} = \left(P_p^{in} \, e^{-\alpha_p L} e^{P^{in}/P_p^{sat}}\right) e^{-P^{out}/P_p^{sat}} + \left(P_s^{in} \, e^{-\alpha_s L} e^{P^{in}/P_s^{sat}}\right) e^{-P^{out}/P_s^{sat}} \tag{5.69}$$

Equation 5.69 can be solved for P^{out}, and then the individual signal and pump output powers can be directly obtained from equation 5.68. This solution has neglected the presence of ASE and is valid only in the presence of a low ASE level, and thus should be considered to be approximately correct only for gains less than about 20 dB.

One of the immediate results of the analytical solution 5.69 is that it is independent of u, and thus, for a given pump power, the gain is predicted to be independent of the direction of propagation of the pump relative to that of the signal. A second interesting result is that the signal saturation power (defined as the output signal power for which the gain drops by e), is given by the expression

$$P_s(sat) \approx P_p^{out}(0) \left[1 - e^{(P_s^{sat}/P_p^{sat})}\right] \tag{5.70}$$

where $P_p^{out}(0)$ is the output pump power with zero signal input. $P_p^{out}(0)$ can be obtained from 5.68. This again shows, as we previously discussed, that the signal output saturation power is proportional to the pump power and can be increased by increasing the input pump power. The analytical model has been shown to be quite accurate for predicting gain values in erbium-doped fibers for gains of less than about 20 dB, where saturation of the amplifier by ASE is not significant.[16] A more recent addition to this work has generalized the approach to include the propagation of ASE, both forward and backward, and thus allows the determination of the output noise spectrum of the amplifier.[19]

Bibliography

[1] E. Desurvire, J. R. Simpson, and P. C. Becker, *Opt. Lett.* **12**, 888 (1987).

[2] C. R. Giles and E. Desurvire, *J. Light. Tech.* **9**, 271 (1991).

[3] J. R. Armitage, *Appl. Opt.* **27**, 4831 (1988).

[4] L. B. Jeunhomme, *Single-Mode Fiber Optics* (Marcel Dekker, New York, 1990).

[5] T. J. Whitley and R. Wyatt, *IEEE Phot. Tech. Lett.* **5**, 1325 (1993).

[6] D. E. McCumber, *Phys. Rev.* **134**, A299 (1964).

[7] P. W. Milonni and J. H. Eberly, *Lasers* (Wiley, New York, 1988).

[8] A. Lidgard, J. R. Simpson, and P. C. Becker, *Appl. Phys. Lett.* **56**, 2607 (1990).

[9] B. Pedersen, A. Bjarklev, J. H. Povlsen, K. Dybdal, and C. C. Larsen, *J. Light. Tech.* **9**, 1105 (1991).

[10] C. R. Giles, C. A. Burrus, D. J. DiGiovanni, N. K. Dutta, and G. Raybon, *IEEE Phot. Tech. Lett.* **3**, 363 (1991).

[11] K. Shimoda, H. Takahashi, and C. H. Townes, *J. Phys. Soc. Jpn.* **12**, 686 (1957).

[12] A.E. Siegman, *Lasers* (University Science Books, Mill Valley, CA, 1986).

[13] M. L. Dakss and P. Melman, *J. Light. Tech.* **LT-3**, 806 (1985).

[14] R. G. Smith, *Appl. Opt.* **11**, 2489 (1972).

[15] J. Auyeung and A. Yariv, *IEEE J. Quant. Elect.* **QE-14**, 347 (1978).

[16] A. A. M. Saleh, R. M. Jopson, J. D. Evankow, and J. Aspell, *IEEE Phot. Tech. Lett.* **2**, 714 (1990).

[17] J. R. Simpson, L. F. Mollenauer, K. S. Kranz, P. J. Lemaire, N. A. Olsson, H. T. Shang, and P. C. Becker, "A distributed erbium-doped fiber," in *Optical Fiber Communication Conference*, Vol. 1, 1990 OSA Technical Digest Series (Optical Society of America, Washington, D.C., 1990), pp. 313–316.

[18] J. R. Simpson, H. T. Shang, L. F. Mollenauer, N. A. Olsson, P. C. Becker, K. S. Kranz, P. J. Lemaire, and M. J. Neubelt, *J. Light. Tech.* **9**, 228 (1991).

[19] R. M. Jopson and A. A. M. Saleh, "Modeling of gain and noise in erbium-doped fiber amplifiers," in *Fiber Laser Sources and Amplifiers III*, M. J. F. Digonnet and E. Snitzer, Eds., *Proc. SPIE* **1581**, pp. 114–119 (1992).

Chapter 6

Erbium-Doped Fiber Amplifiers – Modeling and Complex Effects

6.1 INTRODUCTION

In this chapter we will discuss the more complex aspects involved in modeling the gain and noise behavior of erbium-doped fiber amplifiers. The basic tools for complex modeling are developed, using a full spectral model for the ASE. This will allow us to determine the spectral characteristics of the ASE at the output of the amplifier, as well as to determine the noise figure. The modeling allows us to follow the signal, pump, and ASE as they propagate along the fiber and determine the population inversion. This leads to a better intuitive understanding of the amplification process and how one might design an amplifier for a specific application. We will see that both forward- and backward-propagating ASE are important in determining the overall gain and noise figure.

We will then study the transverse mode aspect of the amplification process, for the purpose of optimizing the amplifier efficiency. We will reexamine the simple three-level system in terms of the more complex effects that can occur, such as excited-state absorption. We will show how the modeling equations can be modified to incorporate these effects and illustrate their practical effect. The 0.80 μm pump band will be discussed in this context. Finally, we will discuss Er^{3+} ion-Er^{3+} ion interactions and their effect in the degradation of amplifier efficiency, through upconversion and pair-induced quenching. These latter effects can be particularly prevalent in high-concentration, Er^{3+}-doped waveguide amplifiers.

6.2 ABSORPTION AND EMISSION CROSS SECTIONS

In general, as discussed in Chapter 4, the absorption and emission cross sections will depend on the particular host glass in which the Er^{3+} ion is embedded. The spectral

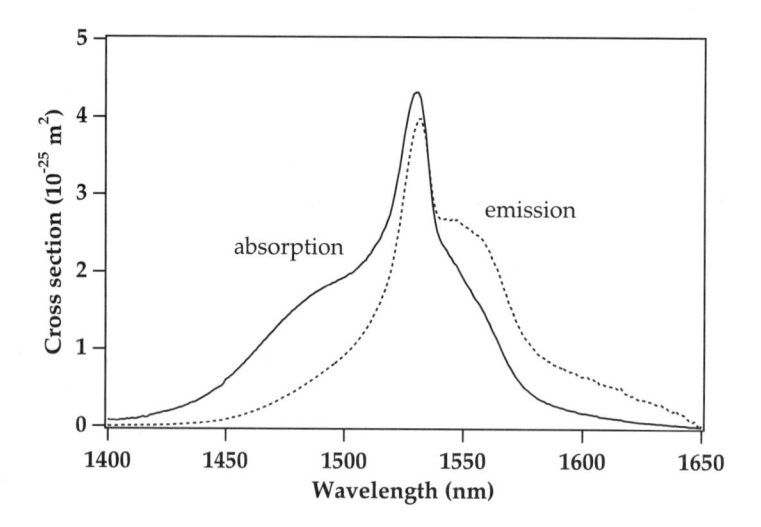

Figure 6.1: Absorption (solid line) and emission (dashed line) cross sections of Er^{3+} near 1.5 μm, for an Al-Ge-Er-doped silica fiber (fiber A). These cross sections are used in the amplifier simulations of Chapter 6.

shape of the cross sections in the 1.5 μm region, as well as the absolute magnitudes of the cross sections, vary with the host environment. While it can be relatively straightforward to obtain the absorption cross sections—by simple normalized absorption measurements on a homogeneous bulk glass sample—the emission cross sections are more difficult to obtain. A standard technique involves either calculating the emission cross sections from the absorption cross sections or directly measuring the small signal gain in a short piece of uniformly inverted Er^{3+}-doped fiber.

We saw in Chapter 4 that an accurate way of obtaining the emission cross sections from the absorption cross sections is through the McCumber relationship. Let us consider the absorption cross section of Er^{3+} in the 1.5 μm region shown in Figure 6.1, determined for an actual Al-Ge-Er-doped silica fiber, which we will refer to as fiber A. We will use the cross sections determined for fiber A for amplifier simulations in the remainder of this chapter to illustrate various properties of the amplification process. The absolute values of the absorption cross section were obtained by measuring the attenuation of the light in the fiber and then dividing by the average erbium ion density and overlap parameter at 1.5 μm for fiber A. The McCumber relationship then yields the emission cross section in terms of the absorption cross section:

$$\sigma^{(e)}(\nu) = \sigma^{(a)}(\nu) \, e^{(\varepsilon - h\nu)/kT} \qquad (6.1)$$

(where ϵ represents the mean energy of the $^4I_{13/2}$ to $^4I_{15/2}$ transition). The emission cross sections obtained with the McCumber relationship with $\varepsilon = 1535$ nm are shown in Figure 6.1. These emission cross sections agree well with those obtained from the small signal gain spectrum measured with this fiber.

For the 980 nm wavelength, the absorption cross section for the case of fiber A is estimated to be equal to 2.7×10^{-21} cm^2 from fiber attenuation measurements. The

6.2. ABSORPTION AND EMISSION CROSS SECTIONS

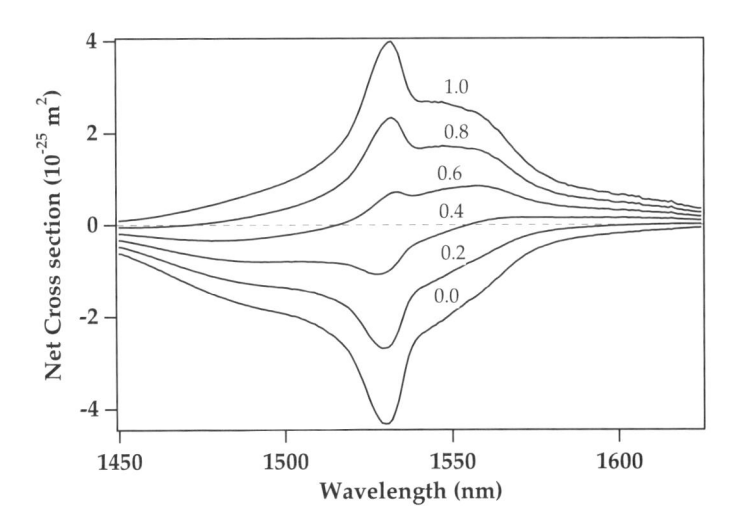

Figure 6.2: Net cross section ($\sigma_e(N_2/N) - \sigma_a(N_1/N)$) of Er^{3+} near 1.5 μm, for different values of the fractional upper-state population (N_2/N) (indicated adjoining each curve) for the cross sections of Figure 6.1 (fiber A).

emission cross section at 980 nm is set equal to zero for the purpose of using the equivalent two-level system amplifier model (see Chapter 5, section 5.3). For 1480 nm pumping, the absorption and emission cross sections were determined to be 1.5×10^{-21} cm^2 and 0.5×10^{-21} cm^2, respectively. The pump and signal cross sections just listed will be used in our numerical modeling of the amplifier properties of fiber A.

A fully accurate model of the gain of a signal at a particular frequency, as well as the output spectrum of the ASE, both forward (copropagating with the signal) and backward (counterpropagating with the signal) traveling, needs to take the spectral shape of the ASE into account.[1, 2] The ASE signal, as discussed in Chapter 5 (section 5.4), originates from spontaneous emission.

The ratio between the absorption and emission cross sections at a particular frequency is critical in determining the gain for a piece of fiber with a given inversion. Figure 6.2 shows the net cross section, on a per-ion basis as a function of wavelength, for different values of the upper-state population and hence inversion. Note that the peak at 1530 nm is very sensitive to the population inversion and suffers relatively more than 1550 nm as the inversion is lowered. A range of inversions exists over which the longer wavelengths will still be amplified while the shorter wavelengths are attenuated.

A very simple amplifier model consists in merely modeling the ASE as a single signal with an effective bandwidth, centered at a frequency corresponding, for example, to the peak of the 1.5 μm emission spectrum. In the effective bandwidth case, the problem is relatively simple to treat numerically as there are only two ASE signals to track, one for the forward direction and one for the backward direction. Further refinements can be made by, for example, dividing the ASE spectrum into two portions that have different center frequencies and effective bandwidths and that correspond to the two peaks at 1.53 μm and 1.55 μm, clearly distinct in the fluorescence spectrum of Er^{3+} at 1.5 μm in silica glass at room temperature.[3] This model is relatively accurate in terms of gain prediction; however, it will not yield the spectrum of the ASE at various pump powers, nor does it take into account the fact that the effective ASE bandwidth can vary with pump power. Additionally, it does not allow a calculation of the noise figure as a function of signal wavelength. The simulations presented in this chapter include the full ASE spectrum, spanning from 1450 to 1650 nm, with channels spaced by 1 nm in the most detailed cases.

In the amplifier simulations of this chapter we will assume that the Er^{3+} $^4I_{13/2}$ to $^4I_{15/2}$ transition in glass (at room temperature) is homogeneously broadened. This implies that a signal at any frequency within the $^4I_{13/2}$ to $^4I_{15/2}$ transition spectrum will increase or decrease the population inversion across the entire spectral range of the transition, and the gain (or loss) spectrum will change in a uniform fashion. In other words, the population of Er^{3+} ions is assumed to be all of one species, and the populations are tracked by means of the two quantities N_2 (upper-state population) and N_1 (lower-state population). Inhomogeneous broadening can be simulated also, although it renders the calculation somewhat more complex. The degree of inhomogeneous broadening needs to be known for an accurate modeling of the behavior of the amplifier.

For the example used in the following sections, we will simulate an experimentally measured fiber, fiber A, which is an Al-Ge-Er-doped silica fiber. Fiber A has a core radius of 1.4 μm, a Δn of 0.026 (refractive index difference between core and cladding), a numerical aperture of 0.28, and a cutoff wavelength of 1.0 μm. For fiber A, the radius of the erbium distribution (assumed to be flat top in shape) is R = 1.05 μm. The overlap parameters (between mode intensity and erbium distribution) are $\Gamma = 0.40$ at 1530 nm and 1550 nm, $\Gamma = 0.43$ at 1480 nm, and $\Gamma = 0.64$ at 980 nm, computed using the Bessel function solutions for the transverse mode shapes. This fiber has an Er^{3+} concentration of $N = 0.7 \times 10^{-19}$ cm^{-3}, typical of erbium-doped fibers with a few hundred ppm of erbium dopant. We will first model our amplifier using the overlap factor approach outlined in Chapter 5, section 5.2.2. In a subsequent section we will then perform a gain simulation with full transverse mode modeling, so as to bring to light issues such as erbium dopant confinement.

6.3 GAIN AND ASE MODELING

6.3.1 Model Equations – Homogeneous Broadening

We saw in Chapter 5, section 5.3, that the gain of a simple three-level system, such as Er^{3+} at 1.5 μm, can be reduced (with certain assumptions) to a two-level model. We will use the absorption and emission cross sections of Figure 6.1. We will also need the erbium ion concentration, the lifetime of the upper state of the amplifier transition, and the overlap integrals for the pump and signal modes. We will assume, for the present case, a very simple erbium ion distribution where the distribution is uniform from r = 0 to r = R, where R is equal to or less than the core radius. In the so-called confined erbium-doped fibers, R will be smaller than the core radius. The intensities of the light fields will then be averaged from r = 0 to r = R to obtain an effective intensity that can be used in the one-dimensional model. This simplification of the transverse variation in

6.3. GAIN AND ASE MODELING

light intensity and erbium ion distribution will allow us to obtain a reasonably accurate model of the spectral shape of the ASE at the output ends of the fiber. We will discuss in the transverse mode modeling section how to modify this treatment to include the transverse variation in the light-field intensity profile.

Following the discussion of Chapter 5, we write first the ion population density in the upper level, as a function of the light-field intensities and the total ion density N.

$$N_2 = \frac{\frac{\tau\sigma_s^{(a)}}{h\nu_s} I_s + \sum_j \frac{\tau\sigma_{\nu_j}^{(a)}}{h\nu_j} I_A(\nu_j) + \frac{\tau\sigma_p^{(a)}}{h\nu_p} I_p}{\frac{\tau\left(\sigma_s^{(a)}+\sigma_s^{(e)}\right)}{h\nu_s} I_s + \sum_j \frac{\tau\left(\sigma_{\nu_j}^{(a)}+\sigma_{\nu_j}^{(e)}\right)}{h\nu_j} I_A(\nu_j) + \frac{\tau\left(\sigma_p^{(a)}+\sigma_p^{(e)}\right)}{h\nu_p} I_p + 1} N \quad (6.2)$$

where $I_A(\nu_j)$ is the ASE signal intensity at frequency ν_j and I_s and I_p are the signal and pump intensities, respectively. We include one signal and one pump, which can be either copropagating or counterpropagating with the signal. We assume that the fiber is single mode and consider the lowest-order mode, for a step index fiber. The ASE is divided into components representing the power present in bandwidth $\Delta\nu_j$ centered at ν_j.

The population equation can now be rewritten in terms of the field powers.

$$N_2 = \frac{\frac{\tau\sigma_s^a}{Ah\nu_s} \Gamma_s P_s + \sum_j \frac{\tau\sigma_{\nu_j}^a}{Ah\nu_j} \Gamma_{\nu_j} P_A(\nu_j) + \frac{\tau\sigma_p^a}{Ah\nu_p} \Gamma_p P_p}{\frac{\tau(\sigma_s^a+\sigma_s^e)}{Ah\nu_s} \Gamma_s P_s + \sum_j \frac{\tau(\sigma_{\nu_j}^a+\sigma_{\nu_j}^e)}{Ah\nu_j} \Gamma_{\nu_j} P_A(\nu_j) + \frac{\tau(\sigma_p^a+\sigma_p^e)}{Ah\nu_p} \Gamma_p P_p + 1} N \quad (6.3)$$

where the Γ_i are the overlap factors between the light-field modes and the erbium distribution, and $A = \pi R^2$ is the effective area of the erbium distribution, as defined in Chapter 5, section 5.2.2.

Each ASE power, $P_A(\nu_j)$, is composed of a forward-traveling ASE component, $P_A^+(\nu_j)$, and a backward-traveling ASE component, $P_A^-(\nu_j)$, such that

$$P_A(\nu_j) = P_A^+(\nu_j) + P_A^-(\nu_j) \quad (6.4)$$

The field propagation equations can also be written in terms of the field powers. Only that part of the optical mode that overlaps with the erbium ion dopant will experience gain or attenuation, so that the overlap parameters enter the propagation equations also. We also allow for possible intrinsic background loss in the fiber by means of the parameters $\alpha_p^{(a0)}$, $\alpha_s^{(a0)}$, and $\alpha_{\nu_j}^{(a0)}$:

$$\frac{dP_p}{dz} = \left(N_2\sigma_p^{(e)} - N_1\sigma_p^{(a)}\right)\Gamma_p P_p - \alpha_p^{(a0)} P_p \quad (6.5)$$

$$\frac{dP_s}{dz} = \left(N_2\sigma_s^{(e)} - N_1\sigma_s^{(a)}\right)\Gamma_s P_s - \alpha_s^{(a0)} P_s \quad (6.6)$$

$$\frac{dP_A^+(\nu_j)}{dz} = \left(N_2\sigma_{\nu_j}^{(e)} - N_1\sigma_{\nu_j}^{(a)}\right)\Gamma_s P_A^+(\nu_j) + N_2\sigma_{\nu_j}^{(e)}\Gamma_s h\nu_j \Delta\nu_j$$
$$\quad -\alpha_{\nu_j}^{(a0)} P_A^+(\nu_j) \quad (6.7)$$

$$\frac{dP_A^-(\nu_j)}{dz} = -\left(N_2\sigma_{\nu_j}^{(e)} - N_1\sigma_{\nu_j}^{(a)}\right)\Gamma_s P_A^-(\nu_j) - N_2\sigma_{\nu_j}^{(e)}\Gamma_s h\nu_j \Delta\nu_j$$

$$+\alpha_{\nu_j}^{(a0)} P_A^-(\nu_j) \tag{6.8}$$

We assume that the same overlap parameter can be used for both the signal and ASE as they are both near 1.5 μm. The ASE power in the channel centered at ν_j is propagated as one signal with an input power of 0 at z = 0 for forward-propagating ASE and another signal with an input power of 0 at z = L (L is the length of the fiber) for the backward-propagating ASE. The model becomes more accurate, but obviously more complex to treat numerically, when the frequency channels $\Delta\nu_j$ are chosen to be very small (i.e., 1 nm or 125 GHz), such that the cross sections are essentially constant across $\Delta\nu_j$. The ASE power in the channnel with a center frequency equal to that of the signal channel will be used to compute the noise figure, using the relationship derived in Chapter 7.

6.3.2 Average Inversion Relationship

The propagation equation for the signal, equation 6.6, can be used to derive a convenient relationship for the signal gain. The signal gain can be expressed in terms of the average inversion of the erbium ions along the fiber. Assuming zero background loss, equation 6.6 can be written

$$\frac{dP_s(z)}{dz} = (N_2(z)\sigma_s^{(e)} - N_1(z)\sigma_s^{(a)})\Gamma_s P_s(z) \tag{6.9}$$

The integration of equation 6.6 then yields the gain

$$G = \exp\left[\int_0^L (N_2(z)\sigma_s^{(e)} - N_1(z)\sigma_s^{(a)})\Gamma_s dz\right] \tag{6.10}$$

Equation 6.10 appears to suggest that the gain is a complex function of the shape of the inversion along the length of the fiber. In fact, the gain can be related simply to the average inversion. Define an average upper-state population, and similarly for the lower-state population, as

$$\overline{N}_2 = \frac{1}{L}\int_0^L N_2(z)dz \tag{6.11}$$

$$\overline{N}_1 = \frac{1}{L}\int_0^L N_1(z)dz \tag{6.12}$$

Equation 6.10 can then be simplified to

$$G = \exp\left[(\overline{N}_2\sigma_s^{(e)} - \overline{N}_1\sigma_s^{(a)})\Gamma_s L\right] \tag{6.13}$$

This shows that the signal gain after traversal of the fiber is dependent only on the average inversion of the erbium ions in the fiber, and does not depend on the details of the shape of the inversion as a function of position along the fiber length. The ASE, on the other hand, will depend on the detailed shape of the inversion, due to the local nature of ASE generation. This also implies that the noise figure will depend on the profile of the inversion, and not just the average inversion. The average inversion

6.3. GAIN AND ASE MODELING

relationship for the gain, equation 6.13, is a result of the assumption of homogeneous broadening for the erbium ion transition at 1.5 μm. As a consequence, once the gain is known at one wavelength, the average inversion can in principle be deduced, and the gain at other wavelengths computed.

6.3.3 Inhomogeneous Broadening

Since the energy levels of erbium ions in glass are inhomogeneously broadened, a full modeling of the erbium-doped amplifier should, strictly speaking, include this fact.[4] Modeling that considers only homogeneous broadening has, however, been very successful in predicting the gain and noise performance of both individual amplifiers and systems of cascaded amplifiers. For this reason, the vast majority of studies has modeled the erbium energy level system as homogeneously broadened, thus avoiding the extra complexity of modeling inhomogeneous broadening. This extra complexity comes at several levels. First, the complexity of the system of equations for the populations and propagation of the light fields increases since one must now include the fact that there are different species of ions, which amplify independently. Second, one needs to quantify the degree of homogeneous and inhomogeneous broadening to be used in the modeling. These values are not always that easy to determine unambiguously. Nevertheless, to model very accurately the saturation behavior of the amplifier, and effects such as hole burning (which are important for WDM systems), inhomogeneous broadening needs to be included in the model.

Consider a collection of sites k over which the erbium ions are distributed, with a distribution function f_k to represent the probability that a given ion occupies site k. This distribution function is, in most situations, a Gaussian function. The ions occupying site k have energy level transitions characterized by a central frequency ν_k. Since the overall transition is constituted by a number of transitions between Stark sublevels of the upper and lower states, ν_k is really a label for the entire collection of transitions. The function f_k can then be written as $f(\nu_k)$ and is normalized such that

$$\int_{-\infty}^{\infty} f(\nu_k)\, d\nu_k = 1 \tag{6.14}$$

The total concentration of erbium ions is denoted by N. The number of ions dN_{ν_k} in a spectral packet $d\nu_k$ with a central transition frequency ν_k is given by the expression

$$dN_{\nu_k} = f(\nu_k) N d\nu_k \tag{6.15}$$

For a site k, we denote by $\sigma_k^{H,a}(\nu, \nu_k)$ the homogeneously broadened absorption cross section as a function of the frequency ν. Similarly, $\sigma_k^{H,e}(\nu, \nu_k)$ is the homogeneously broadened emission cross section. The inhomogeneously broadened cross sections, $\sigma^{I,a}(\nu)$ and $\sigma^{I,e}(\nu)$, for absorption and emission, respectively, of the ensemble of ions are given by

$$\sigma^{I,a}(\nu) = \frac{1}{N} \int_{-\infty}^{\infty} \sigma^{H,a}(\nu, \nu_k) dN_{\nu_k} = \int_{-\infty}^{\infty} f(\nu_k)\, \sigma^{H,a}(\nu, \nu_k)\, d\nu_k \tag{6.16}$$

$$\sigma^{I,e}(\nu) = \frac{1}{N} \int_{-\infty}^{\infty} \sigma^{H,e}(\nu, \nu_k) dN_{\nu_k} = \int_{-\infty}^{\infty} f(\nu_k)\, \sigma^{H,e}(\nu, \nu_k)\, d\nu_k \tag{6.17}$$

Consider now a monochromatic light field traversing the ensemble of erbium ions. The light field will experience amplification and absorption from the entire collection of ions, depending on the homogeneous cross sections of the ions and on the separation between the frequency ν of the light and the central frequency ν_k of each species of ions. The light can have a frequency that is coincident with the center frequency of the upper- to lower-state transition of one species of ions and that is the wing of the transition for other ions. The light will interact more strongly with the resonant set of ions (assuming that the peak cross sections of the homogeneous transitions for different species of ions are all equal). This is the direct cause of the spectral hole-burning phenomenon. Each spectral packet will have population densities in the upper and lower states that are specific to that packet. We denote by dN_{2,ν_k} and dN_{1,ν_k} the upper- and lower-state population densities, respectively, of the packet with dN_{ν_k} total ions per unit volume. We introduce the normalized upper- and lower-state population densities $N_2(\nu_k)$ and $N_1(\nu_k)$, such that

$$dN_{2,\nu_k} = f(\nu_k)N_2(\nu_k)d\nu_k \qquad (6.18)$$

$$dN_{1,\nu_k} = f(\nu_k)N_1(\nu_k)d\nu_k \qquad (6.19)$$

We then have

$$N_1(\nu_k) + N_2(\nu_k) = N \qquad (6.20)$$

We can write the propagation equations for the light fields interacting with the ensemble of erbium ions in terms of the homogeneous cross sections and the population densities $N_2(\nu_k)$ and $N_1(\nu_k)$. The signal intensity, $I_s(\nu)$, for example, will have the propagation equation

$$\frac{dI_s(\nu)}{dz} = \int_{-\infty}^{\infty} f(\nu_k)\left[N_2(\nu_k)\sigma^{H,e}(\nu,\nu_k) - N_1(\nu_k)\sigma^{H,a}(\nu,\nu_k)\right]I_s(\nu)d\nu_k \qquad (6.21)$$

We can write a similar equation for the pump intensity and for a collection of signals representing the ASE (with the inclusion of a term for the spontaneous emission). The population equation for the upper-state population of the subset of ions represented by ν_k is derived from the homogeneous cross sections for that set and their overlap with the collection of monochromatic light signals at frequencies ν_i (pump, signal, or ASE) impinging on the ions:

$$N_2(\nu_k) = \frac{\sum_i \frac{\tau \sigma^{H,a}(\nu_i,\nu_k)}{h\nu_k}I(\nu_i)}{\sum_i \frac{\tau(\sigma^{H,e}(\nu_i,\nu_k)+\sigma^{H,a}(\nu_i,\nu_k))}{h\nu_k}I(\nu_i) + 1} N \qquad (6.22)$$

The lower-state population is derived from equation 6.20. In practice, a numerical simulation of an amplifier constituted by an inhomogeneously broadened ensemble of erbium ions can be performed by dividing the inhomogeneous profile into a finite set of spectral packets.

In the sections that follow, we will model the erbium-doped fiber amplifier using the homogeneous model. That is, we will assume that the erbium ions are all of one species and that they respond in an identical way to the action of light fields. Thus, changes in the upper- and lower-state populations of the amplifier transition result in the same change across the spectral profile of the absorption or gain, regardless of the ion involved.

6.4 AMPLIFIER SIMULATIONS

6.4.1 Signal Gain, ASE Generation, and Population Inversion

Signal Gain in a "Typical" Fiber

To calculate the gain of the fiber amplifier as well as the forward and backward ASE spectral profiles, we will use the model equations outlined in the previous section for the homogeneously broadened system of erbium ions, along with the erbium cross sections of Figure 6.1. The computation involves a dual boundary value problem for the system of differential equations to solve.[1] The forward ASE is zero at $z = 0$, for all $P_A^+(\nu_j)$. The backward ASE is zero at $z = L$, for all $P_A^-(\nu_j)$, where L is the length of fiber considered.

We will first consider a specific fiber length and investigate in depth the mechanics of the gain process for this length. We then shift our attention to two other lengths, representing a "short" fiber and a "long" fiber. This will allow us to explore further the sometimes subtle interplay between pump, signal, forward and backward ASE, and the population inversion in the fiber.

We start by considering a 14 m length of fiber A. The signal and pump are taken to be copropagating and injected at $z = 0$. The gains at the two signal wavelengths of 1530 nm and 1550 nm are computed as a function of pump power. We choose these two signal wavelengths because they are representative of the two main regions of the erbium-doped fiber amplifier, the strong narrow peak near 1530 nm, and the flatter plateau around 1550 nm. They are also quite different in terms of the ratio of emission-to-absorption cross section. The injected signal power was taken to be a small signal value, −40 dBm (0.1 μW). The two pump wavelengths considered were 980 nm and 1480 nm. As we have seen before, the emission cross section is zero at 980 nm; for 1480 nm it is finite. The overlap factor is also quite different for these two wavelengths given their differing mode shapes. Given these differences, the gain process for 980 nm and 1480 nm pump wavelengths can be contrasted on a number of levels.

The results are summarized in Figure 6.3 for 980 nm and 1480 nm pumping of a 14 m piece of the Al-Ge erbium-doped fiber being simulated (fiber A). To help interpret Figure 6.3, Figure 6.4 shows the fractional population in the upper state as a function of pump power for both 980 nm and 1480 nm pumping, averaged across the length of the 14 m fiber. At low pump powers, the 1480 nm pump achieves a better average

---

[1]The numerical solution of the coupled differential equations can be done in two ways, using either Runge-Kutta methods or the relaxation method. The Runge-Kutta method involves solving the equations by propagating the light fields forward and backward along the fiber, using the boundary conditions for the signal, pump, forward ASE, and backward ASE, until a self-consistent solution is found. The presence of counterpropagating backward ASE creates the necessity for this back-and-forth simulation. For certain parameter choices this method does not converge to the desired precision, in which case one can use the relaxation method instead.

162 CHAPTER 6. MODELING AND COMPLEX EFFECTS

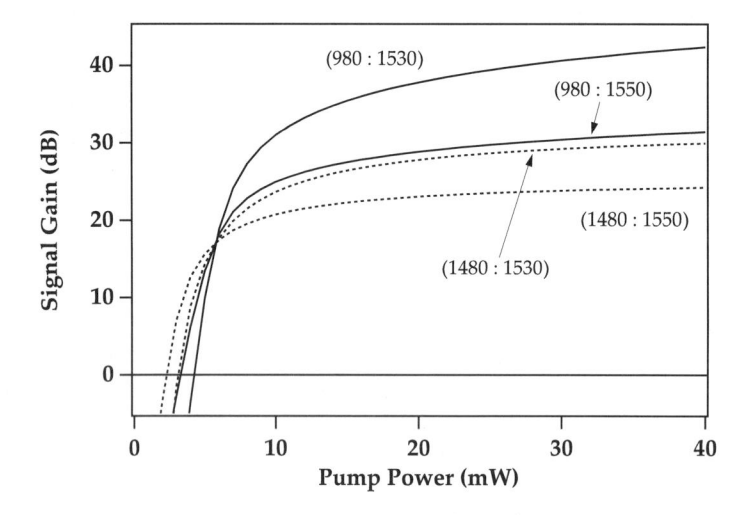

Figure 6.3: Gain as a function of pump power for a 14 m length of erbium-doped Al-Ge silica fiber (fiber A) pumped at 980 nm and 1480 nm. The first number in the parentheses indicates the pump wavelength for the adjoining curve, and the second number indicates the signal wavelength in nm (solid curves: 980 nm pumping; dashed curves: 1480 nm pumping). From numerical simulations.

inversion across the fiber as compared to a 980 nm pump. This situation is reversed at high pump powers as the 980 nm pump's inherent ability to achieve a higher inversion than 1480 nm takes over. Several points are worth noting with respect to Figure 6.3.

- The 980 nm pump wavelength yields higher gains than a 1480 nm pump at high powers. This comes from the fact that 980 nm has achieved a higher inversion than 1480 nm. There is incomplete inversion for 1480 nm pumping, even at high pump powers, due to the nonzero emission cross section at 1480 nm, which drains population out of the upper state.

- For high pump powers, the gain for 1530 nm is higher than that at 1550 nm, due to its higher emission cross section as well as the fact that the fiber is well inverted. At high inversions, with most of the population in the upper state, the gain factor is simply proportional to the emission cross section.

- At low pump powers, the gain factor becomes strongly dependent on the ratio between the emission and absorption cross sections as well as to the exact distribution of the population between the upper and lower states. When the upper-state and lower-state populations are roughly equal, the small signal gain coefficient becomes proportional to the difference between the absorption and emission cross sections. The absorption at 1550 nm is significantly lower than at 1530 nm, so that a lesser amount of pump power is needed to reach threshold for a 1550 nm signal than for a 1530 nm one. This is visible in Figure 6.3.

6.4. AMPLIFIER SIMULATIONS

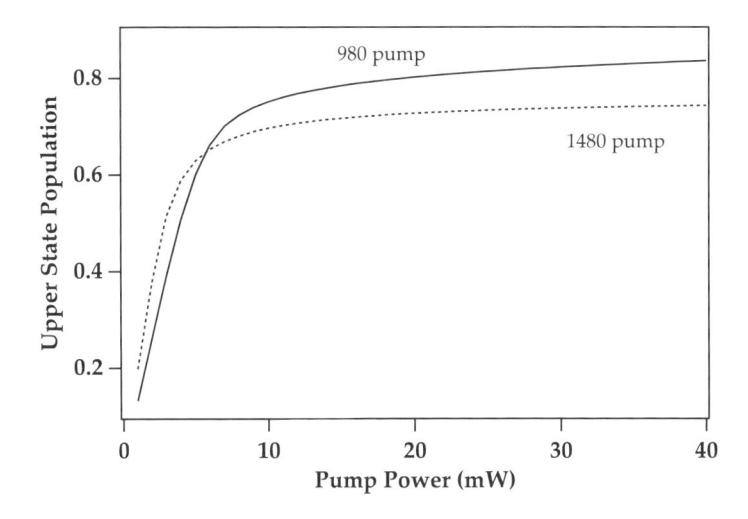

Figure 6.4: Population in the upper state (in units of the fraction of the erbium ion population density), as a function of pump power for a 14 m length of erbium-doped Al-Ge silica fiber (fiber A) pumped at 980 nm and 1480 nm. From numerical simulations.

- Note, finally, that for the 14 m fiber length, the threshold power (pump power required for fiber transparency) for 980 nm pumping is higher than that for 1480 nm pumping. The major reason for this is that 1480 nm pumping has better quantum efficiency than 980 nm pumping. A given amount of pump power represents roughly 50% more photons at 1480 nm as compared to 980 nm. At low pump powers, where ASE is negligible, this translates directly into population inversion, as seen in Figure 6.4.

- When the ASE starts growing as the pump power is increased, the higher inversion created at the beginning of the fiber by the 980 nm pump creates a better seed for the forward ASE than that created by the 1480 nm pump. This robs gain from the signal and increases the transparency point for 980 nm pumping when the fiber is long enough to allow the ASE to build up. We will see this in more delatil in the discussion of ASE generation, in the context of Figure 6.7.

It can be instructive to view at a more microscopic level both the evolution of the light fields as they propagate along the fiber as well as the population inversion at each point along the fiber. Figure 6.5 shows the fractional amount of the population that resides in the upper state as a function of the position along the fiber. At low pump powers, most of the 980 nm pump is absorbed in the beginning of the fiber, thereby locally creating a better inversion than for the 1480 nm pump but leaving most of the remainder of the fiber underpumped. The population determined by the 1480 nm pump is much flatter. The finite emission cross section at 1480 nm serves to flatten both the effective 1480 nm absorption and the upper-state population distribution across the fiber. The peak in the population inversion for 50 mW of pump at 980 nm, near 6 to 8 m, is due to the fact that at this position the total ASE (foward and backward) is

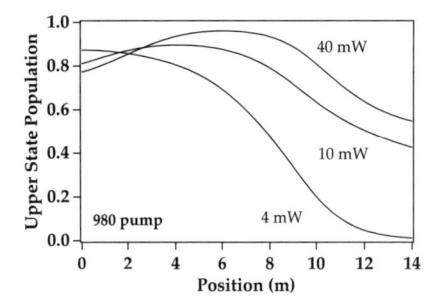

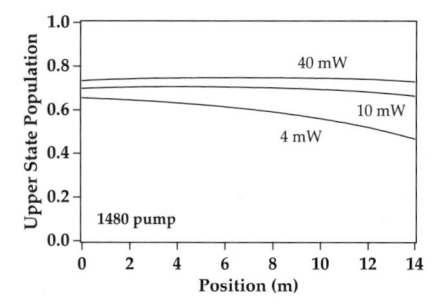

Figure 6.5: Population in the upper state (in units of the fraction of the total erbium ion population) as a function of position along a 14 m length of erbium-doped Al-Ge silica fiber (fiber A), pumped at 980 nm in one case (left) and 1480 nm in the other (right), for the pump powers indicated on the graphs.

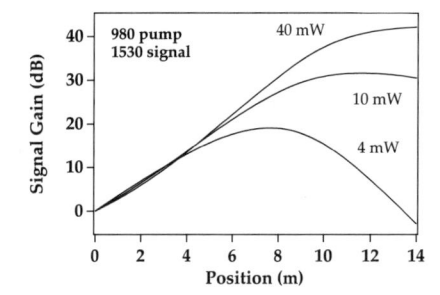

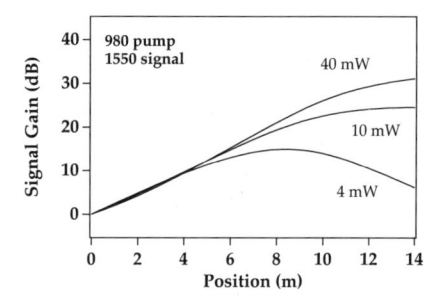

Figure 6.6: Signal gain at 1530 nm (left) and 1550 nm (right) as a function of position along the fiber for a 14 m length of erbium-doped fiber (fiber A), for injected pump powers at z = 0 of 4, 10, and 40 mW at 980 nm and an injected signal power of −40 dBm. From numerical simulations.

the lowest. The inversion is actually lower at the beginning portion of the fiber, even though the 980 nm pump power is highest there, because the backward propagating ASE has reached its highest power there and is significantly depleting the inversion.

It is also instructive to plot the signal gain, as a function of the position along the fiber axis, for the 14 m length of fiber pumped at 980 nm with the small signal −40 dBm input at 1530 nm and 1550 nm, as shown in Figure 6.6. The signal gain is governed by the population inversion shown in Figure 6.5. For the lower pump powers, the fiber is actually too long; in the last section of the fiber the inversion is too low and there is attenuation of the signal rather than gain. The signal gain is proportional to $(\sigma_e N_2 - \sigma_a N_1)$. Thus when N_2/N drops below $(1 + \sigma_e/\sigma_a)^{-1}$, the signal experiences loss rather than gain. For the cross sections of Figure 6.1, the transparency population inversion is 0.53 at 1530 nm and 0.40 at 1550 nm. Thus, at low inversion levels (between 0.40 and 0.53), the 1550 nm signal will continue being amplified while the 1530

6.4. AMPLIFIER SIMULATIONS

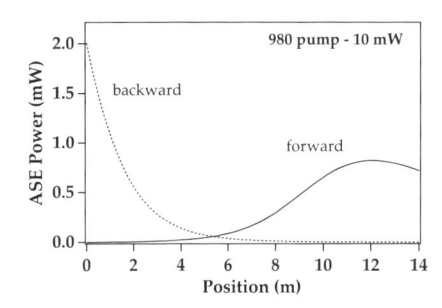

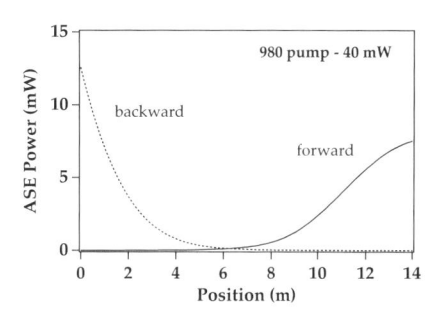

Figure 6.7: Forward- and backward-traveling ASE in a 14 m long erbium-doped fiber pumped at 980 nm (fiber A), for the two pump powers 10 mW (left) and 40 mW (right). From numerical simulations.

nm signal is attenuated. This can be seen, for example, in the 10 mW pump curves of Figure 6.6.

ASE Generation

Figure 6.7 shows the total foward and backward ASE as a function of position, for 980 nm pumping slightly above threshold and well above threshold. The backward ASE benefits from traveling over a well-inverted piece of fiber, as compared to the forward ASE, and thus reaches a higher level. From the point of view of the backward ASE, the 14 m length acts as a source of spontaneous emission followed by a highly inverted piece of fiber a few meters in length. The forward ASE travels over a moderately inverted length of fiber and thus builds up more slowly than the backward ASE. Figure 6.8 shows the forward and backward ASE output at the fiber ends, for both 980 nm and 1480 nm pumping. Consistent with 980 nm providing higher gains than 1480 nm, the ASE is higher for a 980 nm pump than for 1480 nm pump. The backward ASE is higher given that the inversion in the beginning lengths of the fiber is higher than that at the end. The backward ASE is amplified along a well-inverted piece of fiber before exiting the fiber, while the forward ASE travels along a piece of fiber that is progressively less inverted and thus has less gain per unit length than does the backward ASE. For 1480 nm, given that the population inversion is relatively flat across the fiber (for high enough powers), the backward and forward ASE are roughly equal.

The plot of pump power at 980 nm, as a function of position along the fiber axis, is shown in Figure 6.9. At higher pump powers, the pump drops very rapidly, due to depletion of the pump by the backward ASE, before settling to a plateau in the center of the fiber where the ASE is lowest. At the output end, where the forward ASE is strong, the pump is again depleted.

Signal Gain and ASE in "Short" and "Long" Fibers

Figure 6.10 and 6.11 contrast the case of an 8 m piece and a 25 m piece of fiber A pumped at 980 nm and 1480 nm. The 8 m fiber is a "short" fiber given the concentration of erbium ions and fiber geometry for this particular fiber. The 980 nm pump is able to

166   CHAPTER 6. MODELING AND COMPLEX EFFECTS

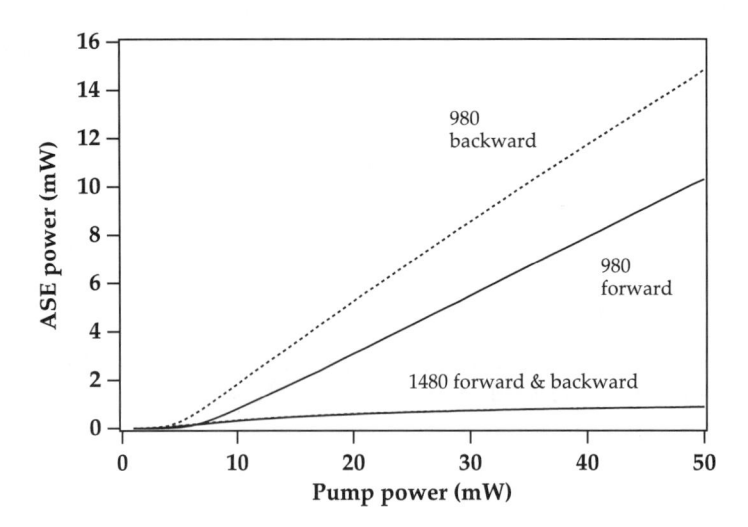

Figure 6.8: Forward- and backward-traveling ASE in a 14 m long erbium-doped fiber (fiber A), as a function of pump power, for both 980 nm and 1480 nm pump wavelengths. From numerical simulations.

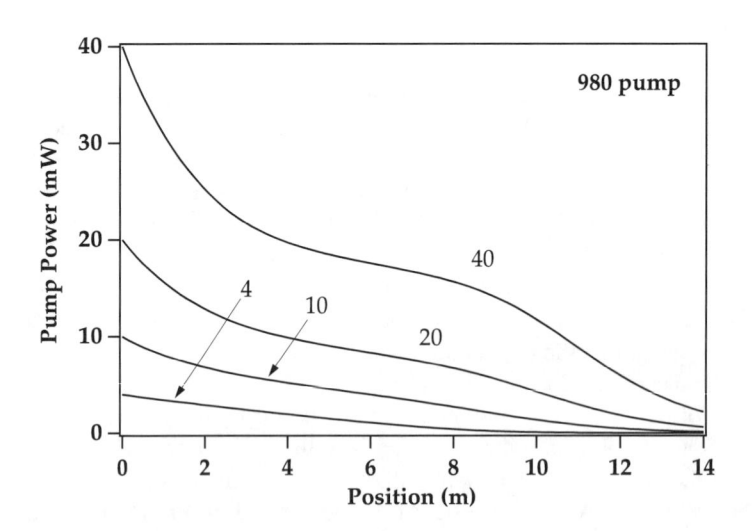

Figure 6.9: Pump power as a function of position along the fiber for a 14 m length of erbium-doped fiber (fiber A), for injected pump powers at z = 0 of 4, 10, 20, and 40 mW at 980 nm, and with an injected signal of power −40 dBm at 1550 nm. From numerical simulations.

6.4. AMPLIFIER SIMULATIONS

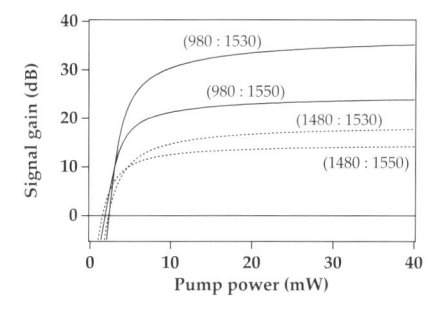

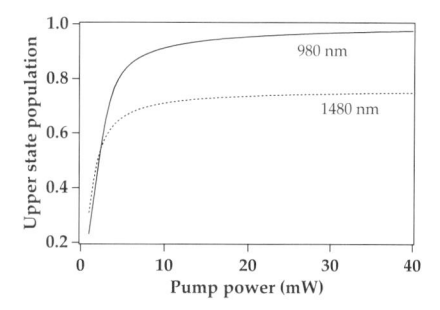

Figure 6.10: Signal gain (left) and upper-state population averaged along the fiber length (right) as a function of pump power for an 8 m length (a "short" length) of erbium-doped Al-Ge silica fiber (fiber A) pumped at 980 and 1480 nm. The first number in the parentheses indicates the pump wavelength for the adjoining curve, and the second number indicates the signal wavelength in nm. From numerical simulations.

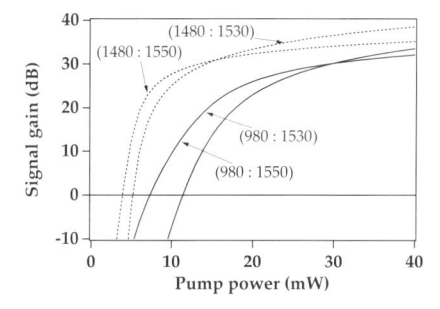

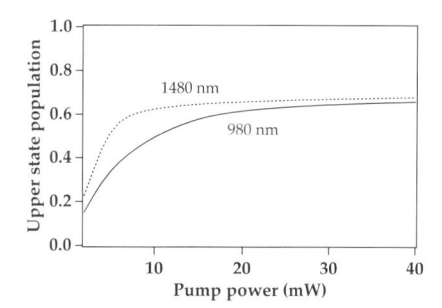

Figure 6.11: Signal gain (left) and upper-state population averaged along the fiber length (right) as a function of pump power for a 25 m length (a "long" length) of erbium-doped Al-Ge silica fiber (fiber A) pumped at 980 and 1480 nm. The first number in the parentheses indicates the pump wavelength for the adjoining curve, and the second number indicates the signal wavelength in nm. From numerical simulations.

almost completely invert the fiber, since the ASE does not have enough fiber length to grow to a value where it is significantly depleting the population.

The 25 m fiber is a "long" one, at least insofar as the 980 nm pump is concerned. The thresholds for transparency at 1550 nm and 1530 nm, for 980 nm pumping, have moved significantly to higher powers, as compared to the thresholds for the 1480 nm pump. The reason for this can be understood by imagining the 25 m fiber as a sequence of shorter length fibers. Higher pump powers are now required at the input to invert the entire fiber, especially toward the end of the fiber. Such a high pump power at 980 nm generates a large amount of ASE, in particular backward ASE, which reduces the pump efficiency in producing signal gain and results in higher thresholds. This can be understood by looking more closely at the growth or decay of the various light fields along the length of the fiber. Consider the 14 m length pumped with 20 mW at 980 nm

168   CHAPTER 6. MODELING AND COMPLEX EFFECTS

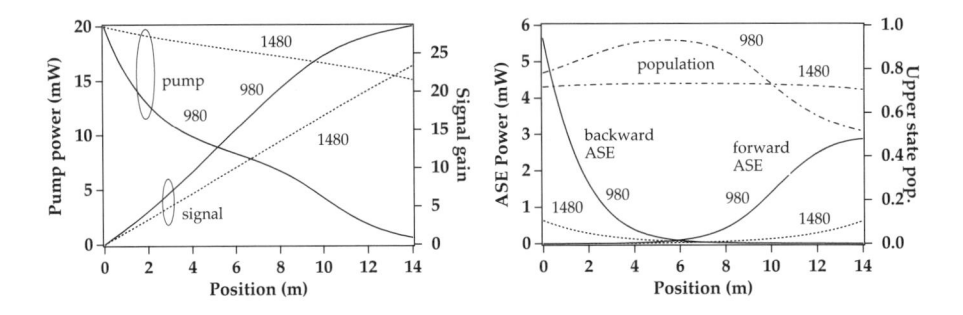

Figure 6.12: Left: pump absorption and signal gain as a function of position in a 14 m erbium-doped fiber (fiber A). Right: upper-state population and forward and backward ASE for 980 nm and 1480 nm pumping with a launched pump power of 20 mW. The signal is at 1550 nm with a launched power of −40 dBm. From numerical simulations.

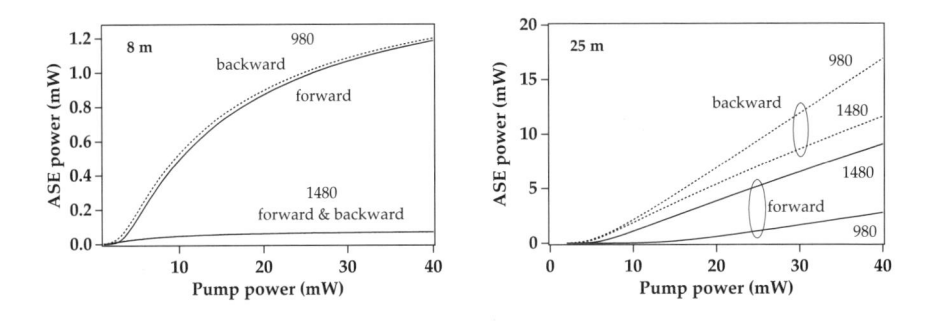

Figure 6.13: Forward (solid curves) and backward (dashed curves) ASE output from the 8 m length of fiber A (left) and the 25 m length of fiber A (right) for 1480 and 980 nm pumping, as a function of pump power. From numerical simulations.

and 1480 nm, as shown in Figure 6.12. Almost all of the 980 nm pump is absorbed in the 14 m length, while a significant fraction of the 1480 nm pump is still available at the output. Nevertheless, the 980 nm pump has yielded only a few dB more gain for the 1550 nm signal than the roughly 25 dB obtained with 1480 nm. The reason for this can be seen in the plot of the upper-state population and the forward and backward ASE. The 980 nm pump gives rise to much higher levels of both forward and backward ASE. The backward ASE is especially strong as the beginning of the fiber is very well inverted and thus allows the 1530 nm band of the ASE to grow rapidly. The ASE depletes the inversion and robs gain at the expense of the signal. One can intuitively extrapolate what situation will arise when an additional 10 m of fiber is added to the 14 m piece for the 20 mW pump power condition.

Figure 6.13 shows the ASE output from the 8 m and 25 m fibers as a function of pump power, for both 980 nm and 1480 nm pumping. The 8 m fiber is short enough that it is highly inverted, uniformly across its length. As a result, the forward and backward ASE at the ends of the fiber, for a given pump power and wavelength, are essentially the

6.4. AMPLIFIER SIMULATIONS

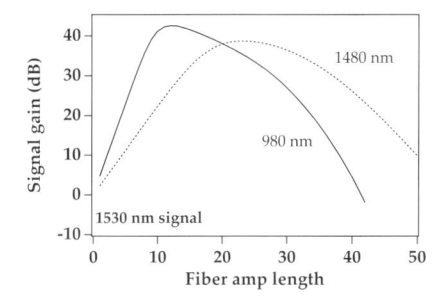

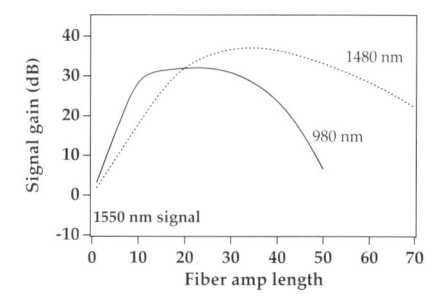

Figure 6.14: Signal gain at 1530 nm (left) and 1550 nm (right) for 1480 and 980 nm pumping of fiber A, as a function of fiber amplifier length. The launched pump power is 40 mW and the launched signal power is −40 dBm.

same. The 25 m fiber is so long that the backward ASE sees a much more propitious inversion than the forward ASE. The forward ASE has to travel over the last portion of the fiber, which has been much less inverted than the beginning portion.

6.4.2 Gain as a Function of Fiber Length

Finally, we can study the small signal gain obtained as the length of fiber is varied between a few m to 50 m. Figure 6.14 shows the signal gain at 1550 nm, for both 980 nm and 1480 nm pumping and 40 mW of pump power, as a function of the erbium-doped fiber length, for fiber A. Interestingly, for the 1550 nm signal, the 1480 nm pump provides a higher gain than the 980 nm pump at equal pump powers. This was noted in previous simulations.[5] The 1480 nm pump can maintain the necessary inversion levels (≥ 0.40 for this particular fiber) over significantly longer lengths than the 980 nm pump. The reason for this is the higher quantum efficiency of 1480 nm pumping. This allows the signal to grow to a higher maximum value. The situation is reversed in the case of the 1530 nm signal. Since the 1530 nm signal requires high inversions (≥ 0.53) and benefits from very high inversion levels, the 980 nm pump with its higher inversion capabilities can reach higher small signal gains than a 1480 nm pump, notwithstanding the greater quantum efficiency of the latter. For higher input powers (e.g., −10 dBm), the fiber will be in a saturation regime where the inversion is reduced. In that case the higher quantum efficiency of the 1480 nm pump is the dominant factor and will give a higher gain at 1530 nm than the 980 nm pump. At high pump powers (for this particular fiber geometry) such as 40 mW, the maximum gain is achieved within a relatively broad range of fiber lengths. At low powers, the optimal length is in a much narrower range, as is shown in Figure 6.15.

6.4.3 Spectral Profile of the ASE

The forward and backward ASE profiles for various 980 nm pump powers are shown in Figure 6.16, for a 14 m length of fiber A. For these computations, the ASE spectrum was divided into channels of 1 nm width, ranging from 1450 nm to 1580 nm. Thus 130

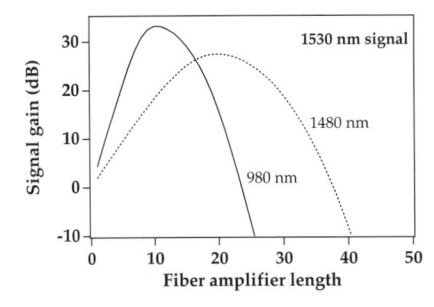

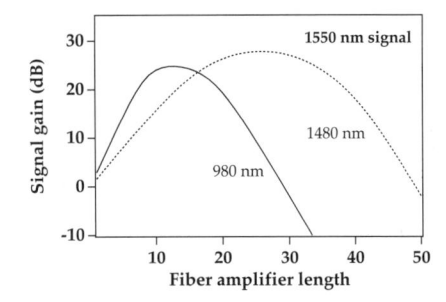

Figure 6.15: Signal gain at 1530 nm (left) and 1550 nm (right) for 1480 and 980 nm pumping of fiber A, as a function of fiber amplifier length. The launched pump power is 10 mW and the launched signal power is −40 dBm. From numerical simulations.

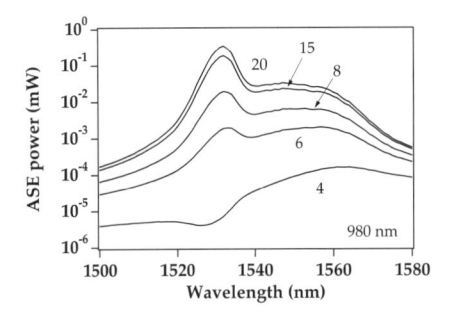

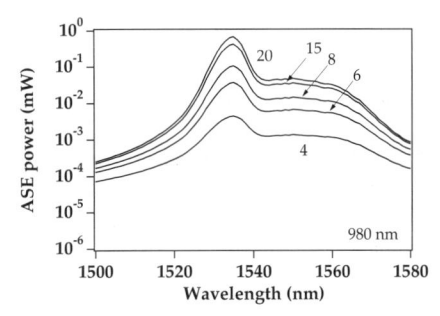

Figure 6.16: Forward- (left) and backward- (right) propagating ASE power spectra (1 nm resolution) for a 14 m length of fiber A. The pump powers are 4, 6, 8, 15, and 20 mW at 980 nm and are indicated on the figure. From numerical simulations.

separate ASE channels were propagated through the fiber amplifier, each one composed of a forward and backward traveling signal.

The forward ASE peak at 1530 nm is very sensitive to the pump power. Similarly to the behavior of the gains for the 1530 nm and 1550 nm signals, the 1530 nm ASE band is at first weaker than the 1550 nm band. As the pump power is increased and the inversion rises, the 1530 nm catches up with and then surpasses the 1550 nm ASE band. At the same pump powers where the small signal gain starts saturating (i.e., 15 to 20 mW), the ASE begins saturating as well. The backward ASE travels over a much better inverted section of fiber, even at low pump powers, and thus the peak at 1530 nm is strong compared to that at 1550 nm, although the overall backward ASE level is low. The fact that the 1480 nm pump produces a different inversion level than the 980 nm pump is reflected in the ASE profiles of Figure 6.17. At low pump powers, the 1480 nm pump produces a higher overall inversion than 980 nm, so that the forward ASE is higher, including the 1530 nm peak. Nevertheless, since 1480 nm does not produce high local inversions, it does not favor the buildup of the 1530 nm peak, as reflected if

6.4. AMPLIFIER SIMULATIONS

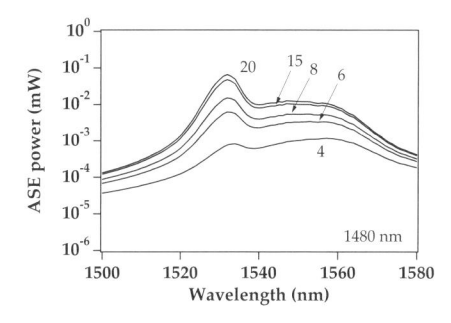

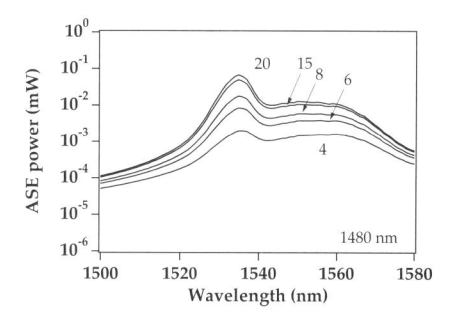

Figure 6.17: Forward (left) and backward (right) propagating ASE power spectra (1 nm resolution) for a 14 m length of fiber A. The pump powers are 4, 6, 8, 15, and 20 mW at 1480 nm and are indicated on the figure. From numerical simulations.

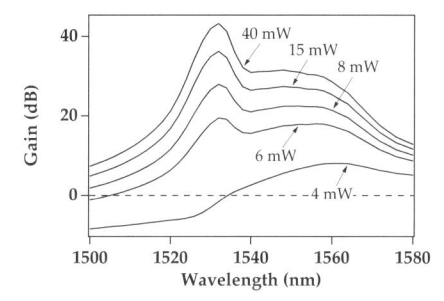

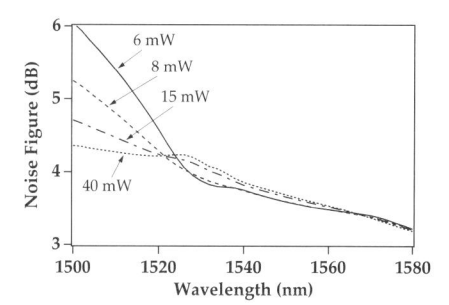

Figure 6.18: Gain (left) and noise figure (right) as a function of signal wavelength (signal power input −40 dBm), for the pump powers indicated on the graphs, for a 14 m length of erbium-doped fiber (fiber A) pumped at 980 nm. From numerical simulations.

one compares the peak ASE powers in the forward and backward ASE curves for 980 nm and 1480 nm at the higher pump powers.

6.4.4 Small Signal Spectral Gain and Noise Modeling

Modeling can also predict the gain and noise as a function of wavelength. The noise figure is computed using the relationships derived in Chapter 7, section 7.3.3. Given the differences in the ratios of the absorption and emission cross sections, as a function of wavelength, the spectral shape of the gain will change nonuniformly with changes in pump power. In particular, as pump powers decrease, signals near 1530 nm will experience a drop in gain much more significant than that for signals near 1550 nm, as shown in Figure 6.18. The corresponding noise figures also vary with wavelength, because of the change in the absorption and emission cross sections. In the 1530 nm region, the higher absorption-to-emission cross section ratio results in a lower inversion parameter n_{sp} than in the 1550 nm region. This is shown in Figure 6.18 for a 14 m fiber with a −40 dBm signal input, swept in wavelength. Several 980 nm pump powers are contrasted: 6, 8, 15, and 40 mW. The noise figure improves monotonically from 1500

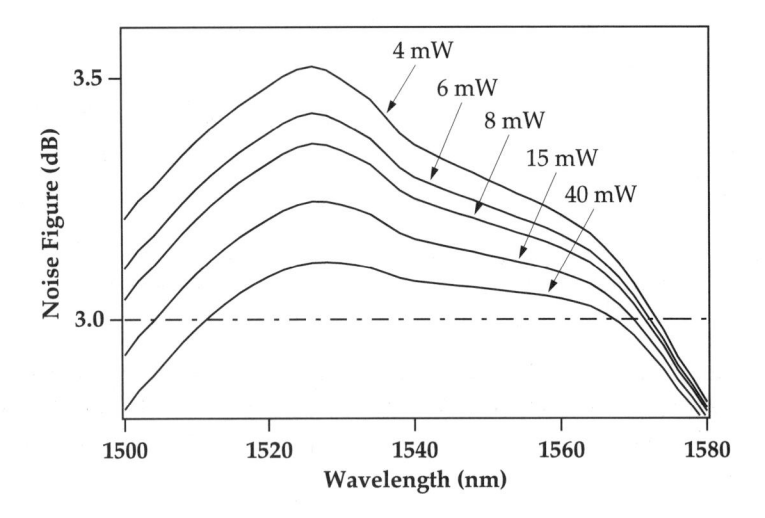

Figure 6.19: Noise figure as a function of signal wavelength (signal power input −40 dBm), for the pump power values 4, 6, 8, 15, and 40 mW, for an 8 m length of erbium-doped fiber (fiber A) pumped at 980 nm. The dashed line indicates the 3 dB signal-spontaneous quantum limit for high gain amplifiers. From numerical simulations.

nm toward the longer signal wavelengths. Surprisingly, the noise figure is actually higher at the higher pump values, and more so in the vicinity of 1530 nm than 1550 nm. The reason for this can be understood by looking again at Figure 6.5. At higher pump powers, the inversion is depleted in the front portion of the fiber by the backward ASE. This is effectively a loss that the signal encounters immediately upon entering the fiber, thus increasing the noise figure. Since the absorption is stronger at 1530 nm than at 1550 nm, the degradation in noise figure is stronger for the former. This effect disappears with shorter fiber lengths.

Figure 6.19 traces the noise figures for an 8 m length of fiber, as a function of wavelength and pump power. Now the noise figures drop smoothly as the pump power is increased, since the backward ASE is not strong enough to significantly affect the inversion at the input of the fiber. Another observation to be made is that the noise figure drops below 3 dB near the ends of the spectrum (below 1510 nm and above 1570 nm, approximately). This is a result of the fact that the fiber considered is very short and highly inverted. In a wavelength range where the inversion parameter n_{sp} is very close to 1 and the gain is relatively low, the noise figure can indeed be less than 3 dB. The 3 dB signal-spontaneous noise quantum limit is derived in the high-gain regime. This issue is discussed further in Chapter 7, section 7.3.3.

Figure 6.20 shows the amplifier spectral gain and noise figures for 1480 nm pumping, for a 14 m length of fiber A. Since pumping at 1480 nm results in flatter population distributions (see Figure 6.5), the noise figure as a function of wavelength drops smoothly with pump power for a 14 m length of fiber. In this regard, Figure 6.20 and Figure 6.18 contrast the difference between 980 nm and 1480 nm pumping.

6.4. AMPLIFIER SIMULATIONS

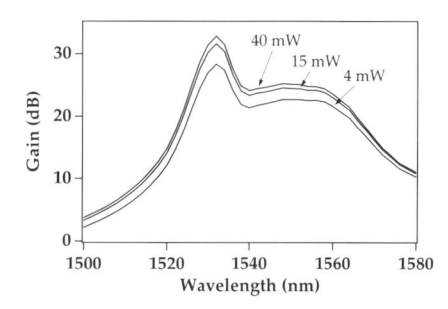

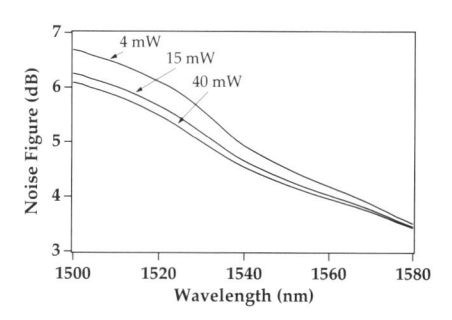

Figure 6.20: Gain (left) and noise figure (right) as a function of signal wavelength (signal power input −40 dBm), for the pump power values 6, 8, 15, and 40 mW, for a 14 m length of erbium-doped fiber (fiber A) pumped at 1480 nm. From numerical simulations.

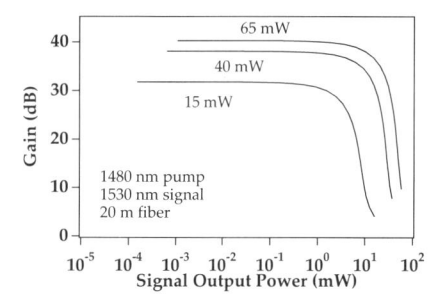

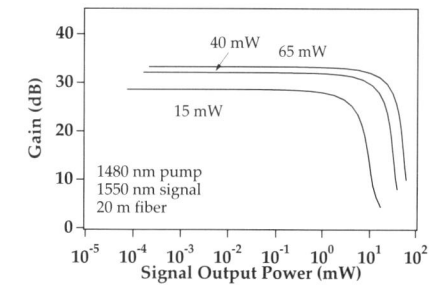

Figure 6.21: Gain as a function of signal output power for the three pump power values 15, 40, and 65 mW, for a 20 m length of fiber A pumped at 1480 nm with a signal at 1530 nm (left) and 1550 nm (right). From numerical simulations.

6.4.5 Saturation Modeling — Signal Gain and Noise Figure

The effects of gain saturation with signal power can also be modeled. Figure 6.21 shows the gain as a function of signal power in a 20 m length of fiber A. The signal wavelengths are 1530 nm and 1550 nm, and the three pump powers are 15, 40, and 65 mW, at a pump wavelength of 1480 nm. As the signal increases in power past the small signal value, the gain decreases since the pump can no longer replenish the inversion as fast as the signal depletes it. The small signal gain and the saturation output power (defined as the output power at which the gain is compressed from its small signal value by 3 dB) both increase with pump power. This is one of the significant advantages of the Er^{3+} three-level amplifier system. The saturation output powers corresponding to the 15, 40, and 65 mW pump powers are, respectively, 2.4 mW, 7.9 mW, and 13.1 mW (1530 nm) and 3.8 mW, 12.4 mW, and 20.5 mW (1550 nm). The saturation output powers are higher for the signal at 1550 nm due to its lower cross section. This has been observed experimentally, as shown in Figure 6.22.

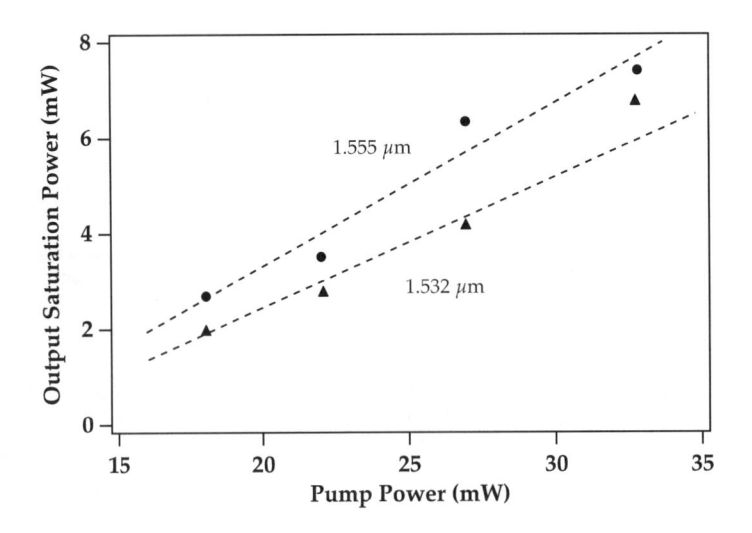

Figure 6.22: Experimentally measured output saturation powers as a function of launched pump power at 975 nm for the signal wavelengths 1532 and 1555 nm, for a Ge-Al fiber with fixed length (40 m). From reference [6]. Reprinted with permission from A. Lidgard, J. R. Simpson, and P. C. Becker, *Appl. Phys. Lett.* Vol. 56, p. 2607 (1990). Copyright 1990 American Institute of Physics.

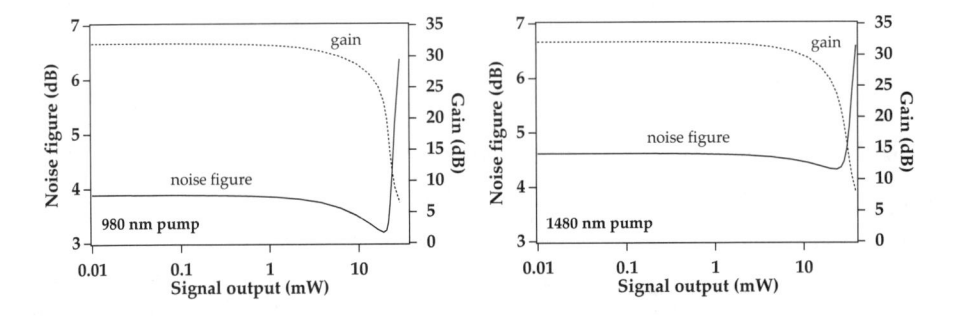

Figure 6.23: Noise figure as a function of input signal power, for a 20 m length of fiber A with a signal input at 1530 nm and 1550 nm, for 980 nm (left) and 1480 nm (right) pumping with 40 mW of power. From numerical simulations.

The noise figure is computed using the relationship derived in Chapter 7, equation 7.76. The noise figure varies with input signal. As the signal level rises, the inversion is depleted, so one expects that the inversion parameter and noise figure will increase with signal level. A simulation of the noise figure is shown in Figure 6.23 for 1530 nm and 1550 nm signals in a 20 m length of fiber A pumped at 980 nm and 1480 nm. The pump power was fixed at 40 mW for this simulation. As expected, as the signal output rises above the saturation values, the noise figure increases dramatically. An interesting feature is the dip in the noise figure at output powers in the 10 to 15

6.4. AMPLIFIER SIMULATIONS

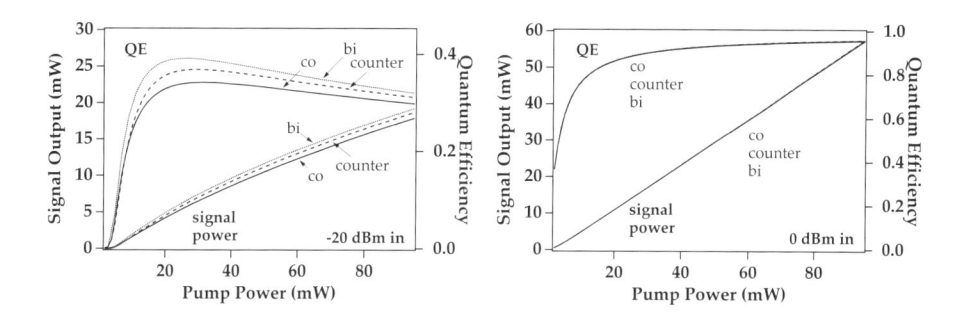

Figure 6.24: Signal output power and quantum conversion efficiency as a function of pump power, for a signal input at 1550 nm with −20 dBm (left) and 0 dBm (right) launched signal power, for a 14 m fiber pumped at 980 nm, with bidirectional (bi), copropagating (co), and counterpropagating (counter) pumping. From numerical simulations.

dBm range (corresponding to input powers in the −15 dBm to −10 dBm range), which are in the range where the amplifier starts saturating with signal input. The noise figure is constant at low input powers as the signal has no significant impact on the level of ASE. Recall that a lower inversion at the input to the amplifier produces some absorption, and loss at the input of an amplifier always degrades the noise figure. When the signal reaches the −10 to −15 dBm input range, it starts affecting the level of ASE and hence the upper-state population depletion in the beginning of the fiber. At these intermediate power levels, the signal robs gain from the backward ASE and thus prevents it from growing and depleting inversion at the input to the amplifier. At higher input power levels the signal significantly depletes the inversion–beyond the pump's ability to replenish it–and the noise figures increase rapidly with signal power.

6.4.6 Power Amplifier Modeling

Power amplifiers are operated with high signal input powers. A characteristic way to measure how well the power amplifier is working is by means of the conversion efficiency, i.e., the conversion of pump light into signal light. This can be measured on a power basis, and one then quotes a power conversion efficiency. The conversion from pump photons into signal photons–the quantum conversion efficiency–is a more direct measure of the efficiency of the amplifier (see section 8.2.2).

Let us consider again the 14 m amplifier fiber and plot the signal output power and quantum conversion efficiency as a function of pump power for a signal input of a 1550 nm and a 980 nm pump. Chapter 8 discusses the choice of fiber length for a power amplifier design for different signal input levels (see Figure 8.11), and the impact of the direction of the pump propagation. Both −20 dBm and 0 dBm signal inputs are contrasted in Figure 6.24. The quantum conversion efficiency is defined as the ratio of the number of output signal photons to the number of input pump photons. The quantum conversion efficiency has also been computed for both a counterpropagating pump and a bidirectional pump (pump power equally split between forward and backward propagating directions). Note that quantum efficiencies are predicted to be

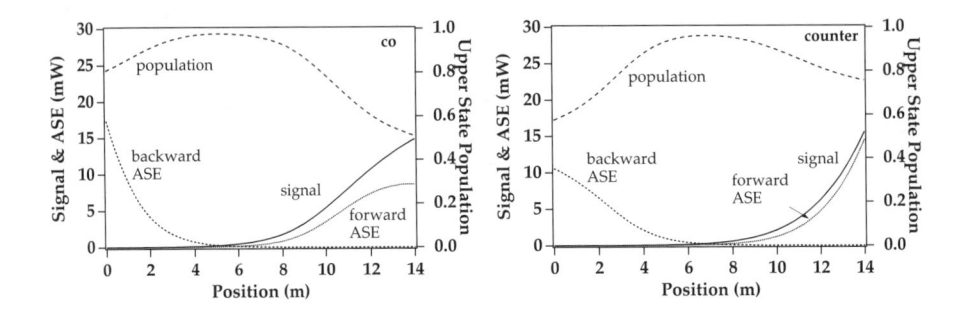

Figure 6.25: Signal gain, forward ASE, and upper-state population as a function of fiber position, for a 14 m fiber with a signal input at 1550 nm with −20 dBm and 75 mW of pump power at 980 nm, for a copropagating pump (left) and a counterpropogating pump (right). From numerical simulations.

near unity, making for very efficient amplifiers. For pumping at 1480 nm as well, the quantum conversion efficiencies are computed to be very high and can exceed 90%.[1] Experimental measurements have established that such conversion efficiencies can be very nearly approached.[7] The counterpropagating pump is more efficient than the copropagating pump and the bidirectional pump is more efficient then both of them. The reason that the counterpropagating pump is more efficient than the copropagating pump is that in the counterpropagating case the pump is strongest in the region where the signal is strongest, Thus the signal can most efficiently deplete the inversion, as opposed to allowing the ASE to deplete the inversion. Consider, for example, Figure 6.25, which shows the 14 m length of fiber and the signal gain, ASE, and population inversion as a function of length for a pump power of 75 mW. The signal input is −20 dBm at a wavelength of 1550 nm. In the copropagating case, the backward ASE has benefited from the high pump powers at the input of the fiber, depleting the inversion and resulting in a backward ASE output that is higher than the signal output. The lower levels of pump power available in the last 5 m of the fiber have produced lower inversion levels, and the growth of the signal is relatively slow. On the other hand, with the counterpropagating pump, the backward ASE is relatively weak and the signal is higher in magnitude than either backward or forward ASE. The signal grows exponentially in the last 5 m where the inversion is strong.

The difference between the copropagating and counterpropagating pump configurations is a result of the moderate signal input of −20 dBm. When the signal input is weak (i.e., less than −40 dBm), the copropagating and counterpropagating pump configurations yield the same conversion efficiency. The signal never gets strong enough to seriously affect the inversion, for the fiber length considered (14 m). On the other extreme, when the signal is strong (0 dBm), then for the pump level and fiber length chosen in this example the copropagating and counterpropagating pump configurations yield again the same conversion efficiency. Here the reason is that the signal is so strong that it dominates the ASE in both pumping cases and does not allow the ASE to favor one of the pump configurations.

6.4. AMPLIFIER SIMULATIONS

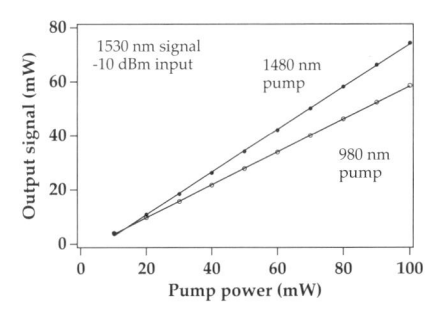

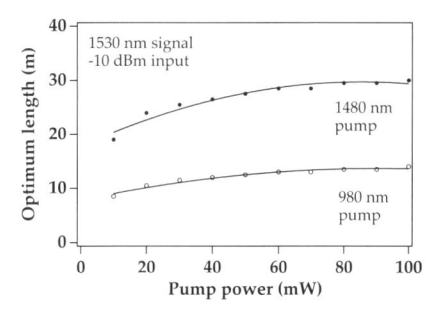

Figure 6.26: Signal output power, for a length optimized (at each pump power) piece of fiber A with a signal input at 1530 nm with −10 dBm input and for a copropagating pump at either 1480 nm or 980 nm (left) and the corresponding optimal lengths (right). From numerical simulations.

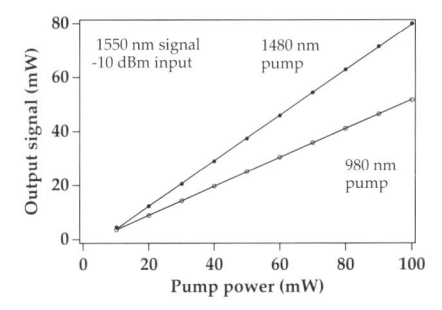

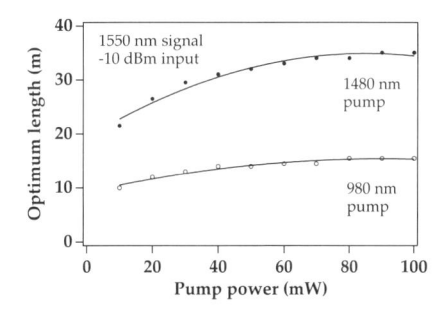

Figure 6.27: Signal output power, for a length optimized (at each pump power) piece of fiber A with a signal input at 1550 nm with −10 dBm input and for a copropagating pump at either 1480 nm or 980 nm (left) and the corresponding optimal lengths (right). From numerical simulations.

To summarize, the most advantageous situation is to have the signal strong where the inversion is strong, so that the signal, not the ASE, will deplete the gain. The pump configuration that yields the highest conversion efficiency is bidirectional pumping. The resulting flattening of the upper-state population distribution results in an optimum situation with respect to the competition between signal and ASE. Bidirectional pumping yields a few dB more in gain (as compared to the counterpropagating pump) for moderate input powers in the −20 dBm range.[8]

Another issue is the choice of a 1480 nm or 980 nm pump, for a power amplifier (either single stage or the second stage of a dual stage amplifier). The choice of a 1480 nm pump yields a better power conversion efficiency, given the larger number of photons available at 1480 nm for a specified pump power, compared to that at 980 nm. This is illustrated in Figures 6.26 and 6.27. For both 1530 nm and 1550 nm signals, 1480 nm provides higher output powers compared to 980 nm. Since current diode laser pumps at 1480 nm currently provide comparable power to 980 nm diode lasers (see Chapter 3, section 3.13), 1480 nm is the preferred pump choice for a power amplifier second stage of a dual stage amplifier, when high output power is desired.

6.4.7 Effective Parameter Modeling

The modeling of the gain and ASE properties of a given length of fiber can be computed, given certain assumptions, without a direct determination of the cross sections. For example, notice that in the power propagation equations 6.5 through 6.8, the gains and attenuations are all proportional to the product of the cross sections, erbium ion density, and overlap parameter. This is nothing more than the small signal absorption and small signal gain of the fiber, depending on whether the cross section involved is that for absorption or emission. The population equation can also be manipulated to contain the small signal absorption and gain coefficients. These can be directly measured from the erbium-doped fiber. One measures the small signal attenuation and small signal gain for a short length, as well as the fiber saturation parameter, and one can then compute the gain for various lengths. Note that when only the fiber attenuation is measured, the fiber small signal gain can be computed from the McCumber relationship. This experimental approach, which avoids a direct computation of the cross sections, has been described by Giles et al.[9] We summarize the key equations here.

The derivation of the effective parameter model is as follows. We consider an average density of erbium ions in the transverse dimension, along with an equivalent radius for the erbium-doped region. The average densities of erbium ions involved in the calculation are the total density, $\bar{n}_t$, and the densities of ions in the ground state and excited state, written respectively as $\bar{n}_1$ and $\bar{n}_2$. The average density of ions in the ground and excited states, at a given axial point z, are computed by integrating over the transverse dimensions at each point z along the fiber. The overlap integrals are calculated also for the overlaps of the light fields with the ion distribution in the ground and upper states. These overlap integrals are assumed to be equal and independent of pump power for the both upper- and lower-state overlaps. This assumption is justified in the case of confined erbium-doped fibers.[2] This then leads to the two basic equations that govern the amplification properties: the population equations for the ground- and excited-state populations, and the light-field propagation equations, which are now expressed in terms of the measured fiber spectral attenuation, α_k, and the measured gain of a fully inverted short length of fiber, g_k^* (both quantities per unit length of fiber). The index k is used to keep track of the wavelength, λ_k. α_k and g_k^* can be written in terms of the overlap parameter (see Chapter 5, section 5.2.2), considering a situation where the erbium ions are uniformly doped in the core of the fiber [2]

$$\alpha_k = \sigma_{a,k} \Gamma_k n_t \quad (6.23)$$

$$g_k^* = \sigma_{a,k} \Gamma_k n_t \quad (6.24)$$

where Γ_k is the overlap integral between the erbium distribution and the transverse optical mode, as defined in Chapter 5, and n_t is the erbium ion density.

In effect, the measurement of the spectral gain over a small piece of fiber allows the prediction of the gain and modeling of the ASE for arbitrary lengths of fiber. Also

6.4. AMPLIFIER SIMULATIONS

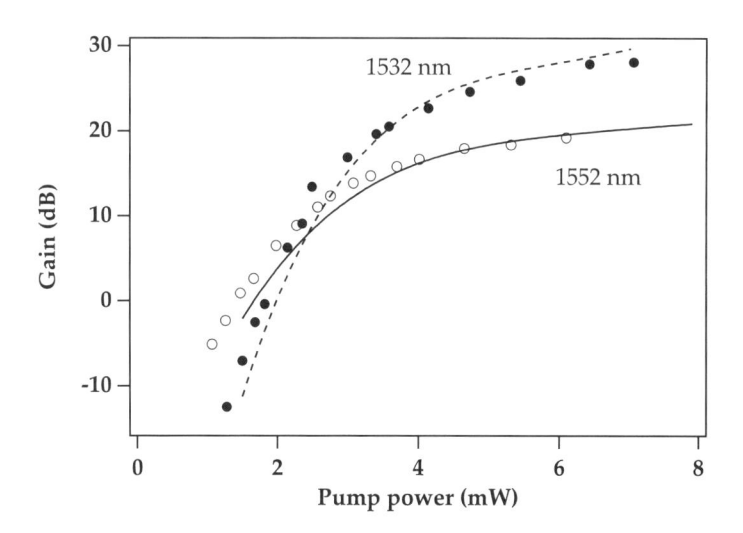

Figure 6.28: Comparison of experimental (points) and calculated with the effective parameter model (solid and dashed curves) small signal gains for a 13.6 m long erbium-doped fiber amplifier pumped at 978 nm, for a signal at 1532 nm (1 μW input) and a signal at 1552 nm (0.5 μW input). From reference [9] (©1991 IEEE).

needed is the fiber saturation parameter, $\zeta = A\bar{n}_t/\tau$, where A is the effective area of the doped region and τ is the lifetime of the Er^{3+} $^4I_{13/2}$ upper-state level. The saturation parameter can be obtained from a measurement of the fiber saturation power as $\zeta = P_k^{sat}(\alpha_k + g_k^*)/h\nu_k$. The equations are then

$$\frac{\bar{n}_2}{\bar{n}_t} = \frac{\sum_k \frac{P_k(z)\,\alpha_k}{h\nu_k \zeta}}{1 + \sum_k \frac{P_k(z)\,(\alpha_k + g_k^*)}{h\nu_k \zeta}} \quad (6.25)$$

for the population in the upper state, and

$$\frac{dP_k}{dz} = u_k(\alpha_k + g_k^*)\frac{\bar{n}_2}{\bar{n}_t}P_k(z) + u_k g_k^* \frac{\bar{n}_2}{\bar{n}_t} mh\nu_k \Delta \nu_k - u_k(\alpha_k + \alpha_k^0)P_k \quad (6.26)$$

for the propagation equation for each light field of index k. u_k is equal to +1 for a forward-propagating field and −1 for a backward-propagating field. The α_k^0 represent background loss. This approach has led to good agreement between measured and predicted gain in erbium-doped fibers, as shown in Figure 6.28.[9] Note also that the average inversion relationship of Section 6.3.2 can be conveniently written in terms of α_k and g_k^* as

$$\ln(G_k) = \left[(\alpha_k + g_k^*)\frac{\bar{n}_2}{\bar{n}_t} - \alpha_k\right] L \quad (6.27)$$

where G_k is the gain of the fiber at wavelength λ_k, L is the length of the amplifier fiber, and $\bar{n}_2$ is the average population in the upper state along the fiber length, defined as in Section 6.3.2.

6.5 TRANSVERSE MODE MODELS — ERBIUM CONFINEMENT EFFECT

It has long been known that for lasers the effective volume occupied by the pump and signal modes, and their overlap with the region containing the active medium, is important to their efficient operation. The main finding of the transverse mode studies for erbium-doped fiber amplifiers is that the amplifier is most efficient when the erbium is confined to the center of the core region of the fiber, in the central 50% of the core area. It is also found, as one might expect, that fiber structures that result in smaller-mode field diameters lead to higher intensities and thus better inversion of the erbium by the pump, resulting in high gain efficiencies.

Kubodera et al introduced the concept of an effective mode volume for a slab waveguide Nd^{3+} laser, defined as a three-dimensional overlap integral between the signal mode and the pump mode over the entire laser cavity.[10, 11] The lasing threshold for single transverse mode conditions was then explicitly calculated as a function of the effective mode volume and compared for side pumping versus end pumping.

This approach was expanded on by Digonnet and Gaeta, who calculated lasing thresholds, slope efficiencies, and single-pass gains for optical fibers containing an active medium.[12] Considering a fiber such that the fiber core is uniformly doped with the active medium, it can be shown that in the low pump power regime the optical gain (for a four-level system as well as a three-level one) is proportional to the ratio of the overlap integral of the pump and signal modes in the core, to the relative amount of pump power in the core.[13] This parameter increases with the V number of the fiber. The small signal gain is also proportional to the pump intensity in the core. Although this treatment does not involve the effects of saturating ASE or signals, and involves only a uniformly doped core, it points out the importance of the choice of the fiber parameters (core size and V number) in maximizing the gain. Figure 6.29 shows, for example, the single-pass gain for a three-level system, as a function of core radius, for various pump modes where it is assumed that the signal mode propagates in the lowest-order mode.[13].

Clearly, the lowest order pump mode provides the highest gain as it offers the highest overlap with the signal mode. There is an optimum core radius in the 2–3 μm range. For larger core sizes the decrease of pump intensity in the core reduces the gain, whereas at very small core sizes a sizable portion of the pump mode extends into the cladding, which reduces the geometric overlap factors and thus the gain. This theory also demonstrates that higher small signal gains are obtained by using a high NA fiber with a small core, as compared to using a low NA fiber with a larger core (e.g., a factor of three for an NA of 0.4 with a core radius of 1.3 μm compared to an NA of 0.1 with a core radius of 4.5 μm).[14]

In the previous section, the transverse variation in the erbium ion distribution and the light-field intensities was simplified by calculating an effective intensity in the erbium-doped region with the help of the overlap integral. This allowed for a one-dimensional model of the amplifier where the various pump and signal fields are propagated along the fiber axis z. This facilitates the calculation of the spectral properties of the amplifier—for example, the spectral shape and strength of the ASE.

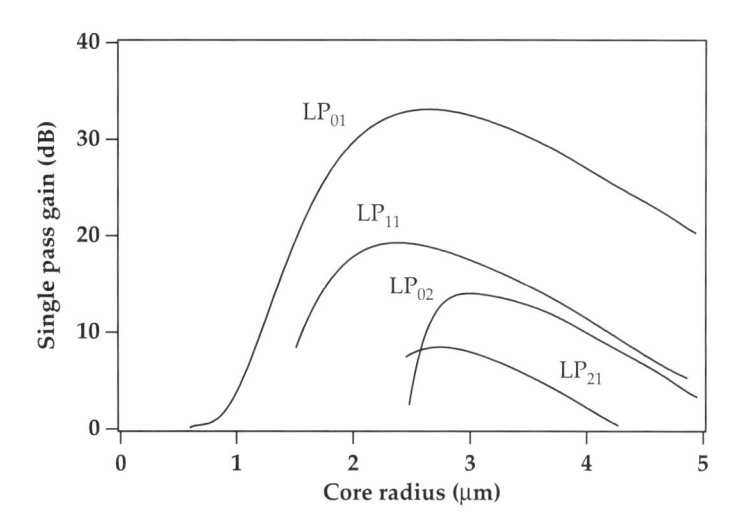

6.5. TRANSVERSE MODE MODELS — ERBIUM CONFINEMENT EFFECT

Figure 6.29: Single-pass gain in a three-level system versus core radius for different pump modes (pump wavelength 800 nm, 10 m fiber length, 50 mW pump power, NA = 0.2, signal propagating in the lowest-order mode, uniformly doped core). From reference [13].

A full calculation that involves an accurate modeling at each point (r, ϕ, z) is needed if one wishes to explore the impact on the amplifier properties of various kinds of fiber waveguide structure as well as the effects of the erbium confinement to the center of the core. In this case, it is sometimes computationally easier to then reduce the ASE signal to a single signal that corresponds to an effective bandwidth of spontaneous emission.[15] One can calculate the full model with both (r, ϕ, z) and spectral dependence of the ASE, as done, for example, by Pedersen et al for the case of a step index fiber.[1] The number of equations to solve simultaneously is then very large. The method involves calculating the light intensity and population densities at each point (r, ϕ, z), then integrating over the transverse variables (r, ϕ) to obtain the total power in each light field at the axial point z. For the ASE, the spontaneous emission contribution $2h\nu\Delta\nu$ is added to the power as a whole. The relevant gain and absorption factors for the signal mode at frequency ν propagating in mode $I^{(n)}(r, \phi)$ at a position z along the fiber are, respectively,

$$\gamma_e(z, \nu) = \sigma_e(\nu) \int_0^{2\pi}\int_0^\infty N_2(r, \phi, z) I^{(n)}(r, \phi) r \, dr d\phi \qquad (6.28)$$

$$\gamma_a(z, \nu) = \sigma_a(\nu) \int_0^{2\pi}\int_0^\infty N_1(r, \phi, z) I^{(n)}(r, \phi) r \, dr d\phi \qquad (6.29)$$

The change in power of the signal field at ν is given by

$$\frac{dP(\nu, z)}{dz} = [\gamma_e(z, \nu) - \gamma_a(z, \nu)] P(\nu, z) \qquad (6.30)$$

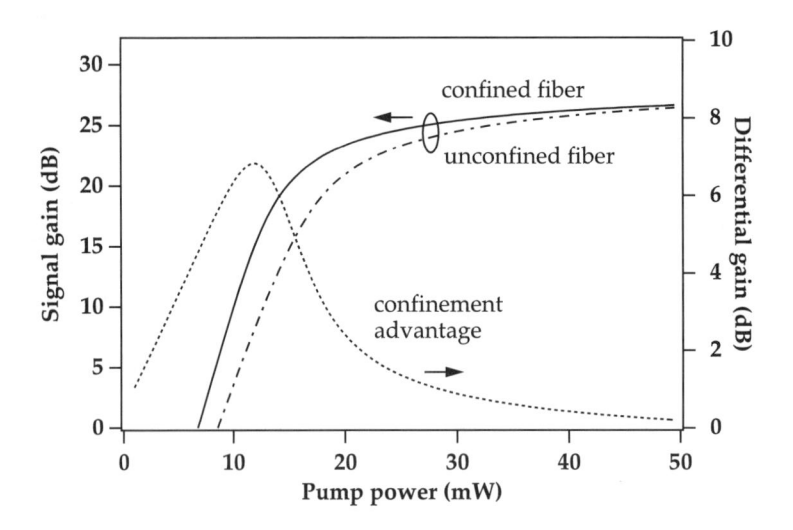

Figure 6.30: Gain, from numerical simulations, as a function of pump power at 980 nm for two erbium-doped fibers, one with a 2 μm core uniformly doped with erbium (dashed curve) and the other with the erbium doped in the center of the core with a distribution radius of 1 μm (smooth curve). The two fibers have lengths of 12 m (unconfined fiber) and 37 m (confined fiber). Also plotted is the difference between the two gain curves.

Similar equations can be written for the pump field and ASE fields (with the addition of the local spontaneous emission). In turn, the population densities in the upper and lower states, $N_2(r, \phi, z)$ and $N_1(r, \phi, z)$ are determined by the pump, signal, and ASE intensities at point (r, ϕ, z). The set of coupled equations is then solved numerically.

The importance of confining the erbium to a small cross-sectional region in the center of the fiber core area was pointed out by Armitage.[16] Intuitively, this effect can be understood as coming from the fact that population inversion will be easier to obtain in the region where the pump intensity is the strongest, i.e., the center of the core for the lowest-order pump mode. Removing the erbium from the outer portion of the core is beneficial as the pump intensity there is lower (for the lowest-order pump mode) and the ions there are not pumped efficiently.

The effect of erbium confinement is illustrated in Figure 6.30 where the gain is plotted as a function of pump power for a given length. Two fibers were used in the simulation: one with no erbium confinement (erbium uniformly distributed throughout the 2 μm core) and one with erbium confined within the core to a cylinder of radius 1 μm. The signal wavelength is 1550 nm with a launched signal power of −40 dBm. The fiber NA is 0.15. The mode field profiles, as a function of the radial distance from the center of the fiber, are plotted in Figure 6.31. The transverse mode profiles were computed from the Bessel function solution for a step-index fiber.[17, 18] The normalized transverse mode profile for the fundamental mode of a step index fiber,

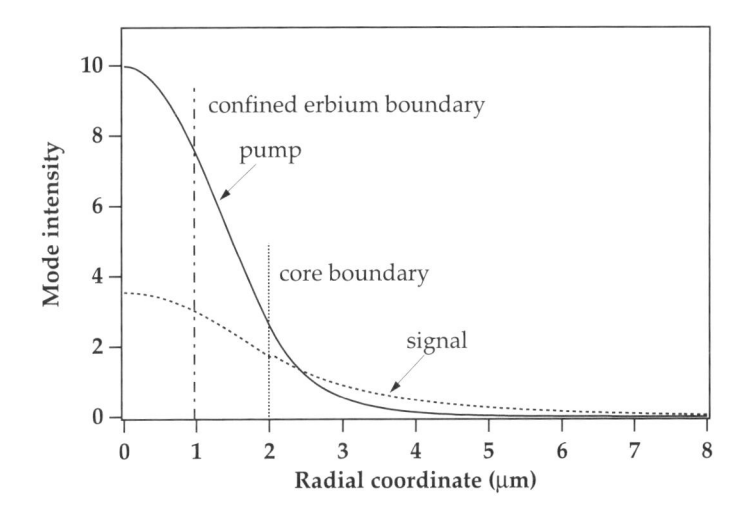

6.5. TRANSVERSE MODE MODELS — ERBIUM CONFINEMENT EFFECT

Figure 6.31: Pump and signal mode intensity profiles as a function of radial distance in μm from the fiber center, for a step-index silica fiber with a 0.15 numerical aperture and a 2 μm core radius. The core boundary is indicated, as well as the boundary for the erbium distribution when it is confined to the center 1 μm of the core.

$I^0(r)$, is calculated from the expressions

$$\begin{align} I^0(r) &= \left(\frac{v}{\pi^2 aV} \frac{J_0(ur/a)}{J_1(u)}\right)^2 \quad \text{for } r < a \\ I^0(r) &= \left(\frac{v}{\pi^2 aV} \frac{K_0(vr/a)}{K_1(v)}\right)^2 \quad \text{for } r \geq a \\ v &= 1.1428V - 0.9960 \\ u &= (V^2 - v^2)^{1/2} \end{align} \tag{6.31}$$

where $V = 2.405\lambda_c/\lambda$ is the fiber V number (λ_c is the cutoff wavelength), a is the core radius, $J_{0,1}$ and $K_{0,1}$ are the Bessel and modified Bessel functions, respectively, and the numerical expressions for u and v in equations 6.31 are valid to within 1% of the exact values for $1 \leq V \leq 3$.

When the erbium is confined to the center 1.0 μm of the core, the overlap factors between the pump and signal modes with the erbium distribution are 0.10 and 0.26, respectively. When the erbium distribution is over the entire 2 μm core, these overlap factors become 0.31 and 0.70, respectively. Part of this increased overlap comes from the lower-intensity wings of the pump and signal transverse intensity distribution.

Clearly, confining the erbium to a region with a radius one-half that of the core radius results in an increase in the small signal gain efficiency for low pump powers. The gain differential is as high as 7 dB at a pump power of 12 mW. This is important in situations where pump power is at a premium, such as is the case for remotely pumped amplifiers. For high pump powers, there is no difference in the overall gain because in the unconfined fiber the pump is strong enough to invert the erbium even in the low-

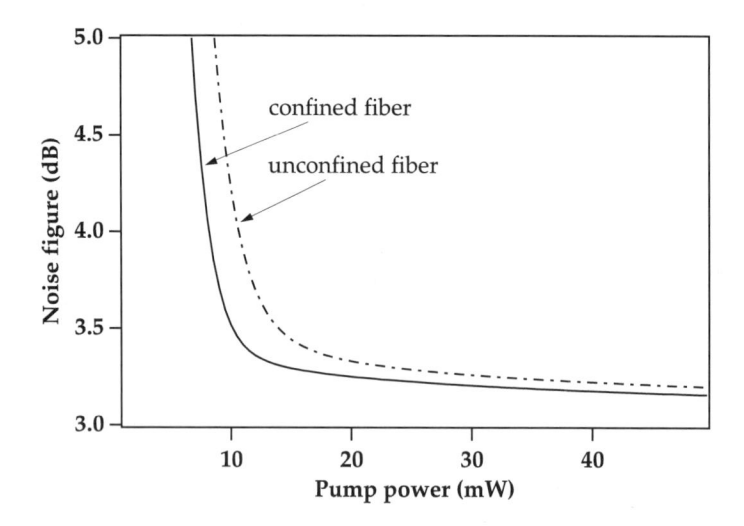

Figure 6.32: Noise figure, from numerical simulations, as a function of pump power at 980 nm for two erbium-doped fibers, one with a 2 μm core uniformly doped with erbium (dashed curve) and the other with the erbium doped in the center of the core with a distribution radius of 1 μm (smooth curve). The two fibers have lengths of 12 m (unconfined fiber) and 37 m (confined fiber).

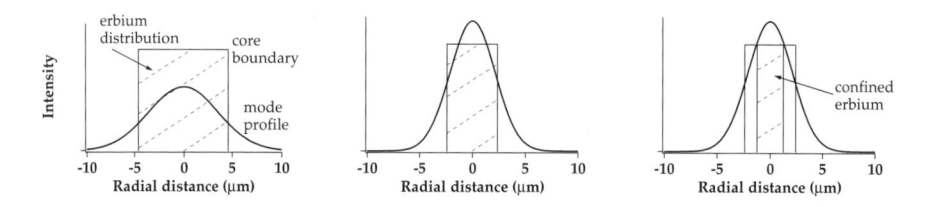

Figure 6.33: Progressive optimization of the design of an efficient erbium-doped fiber, first by reducing the pump mode diameter and increasing its intensity (center), secondly by confining the erbium to the center of the core near the peak pump intensity (right).

intensity wings of the pump transverse intensity profile. The advantage of confinement can also be seen in the noise figure. As shown in Figure 6.32, the confined fiber has a lower noise figure than the unconfined fiber. This is a direct result of the better inversion that can be reached with erbium confined to the region of highest intensity of the pump. The progressive optimization of the geometric design of an erbium-doped fiber is sketched in Figure 6.33.

Figure 6.34 shows the effect of the fiber NA and its cutoff wavelength on the maximum pump gain efficiency (defined as the maximum gain per launched pump power for a given fiber composition, where the fiber length is optimized for that particular pump power and the pump power is varied to find the maximum gain efficiency).[1] The fiber considered was uniformly doped throughout its core. One finds that the optimum cut-

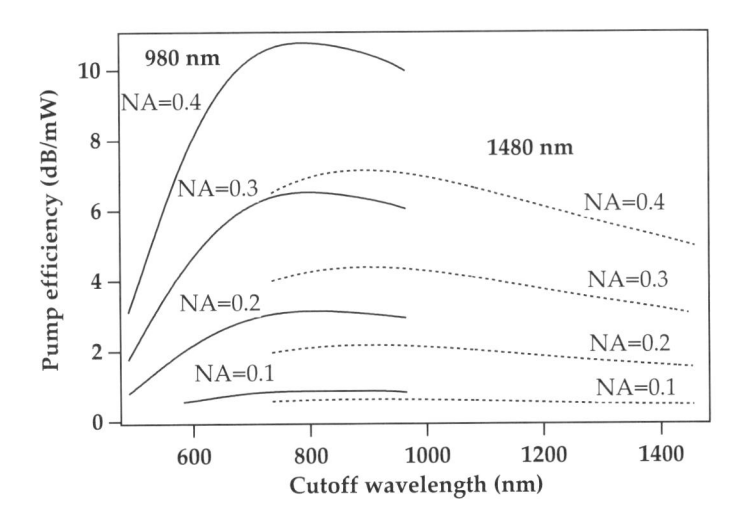

6.5. TRANSVERSE MODE MODELS – ERBIUM CONFINEMENT EFFECT

Figure 6.34: Maximum pump efficiency in dB/mW versus fiber cutoff wavelengths for a step-index fiber where the fiber NA is varied from 0.1 to 0.4 in steps of 0.1. The 980 nm and 1480 nm pumps are compared. From reference [1] (©1991 IEEE).

off wavelength is near 0.8 μm for 0.98 μm pumping and near 0.9 μm for 1.48 μm pumping. Gain efficiency increases with increasing NA, and higher gain efficiencies are obtained for 0.98 μm pumping than for 1.48 μm pumping, due to the presence of stimulated emission at 1.48 μm, which prevents complete population inversion when pumping at 1.48 μm. Similarly, one can use this model to calculate the quantum efficiency of a power amplifier (defined as the ratio of the number of output photons to the number of input photons), which measures the conversion of pump photons into signal photons. Quantum efficiencies that approach unity are predicted.[1]

Gain efficiencies, as measured by the maximum dB-to-mW ratio for a given fiber, are not always the best criterion to use in designing an erbium-doped fiber. For low pump powers, fibers with high dB/mW efficiency will give higher gain; however, at higher pump powers where the erbium ions are all essentially inverted, fibers with a "nonoptimum" design from the dB/mW perspective will give the same gain as an "optimum" fiber (i.e., one where the erbium is confined to the center of a small core region). In particular, in the design of power amplifiers, the dB/mW can be a poor criterion to use. When the erbium dopant is confined to the center of a small core, the Er^{3+}-Er^{3+} interactions start becoming important as the local concentration of Er^{3+} can be quite high (for devices of practical length). This effectively results in a "background" loss that degrades the amplifier performance.[19] The small core fibers, which will be more subject to loss than the large core fibers, drop rapidly in power conversion efficiency as the background loss increases. The large core fiber, with a lower local erbium concentration, is assumed not to be subject to the same high-concentration loss mechanisms as the small core fiber. Thus for power, amplifier applications, the larger core fiber will be more suitable than a small core fiber, which will suffer from background loss

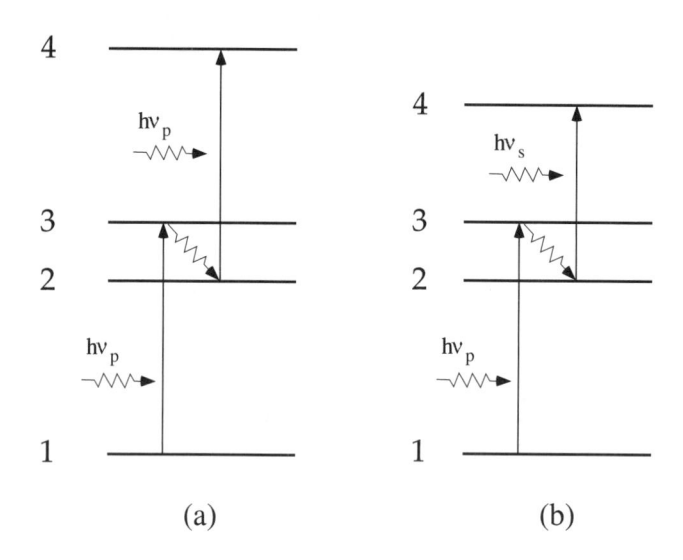

Figure 6.35: (a) Pump excited-state absorption; (b) signal excited-state absorption.

mechanisms for practical device lengths.[19] The choice of fiber types, lengths, and geometries, are discussed further in section 8.3.

6.6 EXCITED−STATE ABSORPTION EFFECTS

6.6.1 Model Equations

We have assumed so far a very simple model of the erbium ion as a three-level system or a two-level system, with the only added complication being possible stimulated emission at the pump wavelength. In fact, there are many more energy levels in the Er^{3+} ion, and the effects of pump or signal excited-state absorption (ESA) are also possible. In these effects, either a pump or signal photon, respectively, is absorbed by an erbium ion in an excited state, thereby promoting it to an even higher energy state. These processes are sketched in Figure 6.35. Since the intermediate level in which the erbium ion is excited is usually either the upper state of the amplifying transition, or, more generally, is important to efficient operation of the amplifier, ESA is deleterious to the high performance operation of an amplifier. It is a wavelength-dependent effect since it depends on matching photon energies to the transition energies between excited states of the Er^{3+} ion.

A model of amplification in the presence of pump ESA includes a population rate equation for a fourth level, the upper level for the excited-state transition. We start with the two-level system discussed in Chapter 5, which is a simplification of the three-level model (recall that level 3, the upper level for the pump absorption, is assumed to empty very rapidly into level 2). We add a fourth level with population N_4, which is the upper level for the pump excited-state absorption, as in Figure 6.35 (a):

$$\frac{dN_2}{dt} = -\Gamma_{21}N_2 + \Gamma_{42}N_4 + \left(N_1\sigma_s^a - N_2\sigma_s^e\right)\varphi_s$$

6.6. EXCITED−STATE ABSORPTION EFFECTS

$$\frac{dN_1}{dt} = \Gamma_{21}N_2 + \left(N_2\sigma_s^e - N_1\sigma_s^a\right)\varphi_s + \left(N_2\sigma_p^e - N_1\sigma_p^a\right)\varphi_p + \left(N_1\sigma_p^a - N_2\sigma_p^e\right)\varphi_p - N_2\,\varphi_p\sigma_{esa}$$

$$\frac{dN_4}{dt} = -\Gamma_{42}N_4 + N_2\,\varphi_p\sigma_{esa} \qquad (6.32)$$

where, for simplicity, we assumed that level 4 relaxes back down only to level 2. When there are other energy levels in between levels 4 and 2, then, strictly speaking, the nonradiative relaxation is via these levels. The excited-state absorption is accounted for by the term $N_2\varphi_p\sigma_{esa}$, where $\sigma_{esa} = \sigma_{24}$ is the excited-state absorption cross section. There can also be, in principle, direct transitions from level 4 to level 1. In general, this is a fluorescent transition and is responsible, for example, for the green emission observable from erbium-doped fibers when ESA is present. The green fluorescence phenomenon is especially visible with a pump source at 800 nm due to transitions from the $^4I_{13/2}$ level to the $^2H_{11/2}$ and $^2S_{3/2}$ levels.

To solve the system of equations 6.32, we consider a steady-state situation where the rates of change of the populations are set to zero and the rate equation system can be solved for the populations in terms of the field photon flux densities. The propagation equation for the signal photon flux density remains the same as that in Chapter 5, but the pump propagation equation needs to be amended to include the pump absorption transition from level 2 to level 4:

$$\frac{d\varphi_p}{dz} = (-N_1\sigma_p^a + \sigma_p^e N_2)\varphi_p - \sigma_{esa}N_2\varphi_p \qquad (6.33)$$

The system of equations thus set up can be quite complex to solve, but such systems have been solved numerically in the case of the Er^{3+} system.[20] An attractive simplification arises from the realization that in most practical cases level 4 will deactivate rapidly to the levels below, in particular to level 2 (the metastable level), and one can then write that $N_4 \simeq 0$. The rate equations then become identical to the ones set up beforehand in Chapters 5 and 6, and the only remaining difference, compared to the treatment of signal amplification without ESA, is contained in the pump propagation equation.[14, 16, 21] What this means, essentially, is that pump photons that are absorbed by ions in level 2 are assumed to be "immediately" recycled back into level 2 by means of fast nonradiative transitions with emission of phonons. This effect is deleterious to amplification, however, because a pump photon that is absorbed by an ion in level 2 is no longer available to excite an ion from level 1 to level 2. As the excited-state cross section σ_{esa} grows in magnitude, it becomes increasingly more difficult to obtain a high population inversion on the 1 to 2 transition.

In the case of Er^{3+}, the only low-lying pump band that is impacted in a very significant way by ESA effects is the 800 nm pump band. Weak ESA has been observed when pumping at 980 nm and is believed to arise from direct ESA from the $^4I_{11/2}$ pumping level.[22, 24] This does not seem to have significantly affected the excellent performance reported for erbium-doped fiber preamplifiers and inline amplifiers pumped at 980 nm. However, it has negatively impacted very high power amplifiers pumped at 980 nm.[23]

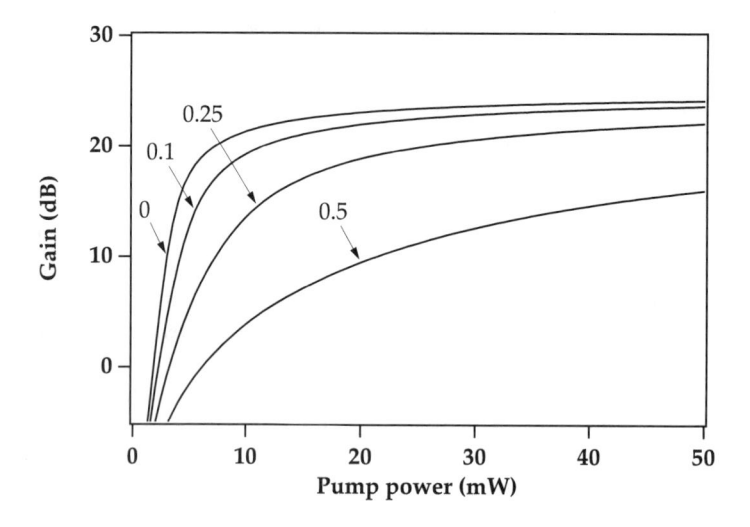

Figure 6.36: Variation of the gain at 1550 nm, from numerical simulations, as a function of pump power for various values of the pump ESA to ground-state absorption cross section ratio σ_{esa}/σ_a (values indicated adjoining the curves). The fiber was a 8 m length of fiber A, and the pump absorption cross section is 2.7×10^{-25} cm^2. The pump emission cross section is taken to be zero.

6.6.2 Modeling Results in the Presence of ESA

In the presence of ESA, the gain obtained is necessarily less than if ESA were absent since some of the pump photons are wasted.[14, 16] Figure 6.36 illustrates this effect, showing that as σ_{esa} increases the small signal gain efficiency decreases dramatically. For the purpose of the simulation, we used the situation of the 980 nm pump and turned on a "fictitious" ESA cross section where the ratio of the pump ESA cross section to pump absorption cross section ranges from 0.1 to 0.5. The noise figure is also affected by the difficulty for the pump to be contributing only to population inversion, in the presence of pump ESA, as shown in Figure 6.37.

The choice of the pumping configuration becomes more important in the presence of ESA.[21] Since a high inversion will result in strong pump ESA and waste of pump photons, it is preferable to mitigate this effect by concentrating the pump where the signal is strongest (i.e., counterpropagating pump for moderate signal inputs). An alternative which is valid both in small signal and moderate signal regimes is to create a more uniform inversion with bidirectional pumping.[25, 26, 27] Simulations of the ESA effect has been explored in some detail for the case of 800 nm pumping where strong pump ESA is present in erbium-doped fibers.

6.6.3 800 nm Band Pumping

Modeling of the 800 nm pumping of erbium-doped fiber amplifiers has been reported. In particular, the choice of an optimum pumping scheme, and the effect of the host glass, have been studied in detail.[25, 26, 27, 28] In the early days of study on such

6.6. EXCITED–STATE ABSORPTION EFFECTS

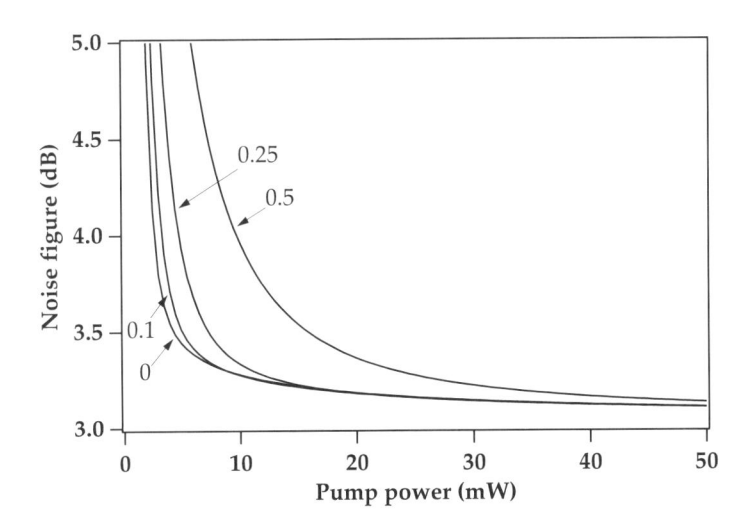

Figure 6.37: Variation of the noise figure at 1550 nm, from numerical simulations, as a function of pump power for various values of the pump ESA to ground-state absorption cross section ratio σ_{esa}/σ_A (values indicated adjoining the curves). The fiber was a 8 m length of fiber A, and the pump absorption cross section is 2.7×10^{-25} cm^2. The pump emission cross section is taken to be zero.

amplifiers, 800 nm pumping of erbium-doped fiber amplifiers was considered to be an attractive option due to the availability of laser diodes at 800 nm. The fact that erbium has a relatively low absorption cross section at 800 nm, however, results in large pump powers being required to invert the erbium ion population when pumping at 800 nm, as compared to 980 nm or 1.48 μm pumping. Figure 6.38 shows the absorption cross section determined for the alumino-germano-silica fiber A, from attenuation measurements. The presence of ESA at 800 nm renders the situation even less attractive. Nevertheless, significant gain has been reported in experiments on 800 nm pumping of erbium-doped fibers.[29, 30, 31] A measurement of gain versus pump power, obtained for an Al co-doped erbium-doped fiber is shown in Figure 6.39.[30] The pump powers required for significant gain are roughly an order of magnitude higher than those needed when pumping at either 980 nm or 1.48 μm, which is one of the main drawbacks of 800 nm pumping and has contributed to the lack of recent interest in that pump band.

The situation regarding ESA with 800 nm pumping is particularly negative in the case of standard Al-Ge co-doped erbium-doped fiber amplifiers. Figure 6.40 shows the ground-state absorption and excited-state cross section on the same scale for an Al-P fiber that has spectral properties very similar to Al-Ge silica fibers.[27] The ESA cross sections are higher than the ground state cross section over most of the 800 nm pump band. From the cross section curves, it can be seen that the most advantageous pump region lies in the vicinity of 820 nm, where the ground-state and excited-state cross sections are roughly equal. This is removed from the peak of the 800 nm band absorption, yet due to the lower ESA results in an improvement of amplifier performance. The op-

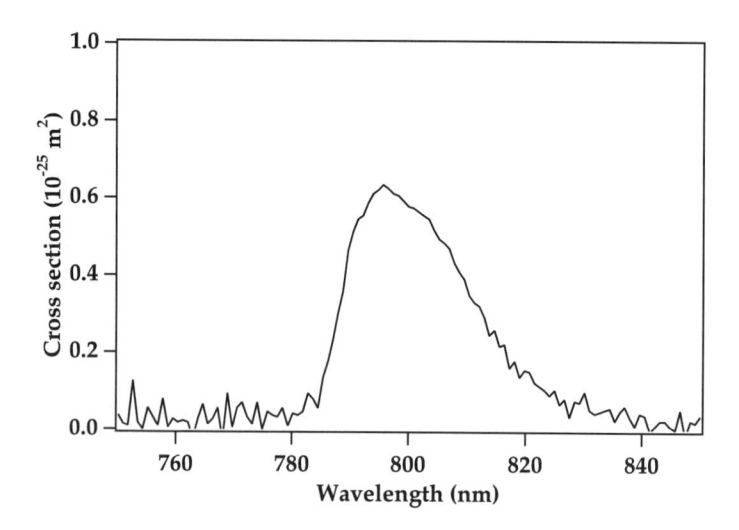

Figure 6.38: Spectral variation of the ground-state absorption cross section near 800 nm, in the alumino-germano-silica fiber A discussed in the text. Absorption measurements courtesy of D. DiGiovanni, Lucent Technologies, Murray Hill, NJ.

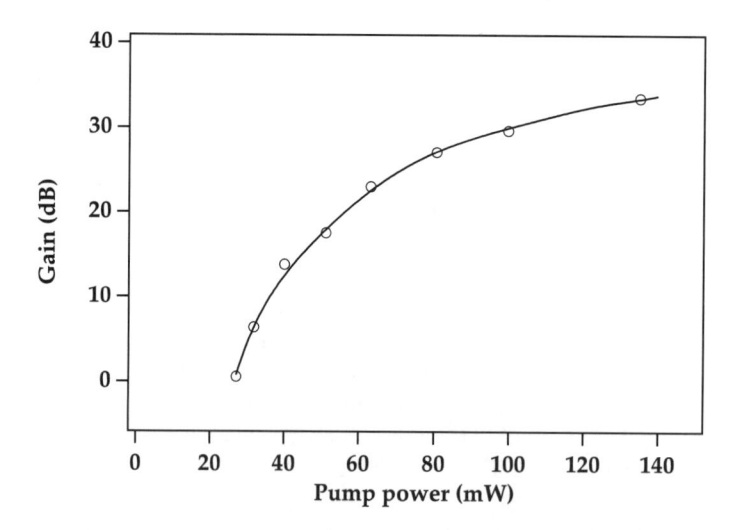

Figure 6.39: Small signal gain at 1535 nm versus pump power for an erbium-doped fiber amplifier pumped at 822 nm (input signal level −41 dBm). From reference [30].

timum pump wavelength will depend on the host glass composition. Figure 6.41 shows the gain obtained in an Al-Ge-P co-doped erbium-doped fiber amplifier as a function of pump wavelength, for a fixed amount of pump power.[27] The optimum pump wavelength shifts with pump power. This results from the fact that ESA is more of a factor

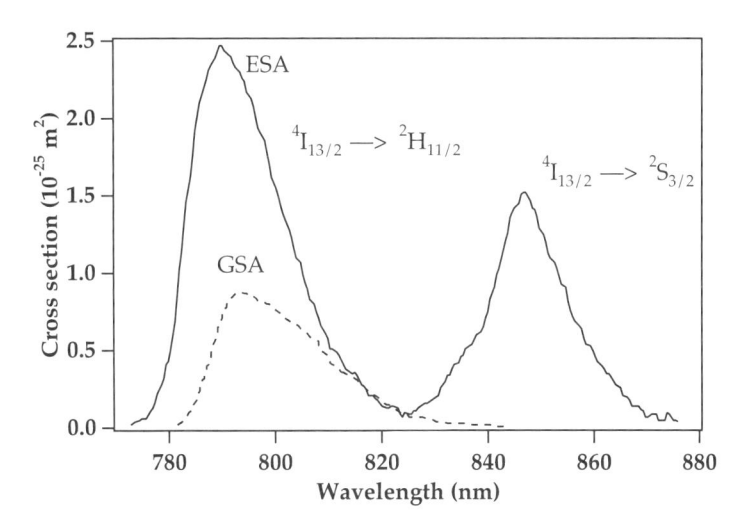

Figure 6.40: Ground-state (dashed line) and excited-state (solid line) absorption cross sections versus wavelength for an Al-P co-doped erbium-doped fiber. From reference [28] (©1991 IEEE).

in long fibers. Thus the optimum wavelength is pushed longer in wavelength as the pump power, and hence optimum fiber length, is increased.[27]

It has been found that compositions such as fluorophosphate-based hosts significantly reduce the ESA. This is illustrated in the ground-state absorption to excited-state absorption cross section ratio of nearly 5 for a fluorophosphate host (see Figure 6.42), whereas this ratio is near unity for an Al-P silica fiber (at the peak wavelength for this ratio).[27] The fluorophosphate hosts have thus been shown theoretically to yield improvements by giving maximum small signal gain efficiencies of between 3 dB/mW and 4 dB/mW compared to efficiencies of around 2 dB/mW for Al-P silica fibers (for bidirectional pumping of high NA fibers).[27] In terms of power amplifier applications, the fluorophosphate host would roughly double the output signal power (as compared to Al-P silica fibers), for a given amount of pump power.[26, 27]

6.7 Er^{3+} - Er^{3+} INTERACTION EFFECTS

The vast majority of erbium-doped fiber amplifiers use fiber lengths that are in the 1 m to 100 m range. It is known that very low concentration erbium-doped fibers are more efficient in terms of overall amplifier performance.[32] It has been found that with short lengths, where the concentration of erbium is high so as to obtain adequate gain, undesirable effects occur that reduce the amplifier performance.[33] These effects have been tied to Er^{3+}-Er^{3+} interaction effects. We previously discussed some aspects of the high concentration effects in Chapter 4, section 4.6. Such effects may also occur with amplifiers of longer lengths where the erbium ions are confined to the center of the core region. The erbium ion density will be locally high in such fibers, and it is the local concentration of Er^{3+} ions that is important in the context of the Er^{3+}-

192   CHAPTER 6. MODELING AND COMPLEX EFFECTS

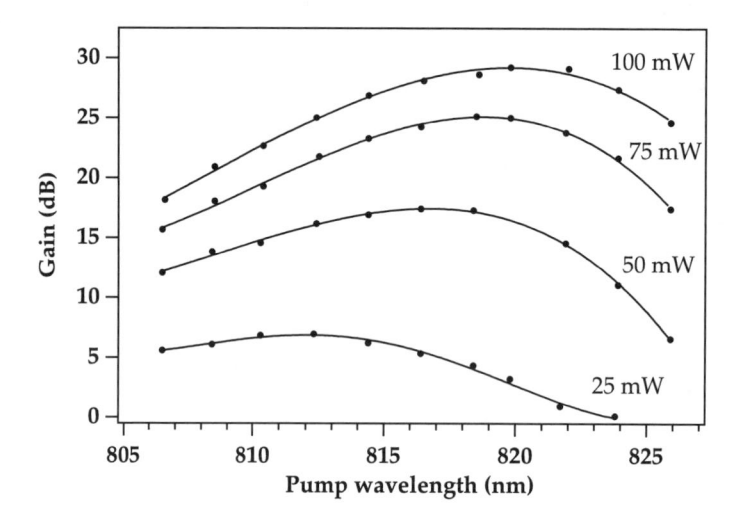

Figure 6.41: Measured small signal gain versus pump wavelength for different pump powers for an erbium-doped Al-Ge-P silica fiber. Each point is for the fiber length that gave the highest gain. From reference [27] (©1992 IEEE).

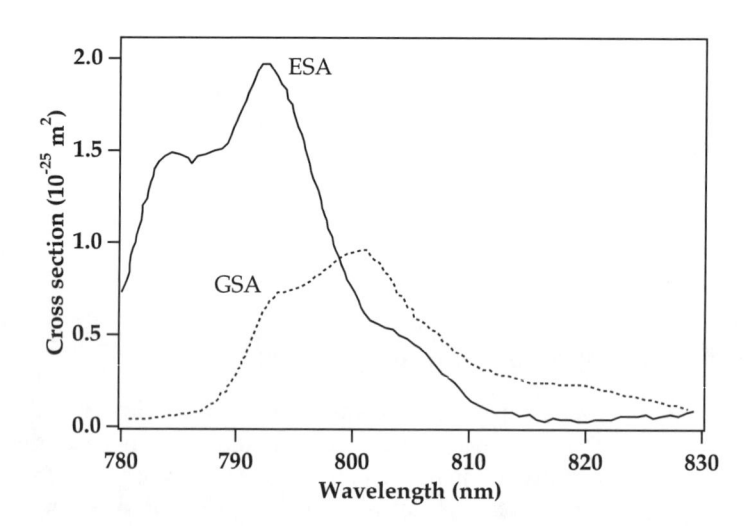

Figure 6.42: Ground-state (dashed line) and excited-state (solid line) absorption cross sections versus wavelength for a low-fluorine fluorophosphate erbium-doped fiber. From reference [27] (©1992 IEEE).

Er^{3+} interaction. We discuss in this section the two mechanisms that are responsible for the decrease in gain from erbium ion interaction effects: upconversion and pair-induced quenching. Both of these effects can be modeled and their effect on amplifier performance computed.

6.7.1 Upconversion Effects on Amplifier Performance

We consider first the simplest picture of upconversion between erbium ions: the case where the erbium density is locally uniform and the ions are independent. The independence is in the sense that if one ion is excited to the $^4I_{13/2}$ state this does not prevent a neighboring ion from also being excited to the $^4I_{13/2}$ state. Upconversion between two Er^{3+} ions will depend on whether both ions are excited to the $^4I_{13/2}$ state.

We consider the two-level model of the Er^{3+} amplifying transition where the $^4I_{13/2}$ state is level 2 and the ground state is level 1, as discussed in Chapter 4. The rate of decay of the fractional population in level 2, for a given ion has a component proportional to the probability that it occupies level 2, multiplied by the probability that a nearby ion is also in level 2. As a result, the rate of decay of the population density in level 2 will be proportional to the square of the population in level 2, $N_2^2(r, \varphi, z)$.

This model was first elaborated by Blixt et al.[34] We make the simplification that the upconversion process results in one ion being deactivated to the ground state while the other is excited to the $^4I_{9/2}$ state, from whence it rapidly cascades back down to the $^4I_{13/2}$ level. We can then write the new equations for the population dynamics in the Er^{3+} two-level model as

$$\frac{dN_2}{dt} = -\frac{N_2}{\tau} + \left(\sigma_p^{(a)}N_1 - \sigma_p^{(e)}N_2\right)\varphi_p + \left(\sigma_s^{(a)}N_1 - \sigma_s^{(e)}N_2\right)\varphi_s - CN_2^2 \quad (6.34)$$

and

$$\frac{dN_2}{dt} = -\frac{dN_1}{dt} \quad (6.35)$$

where φ_s and φ_p are the signal and pump photon fluxes, respectively. Here τ is the lifetime of level 2, both pump and signal absorption and emission have been accounted for, and a new term, CN_2^2, has been added to account for loss of population through upconversion, with C being the upconversion coefficient. The upconversion lifetime τ_{up} is defined as $\tau_{up} = 1/NC$, where N is the concentration of erbium ions.

Blixt et al. studied erbium-doped fibers pumped at 1.48 μm.[34] They observed that the emission at 980 nm is quadratic in the pump power, at low pump powers, as expected for a N_2^2 process. One also expects that as the $^4I_{13/2}$ level gets saturated (fully occupied), the upconversion should no longer grow quadratically with the pump power but merely linearly, or with another power dependence if other nonlinear effects come into play, such as multiphoton absorption. Blixt et al. also observed that the maximum amplification from a given length of fiber decreased significantly as the upconversion increased (as measured from the 980 nm luminescence) for different concentration fibers. The upconversion coefficient was calculated to be roughly 10^{-22} m^3/s, and roughly 1 out of 10,000 ions excited to the $^4I_{11/2}$ level decay by emitting a 980 nm photon, as opposed to decaying to the $^4I_{13/2}$ level (most probably by a nonradiative route). These values were determined for Ge-Al-P erbium-doped fibers.[34]

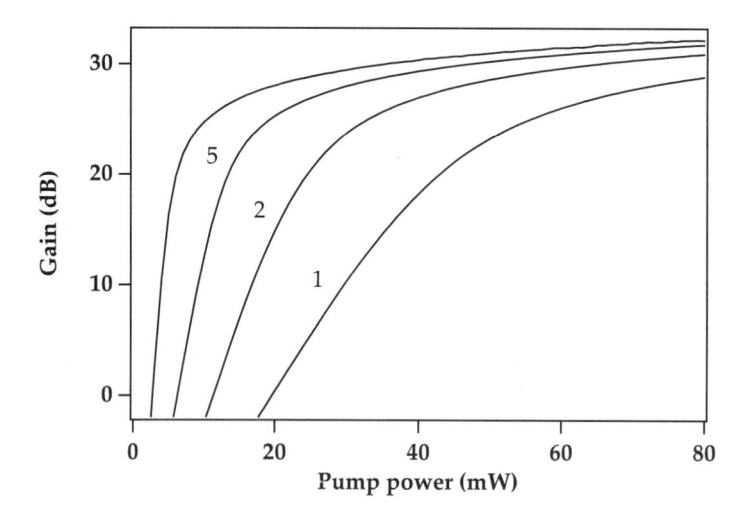

Figure 6.43: Gain at 1550 nm, from numerical simulations, in a 12 m length of erbium-doped fiber A pumped at 980 nm, contrasting fibers where the upconversion lifetime is infinity (no upconversion), and where the upconversion lifetime is 5, 2, and 1 ms. The radiative lifetime of the $^4I_{13/2}$ level is 10 ms.

The effect of upconversion is to decrease the pumping efficiency of erbium-doped fiber amplifiers. Numerical simulations of the amplifier, where the CN_2^2 term is taken into account in the population equation, readily establish the increase in pump needed to achieve a certain gain, as the upconversion rate increases. Figure 6.43 shows the effect on gain resulting from the presence of upconversion.

An experiment involving pulsed laser excitation, with 100 fs pulses at 1.5 μm, has been performed by Thøgersen et al.[35] With such short pulses it is possible to measure the lifetimes of the excited states (which are in the 0.1 to 10 μs range), as well as separate the effect of energy-transfer-mediated upconversion between neighboring ions from stepwise multiphoton absorption by a single ion. It was found that in low-concentration erbium-doped silica fibers, multiphoton absorption by a single ion is the main mechanism responsible for upconversion and the resulting visible fluorescence observed from the fiber. Since the pulses used have very high peak intensities (20 GW/cm^2), some of the multiphoton processes (two- and three-photon) observed in this experiment are not expected to occur with the same strength in cw-pumped fibers where the intensities are much lower. In-high concentration fibers, this experiment confirmed again that energy-transfer-mediated upconversion between neighboring ions is responsible for the promotion of excited ions to states higher than the $^4I_{13/2}$ metastable level.[35]

It should be noted that energy transfer or upconversion does not necessarily require a precise matching of the photon and transition energies involved. Auzel has demonstrated that multiphonon-assisted effects are possible such that photons of lower energy than those needed to bridge the gap between two states of the Er^{3+} ion can effectively

6.7.2 Pair Induced Quenching

Later studies, in particular at high pump powers, have found that the N_2^2 upconversion process as just outlined does not account fully for experimentally observed lifetimes and pump attenuation behaviors.[38] The key experimental observation is that even at high pump powers (with a 980 nm pump) there remains an unsaturable absorption at the pump wavelength. This has led to the concept of *pair-induced quenching*. In this model, originally proposed by Delevaque et al, two ions are considered to be so closely coupled that they can "never" both be simultaneously excited to the $^4I_{13/2}$ level.[38] In practice, this means that the energy transfer rate between the two ions is on a time scale significantly faster than that of the pumping rate, so that at the pump powers considered the pump is unable to keep both ions excited. The pair energy transfer rate has been estimated to be in the range of 5 μs to 50 μs.[38] Presumably, this pair would be a likely candidate for the cooperative luminescence process discussed in Chapter 4. The difference between the pair model and the previously described N^2 model is that in the latter the ions are assumed to be independent in the sense that if a given ion is excited to the $^4I_{13/2}$ level then this does not prohibit one of its neighbors from also being excited to the $^4I_{13/2}$ level, leading to a N^2 dependence. In the pair model, the ions are no longer considered to be independent, and, when one of the ions of the pair is in the $^4I_{13/2}$ level, this virtually assures that the other ion must be in its ground state. Following Delevaque et al, we consider that the erbium population can be broken up into two parts: one made up of isolated ions and the other made up of pairs.[38] The pump is considered to be at 980 nm. $N_1^{(s)}$ and $N_2^{(s)}$ refer to the populations in levels 1 and 2 for single ions, respectively, and $N_1^{(p)}$ and $N_1^{(p)}$ refer to the populations of the pair states in pair state 1 (where neither ion is excited) and pair state 2 (where one ion of the pair is excited), respectively. Pair state 3, where both ions are excited to the $^4I_{13/2}$ level, is assumed to have negligible population.

Studying the effect only of the pump absorption on the combined isolated ions and ion pair populations, we can then write

$$\frac{dN_2^{(s)}}{dt} = -\frac{N_2^{(s)}}{\tau} + \sigma_p^{(a)} \varphi_p N_1^{(s)} - C(N_2^{(s)})^2 \tag{6.36}$$

$$\frac{dN_2^{(p)}}{dt} = -\frac{N_2^{(p)}}{\tau} + 2\sigma_p^{(a)} \varphi_p N_1^{(p)} \tag{6.37}$$

where C is the upconversion coefficient. We also write

$$\frac{dN_2^{(s)}}{dt} = -\frac{dN_1^{(s)}}{dt} \tag{6.38}$$

$$\frac{dN_2^{(p)}}{dt} = -\frac{dN_1^{(p)}}{dt} \tag{6.39}$$

where C is the upconversion coefficient. The factor of 2 appears in the first term on the right-hand side of equation 6.37 due to the fact that the probability of an ion pair in its ground state absorbing a pump photon is twice that of a single ion since a ground state ion pair is made up of two ground state single ions. With 2k as the fraction of ions in pairs, the populations can be decomposed as

$$N_1^{(s)} + N_2^{(s)} = (1 - 2k) N \qquad (6.40)$$

$$N_1^{(p)} + N_2^{(p)} = kN \qquad (6.41)$$

The pump power absorption rate predicted from this model is

$$\frac{dP(z)}{dz} = -\sigma_p^{(a)} \left(N_1^{(s)} + 2N_1^{(p)} + N_2^{(p)} \right) P(z) \qquad (6.42)$$

In the pump absorption in the high pump power regime, it is the pair absorption that is responsible for the unsaturable part of the pump absorption, once the isolated ions absorption has been saturated. This model was used by Delevaque et al in a study of six different Ge-Al-P erbium-doped fibers with varying concentrations of Er^{3+} from 350 to 2000 wt %. It was found from an investigation of the 980 nm pump transmission that the best fits were obtained with C = 0 and k varying between 1 and 20% and increasing with erbium concentration.[38] The incorporation of Al tends to reduce the pair fraction k.

Further evidence for the validity of the pairing model comes from the study of the green fluorescence from erbium-doped fibers pumped at 980 nm.[39] The green fluorescence originates from the $^4S_{3/2}$ level and is caused by excited state absorption from the short-lived $^4I_{11/2}$ level (10 μs lifetime), which absorbs the first 980 nm photon. Since paired ions always provide an ion in the ground state, the paired ion model predicts that the green fluorescence should always have a quadratic dependence on the 980 nm pump power, as observed, at pump powers well beyond the level at which the single ion model saturates.[39] Green fluorescence has also been observed for a high concentration erbium-doped fiber (1250 wt ppm) pumped at 1.48 μm and has been attributed there to a multistep process.[40]

The effect of pair-induced quenching on erbium-doped fiber amplifier performance has been studied by Nilsson et al and Wysocki et al As an example, Nilsson et al have calculated a gain reduction from 30 to 15 dB for an increase in pair fraction from 0 to 20%, in a 15 m long fiber pumped with 17 mW of launched pump power at 980 nm. The gain can be recovered by pumping harder, i.e., for the 15 m fiber length with 20% pairs, a pump power of roughly 30 mW is needed to recover the 30 dB gain.[41] The noise figure suffers also, since the maximum inversion (and thus n_{sp}) is reduced by the inescapable presence of ions in their ground state. Wysocki et al, in a study of superfluorescent fiber lasers, have shown that pump and signal ESA, or an N^2 upconversion process, are unable to account for the observed degradation in fiber laser slope efficiency with increasing erbium ion concentration.[42] The pair-induced quenching model, however, is consistent with this observation. Wysocki et al. determined that in an Al co-doped erbium-doped fiber with an Er^{3+} concentration of 5×10^{19} ions per cm^3 the pair fraction is roughly 20%.[42] Figure 6.44 shows the gain reduction with the increase in pair fraction, using a complete modeling of an erbium-doped fiber amplifier pumped at 979 nm.[42]

6.7. ER^{3+} - ER^{3+} INTERACTION EFFECTS

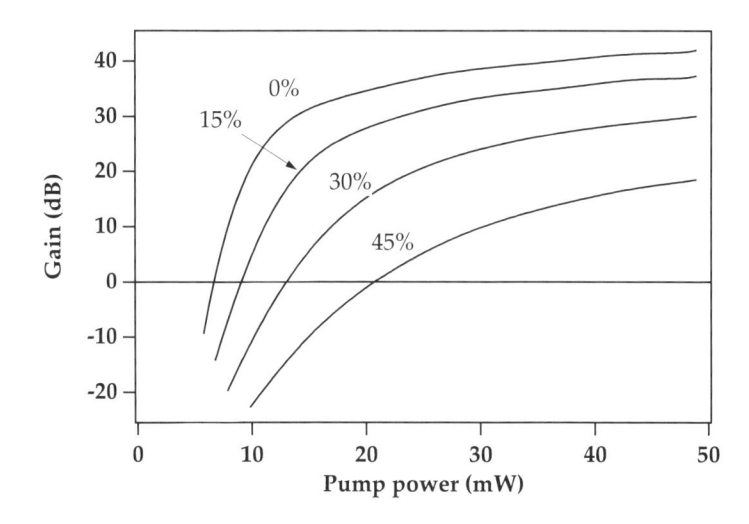

Figure 6.44: Amplifier gain at 1529 nm for a 1 m long erbium-doped fiber amplifier pumped at 979 nm versus pump power, as a function of the erbium ion pair percentage (indicated next to each curve). From reference [42].

Various studies have sought to elucidate the relative importance of upconversion and pair induced quenching, depending on the concentration of Er^{3+}. Some investigations suggest that pair induced quenching is responsible for performance degradation in low concentration Er^{3+} fibers while upconversion is mainly reponsible for that in high concentration Er^{3+} fibers.[43, 44] Other have found a more dominant role for pair induced quenching at high concentrations.[45] Another result of these studies is that in the presence of pair induced quenching, a counterpropagating pump gives a higher quantum efficiency (3 to 10 % more) than a copropagating pump, in the strong signal regime.[45] The reason for this is the same as in the case of ESA: it is preferable for the region of high pump intensity to overlap with that of high signal intensity, so that the signal may optimally capture pump photons at the expense of dissipating mechanisms. One recommendation from a study of pair induced quenching is that in Ge-Al co-doped erbium-doped fibers, the erbium concentration be less than 900 mole ppm in Al co-doped fibers and less than 90 mole ppm in Ge co-doped fibers.[45]

In conclusion, there are a multitude of complex Er^{3+}-Er^{3+} interaction effects that depend on concentration, host composition, and pump wavelength, and all act so as to degrade amplifier performance. In practice, it is better to use as low a concentration of Er^{3+} as is practical for the host composition used and the desired application.

Bibliography

[1] B. Pedersen, A. Bjarklev, J. H. Povlsen, K. Dybdal, and C. C. Larsen, *J. Light. Tech.* **9**, 1105 (1991).

[2] C. R. Giles and E. Desurvire, *J. Light. Tech.* **9**, 271 (1991).

[3] S. Yamashita and T. Okoshi, *Elect. Lett.* **28**, 1323 (1992).

[4] E. Desurvire, J. W. Sulhoff, J. L. Zyskind, and J. R. Simpson, *IEEE Phot. Tech. Lett.* **2**, 653 (1990).

[5] J. H. Povlsen, A. Bjarklev, O. Lumholt, H. Vendeltorp-Pommer, K. Rottwitt, and T. Rasmussen, "Optimizing gain and noise performance of EDFA's with insertion of a filter or an isolator," in *Fiber Laser Sources and Amplifiers*, M. J. F. Digonnet and E. Snitzer, Ed., *Proc. SPIE* **1581**, pp. 107–113 (1991).

[6] A. Lidgard, J. R. Simpson, and P. C. Becker, *Appl. Phys. Lett.* **56**, 2607 (1990).

[7] B. Pedersen, M. L. Dakss, B. A. Thompson, W. J. Miniscalco, T. Wei, and L. J. Andrews, *IEEE Phot. Tech. Lett.* **3**, 1085 (1991).

[8] R. G. Smart, J. L. Zyskind, J. W. Sulhoff, and D. J. DiGiovanni, *IEEE Phot. Tech. Lett.* **4**, 1261 (1992).

[9] C. R. Giles, C. A. Burrus, D. J. DiGiovanni, N.K. Dutta, and G. Raybon, *IEEE Phot. Tech. Lett.* **3**, 363 (1991).

[10] K. Kubodera and K. Otsuka, *J. Appl. Phys.* **50**, 653 (1979).

[11] K. Kubodera and K. Otsuka, *J. Appl. Phys.* **50**, 6707 (1979).

[12] M. J. F. Digonnet and C. J. Gaeta, *Appl. Opt.* **24**, 333 (1985).

[13] M. J. F. Digonnet, "Theory of operation of three- and four-level fiber amplifiers and sources," in *Fiber Laser Sources and Amplifiers*, M. J. F. Digonnet, Ed., *Proc. SPIE* **1171**, pp. 8–26 (1990).

[14] M. J. F. Digonnet, *IEEE. J. Quant. Elect.* **26**, 1788 (1990).

[15] E. Desurvire, J. L. Zyskind, and C. R. Giles, *J. Light. Tech.* **8**, 1730 (1990).

[16] J. R. Armitage, *Appl. Opt.* **27**, 4831 (1988).

[17] D. Gloge, *Appl. Opt.* **10**, 2252 (1971).

[18] L. B. Jeunhomme, *Single-Mode Fiber Optics* (Marcel Dekker, NY, 1990).

[19] D. DiGiovanni, P. Wysocki, and J. R. Simpson, *Laser Focus World*, pp. 95–106 (June, 1993).

[20] P. F. Wysocki, R. F. Kalman, M. J. F. Digonnet, and B. Y. Kim, "A comparison of 1.48 μm and 980 nm pumping for Er-doped superfluorescent fiber sources," in *Fiber Laser Sources and Amplifiers III*, M. J. F. Digonnet and E. Snitzer, Ed., *Proc. SPIE* **1581**, pp. 40–58 (1992).

[21] P. R. Morkel and R. I. Laming, *Opt. Lett.* **14**, 1062 (1989).

[22] P. A. Krug, M. G. Sceats, G. R. Atkins, S. C. Guy, and S. B. Poole, *Opt. Lett.* **16**, 1976 (1991).

[23] J. C. Livas, S. R. Chinn, E. S. Kintzer, and D. J. DiGiovanni, "High power erbium-doped fiber amplifier pumped at 980 nm," in *Conference on Lasers and Electro Optics*, Vol. 15, 1995 OSA Technical Digest Series (Optical Society of America, Washington, D.C., 1995), pp. 521–522.

[24] A. Lidgard, Doctoral Thesis, Royal Institute of Technology, Stockholm, Sweden, TRITA FYS2106, 1991.

[25] B. Pedersen, A. Bjarklev, H. Vendeltorp-Pommer, and J. H. Povlsen, *Opt. Comm.* **81**, 23 (1991).

[26] S. P. Bastien and H. R. D. Sunak, *IEEE Phot. Tech. Lett.* **3**, 1088 (1991).

[27] B. Pedersen, W. J. Miniscalco, and S. A. Zemon, *J. Light. Tech.* **10**, 1041 (1992).

[28] S. A. Zemon, B. Pedersen, G. Lambert, W. J. Miniscalco, L. J. Andrews, R. W. Davies, and T. Wei, *IEEE Phot. Tech. Lett.* **3**, 621 (1991).

[29] A. Lidgard, D. J. DiGiovanni, and P. C. Becker, *IEEE J. Quant. Elect.* **28**, 43 (1992).

[30] M. Nakazawa, Y. Kimura, and K. Suzuki, *Elect. Lett.* **26**, 548 (1990).

[31] M. Horiguchi, M. Shimizu, M. Yamada, K. Yoshino, and H. Hanafusa, *Elect. Lett.* **26**, 1758 (1990).

[32] N. Kagi, A. Oyobe, and K. Nakamura, *IEEE Phot. Tech. Lett.* **2**, 559 (1990).

[33] M. Shimizu, M. Yamada, M. Horiguchi, and E. Sugita, *IEEE Phot. Tech. Lett.* **2**, 43 (1990).

[34] P. Blixt, J. Nilsson, T. Carlnas, and B. Jaskorzynska, *IEEE Phot. Tech. Lett.* **3**, 996 (1991).

[35] J. Thøgersen, N. Bjerre, and J. Mark, *Opt. Lett.* **18**, 197 (1993).

[36] F. Auzel, *Phys. Rev. B* **13**, 2809 (1973).

[37] F. Auzel, *Electron. Lett.* **29**, 337 (1993).

[38] E. Delevaque, T. Georges, M. Monerie, P. Lamouler, and J. F. Bayon, *IEEE Phot. Tech. Lett.* **5**, 73 (1993).

[39] R. S. Quimby, "Upconversion and 980-nm excited-state absorption in erbium-doped glass," in *Fiber Laser Sources and Amplifiers IV*, M. J. F. Digonnet and E. Snitzer, Ed., *Proc. SPIE* **1789**, pp. 50–57 (1993).

[40] S. Arahira, K. Watanabe, K. Shinozaki, and Y. Ogawa, *Opt. Lett.* **17**, 1679 (1992).

[41] J. Nilsson, B. Jaskorzynska, and P. Blixt, "Implications of pair induced quenching for erbium-doped fiber amplifiers," in *Optical Amplifiers and Their Applications*, Vol. 14, 1993 OSA Technical Digest Series, (Optical Society of America, Washington, D.C., 1993), pp. 222–225.

[42] P. F. Wysocki, J. L. Wagener, M. J. F. Digonnet, and H. J. Shaw, "Evidence and modeling of paired ions and other loss mechanisms in erbium-doped silica fibers," in *Fiber Laser Sources and Amplifiers IV*, M. J. F. Digonnet and E. Snitzer, Ed., *Proc. SPIE* **1789**, 66 (1993).

[43] M. P. Hehlen, N. J. Cockcroft, T. R. Gosnell, A. J. Bruce, G. Nykolak, and J. Shmulovich, *Opt. Lett.* **22**, 772 (1997).

[44] G. N. van den Hoven, E. Snoeks, A. Polman, C. van Dam, J. W. M. van Uffelen, and M. K. Smit, *J. App. Phys.* **79**, 1258 (1996).

[45] P. Myslinski, D. Nguyen, and J. Chrostowski, *J. Light. Tech.* **15**, 112 (1997).

Chapter 7

Optical Amplifiers in Fiber Optic Communication Systems – Theory

7.1 INTRODUCTION

Most, if not all, applications of photons and lightwave signals in communications, sensors, signal processing, etc., require the detection and subsequent conversion of the light to an electrical signal. In this process, the useful signal will be corrupted by noise and the ultimate sensitivity and performance of the system is limited by the noise characteristics. The ultimate performance is achieved when the dominant noise source is given by the inherent noise in a coherent light source, i.e., shot noise. However, many systems – digital optical communication links, for example – are not limited by shot noise but rather by thermal and circuit noise in the receiver.

For example, a shot-noise-limited digital lightwave receiver has an error rate of $\frac{1}{2}e^{-N}$, where N = number of photons per bit. Thus an error rate of 10^{-9} requires 20 photons for a 1 bit or an average of 10 photons per bit (assuming equal numbers of 1s and 0s).[1] A typical commercial receiver requires a few thousand photons per bit and the best research receivers a few hundred photons per bit for the same error rate.

It is because of this discrepancy between the ultimate shot noise limit and the real case of thermal and circuit noise that optical preamplifiers can be used to improve the effective receiver sensitivity in lightwave systems. Because optical amplifiers and preamplifiers add noise to the signal amplified, and at some point the amplifier-added noise becomes the dominant noise source, the understanding of the noise properties of optical amplifiers and the resultant effects on systems is of crucial importance to the engineering of lightwave systems.

The basic manifestation of noise in optical amplifiers is in the form of spontaneous emission. Spontaneous emission is present in a spectral interval corresponding to the gain spectrum of the amplifier, and the spectral density of the noise is proportional to the gain. In the optical domain, the spontaneous emission is simply added to the signal as background radiation and, in principle, all of the noise can be filtered out except for that in the same spectral band as the signal. The spontaneous emission not only affects the noise performance of a system but, because of the added optical power in the channel, also limits the maximum gain that can be achieved in high-gain amplifiers, through gain saturation. When an amplified optical signal and the accompanying spontaneous emission are detected in a PIN or an avalanche photo detector (APD), the noise is transformed into the electrical domain and appears along with the signal-induced photocurrent as a noise current.

Photodetection is a nonlinear square-law process; the photocurrent is therefore composed of a number of beat signals between the signal and noise optical fields in addition to the squares of the signal field and spontaneous emission field. Writing the total optical field as the sum of a signal field $\vec{E}_s$ and spontaneous emission field $\vec{E}_n$,

$$\vec{E}_{tot} = \vec{E}_s + \vec{E}_n \tag{7.1}$$

we find for the photo-generated detector current I,

$$I \sim (\vec{E}_{tot})^2 = (\vec{E}_s + \vec{E}_n)^2 = E_s^2 + E_n^2 + 2\vec{E}_s \cdot \vec{E}_n \tag{7.2}$$

The mixing of the noise with itself and with the signal produces frequencies that extend down to zero. A portion of this noise will fall within the bandwidth of the receiver and degrade the signal to noise ratio. On the right-hand side of equation 7.2 we can identify the first term as pure signal, the second term as pure noise, and the third term as a mixing component between signal and noise. The last two terms are commonly referred to as the spontaneous-spontaneous (sp-sp) beat noise and signal-spontaneous (s-sp) beat noise, respectively. These are the two new noise terms that appear (in addition to the shot noise of the total photo current) when optical amplifiers are used.

In this chapter we will examine the noise properties of erbium-doped fiber amplifiers and its system consequences. In Section 7.2 we will concentrate on the device properties, deriving how much spontaneous emission is generated and its power spectrum. A crucial parameter is the inversion parameter, n_{sp}, which is the proportionality coefficient between the gain and the spectral density of the spontaneous emission. Using n_{sp} as the device parameter, in Section 7.3 we will derive the noise characteristics of a square-law receiver and derive electrical signal to noise ratios as a function of all relevant parameters. In Section 7.3 we will determine the bit error rate (BER) characteristics of digital lightwave systems and determine the optimum and maximum system performance. We will also cover nonlinearities in section 7.3.6.

7.2 OPTICAL NOISE: DEVICE ASPECTS

7.2.1 Classical Derivation of Optical Amplifier Noise

We can derive an expression for the spontaneous emission power as a function of the gain available from a two-level system. The result can be obtained semiclassically, by using the Einstein coefficients.

We consider a section of length L of an optical amplifier that is represented by a two-level system, where the population density in the upper state is N_2 and the population density in the lower state is N_1 (see Figure 7.1). The energy separation between

7.2. OPTICAL NOISE: DEVICE ASPECTS

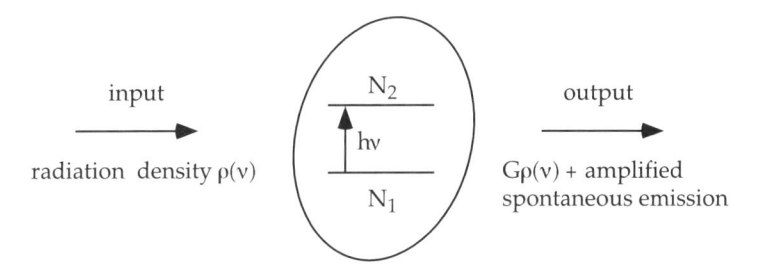

Figure 7.1: Interaction of light at frequency ν with a two-level amplifier with gain G at frequency ν.

the two states is $h\nu$, where h is Planck's constant and ν is the optical frequency of the transition. We assume that the optical line has some broadening so that the distribution of excited ions is given by a normalized distribution function $g(\nu)$ such that

$$\int g(\nu)d\nu = 1 \tag{7.3}$$

We start by considering the Einstein coefficients. We know from Chapter 4 that

$$A_{21} = \left(\frac{8\pi \nu^2}{c^3}\right) h\nu B_{21} \tag{7.4}$$

The factor

$$p_n = \left(\frac{8\pi \nu^2}{c^3}\right) \tag{7.5}$$

is the number of modes per unit volume per unit frequency interval. The spontaneous emission rate at frequency ν is equal to the stimulated emission rate times the number of modes per unit volume per unit frequency at frequency ν times the photon energy. Thus, the spontaneous emission probability into a single mode is equal to the stimulated emission rate into that mode with a single photon occupying that mode.

We now use this fact to calculate the growth in photon number of the mode as it traverses the amplifier. We consider a mode at frequency ν and calculate the number of photons in the mode after it has traversed the given section of length L. The number density of photons in the mode (in number of photons per unit volume), N_m – after an incremental section of length dz of the amplifier – is given by the combination of absorption, stimulated emission, and spontaneous emission. We have

$$dN_m = [(N_2 B_{21}\rho(\nu) - N_1 B_{12}\rho(\nu)) + h\nu N_2 B_{21}] g(\nu) dz \tag{7.6}$$

where $\rho(\nu)$ is the radiation density (in energy per unit volume). The first term on the right-hand side of equation 7.6 represents stimulated emission, the second absorption, and the third spontaneous emission. The radiation density is simply equal to the photon number density times the photon energy $h\nu$. We have, assuming that $B_{21} = B_{12}$ (no degeneracy),

$$dN_m = B_{21} h\nu \left[(N_2 - N_1) N_m + N_2 \right] g(\nu) dz \tag{7.7}$$

We can write this as

$$\frac{dN_m}{[(N_2 - N_1)N_m + N_2]} = B_{21}h\nu g(\nu)dz \tag{7.8}$$

or

$$\frac{dN_m}{N_m + n_{sp}} = (N_2 - N_1)B_{21}h\nu g(\nu)dz \tag{7.9}$$

where we define the inversion parameter n_{sp} as

$$n_{sp} = \frac{N_2}{N_2 - N_1} \tag{7.10}$$

which is valid for a nondegenerate atomic transition where the absorption cross section equals the emission cross section.

Consider now that the erbium ion transitions in the amplifier model we developed in Chapter 5 are modeled using different emission and absorption cross sections to reflect the thermal distribution of population in the upper and lower states (see in particular the discussion in Section 4.4.1). The growth in the radiation density is then calculated as in equation 7.6 except that the stimulated emission probability is now proportional to $\sigma_e(\lambda)N_2$ and the absorption probability is now proportional to $\sigma_a(\lambda)N_1$, where $\sigma_e(\lambda)$ is the emission cross section, $\sigma_a(\lambda)$ the absorption cross section, and λ the wavelength of the transition. The inversion parameter is then given by the expression

$$n_{sp} = \frac{\sigma_e(\lambda)N_2}{\sigma_e(\lambda)N_2 - \sigma_a(\lambda)N_1} \tag{7.11}$$

We now integrate both sides of equation 7.9. The right-hand side length integration goes from 0 to L, and the left-hand side integration goes from $N_m(0)$ to $N_m(L)$, the input and output number density of photons, respectively. Assuming small signal conditions – i.e., no gain depletion – we obtain

$$\ln\left[\frac{N_m(L) + n_{sp}}{N_m(0) + n_{sp}}\right] = B_{21}h\nu g(\nu)(N_2 - N_1)L \tag{7.12}$$

so that

$$\frac{N_m(L) + n_{sp}}{N_m(0) + n_{sp}} = e^{B_{21}h\nu g(\nu)(N_2-N_1)L} = e^{gL} = G \tag{7.13}$$

where g is the incremental gain of the amplifier and G the overall gain. We then find that the output number density of photons is equal to

$$N_m(L) = GN_m(0) + n_{sp}(G - 1) \tag{7.14}$$

where the first term on the right-hand side is the amplified signal, and the second term is the amplified spontaneous emission (or noise) output. We calculate the noise power output for this mode and compute the number of modes in the frequency band $\Delta\nu$ (as in Chapter 5, section 5.4) to obtain the noise output power (resulting from amplified

7.2. OPTICAL NOISE: DEVICE ASPECTS

spontaneous emission) in the bandwidth $\Delta \nu$ around the frequency ν where the gain of the amplifier is G

$$P_{ASE} = n_{sp}(G-1)h\nu\Delta\nu \qquad (7.15)$$

Equation 7.15 is a fundamental result in understanding noise in optical amplifier systems. It is important to realize that the expression for P_{ASE} in equation 7.15 is per spatial and polarization mode of the amplifier. Single-mode fiber amplifiers have two polarization modes and multimode fiber amplifiers can have a large number of spatial modes. For a single-mode fiber, one should multiply the right-hand side of equation 7.15 by a factor of 2 to get the total ASE power.

7.2.2 Noise at the Output of an Optical Amplifier

In lightwave communication systems that employ optical amplifiers, the optical signal is converted to an electrical signal at the end of the transmission path. We must consider the conversion of the light field into an electrical signal, whereby photons are converted into electrons. Since the detector will convert all photons into electrons, spontaneous emission exiting the optical amplifier will give rise to an electrical signal that must be considered as noise, because it is random and contains no information. The noise can be restricted by a spectral filter placed in front of the detector so that the detector will accept only the bandwidth of light corresponding to the signal. There will still be, however, some spontaneous emission noise in that bandwidth and signal-spontaneous noise will still be present. The total spontaneous emission power is given by Expression 7.15.

The sources of noise can be detailed by considering more closely the total electric field intensity at the square-law detector.[2] We will compute in the present section the optical noise components. Note that another effect of the electrical receiver is to down-convert the signal from optical carrier frequencies to frequencies near dc. This gives the opportunity to use an electrical filter in the electrical receiver circuit, in addition to the optical filter which is in front of the optical receiver. As we shall see later, this electrical filter is instrumental in reducing the noise detected by the receiver.

We write the total electric field at the detector as the sum of the fields from the spontaneous emitted light and the signal light

$$\vec{E}_{tot} = \vec{E}_{sig} + \vec{E}_{spont} \qquad (7.16)$$

Because the detector response is proportional to $\vec{E}_{tot} \times \vec{E}_{tot}$, the photocurrent will depend on the polarization state of the total field $\vec{E}_{tot}$. The photocurrent generated will be proportional to the light intensity since the total number of photons is proportional to the light intensity and the photocurrent is proportional to the total number of photons (each photon absorbed is assumed to generate one electron).

We square the previous sum to obtain the total intensity (only the field components along the signal polarization are considered, the other polarization direction can be rejected by a polarizing filter)

$$I \sim \left[E_{sig}^2 + E_{spont}^2 + (E_{sig}E_{spont}^* + E_{sig}^*E_{spont}) \right] \left(\frac{e}{h\nu}\right) \qquad (7.17)$$

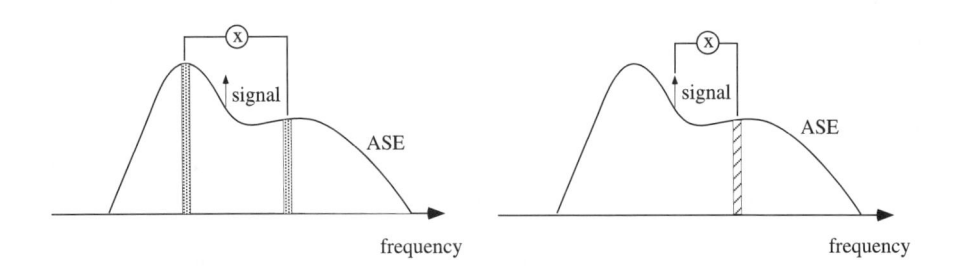

Figure 7.2: Spectral density of spontaneous-spontaneous noise (left) and signal-spontaneous noise (right). The frequency position of the signal is indicated by the arrow and the hatched vertical areas indicate examples of frequency bands of ASE which contribute to the noise generation process via beating with the signal.

where ∗ denotes the complex conjugate and I is the intensity of the light at the receiver (or equivalently the photocurrent).

The first term represents the expected signal intensity. The other terms all correspond to noise. The second term corresponds to the product of the spontaneous emission electric field with itself and is called the *spontaneous-spontaneous* (sp-sp) beat noise term. The last term, the product of the signal electric field and the spontaneous emission electric field, is known as the *signal-spontaneous* (s-sp) beat term. The two noise terms are pictorially shown in Figure 7.2.

Let's examine each term a little more closely by calculating the power spectrum of each noise term. The power spectrum can be obtained from a Fourier decomposition of each noise term. We will consider an optical amplifier with unity coupling efficiency, uniform gain G (and thus uniform ASE over an optical bandwidth B_o), and an average input signal power of P_{in} at optical frequency ω_0 centered in the optical passband B_o.

The spontaneous emission power in the optical bandwidth B_o, in one polarization mode, is given by

$$P_{sp} = n_{sp} (G-1) h\nu B_o \tag{7.18}$$

Writing the electric field E_{sp} (the spontaneous emission), as a sum of cosine terms spaced $\delta\nu$ apart in frequency, we have

$$E_{sp}(t) = \sum_{k=(-B_0/2\delta\nu)}^{B_0/2\delta\nu} \sqrt{2n_{sp}(G-1)h\nu\delta\nu} \, \cos((\omega_0 + 2\pi k\delta\nu)t + \Phi_k) \tag{7.19}$$

where Φ_k is a random phase for each component of spontaneous emission. In equation 7.19, $\delta\nu$ is an arbitrarily small frequency width and is chosen so that $B_0/2\delta\nu$ is an integer value (for ease of analysis). Thus the sum of monochromatic waves over the frequency bandwidth B_o accurately represents the total electric field of the spontaneous emission. The amplitude factor for each electric field component is the same since we assumed a uniform gain.

With

$$N_0 = n_{sp} (G-1) h\nu \tag{7.20}$$

7.2. OPTICAL NOISE: DEVICE ASPECTS

and
$$M = \frac{B_0}{2\delta\nu} \tag{7.21}$$

the total electric field at the output of the amplifier is given by

$$E(t) = \sqrt{2GP_{in}}\cos(\omega_0 t) + \sum_{k=-M}^{M} \sqrt{2N_0\delta\nu}\cos((\omega_0 + 2\pi k\delta\nu)t + \Phi_k) \tag{7.22}$$

The photocurrent i(t) generated by a unity quantum efficiency photodetector is proportional to the intensity

$$i(t) = \overline{E^2(t)}\,\frac{e}{h\nu} \tag{7.23}$$

where the bar indicates time averaging over optical frequencies by the photodetector. In the following we will ignore all terms at frequencies $\simeq 2\omega_0$, which average to zero and are outside the system passband. Hence

$$\begin{aligned}
i(t) = {} & GP_{in}\frac{e}{h\nu} + \frac{4e}{h\nu}\overline{\sum_{k=-M}^{M}\sqrt{GP_{in}N_0\delta\nu}\cos(\omega_0 t)} \\
& \cdot \cos((\omega_0 + 2\pi k\delta\nu)t + \Phi_k) \\
& + \frac{2eN_0\delta\nu}{h\nu}\overline{\left[\sum_{k=-M}^{M}\cos((\omega_0 + 2\pi k\delta\nu)t + \Phi_k)\right]^2}
\end{aligned} \tag{7.24}$$

The three terms in equation 7.24 represent signal, signal-spontaneous beat noise, and spontaneous-spontaneous beat noise, respectively, for one polarization.

Signal-Spontaneous Beat Noise

We will examine the signal-spontaneous beat noise term first.

$$\begin{aligned}
i_{s-sp}(t) &= \frac{4e}{h\nu}\overline{\sum_{k=-M}^{M}\sqrt{GP_{in}N_0\delta\nu}\cos(\omega_0 t)\cos((\omega_0 + 2\pi k\delta\nu)t + \Phi_k)} \\
&= \frac{2e}{h\nu}\sqrt{GP_{in}N_0\delta\nu}\sum_{k=-M}^{M}\cos(2\pi k\delta\nu\, t + \Phi_k)
\end{aligned} \tag{7.25}$$

where terms that are proportional to $\cos(2\omega_0 t)$ and that are averaged to zero by the photodetector have been neglected. For each non-optical difference frequency $2\pi k\delta\nu$, the sum has two components – a positive frequency component and a negative frequency component – but a random phase. Therefore, the power spectrum of $i_{s-sp}(t)$ is uniform in the frequency interval $-B_o/2$ to $B_o/2$ and has an equivalent one-sided power density of

$$\begin{aligned}
N_{s-sp} &= \frac{4e^2}{(h\nu)^2}GP_{in}N_0 \cdot \frac{1}{2} \cdot 2 \\
&= \frac{4e^2}{h\nu}P_{in}\,n_{sp}\,(G-1)\,G
\end{aligned} \tag{7.26}$$

Spontaneous-Spontaneous Beat Noise

Now let's consider the spontaneous-spontaneous beat noise. From equation 7.24 we have

$$i_{sp-sp}(t) = 2N_0 \frac{\delta ve}{hv} \left[\sum_{k=-M}^{M} \cos((\omega_0 + 2\pi k\delta v)t + \Phi_k) \right]^2$$

$$= 2N_0 \frac{\delta ve}{hv} \left[\sum_{k=-M}^{M} \cos(\beta_k) \right] \left[\sum_{j=-M}^{M} \cos(\beta_j) \right] \quad (7.27)$$

where

$$\beta_k = (\omega_0 + 2\pi k\delta v)t + \Phi_k \quad (7.28)$$
$$\beta_j = (\omega_0 + 2\pi j\delta v)t + \Phi_j \quad (7.29)$$

Equation 7.27 can be written as

$$i_{sp-sp}(t) = 2N_0 \frac{\delta ve}{hv} \sum_{k=-M}^{M} \sum_{j=-M}^{M} (\frac{1}{2}\cos(\beta_k - \beta_j) + \frac{1}{2}\cos(\beta_k + \beta_j)) \quad (7.30)$$

The terms proportional to $\cos(\beta_k + \beta_j)$ have frequencies $\sim 2\omega_0$ and average to zero. Rewriting equation 7.30 gives the remaining nonoptical frequency terms, which can be written as

$$i_{sp-sp}(t) = \frac{N_0 \delta ve}{hv} \sum_{k=0}^{2M} \sum_{j=0}^{2M} \cos((k-j)2\pi \delta vt + \Phi_k - \Phi_j) \quad (7.31)$$

In the double summation of equation 7.30 we note that the limits of the summations are $-M$ and M, and since $\cos(-\alpha t) = \cos(\alpha t)$, the limits of the summation can be shifted to 0 and 2M, as in equation 7.31.

The dc term is obtained for $k = j$, and there are $2M + 1$ such terms

$$I_{sp}^{dc} = \frac{e}{hv} N_0 \delta v (2M+1) = n_{sp}(G-1)eB_0 \quad (7.32)$$

where we used the fact that as M becomes large $2M = 2M+1$. I_{sp}^{dc} in equation 7.32 does not represent a noise term in itself; it is simply the dc value of the detected spontaneous emission (per polarization state). This dc current will generate a shot noise that we will take into consideration in Section 7.3. Organizing the terms in equation 7.31 according to their frequency we obtain

7.2. OPTICAL NOISE: DEVICE ASPECTS

frequency	#terms
$-2M\delta\nu$	1
$(-2M+1)\delta\nu$	2
$(-2M+2)\delta\nu$	3
.	.
.	.
.	.
$-m\delta\nu$	$2M-m+1$
.	.
.	.
.	.
$-\delta\nu$	$2M$
$\delta\nu$	$2M$
.	.
.	.
.	.
$m\delta\nu$	$2M-m+1$
.	.
.	.
.	.
$(2M-2)\delta\nu$	3
$(2M-1)\delta\nu$	2
$2M\delta\nu$	1

The terms with the same absolute frequency but of opposite sign add in phase. Therefore, the power spectrum of the spontaneous-spontaneous beat noise extends from 0 to B_o with a triangular shape and a single-sided power density near dc of

$$N_{sp-sp} = \frac{1}{2}\left(\frac{N_0 \delta\nu e}{h\nu}\right)^2 \cdot \frac{1}{\delta\nu} \cdot 2M \cdot 2 = 2n_{sp}^2(G-1)^2 e^2 B_0 \quad (7.33)$$

Equation 7.33 is for one polarization mode only. Since there are two polarization states in a single-mode fiber, the total sp-sp power will be twice as large as the amount given in equation 7.33 when no polarizer is used. Note that the two orthogonal polarizations do not beat with each other so the total increase in sp-sp noise is a factor of 2 rather than a factor of 4 for a doubling of the ASE power in the *same* polarization. Also, the amount of s-sp noise is unaffected by the presence of a second noise mode. This follows from the fact that the signal itself is polarized and beats with only one noise polarization.

Figure 7.3 summarizes the results of equations 7.26 and 7.33 and shows the electrical power spectrum of the beat noise. Note that reducing B_o, by using an optical filter with a smaller passband, decreases the spontaneous-spontaneous noise power density but leaves the signal-spontaneous noise power density unaffected. Once the optical signal has been down-shifted by the receiver, an electrical filter can be used advantageously in the electrical receiver circuit. This electrical filter has a very large effect

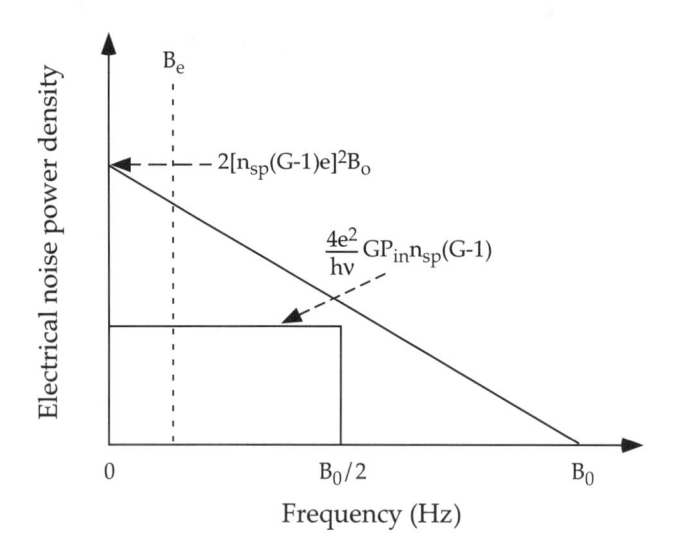

Figure 7.3: Electrical single-sided noise power spectrum of spontaneous-spontaneous (triangular) and signal-spontaneous (rectangular) beat noise (per polarization mode). Also indicated is the electrical bandwidth B_e. The noise accepted by the receiver is the noise power density integrated from dc up to B_e. Note that the absolute level of the spontaneous-spontaneous noise can be reduced by reducing B_o while this is not the case for the signal-spontaneous noise.

in reducing the noise from the optical amplifier. In Section 7.3 we will use the basic results of equations 7.26 and 7.33 to calculate the system-level noise performance of amplified systems, calculated at the system receiver. In particular we will integrate the areas under the noise density curves of Figure 7.3 to obtain the total electrical noise terms at the receiver.

7.2.3 Comparison of Optical Amplifier Devices

Various optical amplifiers have been proposed and characterized since the advent of optical communications. The ideal goal is to have an amplifier that has a very large bandwidth, high, polarization independent, gain with modest power requirements, and that adds no noise to the actual signal or signals. Of course, the ideal amplifier does not exist, although the erbium-doped fiber amplifier comes close. We take a brief detour here to compare here the most prevalent optical amplifiers to date, in particular from the point of view of their noise performance.

Erbium-Doped Fiber Amplifiers

The erbium-doped fiber amplifier has a theoretical quantum limited noise figure due to spontaneous emission, as discussed in the previous sections. This noise figure of 3 dB has been achieved experimentally, although realistic values lie in the range of 3.5 to 6 dB. The gains accomplished with the erbium-doped amplifier are extremely high. Gains of up to 45 dB have been achieved with two-stage erbium-doped fiber amplifiers.

However, the erbium-doped fiber amplifier exhibits some very slight polarization sensitivity, which can be a factor in system performance when many (e.g., hundreds of) amplifiers are cascaded in series. The saturation characteristics are very good, because the saturation power can be linearly increased with the pump power.

The long 10 ms upper-state lifetime is unique to Er^{3+} and dramatically distinguishes the erbium-doped fiber amplifier from all the other amplifiers. It is responsible for several of the key advantages of the erbium-doped fiber amplifier. Because the upper state can integrate the pump power for such long time, the pump power required to store sufficient energy in the amplifier is typically very low– $\simeq$ 10 mW to 20 mW for a 30 dB small signal gain. For signals at different wavelengths, multichannel crosstalk in the amplifier is also very low due to the long upper-state lifetime, which ensures that the population in the upper state cannot respond to fast signal changes that would otherwise be carried over to another signal via the population inversion. For the same reason, the erbium-doped fiber amplifier is still a non distorting amplifier even when operating deep in saturation. Obviously, the erbium-doped fiber amplifier can obviously easily be integrated in a fiber network since it is already a silica-based fiber component. The major drawback of the erbium-doped fiber amplifier is that the gain spectrum is not inherently flat but peaked, so the gain is not equal at different wavelengths.

Raman Amplifiers

In Raman scattering a photon is annihilated to create an optical phonon and a lower frequency photon (Stokes photon). For sufficiently high optical powers, the process becomes stimulated (stimulated Raman scattering or SRS), and optical gain is generated at the Stokes frequency.[3, 4] We will discuss the possible negative consequences of SRS in an optically amplified long-haul system in section 7.3.6. In Raman amplifiers utilizing silica fiber, the Stokes shift between the pump and the signal frequency where gain occurs is approximately 15 THz. Since the gain peak can be shifted by varying the center wavelength of the pump, a Raman amplifier is tunable. The gain is proportional to the pump power, so that gain flatness to the extent of the gain width of the optical phonon band can be ensured. The noise for a Raman amplifier is derived from an equivalent n_{sp}, given by

$$n_{sp} = \frac{1}{1 - e^{-h\nu_s/kT}} \qquad (7.34)$$

where ν_s is the Stokes shift.[5] Typically, $h\nu_s \gg kT$ so that $n_{sp} \simeq 1$ and the noise is close to the quantum limit. The pump powers needed for Raman amplifiers are quite high, on the order of 0.5 W to 1 W. In addition, Raman amplifiers can exhibit large crosstalk between different wavelength signals, via gain saturation, which, in contrast to erbium-doped fiber amplifiers, is not damped by a slow population response time. Raman amplifiers, in particular for 1.3 μm amplification purposes, have recently witnessed a resurgence of interest with the advent of high-power solid state laser pump sources.[6, 7]

Brillouin Amplifiers

Brillouin amplifiers are analogous to Raman amplifiers with the difference that acoustical phonons are involved in the gain process.[8] In the case of Brillouin amplifiers, the

Stokes wave propagates backward. The Stokes shift for Brillouin scattering is typically very small–$\simeq$ 11 GHz in silica fiber. Therefore, using equation 7.34, the spontaneous emission factor n_{sp} is large, about 500. The Brillouin gain bandwidth is only about 20 MHz, which makes Brillouin amplifiers attractive for narrowband selective amplification but, because of the poor noise performance, Brillouin amplifiers have not found wide use in communication applications.[9] As with Raman amplifiers, Brillouin amplifiers are tunable via the pump wavelength. The threshold power for stimulated Brillouin scattering is only a few mW in single-mode fiber and is one of the main effects that limit the usable optical power in communication systems. We will discuss SBS further in our discussion of nonlinearities in section 7.3.6.

Semiconductor Laser Amplifiers

Semiconductor laser amplifiers have the advantage that they can be directly pumped electrically and require very low electrical pump power. Semiconductor laser amplifiers were the optical amplifiers of choice prior to erbium-doped fiber amplifiers, and much of the early research done on optically amplified systems was performed with semiconductor laser amplifiers. Some of the basic results concerning spontaneous emission noise derived for the fiber amplifier remain valid for semiconductor amplifiers.[2, 10, 11] Semiconductor laser amplifiers can be fabricated at most of the wavelengths of interest (1.5 μm and 1.3 μm). The bandwidth of semiconductor laser amplifiers is very large, and they have large small signal gains.

The intrinsic noise performance of semiconductor laser amplifiers is quite good. However, they suffer from coupling loss to the adjoining fiber transmission line. This coupling problem is a technically difficult problem; it impairs the gain and noise figure and adds components that are not necessary with fiber amplifiers. For a semiconductor laser amplifier, typically $n_{sp} \simeq 2$ and is wavelength dependent. Polarization sensitivity is also an issue.

Another drawback of semiconductor laser amplifiers is multiwavelength crosstalk. It has been shown that four-wave mixing is a particularly strong effect in semiconductor laser amplifiers and gives rise to large intersignal modulation and information distortion, unless the wavelengths are separated by unreasonably large frequency gaps. This restricts the information-carrying capacity in wavelength division multiplexed systems.

Crosstalk from gain saturation is also a concern for these amplifiers. The carrier lifetime in an typical semiconductor laser amplifier is a few ns (compared to 10 ms in erbium-doped fiber amplifiers), which results in crosstalk that extends to several 100 MHz through cross-channel gain saturation. The short lifetime also results in pulse distortion when the amplifier is operated in deep gain saturation.

7.3 OPTICAL NOISE: SYSTEM ASPECTS

The first theoretical treatment of optical amplifiers in optical transmission systems dates back almost to the first development of the laser (and the maser). Some of the more fundamental work has been done by Personick, who calculated the signal to noise ratios in an amplified digital transmission system.[12, 13]

7.3.1 Receivers

The noise is calculated at the optical receiver and is summed from the various sources. The optical receiver is responsible for converting the optical signal into an electrical form suitable for further processing. An ideal receiver converts photons into electrons without adding any noise (consisting of either extra or missing electrons). Independently of the spontaneous emission noise, which was previously considered, a real optical receiver will add noise of its own, usually thermal noise and shot noise. The calculation will depend on the kind of receiver considered–for example, whether it is an APD or a PIN detector.

Detectors used today are either APDs or PINs. Both are made of semiconductor material. An APD works by generating a large number of electrons for each photon impinging on the device to ensure that the receiver will detect the most signal photons possible. This is done, in a manner similar to photomultiplier tube operation, by having the electron generation work in a cascade fashion (hence the term *avalanche*). An APD is a reverse-biased p-n junction in which the applied voltage is very large. The electron-hole pair initially created by the absorbed photon will acquire a large kinetic energy and, as it accelerates through the semiconductor material, will create further electron-hole pairs as it transfers its energy via collisions. An APD provides gain, which is useful in that it reduces the need for electrical preamplification of the signal. However, the avalanche process produces excess noise. At low gains, the noise from an APD is lower than that from an electrical amplifier so the overall sensitivity of the receiver is improved.

A PIN detector relies on the presence of a undoped layer of semiconductor material placed between the p and n sides of the junction to absorb the photons. This so-called depletion region has an electric field across it, and the electron-hole pairs generated in the depletion region are swept across the junction and create an electrical current. The larger the depletion region in a PIN, the larger the responsivity of the device. The speed of the device, however, will decrease as the depletion region increases in size. A PIN detector will add only shot noise since one photon absorbed results in one electron-hole pair. In most cases a PIN detector is preferable to an APD in the case of an optically preamplified receiver.

The thermal noise of a receiver arises from the fact that electrons in a receiver circuit have some probability of generating a current even in the absence of an optical signal. This noise, often referred to as Johnson noise, can be summarized by the variance of the thermal current per unit frequency,

$$\sigma_{th}^2 = \langle i_{th}^2(t) \rangle = \frac{4kT}{R} \tag{7.35}$$

where T is the temperature, k is Boltzmann's constant, and R is the resistance of the detector load resistor. For an electrical filter bandwidth of $\Delta \nu$ the thermal noise power will be the density given by equation 7.35 multiplied by $\Delta \nu$.

The shot noise arises from the Poissonian distribution of the electron-hole generation by the photon stream. The latter is a stochastic process having random arrival times. On average, the number of electron-hole pairs created will be proportional to the number of photons, with a given constant of proportionality. During a given time interval, with a certain number of photons incident upon the detector, the number of electron-hole pairs generated will have fluctuations as determined by Poisson statistics. This is the shot noise and is inescapable. A dc photocurrent of I_{pd} will generate a shot noise power density of $2eI_{pd}$.

7.3.2 Bit Error Rate Calculations - Direct Detection

Consider a digital transmission system where the information is transmitted in bits that represent either a 1 or a 0. We can assume that a 0 bit is represented by the absence of a light pulse in an interval T, whereas a 1 bit is represented by the presence of a light pulse at the communication wavelength. The bit error rate obtained can be computed from the various sources of noise that we have discussed so far. The bit error rate is defined as the relative number of errors made by a receiver in an optical transmission system in interpreting the original transmision of a sequence of bits.

Direct Detection - Q factor calculation

The most fundamental source of noise in a bit error rate measurement is the shot noise discussed previously. When the bit transmitted is a 0 bit, there can be no shot noise. Thus, in first approximation, it is safe to assume that from the point of view of shot noise a 0 bit will always be interpreted as a 0 bit. A 1 bit will be recognized as a 1 bit unless zero electron-hole pairs are generated. The probability for this occurrence is given by Poisson's equation

$$P(0) = \exp(-E_s/h\nu) \tag{7.36}$$

If the probability of missing a 0 is equal to 0 and 1s and 0s are equally probable, then one can tolerate an error rate of 2×10^{-9} for the 1s to have an overall BER of 1×10^{-9}. Thus, $E_s = 20\ h\nu$ or an average of 10 photons per bit is required for a 10^{-9} BER in a shot-noise-limited system.

A more complete calculation of the bit error rate can be made by treating the noise sources in terms of Gaussian noise statistics. One can then show that the BER, for optimum setting of the decision threshold for choosing whether a bit is a 1 or a 0, is given by the Gaussian error function

$$\text{BER} = \frac{1}{2} \operatorname{erfc}\left(\frac{1}{\sqrt{2}} \frac{I_1 - I_0}{\sigma_1 + \sigma_0}\right) \tag{7.37}$$

where I_0 and I_1 are the average photocurrents generated by a 1 bit and a 0 bit, respectively, and σ_0^2 and σ_1^2 are the noise variances for a 0 bit and a 1 bit, respectively.[13] The total noise variance is determined from the thermal, shot noise, and ASE contributions:

$$\sigma = \sqrt{\left(\sigma_{\text{shot}}^2 + \sigma_{\text{thermal}}^2 + \sigma_{\text{s-sp}}^2 + \sigma_{\text{sp-sp}}^2\right)} \tag{7.38}$$

If the receiver is being used without an optical amplifier, then only the shot noise and thermal noise contributions need to be included in equation 7.38.

Note that the I_0, I_1 in the numerator for the BER are the time average of I_S over the duration T of the bit. Note also that the variance of the current I_{noise} is equal to the expression $\sigma_{\text{noise}}^2 = \overline{I_{\text{noise}}^2} - \overline{I_{\text{noise}}}^2$, and $\overline{I_{\text{noise}}} = 0$. The expectation value of the square

7.3. OPTICAL NOISE: SYSTEM ASPECTS

of the noise is equal to the integral of the noise power density over the frequency range of interest from the Wiener-Khinchin theorem,[1]

$$\overline{I_{noise}^2} = \int N_{noise}(\nu)d\nu = N_{tot} \quad (7.39)$$

The noise variances are thus equal to the previously computed noise power densities of Section 7.2.2, integrated over an electrical bandwidth B_e.

In the case of an optically amplified system, we now need to compute the noise currents corresponding to the signal-spontaneous and spontaneous-spontaneous noise previously described. In a square-law detector, the received signal power, assuming an infinite extinction ratio between 1s and 0s, is

$$S = (GI_s)^2 \quad (7.40)$$

I_s is the photocurrent generated by the signal photons,

$$I_s = P_{in}\frac{e}{h\nu} \quad (7.41)$$

where P_{in} is at the input to the optical amplifier with gain G. We will denote by $I_S = GI_s$ the signal current corresponding to the signal light power at the output of the amplifier. From equations 7.26 and 7.33 we obtain the signal-spontaneous and spontaneous-spontaneous noise powers (the integrals of the noise power densities over the electrical bandwidth B_e), respectively:

$$N_{s-sp} = 2GI_s I_{sp}\frac{B_e}{B_o} \quad (7.42)$$

$$N_{sp-sp} = \frac{1}{2}I_{sp}^2 \frac{B_e(2B_o - B_e)}{B_o^2} \quad (7.43)$$

where B_e is the bandwidth of the electrical filter used in the electrical receiver circuit and I_{sp} is the photocurrent generated by the spontaneous emission (including both polarization states) at the output of the amplifier,

$$I_{sp} = 2n_{sp}(G-1)eB_o \quad (7.44)$$

To obtain the two noise terms we computed the areas under the relevant curves of Figure 7.3, bounded in addition by the electrical bandwidth B_e. To get the total noise we need to add the shot noise and the thermal noise of the receiver:

$$N_{shot} = 2B_e(GI_s + I_{sp})e \quad (7.45)$$

$$N_{th} = I_{th}^2 \quad (7.46)$$

where I_{th}^2 is variance of the thermal noise current. The total noise is

$$N_{tot} = N_{shot} + N_{s-sp} + N_{sp-sp} + N_{th} \quad (7.47)$$

Some modifications are necessary to the previous equations when considering coherent systems.[1]

Now consider an amplitude-modulated signal of average power P_{in}, 50% duty cycle, and an extinction ratio of r. The photocurrent equivalents of the amplifier output powers for a 1 bit, $I_S(1)$, and for a 0 bit, $I_S(0)$, are

$$I_S(1) = eGP_{in}2r/(h\nu(r+1)) \tag{7.51}$$

$$I_S(0) = eGP_{in}2/(h\nu(r+1)) \tag{7.52}$$

The bit error rate is given by equation 7.37, which can be well approximated by:

$$\text{BER} = \frac{1}{\sqrt{2\pi}} \frac{\exp\left(-\frac{Q^2}{2}\right)}{Q} \tag{7.53}$$

where Q is given by:

$$Q = \frac{I_S(1) - I_S(0)}{\sqrt{N_{tot}(1)} + \sqrt{N_{tot}(0)}} \tag{7.54}$$

$I_s(1)$, $I_s(0)$, and $N_{tot}(1)$, $N_{tot}(0)$ are the signal and total noise for a 1 bit and a 0 bit, respectively. Figure 7.4 shows the dependence of the BER on the Q factor, from equation 7.53.

A BER of 10^{-9} requires that $Q = 6$ (15.6 dB), from equation 7.53. Additionally, the definition of Q, as given in equation 7.54, can be used with some simplifying assumptions to quantify the SNR (signal to noise ratio) needed to achieve a 10^{-9} BER. This SNR value can often be found quoted in the literature. Assume an infinite extinction ratio, such that $I_S(0) = 0$. Let us assume also that the noise on the 1 rail is the same as that on the 0 rail, i.e. $\sigma_1 = \sigma_0 = \sqrt{N_{tot}}$. This is not necessarily a very good assumption, since the noise contains signal dependent noise (the signal-spontaneous beat noise) and will be different for a 1 bit as compared to a 0 bit. Nevertheless, we then obtain for Q

$$Q = \frac{I_S(1)}{2\sigma_1} = \frac{I_S(1)}{2\sqrt{N_{tot}}} \tag{7.55}$$

where N_{tot} is the total noise power. Squaring both sides of equation 7.55, we obtain:

$$Q^2 = \frac{I_S^2(1)}{4N_{tot}} = \frac{\text{SNR}}{4} \tag{7.56}$$

---

[1] For a coherent system with amplitude shift keying (ASK) the signal term 7.40 is replaced by:

$$S_{coh} = 2I_s I_{lo} LG \tag{7.48}$$

where L is any loss between the output of the amplifier and the detector, and I_{lo} is the photocurrent equivalent of the local oscillator power P_{lo}:

$$I_{lo} = P_{lo} e/(h\nu) \tag{7.49}$$

The local oscillator shot noise needs to be added to the total noise power, $2I_{lo}B_e e$, as well as the local oscillator-spontaneous beat noise:

$$I_{lo-sp} = 4I_{lo}L(G-1)N_{sp}eB_e \tag{7.50}$$

7.3. OPTICAL NOISE: SYSTEM ASPECTS 217

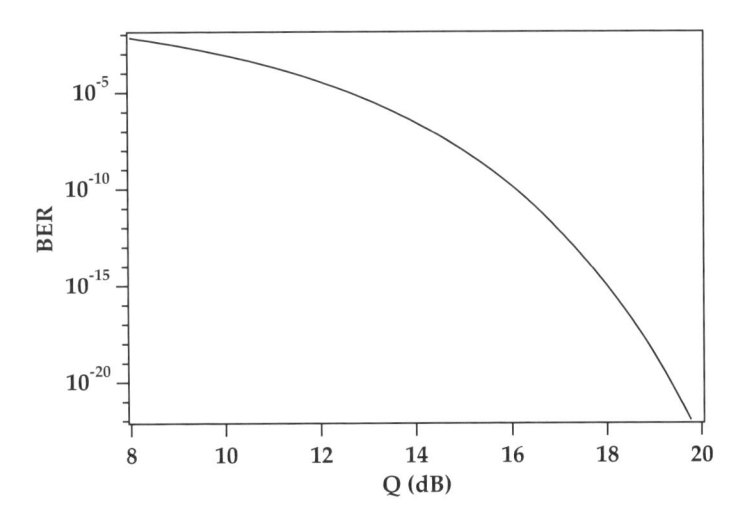

Figure 7.4: Dependence of the BER on the Q factor for Gaussian noise statistics, from equation 7.53. Q(dB) is given by 20log(Q).

where the (electrical) SNR refers to the power level of the 1 bit (as opposed to an average power level of 1 bits and 0 bits). This implies that with the simplifying assumptions listed above

$$\text{SNR} = 4Q^2 \tag{7.57}$$

For a BER of 10^{-9} we have Q=6 (15.6 dB in 20log(Q) units), so that we obtain SNR = 144, or 21.6 dB. We can choose to specify the average SNR, SNR_{avg}, where the average current is

$$I_S(\text{avg}) = 1/2 \times (I_S(1) + I_S(0)) \tag{7.58}$$

and the average signal is

$$1/2 \times (I_S^2(1) + I_S^2(0)) = 2I_S^2(\text{avg}) \tag{7.59}$$

and

$$\text{SNR}_{\text{avg}} = \frac{\text{average signal}}{\text{average noise}} = \frac{2I_s^2(\text{avg})}{N_{\text{tot}}} \tag{7.60}$$

for an infinite extinction ratio. We then have in terms of Q

$$Q = \frac{2I_S(\text{avg})}{2\sigma_1} = \frac{I_S(\text{avg})}{\sqrt{N_{\text{tot}}}} \tag{7.61}$$

by squaring both sides we obtain

$$Q^2 = \frac{I_S^2(\text{avg})}{N_{\text{noise}}} = \frac{\text{SNR}_{\text{avg}}}{2} \tag{7.62}$$

We then obtain for a BER of 10^{-9} that $\text{SNR}_{\text{avg}} = 72$ or 18.6 dB, i.e. (not surprisingly) 3 dB less than the SNR for the 1 bit alone.

A more accurate expression can be derived for Q in terms of the optical signal to noise ratio, SNR_{opt}.[14] This expression takes into account the difference in noise powers corresponding to a 1 bit and a 0 bit. Assume again an infinite extinction ratio such that Q can be written as

$$Q = \frac{I_S(1)}{\sqrt{N_{noise}(1)} + \sqrt{N_{noise}(0)}} = \frac{2I_S(avg)}{\sqrt{N_{noise}(1)} + \sqrt{N_{noise}(0)}} \quad (7.63)$$

In terms of the various noise power components this is written

$$Q = \frac{2I_S(avg)}{\sqrt{N_{s-sp}(1) + N_{sp-sp}} + \sqrt{N_{sp-sp}}} \quad (7.64)$$

where we assumed that the shot noise and receiver thermal noise contributions were negligible compared to the amplifier ASE noise. Using the noise powers for the signal-spontaneous and spontaneous-spontaneous beat noises calculated previously (Equations 7.42 and 7.43), we obtain

$$Q = \frac{2I_S(avg)}{(4\frac{B_e}{B_o}I_S(avg)I_{sp} + \frac{B_e}{B_o}I_{sp}^2)^{\frac{1}{2}} + (\frac{B_e}{B_o}I_{sp}^2)^{\frac{1}{2}}} \quad (7.65)$$

where we also assumed that $B_o \gg B_e$. We note that the average optical SNR at the receiver, SNR_{opt} is given by the ratio of the optical powers, which is equal to the ratio of the photocurrents generated by the signal and noise

$$SNR_{opt} = \frac{I_S(avg)}{I_{sp}} \quad (7.66)$$

The Q factor is then simply given by the expression

$$Q = (\frac{B_o}{B_e})^{\frac{1}{2}} \frac{2SNR_{opt}}{(4SNR_{opt} + 1)^{\frac{1}{2}} + 1} \quad (7.67)$$

Equation 7.67 can be used to derive the optical SNR needed to obtain a given BER, for an ideal system with only amplifier noise and without nonlinearities or inter-symbol interference. Most single channel systems use an optical filter bandwidth of 1-2 nm. For a Q equal to 15.6 dB this leads to a requirement on SNR_{opt} of only a few dB in the bandwidth considered, surprisingly. This is usually quoted on the basis of a bandwidth equal to 1 Å, the typical bandpass of an optical spectrum analyzer (OSA). This then leads to an SNR_{opt} requirement of 10 dB to 13 dB when measured on an OSA set at 1 Å. For very long transmission systems, such as undersea systems, the importance of nonlinearities, dispersion, pattern dependent effects, etc., means that achieving a certain SNR_{opt} does not necessarily imply a certain Q will be achieved. The SNR_{opt} can also be difficult to measure due to pulse spreading or system gain slope effects. System developers for such systems thus usually measure Q directly to estimate the error performance of the system. The desired Q will be significantly higher than 15.6 dB to take into account these effects as well as aging of the system.[15] For terrestrial systems, however, the translation from SNR_{opt} to Q is much more reliable. System

7.3. OPTICAL NOISE: SYSTEM ASPECTS

developers often target an SNR$_{opt}$ in the range of 20-25 dB (relative to 1Å) to obtain error free transmission, for systems ranging from a few 100 km to 1000 km.[16]

The Q factor is a very useful number for characterizing the noise performance of a system. Bergano et al have shown that the Q factor can be accurately obtained even when the BER is too low to be measurable.[17] In this way, the noise performance of various systems can be quantified (as a function of transmission length, bit rate, etc.) even when measuring the actual bit error rate is a daunting task. If the currents that correspond to the 1 and 0 marks are converted to voltages, as would be the case with an oscilloscope trace, the Q factor can be written as

$$Q = \frac{|V_1 - V_2|}{\sigma_1 + \sigma_\Theta} \tag{7.68}$$

where V_1 and V_2 are the mean voltages corresponding to the 1 and 0 marks and σ_1 and σ_2 are the standard deviations of the voltage signals and are equal to the square root of the noise variances $N_{tot}(1)$ and $N_{tot}(0)$ of the voltages, respectively. The bit error rate is measured after setting a voltage decision threshold on the instrument. When this voltage is not set optimally, a large error results. The bit error rate is then given by the expression

$$\text{BER} = \frac{1}{2}\left[\text{erfc}\left(\frac{|V_1 - D|}{\sigma_1}\right) + \text{erfc}\left(\frac{|V_2 - D|}{\sigma_2}\right)\right] \tag{7.69}$$

where D is the voltage threshold and erfc(x) denotes the error function of x. By varying D and plotting the BER over the measurable range, one can then obtain a fit to Expression 7.69 that yields values for V_1, V_2, σ_1, and σ_2. Using these values in Expression 7.68 then gives the Q value. This method has proven to be very advantageous in characterizing the noise performance of long span systems.[18]

BER curves and eye patterns

The bit error rate characteristics of a system are usually displayed as a plot of BER versus signal power. Depending on the signal power dependence of the Q factor, the BER curves take different forms. The BER curve for a system with fixed noise, such as a receiver limited by thermal noise, is shown as curve (a) in Figure 7.5. *Power penalty* is a commonly used term and refers to a horizontal displacement of the BER curve without alteration of its shape. Curve (b) in Figure 7.5 shows the same system as does curve (a) except that we have assumed that the extinction ratio is 20:1, which leads to a power penalty of approximately 1 dB. For a system with signal-dependent noise—such as an amplified system in the signal-spontaneous beat noise limit—the BER curve, shown in curve (c), has a different slope. Curve c was generated for a receiver with an optical preamplifier, and the signal power was measured at the input of the preamplifier. Finally, BER floors are BER curves that reach an asymptotic value for the BER as the signal power is increased. BER floors usually indicate a serious system deficiency and can be caused by anything that limits the SNR of the system or by source laser problems such as mode-partitioning noise. Curve (d) in Figure 7.5 was generated for a capped SNR system; this is the same preamplifer system as in curve (c) but with the signal power measured at the detector and varied by varying the gain of the preamplifier.

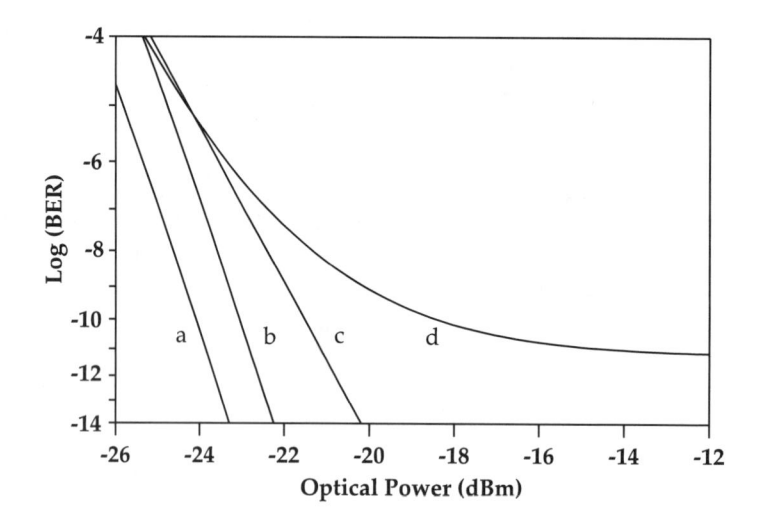

Figure 7.5: Generic bit error rate verses power curves. (a) Fixed noise system; (b) fixed noise system with penalty; (c) signal-dependent noise system; (d) system with BER floor.

Another means for assessing quickly the noise performance of an optically amplified system is through the use of "eye patterns" displayed on an oscilloscope. Such eye patterns are generated by capturing on an oscilloscope the superposition of a random sequence of many 1s and 0 bits. Any noise or intersymbol interference appears as a spreading of the 1 and 0 voltage rails, as well as contributing to a distortion of the transition between a 1 and a 0. This gives a more qualitative view of the noise impressed on the digital transmission pattern and can be used to recognize which kinds of effects are responsible for the transmission degradation. Figure 7.6 is an experimental illustration of eye patterns. These curves were generated with an integrated electro-absorptive modulated laser used as a signal laser and transmited over standard (non dispersion shifted) fiber. The degradation of the eye pattern after transmission over 376 km, with this particular device, is clearly visible. The laser used in this experiment had significant chirp. Thus, a major contribution to the transmission degradation, in this example, is dispersion of the fiber, which leads to a spreading of the laser pulse and a decrease in the slope of the transition between a 1 and 0. Pattern (the particular sequence of 1s and 0s) dependent effects also contribute to the degradation of the eye in this particular case.

7.3.3 Optical Preamplifiers – Noise Figure and Sensitivity

An optical preamplifier is usually placed directly in front of the optical receiver. Its purpose is to amplify an optical signal that has been transmitted (and attenuated) over some distance to effectively increase the sensitivity of the detector. An optical filter can be placed in front of the receiver and after the preamplifier to minimize the noise from the spontaneous emission at frequencies other than the signal frequency.

7.3. OPTICAL NOISE: SYSTEM ASPECTS

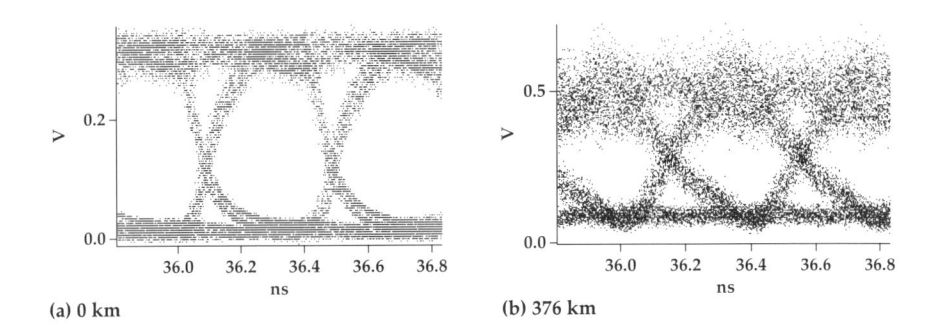

Figure 7.6: Recorded eye patterns, captured for an experimental integrated laser-electro-absorptive modulator transmitter at 1550 nm. Curve (a) is after transmission over 0 km, while curve (b) is after transmission over 276 km of standard (non dispersion shifted) fiber. Courtesy L. Adams, Lucent Technologies, Murray Hill, NJ.

A common way to characterize the performance of a preamplifier is through its noise figure. The noise figure is defined as the ratio of the signal to noise ratio at the input of the preamplifier-receiver to that at the output of the preamplifier-receiver combination:

$$\text{NF} = \frac{(\text{SNR})_{\text{in}}}{(\text{SNR})_{\text{out}}} \tag{7.70}$$

The noise figure will always be greater than one, due to the fact that the amplifier adds noise during the amplification process and the signal to noise ratio at the output is always lower than that at the input. The noise figure value is usually given in dB. A high noise figure implies that the signal to noise ratio has been impaired by the amplification process. One can show that for an ideal amplifier the quantum limit of the noise figure — i.e., the best value achievable — is 3 dB. Noise figures close to 3 dB have been obtained with erbium-doped fiber amplifiers.

The noise figure at the input of the amplifier is computed assuming that the signal is shot noise limited. The shot noise for a current I_s in a bandwidth B_e is $2eI_sB_e$, where the current I_s is related to the optical power P by the relation $I_s = (P/h\nu)e$. The electrical signal to noise ratio at the input of the amplifier is thus

$$(\text{SNR})_{\text{in}} = \frac{I_s^2}{2eI_sB_e} = \frac{I_s}{2eB_e} \tag{7.71}$$

The signal to noise ratio at the output of the amplifier is given by the expression

$$(\text{SNR})_{\text{out}} = \frac{(GI_s)^2}{N_{s-sp} + N_{sp-sp} + N_{\text{shot}}} \tag{7.72}$$

where the noise powers are given by the expressions in Section 7.3.2 (the thermal noise will be negligible at high enough gains). The noise figure can then be calculated to be

$$\text{NF} = \frac{G\frac{I_s^2}{e}I_{sp}\frac{1}{B_0} + \frac{1}{4}\frac{I_s}{e}I_{sp}^2\frac{(2B_0-B_e)}{B_0^2} + I_s(GI_s + I_{sp})}{(GI_s)^2} \tag{7.73}$$

Using the fact that the spontaneous emission power current is given by $I_{sp} = 2n_{sp}(G-1)eB_o$, we can further write the noise figure as

$$\text{NF} = 2n_{sp}\frac{(G-1)}{G} + \frac{1}{G} + \frac{n_{sp}(G-1)^2 e(2B_0 - B_e)}{G^2 I_s} + \frac{2(G-1)n_{sp}eB_0}{G^2 I_s} \quad (7.74)$$

The first term is the dominant term in the right-hand side of equation 7.74. Assuming that $G \gg 1$ and reasonably high input powers, the noise figure is given by $2n_{sp}$. Since n_{sp} is at best 1, when complete inversion is attained, the lowest noise figure achievable is 2 (i.e., 3 dB), given the assumptions just outlined. This is the source of the often quoted 3 dB noise figure limit. Cases where the noise figure is below the 3 dB noise limit can sometimes be encountered. They usually occur when the gain is low and the inversion factor is close to unity, a situation that typically arises when the fiber length is short. Since most practical systems are either high gain or else not well inverted because of saturation, this situation is seldom encountered in practice.

As we saw in Section 7.2.1, n_{sp} depends on the inversion in the amplifier as well as the absorption and emission cross sections $\sigma_a(\lambda)$ and $\sigma_e(\lambda)$ at the signal wavelength λ:

$$n_{sp} = \frac{\sigma_e(\lambda)N_2}{\sigma_e(\lambda)N_2 - \sigma_a(\lambda)N_1} = \frac{N_2}{N_2 - \frac{\sigma_a(\lambda)}{\sigma_e(\lambda)}N_1} \quad (7.75)$$

In an erbium-doped amplifier, complete inversion can only be achieved with 0.98 μm pump sources. Therefore, a 0.98 μm pumped amplifier can achieve a better noise figure than a 1.48 μm pumped amplifier. As an example, Figure 7.7 shows the noise figure at a signal wavelength of 1550 nm as a function of amplifier gain for that signal both for 0.98 μm and 1.48 μm pumping, from amplifier simulations. The fiber lengths are chosen to have approximately the same curves of gain as a function of pump power. Longer fiber lengths are required for 1480 nm pumping due in significant part to the smaller overlap integral between pump and signal, and also the stimulated emission at the pump wavelength.

As a result of the fact that the noise figure is a function of the cross sections, longer signal wavelengths will have a lower noise figure than shorter signal wavelengths. This is a direct result of the decrease in the emission to absorption cross section ratio, as shown in Figure 7.8.

The noise figure can also be very simply written in terms of the ASE power exiting the fiber in a bandwidth $\Delta\nu$. Considering the first two terms of equation 7.74, since the noise power is given by $P_{ASE} = 2n_{sp}h\nu\Delta\nu(G-1)$ we can write

$$\text{NF} = \frac{P_{ASE}}{h\nu\Delta\nu G} + \frac{1}{G} \quad (7.76)$$

In order to calculate the preamplifier receiver BER performance we use equations 7.40 through 7.54 to arrive at an expression for the Q of the pre-amplified receiver. With the simplifying assumption that the dominant noise source is signal-spontaneous beat noise and that the extinction ratio is infinite, equation 7.54 becomes

$$Q = \sqrt{\frac{P_{in}}{h\nu 2B_e n_{sp}}} \quad (7.77)$$

7.3. OPTICAL NOISE: SYSTEM ASPECTS   223

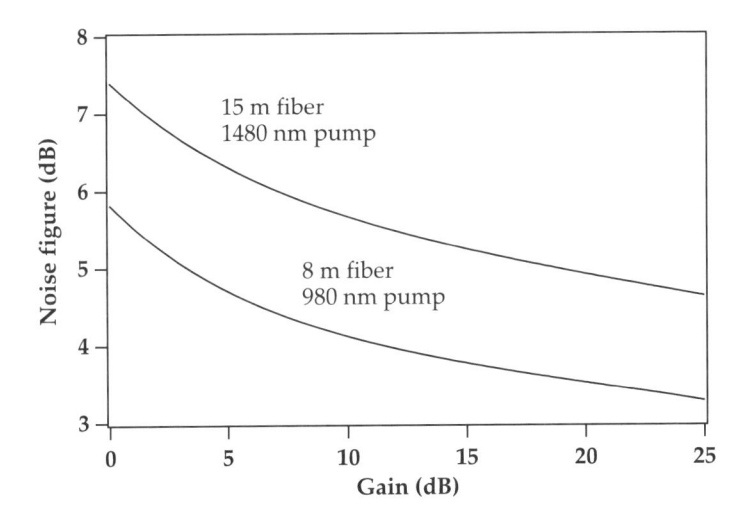

Figure 7.7: Noise figure at 1550 nm as a function of gain, for $1.48\,\mu m$ (15 m fiber length) and $0.98\,\mu m$ (8 m fiber length) pumping of an erbium-doped fiber amplifier (fiber A of Chapter 6). From numerical simulations.

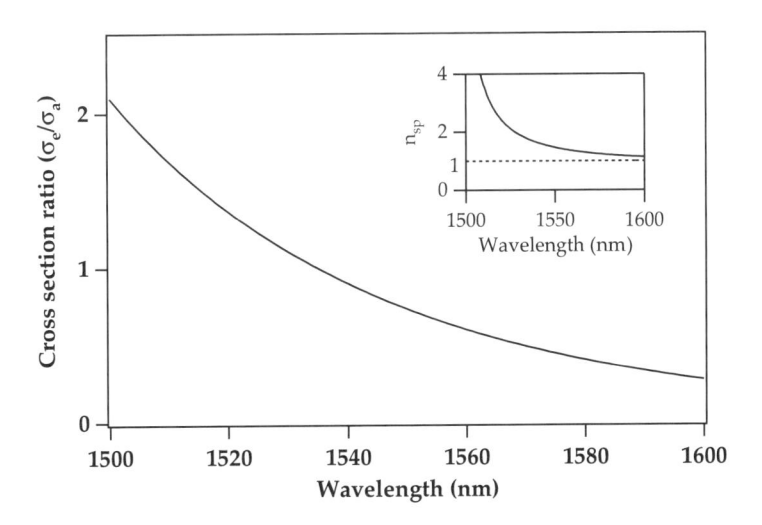

Figure 7.8: Ratio of the emission to absorption cross section as a function of wavelength for an alumino-germano-silica fiber (fiber A of Chapter 6). The inset graph shows the corresponding inversion parameter n_{sp} for the example of an inversion level of 75% of the population in the upper state and 25% in the ground state, calculated using equation 7.75.

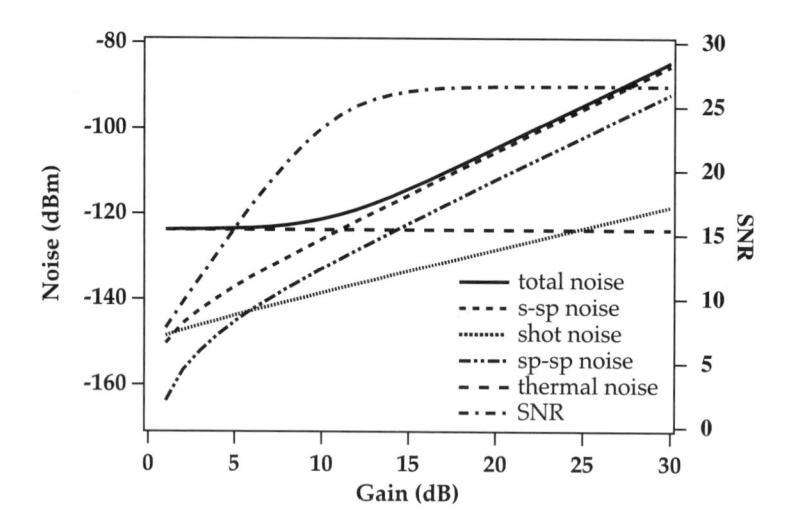

Figure 7.9: Noise powers and signal to noise ratios for an erbium-doped fiber preamplifier, as a function of the gain G of the amplifier, computed from the equations of section 7.3.2 (parameters listed in the text).

Recalling that a BER of 10^{-9} requires Q = 6 and using the fact that $P_{in}/h\nu 2B_e$ = (average number of photons per bit), we find that the ultimate sensitivity, taking only signal-spontaneous noise into consideration, is 36 photons per bit. Including the spontaneous-spontaneous noise into the expression for Q increases the ultimate senisitivity. One finds that the number of photons per bit for a 10^{-9} BER can be, as examples, 39 ($B_o = B_e$, polarizer at detector input), 40 ($B_o = B_e$, no polarizer at detector input), 41 ($B_o = 2B_e$, polarizer at detector input), 43 ($B_o = 2B_e$, no polarizer at detector input). A more complete calculation of the SNR that takes all noise sources into account is shown in Figure 7.9.

Figure 7.9 shows the noise powers and the total SNR due to the contributions of thermal noise, shot noise, signal-spontaneous noise, and spontaneous-spontaneous noise. At low gains, the thermal noise is dominant; at high gains, the signal-spontaneous beat noise becomes dominant. The preamplifier modeled in Figure 7.9 operates at a bit rate of 5 Gb/s, has an electrical filter with a bandwidth $B_e = 2.5$ GHz, an optical filter with a bandwidth of 100 Å, and assumes $n_{sp} = 1.4$ and an input power of -30 dBm. The thermal noise current was chosen to yield a base receiver sensitivity (no optical preamplification) of 5000 photons per bit at 5 Gb/s, which is typical for a PIN receiver. Figure 7.10 shows for the same conditions the relative contributions of the thermal and signal-spontaneous noise terms to the total noise as a function of the preamplifier gain.

Preamplifiers are often characterized by their sensitivity. The sensitivity of a preamplifier is defined as the minimum input signal required to achieve a bit error rate of 10^{-9} at a given bit rate. Figure 7.11 shows the receiver sensitivity, using realistic device parameters, as a function of the gain of the preamplifier. These curves were calculated for a bit rate of 5 Gb/s. The curves can be scaled for other bit rates by using the fact

7.3. OPTICAL NOISE: SYSTEM ASPECTS 225

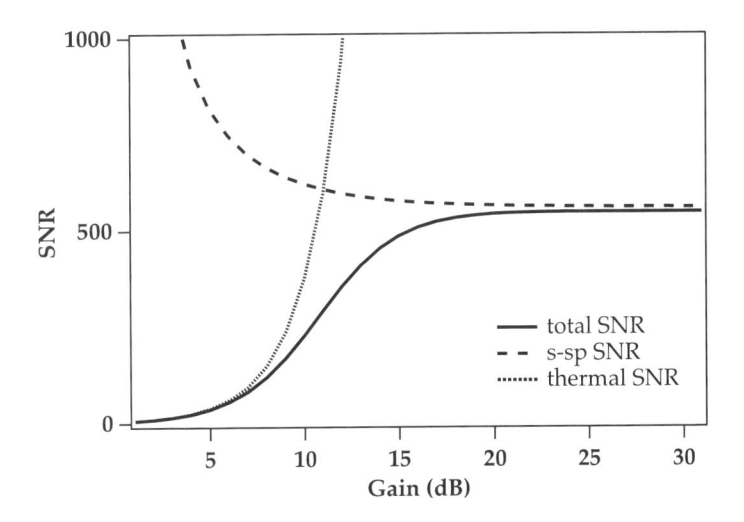

Figure 7.10: SNR of an erbium-doped fiber preamplifier as a function of the gain of the amplifier. The operating parameters of the preamplifier are the same as in Figure 7.9.

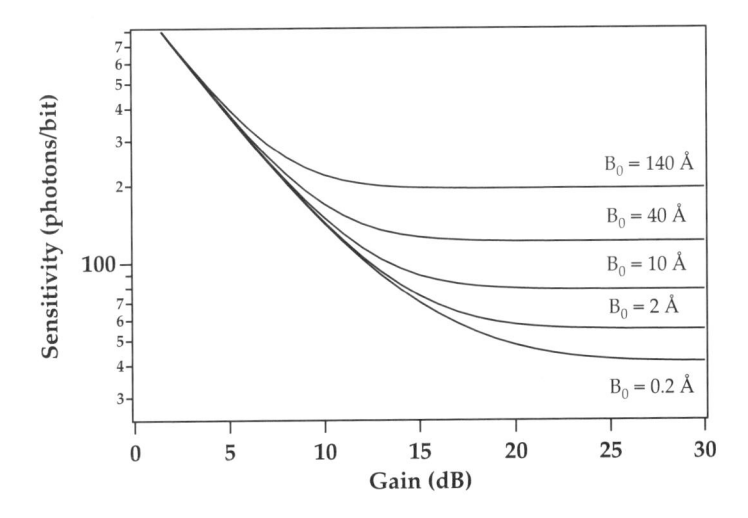

Figure 7.11: Sensitivity of an erbium-doped fiber preamplifier operating at 5 Gb/s, for the indicated optical bandwidths (for an amplifier noise figure of 3 dB).

that these curves are for optical bandwidths equal to the electrical bandwidth (setting it equal to one-half of the bit rate) times a factor of 1, 10, 50, 400, and 700 (respectively). As seen in Figure 7.11, a receiver sensitivity of less than 100 photons per bit can be achieved provided the optical bandwidth is less than $\simeq$ 100 times the electrical bandwidth.

7.3.4 Optical Inline Amplifiers - Amplifier Chains

SNR for a Single Amplifier

Inline amplifiers are used for long haul transmission systems (e.g., transoceanic systems), where the signal needs to be amplified along the way to arrive at the receiver end with some reliable chance of being detected.[18, 19, 20] Inline amplifiers typically operate with moderately large optical input signals. Therefore the signal to noise at the output of the amplifier is dominated by signal-spontaneous beat noise. In this limit the equivalent electrical SNR at the output of the amplifier is given by

$$\text{SNR} = \frac{S}{N_{s-sp}} = \frac{GP_{in}}{4h\nu n_{sp} B_e (G-1)} \tag{7.78}$$

where the received electrical signal power S is given by $S = (GI_s)^2$ where I_s is the current corresponding to the optical signal P_{in} (assumed to be continuous) at the input to the amplifier. Thus, provided G is reasonably high, the SNR is determined only by the input power and the inversion parameter n_{sp}. More specifically, the SNR is independent of the gain. This is an important result that governs the system performance of inline amplifiers. Considering the ratio of the signal-spontaneous and spontaneous-spontaneous noise terms, the above approximation is valid when signal-spontaneous beat noise is the dominant optical noise source, i.e., when

$$\frac{P_{in}}{h\nu} \gg B_o n_{sp} \tag{7.79}$$

where we have used the fact that $G \simeq (G-1)$ and that $B_o \gg B_e$. Equation 7.79 implies that if the signal photon flux is larger than n_{sp} per unit of optical bandwidth, signal-spontaneous beat noise is the dominant noise source. In practice, for an optical bandwidth of 100 Å, the input power must be -30 dBm or larger.

In the absence of gain saturation, equation 7.78 indicates that the most advantageous placement of an amplifier in a fiber link is at the beginning of the link where the input power is the highest, i.e., in a power amplifier configuration. However, in this configuration, gain saturation will reduce the effective gain of the amplifier and reduce the signal strength at the end of the link. In this case, the link performance is degraded by the increased effect of thermal noise. The optimal placement will depend on the detail of the system and amplifier parameters.[21] As a general rule, the amplifier should be placed as close to the input as possible without causing gain saturation. Figure 7.12 shows a detailed calculation of the system SNR as a function of the amplifier placement in a link with a total fiber loss of 60 dB. In this case, the optimum placement is approximately mid-span. For the purposes of this calculation the noise figure was assumed to be constant. The noise figure is actually higher for amplifiers placed near the input end and operating in saturation, but in this situation the SNR is mostly dictated by the large loss of the remaining span of fiber. When the amplifier is moved away from the input and no longer operating in deep saturation, the noise figure is essentially constant.

7.3. OPTICAL NOISE: SYSTEM ASPECTS

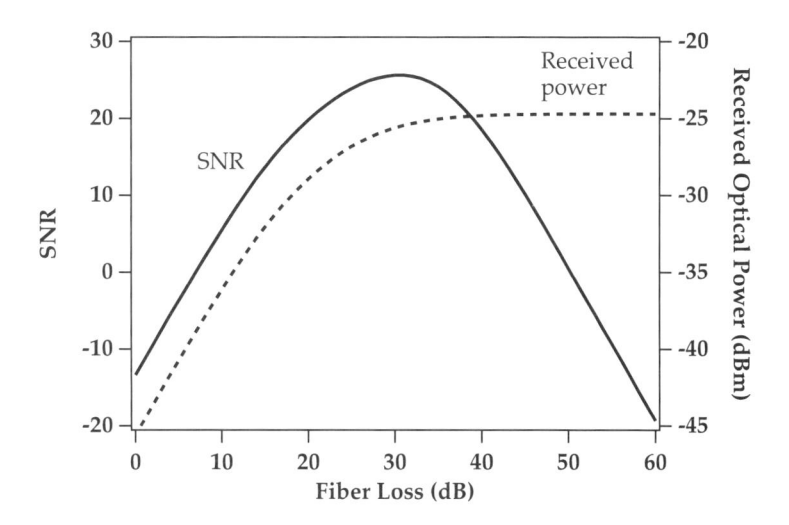

Figure 7.12: Link output SNR and signal power as a function of inline amplifier placement, from numerical simulations. The horizontal scale is the fiber loss preceeding the amplifier, in dB. The total link loss is 60 dB and the calculation was done for a transmitter power of 0 dBm, B_o = 100 Å, amplifier small signal gain G_0 = 35 dB, output saturation power P_{sat} = 10 dBm, and NF = 3.7 dB.

Amplifier Chains - Optimum Amplifier Spacing

The simplest way to analyze a cascade of optical amplifiers is to assume that all amplifiers have the same gain and that the loss between amplifiers exactly matches the amplifier gain. The signal output power at the end of the chain is assumed to be equal to the signal input power at the beginning of the chain, a relatively accurate assumption when the ASE generated is small compared to the signal power. In fact, the ASE is growing at the expense of the signal in saturated amplifier chains, but this does not obviate the main conclusions of the following discussion. Each amplifier generates an equal amount of ASE as the other amplifiers, and this ASE propagates transparently to the output of the chain, much as the signal does. Thus, the ASE at the output of the chain is the linear addition of the ASE generated by each amplifier. The SNR at the output of the amplifier chain is then obtained by replacing n_{sp} in the expression for the SNR for one amplifier by Nn_{sp}, where N is the number of amplifiers in the chain.[2]

We consider the case of the configuration of Figure 7.13A, which corresponds to an amplifier chain where the input power, P_{in}, to the amplifiers is independent of the amplifier gain G or the span loss L.[22] We compute the SNR under the assumption that it is determined by signal-spontaneous beat noise. The SNR at the output of the amplifier chain can be written, as noted above, using the fact that the ASE at the output of the amplifier chain is N times the ASE generated by one amplifier. The signal input to the receiver is P_{in}/L, where $GL = 1$

$$\text{SNR}_A = \frac{P_{in}}{4Nn_{sp}h\nu B_e(G-1)L} \tag{7.80}$$

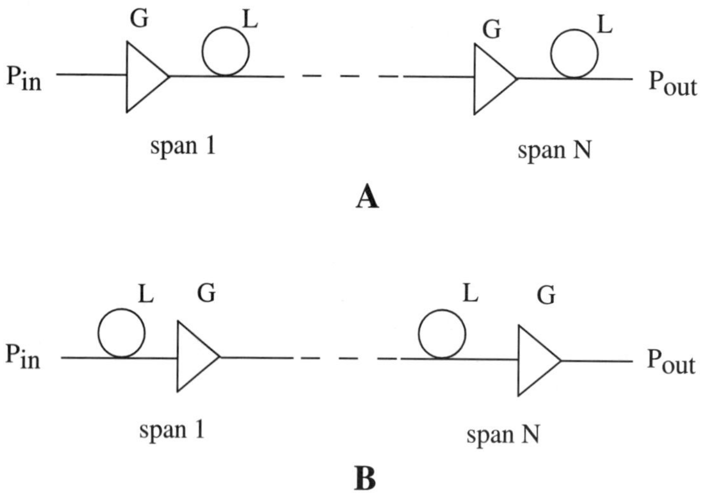

Figure 7.13: Two conceptual configurations for an optical amplifier chain. (A): Amplifier before the lossy fiber span; (B): Amplifier after the lossy fiber span. The amplifier gain is G and the span loss L. While seemingly similar, the two configurations lead to different conclusions as to optimum amplifier spacing.

The total gain G_{tot} and loss L_{tot} of the system are given by

$$G_{tot} = \frac{1}{L_{tot}} = G^N = \frac{1}{L^N} \qquad (7.81)$$

The SNR can then be written as

$$SNR_A = \frac{P_{in}}{4n_{sp}h\nu B_e} \frac{(G_{tot}^{1/N} - 1)N}{G_{tot}^{1/N}} \qquad (7.82)$$

Equation 7.82 yields the result that the SNR of case A is maximized when $N = 1$, i.e. there should be only one amplifier and the amplifier spacing should be the longest possible. From the point of view that each amplifier adds noise, it makes sense that we find that the minimum number of amplifiers gives the best system performance.

We now turn our attention to the system in Figure 7.13B. This configuration corresponds to a situation where the amplifier output power is held constant, since whatever the span loss L, the amplifier will be selected to have a gain G such that the output of each lossy span and amplifier unit is equal to the signal input power P_{in}, to ensure a fully transparent chain. The SNR at the output of the amplifier chain is given by:

$$SNR_B = \frac{P_{in}}{4Nn_{sp}h\nu B_e(G - 1)} \qquad (7.83)$$

The difference with respect to equation 7.78 (the factor of G in the numerator) is that here P_{in} is measured at the beginning of the chain, whereas in 7.78 P_{in} was measured

7.3. OPTICAL NOISE: SYSTEM ASPECTS

at the input to the amplifier. Similarly to the derivation of SNR_A, we can write

$$\text{SNR}_B = \frac{P_{in}}{4n_{sp}h\nu B_e} \frac{1}{(G_{tot}^{1/N} - 1)N} \quad (7.84)$$

Equation 7.84 can be rewritten as

$$\text{SNR}_B = \frac{P_{in}}{4n_{sp}h\nu B_e} \frac{1}{\ln(G_{tot})} \frac{\ln(G)}{(G-1)} \quad (7.85)$$

In this case the SNR is improved by increasing the number of amplifiers and reducing the gain G correspondingly. In the limit where the number of amplifiers goes to infinity (a distributed amplifier) the SNR is maximized and is equal to

$$\text{SNR}_B(\max) = \frac{P_{in}}{4n_{sp}h\nu B_e} \frac{1}{\ln(G_{tot})} \quad (7.86)$$

This result can be understood from the fact that as we decrease the amplifier gain, the amplifier input power increases since the preceding span is now shorter and its loss less. As shown in the previous section, the amplifier output SNR depends only on its input power. Another way of saying this is that (as seen abundantly in the simulations of Chapter 6) any loss prior to the input of an amplifier degrades the noise figure and the output SNR.

How do we resolve the apparent contradiction between the conceptual conclusions of cases A and B? The maximum SNR for case A ($P_{in}/4n_{sp}h\nu B_e$) is the highest SNR for either configuration. In fact, practical limitations render it impossible to construct an amplifier without limitations on output power. For example, a 10,000 km system with span fiber loss of 0.2 dB/km would require a single amplifier with a gain of 2,000 dB! This is clearly not practical. In addition, the onset of nonlinear effects above a certain power level limit the output power to be launched into the chain. Real life systems are similar to the configuration of Figure 7.13B, where the amplifier output power is held constant. In this case, short amplifier spacings are desirable. Given the economic cost of amplifiers, practical systems use amplifier spacings as long as possible while still maintaining a minimum system SNR for low error rate detection.

SNR in Cascaded Amplifier Chains

We consider the expression for the SNR (assuming only signal-spontaneous beat noise) with a continuous signal P_{in}, for configuration B.

$$\text{SNR} = \frac{P_{in}}{4Nn_{sp}h\nu B_e(G-1)} = \frac{P_{in}}{4n_{sp}h\nu B_e \ln(G_{tot})} = \frac{B_o}{2B_e}\text{SNR}_{opt} \quad (7.87)$$

For a random pattern of 1s and 0s, and an infinite extinction ratio, one can derive, given that the signal-spontaneous noise is only present during the 1s, that the electrical SNR is given by:

$$\text{SNR} = \frac{P_{in}}{2Nn_{sp}h\nu B_e(G-1)} = \frac{B_o}{B_e}\text{SNR}_{opt} \quad (7.88)$$

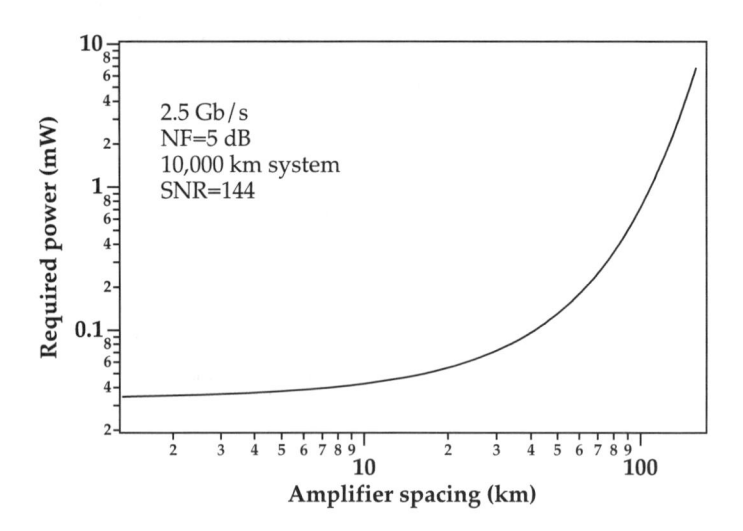

Figure 7.14: Required signal power needed to maintain an electrical SNR of 144 for a 10,000 km link operating at 2.5 Gb/s, with a transmission fiber loss of 0.20 dB/km, as a function of the amplifier spacing, from equation 7.88.

where P_{in} is the average optical power. Although this equation is only approximate, as it neglects the decrease in signal strength in saturated amplifier chains where the ASE is robbing gain at the expense of the signal, it gives valuable insight into the main determinants of noise accumulation in optically amplified systems. By fixing the SNR to the desired value for near error free performance, any one of the remaining quantities in equation 7.87 (e.g., amplifier spacing or amplifier gain) can be computed in terms of the other quantities (e.g., overall system length, electrical bandwidth, signal input power). For example, equation 7.87 shows that when increasing the amplifier spacing, for a given system length and bit rate, the input power needs to increase commensurately. In fact, the signal power needed to maintain a given signal to noise ratio in a system of cascaded amplifiers increases exponentially as the spacing between amplifiers increases, as shown in Figure 7.14. The electrical filter bandwidth was assumed to have an ideal value, i.e. one-half of the bit rate. The power required is roughly constant until the spacing starts increasing past $1/\alpha$ where α is the attenuation coefficient of the transmission fiber.

Similarly, when the system length increases, for a given signal input power and bit rate, the amplifier spacing needs to decrease to maintain a given SNR. The electrical SNR expression, equation 7.87, can be plotted as a function of the amplifier gain to show the advantages of using many small gain amplifiers. Figure 7.15 shows the number of amplifiers that can be cascaded for a given amplifier gain, while maintaining an SNR of 21.6 dB, as well as the corresponding total system length where a fiber loss of 0.20 dB/km was assumed. The curves show the impact of changing the bit rate from 2.5 Gb/s to 5 Gb/s, and also the effect of a degradation in the amplifier noise figure from 4 dB to 6 dB. Note, for example, that with an amplifier gain of 18 dB and a noise figure of 6 dB, simply increasing the bit rate from 2.5 Gb/s to 5 Gb/s reduces the allowed system

7.3. OPTICAL NOISE: SYSTEM ASPECTS

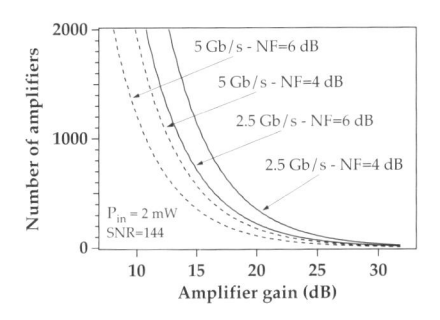

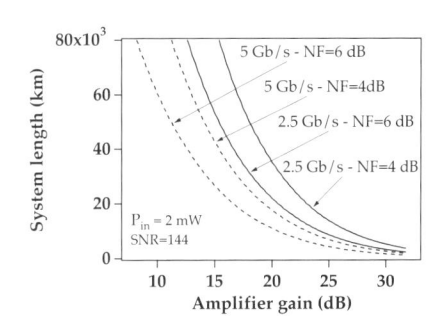

Figure 7.15: Maximum number of amplifiers that can be cascaded in a transparent chain (left), and equivalent total system length allowed (right), while maintaining an electrical signal to noise ratio (SNR) of 21.6 dB, as a function of amplifier gain. Bit rates are 2.5 and 5 Gb/s and amplifier noise figures of both 4 dB and 6 dB were considered; the fiber loss is assumed to be 0.20 dB/km for the right-hand curves. The curves were computed using equation 7.88.

length from 32,000 km to 16,000 km. Practical systems are not as forgiving as those simulated in Figure 7.15 and require an SNR well in excess of 21.6 dB. Nevertheless, Figure 7.15 illustrates the advantage of using many small gain amplifiers, as well as the sensitivity to some of the key system parameters.

The optical SNR is often used to quickly characterize the system properties of a cascaded amplifier chain, since the SNR can be directly measured on an optical spectrum analyzer (typically with a bandwidth of 1 Å). While this gives some good indications as to the noise properties of the system, the SNR alone is not necessarily an accurate predictor of the BER performance. The Q value contains more information than the optical SNR, for example error rates resulting from inter-symbol interference and non-linear effects. Nevertheless, system designers often know heuristically what level of optical SNR is required to obtain a certain BER.

We can plot the optical SNR as a function of the various system parameters. The optical SNR is given by the expression

$$\text{SNR}_{\text{opt}} = \frac{P_{\text{in}}}{2Nn_{\text{sp}}h\nu B_o(G-1)} \qquad (7.89)$$

The impact of using a high input signal is seen in Figure 7.16. We used an optical SNR requirement of 25 dB in 1 Å, a value often used by systems experimentalists.[16] While at low powers it is difficult to cover transoceanic distances with the given assumptions, once the power increases to a few mW the distance can be covered with required amplifier spacings of several tens of km. Note that in a WDM system the power per channel is the total power divided by the number of channels. Since the amplifier output power is limited by amplifier saturation, using a large number of channels will result in the system operating on one of the lower power curves of Figure 7.16, on a per channel basis, thus requiring a short amplifier spacing or a lower bit rate. Finally, reducing the amplifier noise figure has a significant impact in relaxing the requirements on the other system parameters such as amplifier spacing and signal input power.

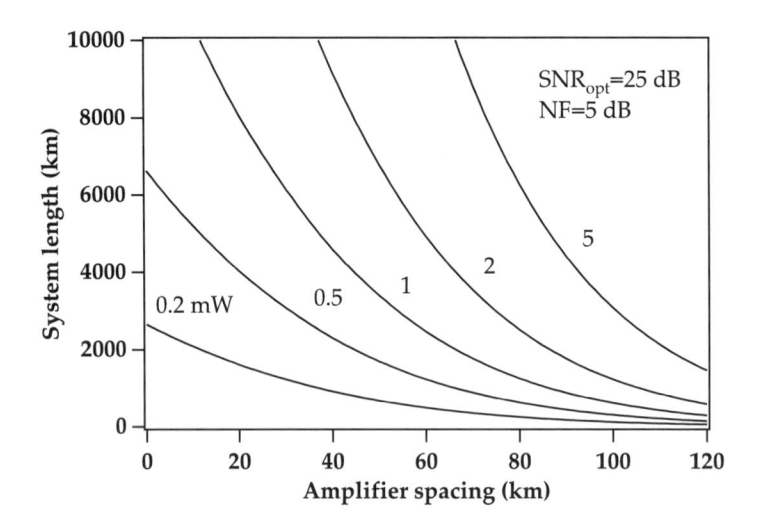

Figure 7.16: Maximum system length in a transparent amplifier chain while maintaining an optical signal to noise ratio (SNR) of 25 dB in 1 Å, as a function of amplifier spacing and input optical power. The amplifier noise figure is 5 dB and the fiber loss is assumed to be 0.20 dB/km.

A more sophisticated analysis of cascaded amplifiers than simply assuming constant signal output power after each amplifier must take gain saturation and unequal gain effects into account. A common approximation to treat gain saturation is to assume that the total output power (signal + ASE) is constant in all amplifiers.[23] One can then show that the signal power after amplifier i is given by

$$P^i_{sig} = P_{in}(GL)^i \tag{7.90}$$

and the corresponding total ASE power after amplifier i is given by

$$P_{sp} = 2n_{sp}h\nu(G-1)B_o \sum_{k=0}^{i}(GL)^k \tag{7.91}$$

where the gain G is given in terms of the input signal power P_{in} and the bandwidth B_o of the optical filter assumed to be placed after each amplifier

$$G = \frac{(P_{in} + 2n_{sp}h\nu B_o)}{(P_{in}L + 2n_{sp}h\nu B_o)} \tag{7.92}$$

and we assumed that all amplifiers have equal gain and the span loss between amplifiers is constant. Because ASE is linearly built up in a cascade of amplifiers, the signal power is decreasing as it propagates through the amplifier chain. In Figure 7.17 we show a calculation of the signal power and ASE power in an amplifier chain. An experimental result is shown in Figure 7.18. Further effects of importance in amplifier chains are spectral narrowing of the gain spectrum (discussed in Chapter 9, section 9.2.7), non linear effects (discussed in Section 7.3.6), and issues related to WDM signal propagation in amplifier chains (discussed in Chapters 8 and 9).

7.3. OPTICAL NOISE: SYSTEM ASPECTS   233

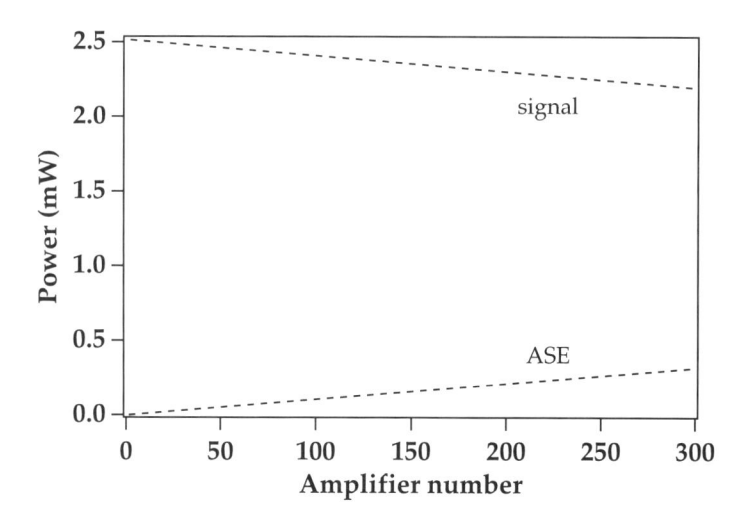

Figure 7.17: Calculated signal power and ASE power in a cascade of amplifiers. The calculation was done using an optical bandwidth of 2 nm, noise figure of 6 dB for the optical amplifiers, signal wavelength of 1558 nm, and a fiber loss of 10 dB between amplifiers.

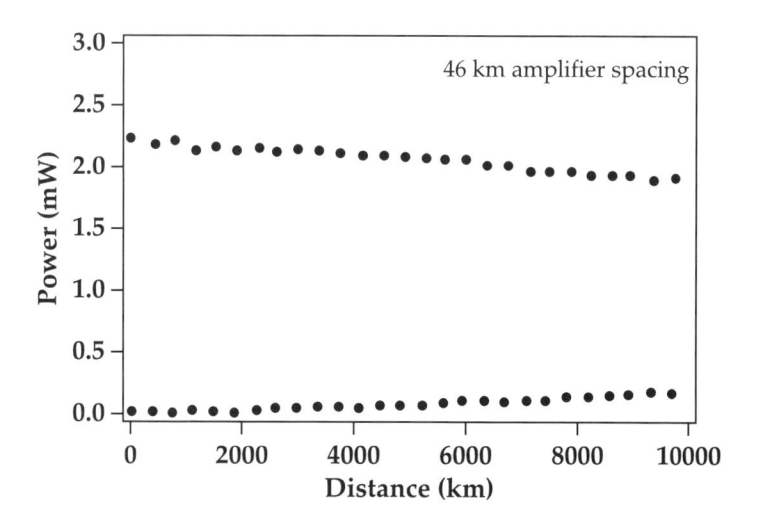

Figure 7.18: Experimental signal power and ASE power in a cascade of amplifiers. From reference [18] (©1995 IEEE).

Noise figure in an amplifier chain

The noise figure for a system consisting of a chain of optical amplifiers can be computed from the noise figure for an individual amplifier.[24] Consider a system of N amplifiers where SNR_i denotes the SNR after amplifier i, and each amplifier provides a gain G to exactly compensate the span loss L. The overall noise figure of the system is given by

$$F_{sys} = \frac{SNR_0}{SNR_N} = \frac{SNR_0}{SNR_1} \frac{SNR_1}{SNR_2} \cdots \frac{SNR_{N-1}}{SNR_N} \quad (7.93)$$

where SNR_0 is the SNR at the input of the system (immediately after the transmitter, and prior to the first span of fiber). The SNR ratios in equation 7.93 are the noise figures of each amplifier multiplied by 1/L since the amplifier noise figure is defined by the SNR's immediately prior and after the amplifier. In equation 7.93, each SNR_i is separated from the following amplifier by a span with loss L, hence we obtain for the system noise figure, in logarithmic units

$$F_{sys} = GF_1 + GF_2 + ... + GF_N = NGF \quad (7.94)$$

assuming all the amplifiers have an equal noise figure and G=L. The SNR degradation in a cascaded amplifier transparent chain is seen to be linear with the number of amplifiers.

An interesting result can be derived when G and L are different, as is the case for a multistage amplifier constructed by piecing together several amplifiers. Equation 7.93 is then written more generally as

$$F_{sys} = \frac{F_1}{L_1} + \frac{F_2}{L_1 G_1 L_2} + \cdots + \frac{F_N}{L_1 G_1 L_2 G_2 \cdots L_N} \quad (7.95)$$

where L_i and G_i refer to the loss prior to amplifier i and the gain of amplifier i, respectively. With $L_i = 1$ for all i this becomes

$$F_{sys} = F_1 + \frac{F_2}{G_1} + \cdots + \frac{F_N}{G_1 G_2 \cdots G_{N-1}} \quad (7.96)$$

which shows that the noise figure of a multi-stage amplifier is dominated by the noise of the first stage.

Power Self-Regulation in Amplifier Chains

One of the main advantages of cascaded erbium-doped fiber amplifiers is their ability for self-regulation of power. When inline amplifiers are run in moderate compression, changes in the input signal power tend to self-correct after a few amplifiers. Figure 7.19 show the typical gain saturation curve of an erbium-doped fiber amplifier. When in saturation, for a nominal signal input power of P_A, the amplifier will be operating at point A. If there is a drop in power of the signal due to a change in the cable conditions preceding the amplifier, the input to the amplifier moves to point B. This point has higher gain and thus increases the signal output power so that it will be closer to P_A at the next amplifier. Correspondingly, if the signal power is higher than it should be (point C), the gain will be lower, thus decreasing the signal power for the next amplifier.

7.3. OPTICAL NOISE: SYSTEM ASPECTS

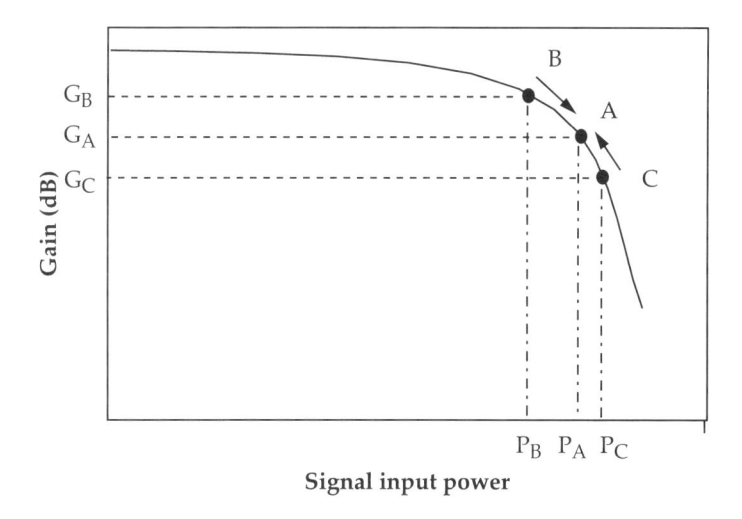

Figure 7.19: Self-regulation effect in cascaded amplifier systems, from the gain saturation profile of an erbium-doped fiber amplifier. The nominal signal input power is P_A. When the signal input power is higher or lower (points B and C), the amplifier tends to return the power to its nominal value in a long haul cascaded amplifier and transmission fiber chain.

This self-regulation is a direct result of the slow gain dynamics of the erbium-doped fiber amplifier, which allows it to regulate the average power of a signal modulated at the high speeds typical of modern high-capacity transmission systems. Note that self-regulation cannot be used as effectively in WDM systems, since it does not regulate the power on a channel-by-channel basis, only on a total power basis.

7.3.5 Noise in Optical Power Amplifiers

Power amplifiers typically boost the signal power emitted directly from a transmission source (typically in the -10 to 0 dBm range) and enable the signal to travel much farther than it might otherwise have without amplification, or to be split among several transmission branches. The typical optical amplifier requirement here is a high ouput power. The noise figure will be that obtained in the saturation regime and can be quite high, typically 5 dB to 10 dB, even with 980 nm pumping. This high noise figure is not a problem from the transmission perspective. Indeed, the signal powers at the input and output of the booster amplifier are so high that the signal to noise ratio, which is the important parameter from the point of view of error-free transmission, is high. This is intuitively clear from the fact that the signal to noise ratio at the output of the amplifier is equal to the signal to noise ratio at the input of the amplifier divided by the noise figure. The high noise figure is counterbalanced by the high signal to noise ratio at the input of the amplifier resulting from the high signal level used.

7.3.6 Nonlinearity Issues

Optical amplifiers provide numerous benefits for optical transmission and allow a significant increase in information capacity of such systems. However, they also open the way to new, previously unimportant system degradations. The most important class of such effects arises from fiber nonlinearities.[25, 26] These effects are becoming better understood, and means to combat them are becoming well known. The nonlinear effects are brought about by the conjunction of two separate system aspects resulting from the use of optical amplifiers. First, the amplifiers increase the pulse energy to the point where nonlinearities (produced by the intensity dependent refractive index) become a cause of information scrambling and loss. Second, many such nonlinearities grow with the distance traveled, and the emergence of optical amplifiers has meant that the distance between transmitter and receiver is large enough that the nonlinear interaction between two pulses can grow significantly over the entire fiber link.

The nonlinear effects to be considered are

- Stimulated Brillouin scattering.
- Stimulated Raman scattering.
- Self-phase modulation.
- Cross-phase modulation.
- Four-wave mixing.

We will consider each of these in turn, giving a description of the effect, and the important orders of magnitude involved, as well as possible ways in which to minimize the problem. The aim is to make the reader aware of the problems and able to realize when they may become important in a particular system.

Two important fiber parameters are needed to estimate the thresholds for the various nonlinearities. These are the effective area and the effective length of a fiber span. Both of these quantities remove the need to integrate the nonlinear interaction effects over the cross section of the fiber or over its length. The power is instead assumed to be constant over the effective area A_{eff} and along the length L_{eff}. The effective area is defined as

$$A_{\text{eff}} = \frac{\left(\int I \, dA\right)^2}{\int I^2 \, dA} \qquad (7.97)$$

Typical values are 80 μm^2 for a conventional single mode fiber and 50 μm^2 for a dispersion shifted fiber. The effective length is defined as

$$L_{\text{eff}} = \frac{1 - e^{-\alpha L}}{\alpha} \qquad (7.98)$$

where α is the attenuation coefficient of the transmission fiber. For typical fiber spans, $\alpha L \gg 1$ and $L_{\text{eff}} \simeq 1/\alpha$. A typical value of L_{eff} is 20 km for $\alpha = 0.22$ dB/km. For a multiple span system the overall L_{eff} is the sum of the individual L_{eff} of each span. In general, nonlinearities increase as the amplifier spacing is increased. This is because

7.3. OPTICAL NOISE: SYSTEM ASPECTS

the launched power needed for a certain SNR increases exponentially as the amplifier spacing is increased, as was shown in Figure 7.14.

Stimulated Brillouin Scattering

Stimulated Brillouin scattering (SBS) arises from the interaction between a photon and an acoustic phonon, as discussed in section 7.2.3.[27, 28] The backscattering event involves the scattering of a photon into a photon of lower frequency with the simultaneous creation of a phonon. The phonons of silica glass involved in the Brillouin scattering events have a typical frequency of 11 GHz, so that the photon is downshifted by this amount. The Brillouin bandwidth $\Delta \nu_B$ is very narrow, typically about 20 MHz. The light created by the SBS process propagates in a direction counter to that of the original lightwave in a single-mode fiber. This results from conservation of momentum. The SBS light is then amplified exponentially with distance as the SBS photons stimulate the creation of more photons of the same frequency. This process robs signal photons and thus increases noise due to both the fact that the signal will not be as powerful at the receiving end and the fact that the random removal of photons is equivalent to the addition of random noise.

The SBS light grows exponentially, with a gain coefficient γ approximately equal to 4×10^{-9} cm/W. The threshold power for the SBS process for a cw narrow linewidth source is given by:[29]

$$P_{th} = \frac{21 \, A_{eff}}{\gamma \, L_{eff}} \qquad (7.99)$$

Because the linewidth of the Brillouin interaction is very narrow, the actual SBS threshold will depend on the actual linewidth of the light source. For an inherently broad (Lorentzian) source, the threshold scales with the factor $(\Delta \nu_L + \Delta \nu_B)/\Delta \nu_B$ where $\Delta \nu_L$ is the Lorentzian linewidth of the source.[26] For modulation-broadened sources, however, the situation is more complicated. Linewidth broadening from ASK and FSK modulation results in an increase in threshold of $2\times$ and $4\times$, respectively, irrespective of the modulation-induced spectral width. PSK modulation, on the other hand, increases the SBS threshold by a factor proportional to the modulation frequency.[30]

For a conventional fiber the threshold for a narrow cw source is about 4.2 mW (6.2 dBm), and for a dispersion-shifted fiber it is about 2.6 mW (4.2 dBm).[31] For a modulated source with an equal number of 1 and 0 bits, on average these numbers would be doubled and would refer to the average power. Figure 7.20 shows the scattered SBS light power and the transmitted power as a function of launched power.

Isolators at each amplifier that are codirectional with the signal prevent the SBS light from crossing from span to span and suppresses the possibility that SBS can grow unabated over an entire transmission link. The SBS process can be eliminated by dithering the central frequency of the signal laser.[32] This broadens out the carrier spectrum and the effective gain profile so that the peak gain is reduced. The dither period needs to be shorter than the time it takes to traverse the effective length, i.e., less than about 100 μs, for a 20 km effective length.

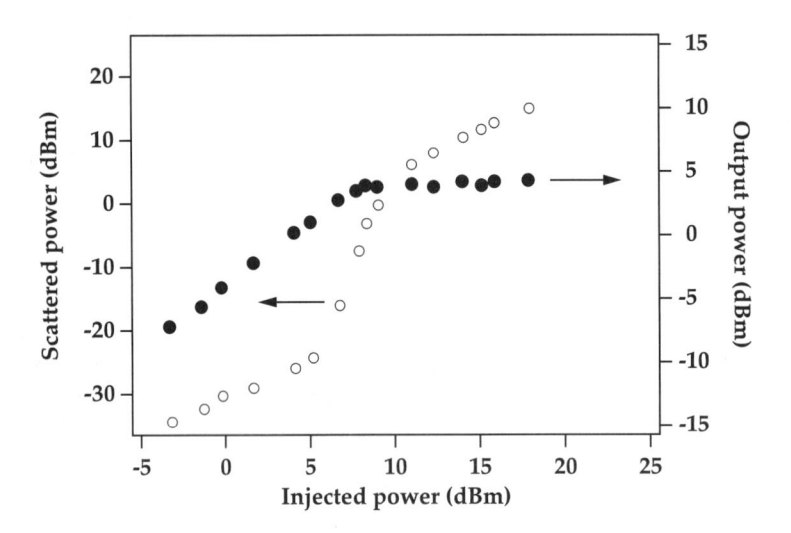

Figure 7.20: Scattered SBS power and transmitted light power as a function of launched power into the fiber. From reference [31].

Stimulated Raman Scattering

Stimulated Raman scattering (SRS) is similar to SBS but differs in that the scattering event is between a photon and an optical phonon.[33] We discussed SRS earlier in section 7.2.3 in the context of its role as an amplifier mechanism. Optical phonons have much higher frequencies and energies than acoustical phonons so that the energy shift between the original and SRS-created photon is quite large, about 15 THz. The bandwidth for the Raman process is also quite large, up to $\sim$ 20THz. The peak Raman gain coefficient in silica is $\gamma \simeq 7 \times 10^{-12}$ cm/W.[3] The threshold power for the SRS process is given by:[29]

$$P_{th} = 32 \frac{A_{eff}}{\gamma L_{eff}} \tag{7.100}$$

For a conventional fiber the threshold is then about 1.8 W and for a dispersion-shifted fiber about 1.1 W (with L_{eff} = 20 km). The SRS light can propagate in the forward direction and is amplified by the signal as signal and SRS light copropagate. In the case of multiple channels, for a WDM system, SRS will impair the system performance by providing an avenue by which energy can be transferred from the short wavelength to the long wavelength channels.[34] For equally spaced channels in a system with N channels with spacing $\delta \nu$ and input power P_0, and for an effective length L_{eff}, the following relationship needs to be respected to obtain a power penalty of less than 1 dB (per channel) from SRS:

$$N \cdot P_0 \cdot (N-1)\delta \nu \cdot L_{eff} \leq 10^4 \text{ W} \cdot \text{GHz} \cdot \text{km} \tag{7.101}$$

This produces a significant limitation in long haul systems where L_{eff} is long.

Self-Phase Modulation

In general, the refractive index of a medium is intensity dependent. Under most conditions this can be neglected except when the pulse intensity becomes large. This occurs with very short pulses, even with weak energies per pulse, as the peak intensity is approximately the pulse energy divided by the pulsewidth. As described in the discussion of solitons (Chapter 9, section 9.3), the varying intensity across the pulse envelope results in a varying index and thus a varying frequency shift of the pulse spectral components. Spectral broadening is thus the main consequence of SPM.

SPM can impair system performance if there is dispersion at the central wavelength of the pulse, since the different spectral components will then walk away from each other and the pulse will broaden temporally. This can lead to overlap between adjacent bits and increased error rates. The walkoff of the SPM broadened spectrum can be minimized if the pulse wavelength and the zero dispersion wavelength of the fiber are very close. This is often not a good solution for a single channel system, however, since operating near zero dispersion leads to significant impairment from the phase-matched mixing between the signal and the amplifier noise. In addition, if any optical filters (or the amplifiers themselves) clip the broadened spectrum then pulse distortion will occur in the time domain. In principle, if the total spectrum is admitted to the receiver then no pulse distortion will occur, although this may lead to impairment of the SNR due to the increased ASE noise admitted. Thus, dispersion compensation at the ends of the transmission link, which null the overall dispersion of the link, can significantly reduce the combined effect of SPM and GVD. For WDM systems, the spectral broadening can lead to spectral overlap between adjacent channels.

Cross-Phase Modulation

Cross-phase modulation (XPM) arises from the same physical principles as SPM and occurs in the context of WDM systems. Intensity variations in one pulse alter the phase of another channel, via the nonlinear refractive index of the glass. This leads to spectral broadening as in SPM. Two pulses from two different channels, and thus possessing different center frequencies, will travel at different velocities along the fiber and can cross each other at some point.[26] As the two pulses traverse each other, one pulse's time varying intensity profile will cause a frequency shift in the other. However, since one pulse will interact both with the rising edge and the falling edge of the other pulse, the net XPM effect will usually be nil since the rising edge and falling edge of a pulse will cause equal and opposite frequency shifts. When the collision length is longer than L_{eff} or the pulses pass through an amplifier during the collision, the symmetry of the collision is broken and XPM will have a larger impact on the pulses.[35] XPM is a major concern for the design of WDM systems. In certain situations, in particular when the channel separation is small, it can be a significant limiting factor.[36, 37, 38, 39]

Four-Wave Mixing

Four-wave mixing (FWM) is a nonlinear interaction that can occur between several different wavelength channels in a WDM system. In the interaction between two distinct wavelength channels, the electromagnetic waves generate an intensity that is proportional to the cube of the sum of their electric fields and thus contains beat terms at

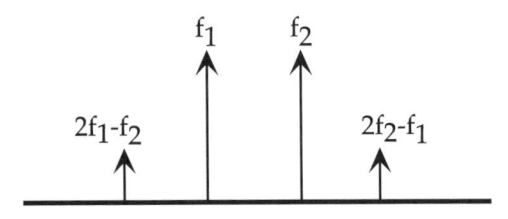

Figure 7.21: Creation of spectral sidebands in a two-channel transmission system (frequencies f_1 and f_2), due to four-wave mixing between the two channels.

various sum and difference frequencies, as shown in Figure 7.21. The index of refraction is modulated at the beat frequencies and generates sidebands of the original frequencies. These sidebands have a significant effect if the original pulses and the sidebands travel together along the fiber for a long enough distance for there to be significant power transferred to the sidebands. Hence for a dispersion-shifted fiber (dispersion zero at 1.55 μm) with channels near 1550 nm where the pulses travel together for long distances, the four-wave mixing efficiency is quite high. For such fibers, significant deterioration of the spectral purity can occur after about 20 km of fiber. The spectral changes can be very pronounced, as shown in Figure 7.22 for the extreme case when phase matching is intentionally maintained in the fiber transmission system. For a fiber with significant dispersion at 1.55 μm, such as a standard fiber with dispersion zero near 1.3 μm, the effect is much smaller. Figure 7.23 shows the four-wave mixing efficiency as a function of channel spacing for several different dispersion values.

For N channels, $N^2(N-1)/2$ sidebands are created. The sidebands deplete power at the expense of the main frequencies. Additionally, the sidebands can fall at the same frequency as one of the signal channels thus causing interference and information loss. One way to combat the four-wave mixing process is to use unequal channel spacings so that the mixing products do not coincide with signal frequencies.[42, 43] Another, and very effective way, is by fiber dispersion management.[44, 45] This method alternates fibers with different dispersions along the transmission path so that the overall dispersion is 0 (on average), but the local dispersion is nonzero. This method has been used with significant success in constructing high capacity long haul WDM systems. Local dispersion with an absolute value greater than 1 ps/km · nm has been shown to effectively reduce the four wave mixing penalty, in the case of an 8 × 10 Gb/s system.[46]

7.3.7 Analog Applications

Introduction

Fiber optics is emerging as a prime vehicle for information transmission in analog systems, heretofore the exclusive domain of electrical transmission systems (e.g., coaxial cable). The underlying reason is the low loss of optical fiber such that the distance between repeaters or nodes can be increased to 10–20 km vs. the 1–5 km usually obtained with coaxial cable. The most important application in this case is CATV trunk line systems. Optical amplifiers can also increase transmission span lengths and thus can play an important role in a fiber-based CATV transmission system. In addition, amplifiers

7.3. OPTICAL NOISE: SYSTEM ASPECTS 241

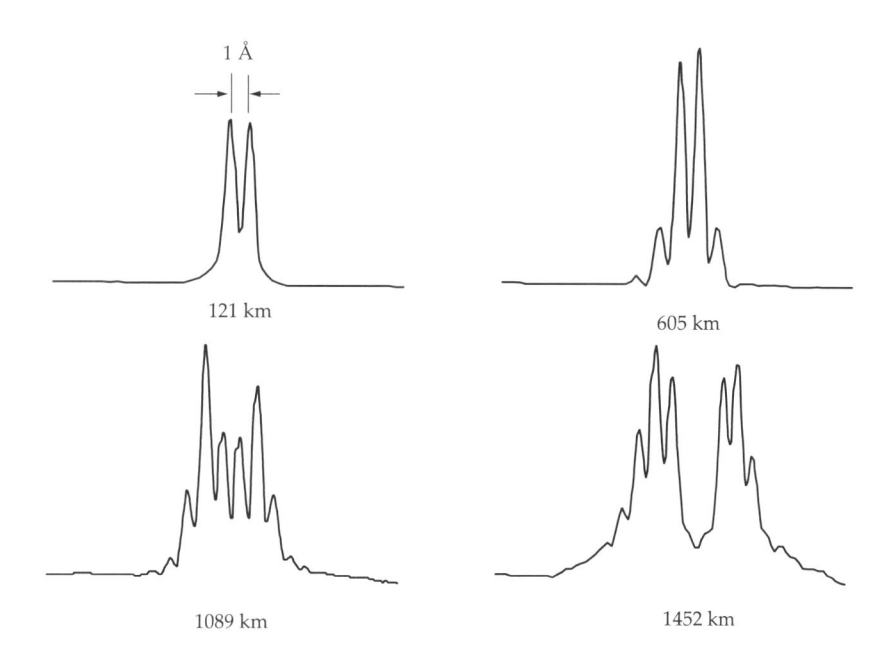

Figure 7.22: Creation of spectral sidebands in a two-channel transmission system near 1.55 μm as a function of distance traveled in a dispersion-shifted fiber (signal wavelengths at the dispersion zero). From reference [40]. The horizontal scale represents wavelength. Amplifiers are spaced every 40 km and the launched signal power is 1 mW per channel.

provide the opportunity to boost a signal before splitting it among many users. This latter application would insert optical amplifiers in CATV distribution networks.

Analog systems such as CATV have a quite different architecture than digital transmission systems. Analog systems are typically part of a large interconnected distribution web serving a very large number of spatially distinct users. Digital systems, in contrast, are typically based on a single long distance transmission line operating at an ultrahigh transmission rate. Analog transmission systems are also different from digital transmission systems in their technical characteristics and specifications. Analog systems have, in principle, a greater information-carrying capacity than digital systems; a digital signal can have only two levels, 1 or 0, whereas an analog signal can have a continuous range of levels. The drawback of analog systems is that to keep this information-carrying capacity error free, the signal to noise level needs to be maintained very high. This places very stringent requirements on the optical amplifiers, much more so than in a digital system. As we shall see shortly, the amplifier must not distort the analog waveform in any way.

Analog Transmission Basics

A typical analog transmission link is depicted in Figure 7.24. A number of electrical RF channels—which usually number 40, 60, 80, or 110 and have frequencies between

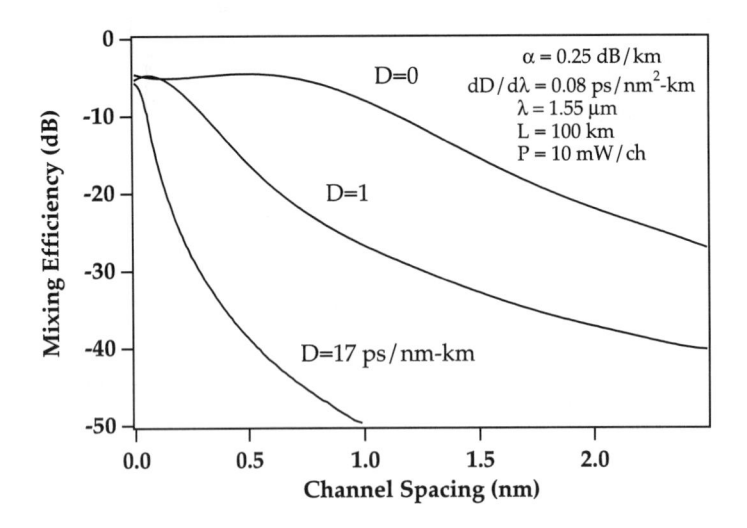

Figure 7.23: Four-wave mixing efficiency as a function of channel spacing, for different values of the fiber dispersion. From reference [41] (©1993 IEEE).

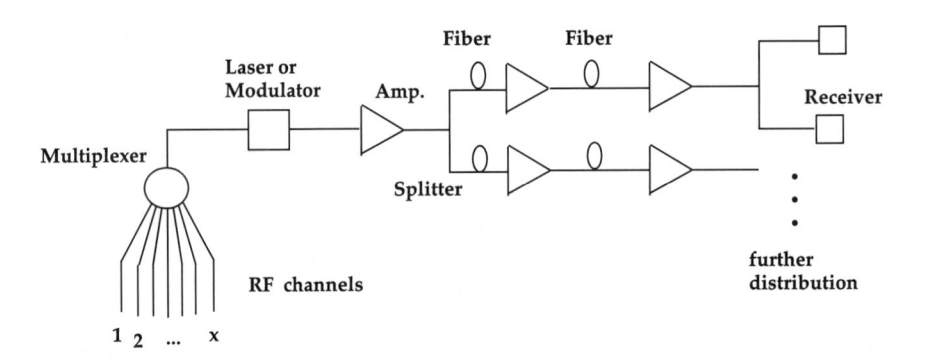

Figure 7.24: Sketch of a typical analog transmission and distribution system delivering x RF channels.

50 and 700 MHz—are impressed with the video information. They are combined, and the resulting electrical signal is then directed to a laser or a modulator. An amplifier boosts the signal, which is then launched into one or more fiber links. The signal can be split off at a variety of points along its path and amplifiers can be used to compensate for the splitting loss. The receiver converts the optical signal into an electrical signal and demultiplexes the signal into its original channels. A large number of variables determine system performance, including laser linearity and noise, dispersion of the fiber link, nonlinear effects, optical reflections in the system, etc. When amplifiers are used, they can be a significant source of degradation of system performance if they do not have the right characteristics.

7.3. OPTICAL NOISE: SYSTEM ASPECTS 243

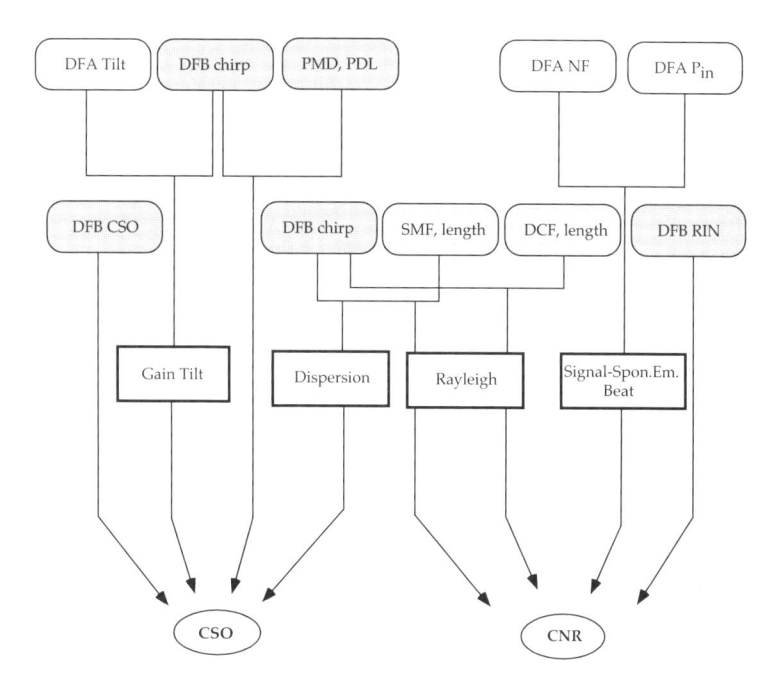

Figure 7.25: Contributions to CNR degradation and appearance of CSO in a real analog system composed of a directly modulated laser (DFB), a doped fiber amplifier (DFA) with noise figure NF, an optical fiber (SMF) with polarization mode dispersion (PMD) and polarization-dependent loss (PDL), and possibly a span of dispersion compensating fiber (DCF). Adapted from reference [48]. The effects are described in more detail in the text.

There are two key issues concerning the noise-free operation of such an analog system: signal to noise (as measured by the SNR) and any distortion of the original waveform. A high signal to noise ratio is required at the receiver. Typical values for the required SNR are 65 dB for broadcast quality video, 55 dB for short span video, and 45 dB for surveillance quality video.[47] The noise is usually characterized by a quantity known as the CNR (carrier to noise ratio) of the system. Distortions of the system are usually characterized by a quantity known as CSO (composite second order) distortion, and CTB (composite triple beat) distortion. The CSO and CTB distortions are nonlinear effects whereby overtones are created by the fundamental frequencies or channels of the system. This degrades the system performance. The CNR and CSO will be described in more quantitative detail below.

A map of the different sources of CNR degradation and contributors to CSO in a real system is depicted in Figure 7.25, in the case of a directly modulated laser transmitter. For an externally modulated laser, due to the absence of laser chirp, a number of these effects do not occur. The noise of the laser, parameterized by its relative intensity noise (RIN), will directly degrade the CNR of the system. The amplifier ASE will interact with the signal traveling through the amplifier, via signal-spontaneous beat noise, to degrade the CNR. Rayleigh backscattering will also degrade the CNR since the double Rayleigh scattered light will interfere with the transmitted light at the detector. The Rayleigh scattering will occur in the single-mode transmission fiber as well as in any dispersion compensating fiber employed. The scattering increases with the length of fiber used as well as with the fiber NA. The resulting CNR is also proportional to the laser chirp since the laser chirp determines the lineshape and coherence properties of the transmitted signal.[49]

A number of different interactions can contribute to CSO. CSO can originate directly from the laser, in particular from the clipping of the laser output when the drive current falls below the laser threshold.[50] The contribution to CSO from the combination of the transmitter chirp and the nonflatness of the gain spectrum will be discussed below. The chromatic dispersion of the single-mode transmission (and also DCF) fiber introduces phase and intensity distortions in the transmission of the signal from a chirped laser where the frequency varies with the intensity of current modulation.[51] Essentially, the frequency modulation of the laser is converted to amplitude modulation due to the dispersion of the fiber. This creates nonlinear distortions in the detected signal and a finite CSO.[52]

When coupled with laser chirp, polarization mode dispersion (PMD) and polarization-dependent loss (PDL) also give rise to nonlinear distortion.[53] The PMD of the fiber converts the frequency modulation of the signal into polarization modulation. The change in polarization is then converted to amplitude modulation when the signal traverses a polarization-dependent loss element, giving rise to nonlinear components in the detected signal. Nonlinear distortion can also arise from PMD alone, when there is significant coupling between the two polarization modes, and is dependent on the polarization-dependent dispersion of the fiber.[53] The CSO from these two polarization-dependent effects interacting with the laser chirp can be minimized by choosing fibers and components with low PMD and PDL.

The CNR is the signal to noise ratio (SNR) of the system when only the pure carrier wave, without modulation by an information-carrying envelope, is used for transmission. The CNR for a single video channel is given as

$$\text{CNR} = \frac{m^2 G^2 P_{in}^2 C^2/2}{\sigma^2} \tag{7.102}$$

with

$$\sigma^2 = (2e(GP_{in} + P_{sp})C + 2GP_{in}P_{sp}C^2/\Delta f \\ + P_{sp}^2 C^2/2\Delta f + i_{th}^2 + P_{in}^2 G^2 C^2 \text{RIN})B \tag{7.103}$$

where m is the modulation depth per channel, B is the video channel electrical bandwidth, i_{th} is the thermal noise current spectral density, RIN is the laser relative intensity noise (dB/Hz), G is the amplifier gain, P_{sp} is the ASE power emitted by the amplifier, and C is the coupling from amplifier output into the photocurrent (including any losses).[47] At high input powers the limiting contribution to the CNR is the signal-spontaneous noise term. A high CNR and SNR require a large signal power at the input of the amplifier to overcome the spontaneous-spontaneous beat noise, as illustrated in Figure 7.26. High output powers are required at the output of the amplifier so that overall it should be capable of high gains with high input powers.

7.3. OPTICAL NOISE: SYSTEM ASPECTS

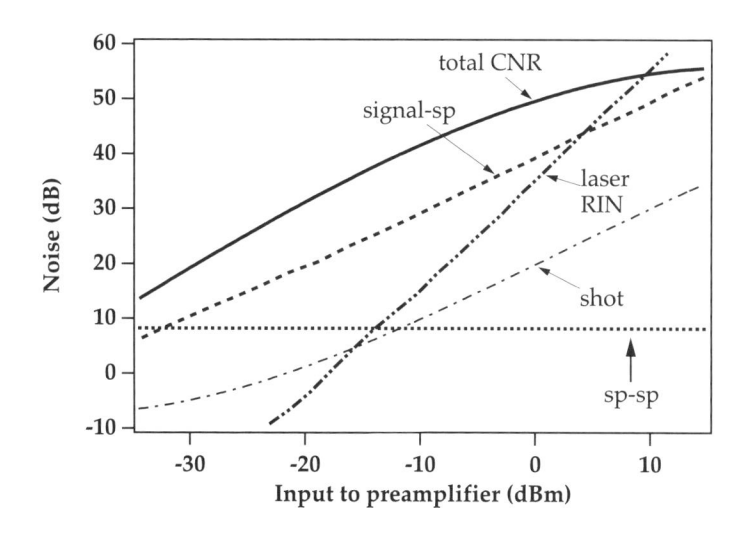

Figure 7.26: Laser RIN and noise components relative to thermal noise, calculated for an FM analog transmission system. From reference [47].

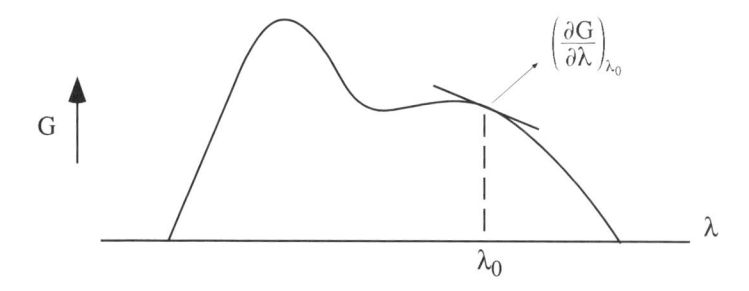

Figure 7.27: Gain as a function of wavelength in an erbium-doped fiber amplifier. The gain in a small region about a central wavelength can be approximated linearly by the tangent to the gain curve.

The distortion introduced by an optical amplifier for a transmitter with chirp such as a directly modulated laser arises from the departure from flatness of the gain spectrum. A change in the signal level, resulting in a change in the drive current applied to the laser, causes a slight change in the laser operating wavelength (chirp). Due to the nonflatness of the gain (gain tilt), this causes a change in gain. Thus different signal levels will undergo different gains and the original waveform will be distorted.[54]

The origin of the CSO effect is the production of overtones by the fundamental frequencies of the system. A single channel at frequency f can produce an overtone at 2f. Similarly, a channel at frequency f_1 and a channel at frequency f_2 can combine to produce overtones at $(f_1 + f_2)$ and $(f_1 - f_2)$. This causes noise on a nearby channel f_n. The appearance of the overtones can be understood with some simple calculations. We

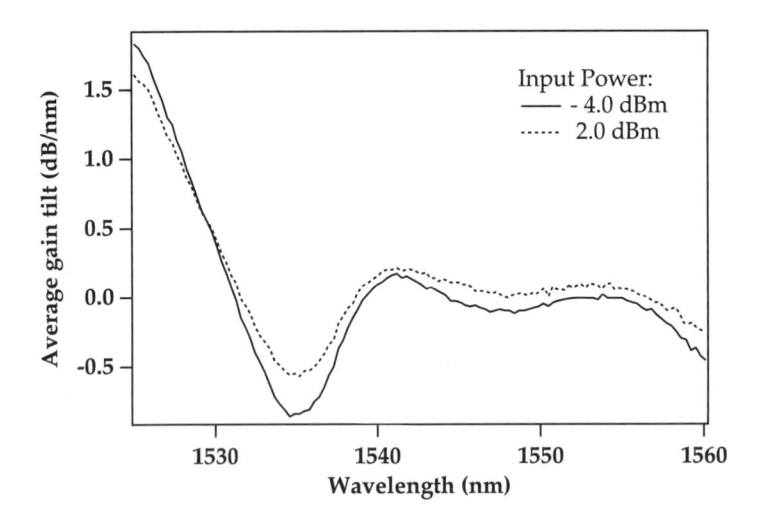

Figure 7.28: Experimental gain tilt of an erbium-doped fiber amplifier. Courtesy T. Nielsen, Lucent Technologies, Murray Hill, NJ.

assume that the gain varies with wavelength as

$$G = G_0 + \Delta G(\lambda - \lambda_0) \quad (7.104)$$

where $\Delta G = \left(\frac{\partial G}{\partial \lambda}\right)_{\lambda_0}$ as shown in Figure 7.27. This is equivalent to expanding the gain to first order around the center wavelength λ_0. The signal input to the amplifier has the amplitude

$$P(t) = P_0(1 + m\sin(\omega t)) \quad (7.105)$$

For a chirped source, such as a directly modulated laser, the wavelength can be written as

$$\lambda(t) = \lambda_0 + \frac{\partial \lambda}{\partial P}P(t) = \lambda_0 + \Delta\lambda(1 + m\sin(\omega t)) \quad (7.106)$$

where $\Delta\lambda = \left(\frac{\partial \lambda}{\partial P}\right)_{\lambda_0} P_0$. The output of the amplifier $P_{out} = GP_{in}$ can then be written as

$$\begin{aligned} P_{out} &= (G_0 + \Delta G(\lambda - \lambda_0))\, P_0(1 + m\sin(\omega t)) \\ &= G_0 P_0 + P_0\, m\,(G_0 + \Delta G\, \Delta\lambda)\sin(\omega t) + \Delta G\, \Delta\lambda\, P_0\, m^2 \sin^2(\omega t) \\ &= \frac{1}{2}P_0(G_0 + \Delta G\, \Delta\lambda\, m^2) + P_0 m\,(G_0 + \Delta G\, \Delta\lambda)\sin(\omega t) \\ &\quad - \frac{1}{2}\Delta G\, \Delta\lambda\, P_0\, m^2 \cos(2\omega t) \end{aligned} \quad (7.107)$$

Thus we see that P_{out} contains a term that varies with 2ω. This simple example shows that when the gain has some slope as a function of wavelength and the source is chirped, there will be higher-order frequencies produced by the interaction of the signal and the amplifier. A first-order expansion of the gain as a function of wavelength yielded the

7.3. OPTICAL NOISE: SYSTEM ASPECTS 247

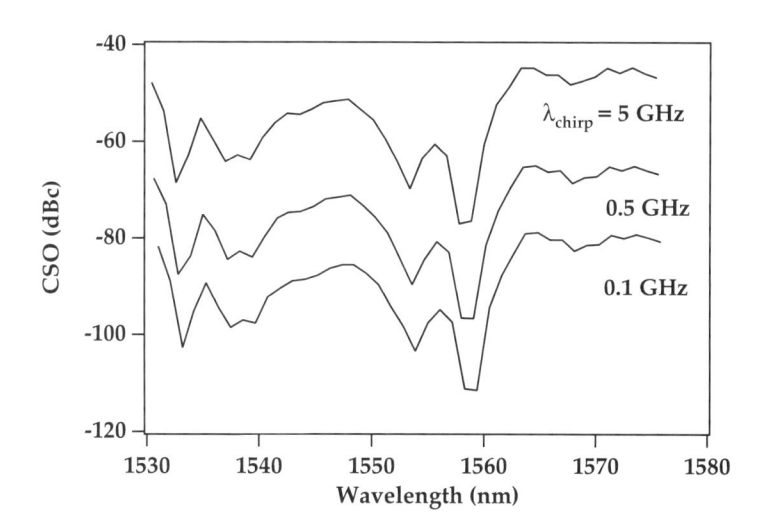

Figure 7.29: Calculated CSO dependence on input wavelength with experimentally measured gain tilt. From reference [56] (©1994 IEEE). The laser chirps assumed are 0.1, 0.5, and 5 GHz, respectively.

CSO term where two frequencies are added or subtracted; similarly, an expansion to second order would yield the CTB term where three frequencies are added or subtracted.

Gain tilts for erbium-doped fiber amplifiers have been measured experimentally.[55, 56] The gain tilt changes with input power due to the change in the shape of the spectrum with the strength of the signal. An experimentally measured gain tilt curve is shown in Figure 7.28. It is important to keep in mind that the gain tilt should be measured as the gain tilt seen by a small signal probe in the presence of a strong saturating signal.[57] The resulting CSO as a function of signal wavelength is shown in Figure 7.29. An amplifier with flatter gain, such as a fluoride-based erbium-doped fiber, leads to a lower CSO.[58] The CSO has also been shown to increase with the input power to the erbium-doped fiber amplifier.[55]

Finally, it has been shown that there are techniques that offer opportunities for cancellation of second-order distortion in an erbium-doped fiber amplifier. Spectral filters, for example, can be inserted after the amplifier, which will flatten the overall gain profile.[59] The most obvious way to mitigate the problems discussed in this section is to use an externally modulated laser. The use of an externally modulated laser eliminates chirp and the interaction with the gain tilt of the amplifier.[60]

Bibliography

[1] G. P. Agrawal, *Fiber-Optic Communication Systems* (John Wiley & Sons, Inc., New York, 1992).

[2] N. A. Olsson, *J. Light. Tech.* **7**, 1071 (1989).

[3] R. H. Stolen, E. P. Ippen, and A. R. Tynes, *Appl. Phys. Lett.* **20**, 62 (1972).

[4] R. H. Stolen and E. P. Ippen, *Appl. Phys. Lett.* **22**, 276 (1973).

[5] N. A. Olsson and J. Hegarty, *J. Light. Tech.* **4**, 396 (1986).

[6] S. G. Grubb, T. Erdogan, V. Mizrahi, T. Strasser, W. Y. Cheung, W. A. Reed, P. J. Lemaire, A. E. Miller, S. G. Kosinski, G. Nykolak, and P. C. Becker, "1.3 μm cascaded Raman amplifier in germanosilicate fibers," in *Optical Amplifiers and Their Applications*, Vol. 14, 1994 OSA Technical Digest Series, (Optical Society of America, Washington, D.C., 1994), pp. 188–190.

[7] A. J. Stentz, "Progress on Raman amplifiers," in *Optical Fiber Communication Conference*, Vol. 6, 1997 OSA Technical Digest Series (Optical Society of America, Washington D.C.,1997), p. 343.

[8] N. A. Olsson and J. P. Van der Ziel, *Elect. Lett.* **22**, 488 (1986).

[9] A. R. Chraplyvy and R. W. Tkach, *Elect. Lett.* **22**, 1084 (1986).

[10] J. C. Simon, *J. Light. Tech.* **5**, 1286 (1987).

[11] T. Mukai, Y. Yamamoto, and T. Kimura, "Optical amplification by semiconductor lasers," in *Semiconductors and Semimetals*, Vol. 22, Part E, W. T. Tsang, Ed. (Academic Press, New York, 1985), pp. 265–319.

[12] S. D. Personick, *Bell Syst. Tech. J.* **52**, 843 (1973).

[13] R. G. Smith and S. D. Personick, in *Semiconductor Devices for Optical Communications (Topics in Applied Physics*, Vol. 39), H. Kressel, Ed. (Springer-Verlag, Berlin, 1982), pp. 89–160.

[14] S. A. Kramer, Lucent Technologies, private communication.

[15] J. Schesser, S. M. Abbott, R. L. Easton, and M. Stix, *AT&T Tech. J.* **74**, 16 (1995).

[16] R. Tench, Lucent Technologies, Breinigsville, PA, private communication.

[17] N. S. Bergano, F. W. Kerfoot, and C. R. Davidson, *IEEE Phot. Tech. Lett.* **5**, 304 (1993).

[18] N. S. Bergano and C. R. Davidson, *J. Light. Tech.* **13**, 879 (1995).

[19] J. Aspell and N. Bergano, "Erbium-doped fiber amplifiers for future undersea transmission systems,"in *Fiber Laser Sources and Amplifiers II*, M. J. F. Digonnet, Ed., *Proc. SPIE* **1373**, pp. 188–196 (1990).

[20] H. Taga, N. Edagawa, S. Yamamoto, and S. Akiba, *J. Light. Tech.* **13**, 829 (1995).

[21] E. Lichtman, *Elect. Lett.* **29**, 2058 (1993).

[22] N. A. Olsson, *Proc. of the IEEE* **80**, 375 (1992).

[23] C. R. Giles and E. Desurvire, *J. Light. Tech.* **9**, 147 (1990).

[24] Y. Yamamoto and T. Mukai, *Opt. and Quant. Elect.* **21**, S1 (1989).

[25] A. R. Chraplyvy, *J. Light. Tech.* **8**, 1548 (1990).

[26] G. P. Agarwal, *Nonlinear Fiber Optics* (Academic Press, San Diego, CA, 1989).

[27] R. Y. Chiao, C. H. Townes, and B. P. Stoicheff, *Phys. Rev. Lett.* **12**, 592 (1964).

[28] E. P. Ippen and R. H. Stolen, *Appl. Phys. Lett.* **21**, 539 (1972).

[29] R. G. Smith, *Appl. Opt.* **11**, 2489 (1972).

[30] Y. Aoki, K. Tajima, and I. Mito, *J. Light. Tech.* **LT-6**, 710 (1988).

[31] A. Chraplyvy, "Systems impact of fiber nonlinearities", Short Course 101, *Conference on Optical Fiber Communications* (Optical Society of America, Washington D.C., 1997).

[32] N. A. Olsson and J. P. van der Ziel, *Elect. Lett.* **22**, 488 (1986).

[33] R. W. Hellwarth, *Phys. Rev.* **130**, 1850 (1963).

[34] A. R. Chraplyvy, *Elect. Lett.* **20**, 58 (1984).

[35] P. A. Andrekson, N. A. Olsson, P. C. Becker, J. R. Simpson, T. Tanbun-Ek, R. A. Logan, and K. W. Wecht, *Appl. Phys. Lett.* **57**, 1715 (1990).

[36] D. Marcuse, A. R. Chraplyvy, and R. W. Tkach, *J. Light. Tech.* **12**, 885 (1991).

[37] N. Kikuchi, K. Sekine, and S. Sasaki, *Elect. Lett.* **33**, 653 (1997).

[38] L. Rapp, *IEEE Phot. Tech. Lett.* **9**, 1592 (1997).

[39] L. E. Nelson, A. H. Gnauck, R. M. Jopson, and A. R. Chraplyvy, "Cross-phase modulation resonances in wavelength-division-multiplexed lightwave transmission," in *24th European Conference on Optical Communication*, Proceedings Vol. 1, pp. 309–310 (1998).

[40] N. S. Bergano and J. C. Feggeler, "Four-wave mixing in long lengths of dispersion-shifted fiber using a circulating loop," in *Optical Fiber Communication Conference*, Vol. 5, 1992 OSA Technical Digest Series (Optical Society of America, Washington D.C.,1992), p. 7.

[41] T. Li, *Proc. of the IEEE* **81**, 1568 (1993).

[42] F. Forghieri, R. W. Tkach, A. R. Chraplyvy, and D. Marcuse, *IEEE Phot. Tech. Lett.* **6**, 754 (1994).

[43] F. Forghieri, A. H. Gnauck, R. W. Tkach, A. R. Chraplyvy, and R. M. Derosier, *IEEE Phot. Tech. Lett.* **6**, 1374 (1994).

[44] A. R. Chraplyvy, A. H. Gnauck, R. W. Tkach, and R. M. Derosier, *IEEE Phot. Tech. Lett.* **5**, 1233 (1993).

[45] N. Henmi, T. Saito, and S. Nakaya, *IEEE Phot. Tech. Lett.* **5**, 1337 (1993).

[46] R. W. Tkach, A. R. Chraplyvy, F. Forghieri, A. H. Gnauck, and R. M. Derosier, *J. Light. Tech.* **13**, 841 (1995).

[47] M. J. Pettitt, *IEE Proc.-J.* **140**, 404 (1993).

[48] N. Park, Lucent Technologies, Murray Hill, NJ, private communication.

[49] T. E. Darcie, G. E. Bodeep, and A. A. M. Saleh, *IEEE J. Light. Tech.* **9**, 991 (1991).

[50] N. J. Frigo, M. R. Phillips, and G. E. Bodeep, *IEEE J. Light. Tech.* **11**, 138 (1993).

[51] G. J. Meslener, *IEEE J. Quant. Elect.* **QE-20**, 1208 (1984).

[52] M. R. Phillips, T. E. Darcie, D. Marcuse, G. E. Bodeep, and N.J. Frigo, *IEEE Phot. Tech. Lett.* **3**, 481 (1991).

[53] C. D. Poole and T. E. Darcie, *J. Light. Tech.* **11**, 1749 (1993).

[54] K. Kikushima and H. Yoshinaga, *IEEE Phot. Tech. Lett.* **3**, 945 (1991).

[55] B. Clesca, P. Bousselet, and L. Hamon, *IEEE Phot. Tech. Lett.* **5**, 1029 (1993).

[56] K. Kikushima, *J. Light. Tech.* **12**, 463 (1994).

[57] S. L. Hansen, P. Thorsen, K. Dybdal, and S. B. Andreasen, *IEEE Phot. Tech. Lett.* **4**, 409 (1993).

[58] H. Ibrahim, D. Ronar'ch, L. Pophillat, A. Madani, J. Moalic, M. Guibert, J. Le Roch, and P. Jaffre, *IEEE Phot. Tech. Lett.* **5**, 540 (1993).

[59] A. Lidgard and N. A. Olsson, *IEEE Phot. Tech. Lett.* **2**, 519 (1990).

[60] W. Muys, J. C. van der Plaats, F. W. Willems, H. J. van Dijk, J. S. Leong, and A. M. J. Koonen, *IEEE Phot. Tech. Lett.* **7**, 691 (1995).

Chapter 8

Amplifier Characterization and Design Issues

8.1 INTRODUCTION

In this chapter, we will dissect the amplifier in terms of its key characteristics. These characteristics are inputs to decisions regarding systems implementation of specific amplifiers. Chapter 9 will then discuss systems issues with the amplifier treated as a building block. We will first discuss here the characterization of amplifiers as they pertain to gain, noise, and conversion efficiency. We will then turn our attention to amplifier design issues for various applications.

8.2 BASIC AMPLIFIER MEASUREMENT TECHNIQUES

8.2.1 Gain Measurements

The gain of an amplifier is expressed as the ratio between the input signal level and the output signal level, typically expressed in dB.

$$\text{Gain(dB)} = 10\log_{10}\left(\frac{P_{\text{signal-out}}}{P_{\text{signal-in}}}\right) \quad (8.1)$$

Conceptually simple, the measurement of the gain of an optical amplifier is complicated by polarization-dependent effects and broadband optical noise (amplified spontaneous emission, or ASE) accompanying the signal at the amplifier output. As an example, the wavelength-dependent optical power of the input signal and amplified output for an erbium-doped fiber amplifier with 14 dB gain is shown in Figure 8.1.

To differentiate the signal from the noise at the output, two general methods are used. In the first method, one filters the output by an optical spectrum analyzer, as in Figure 8.1. In the second method, one modulates the input signal and detects the output signal level by a phase-sensitive detector (such as a lock-in amplifier).

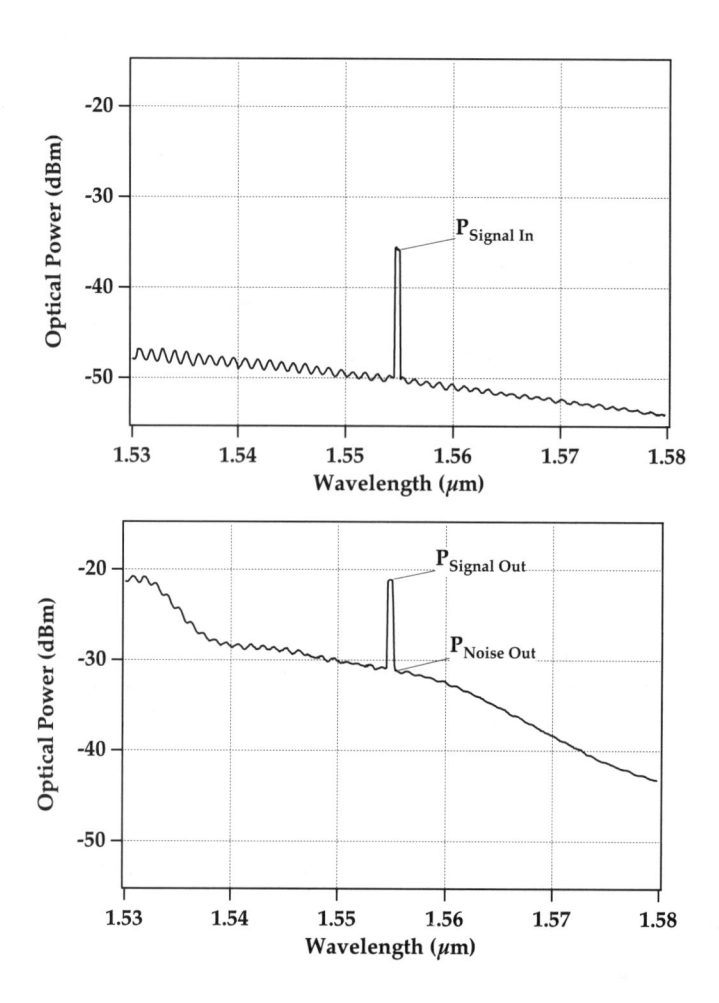

Figure 8.1: Amplifier input (top) and output (bottom) spectra, measured with an optical spectrum analyzer with a 0.5 nm spectral bandpass.

As an example of the optical spectrum analyzer (OSA) method, the gain of the amplifier shown in Figure 8.1 would be as follows.

$$\begin{aligned} \text{Gain(dB)} &= 10\log_{10}\left[\frac{((P_{\text{signal-out}} + P_{\text{noise-out}}) - P_{\text{noise-out}})}{P_{\text{signal-in}}}\right] \\ &= 10\log_{10}\left(\frac{8.04\ \mu\text{W} - 0.76\ \mu\text{W}}{0.27\ \mu\text{W}}\right) \\ &= 14.34\ \text{dB} \end{aligned} \qquad (8.2)$$

For this example the noise power within the 0.5 nm bandpass of the OSA is small (0.76 μW compared with the 8.04 μW signal), although for other amplifier designs, the noise power can be a substantial part of the output. To illustrate the noise power

8.2. BASIC AMPLIFIER MEASUREMENT TECHNIQUES

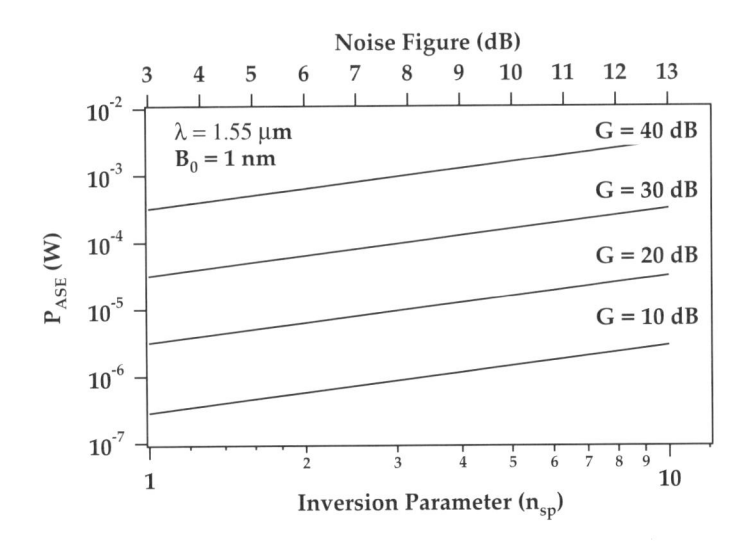

Figure 8.2: Spontaneous noise power (P_{ASE}) in a 1 nm bandwidth about the signal wavelength and noise figure (defined as $10\log_{10}(2n_{sp})$) for amplifier gains of 10, 20, 30, and 40 dB and for inversion parameters n_{sp} between 1 and 10.

levels for a variety of amplifier conditions, Figure 8.2 traces the total noise power P_{ASE} (considering both of the single-mode fiber polarization modes) within a 1 nm bandwidth centered at the signal wavelength, for amplifier gains of 10, 20, 30, and 40 dB at the signal wavelength, and for inversion parameters n_{sp} between 1 and 10 (corresponding to noise figures between 3 and 13 dB). The noise power at the signal wavelength is calculated from the expressions for spontaneous emission noise derived in Chapter 7 (see equation 7.15):

$$P_{ASE} = 2n_{sp}(G-1)h\nu B_o \qquad (8.3)$$

In this expression, B_o is equal to the bandpass of the optical spectrum analyzer at the signal frequency ν, h is Planck's constant, and G is the gain of the amplifier at the signal frequency. The noise figure in dB is defined as $NF = 10\log_{10}(2n_{sp})$, which is valid in the case of high-gain amplifiers.

A complete amplifier evaluation requires measurement of the gain for a range of input signal and pump powers. To accomplish this, a test setup such as the one shown in Figure 8.3 is required. Attenuators, taps, and power meters have been inserted to control and monitor the counterpropagating pump power and input signal levels, and an optical spectrum analyzer is used to detect the output signal and noise components. An isolator is typically placed after the signal laser to reduce backward-propagating light, which could destabilize the signal laser. A 1 nm to 3 nm wide bandpass filter is also included to filter out the signal laser spontaneous emission. In practice, these measurements are complicated by losses introduced by components, connections, and polarization-dependent losses that may vary with time or orientation of the component fiber pigtails. Overall uncertainty in the measured gain of less than 0.45 dB has been claimed for this method.[1]

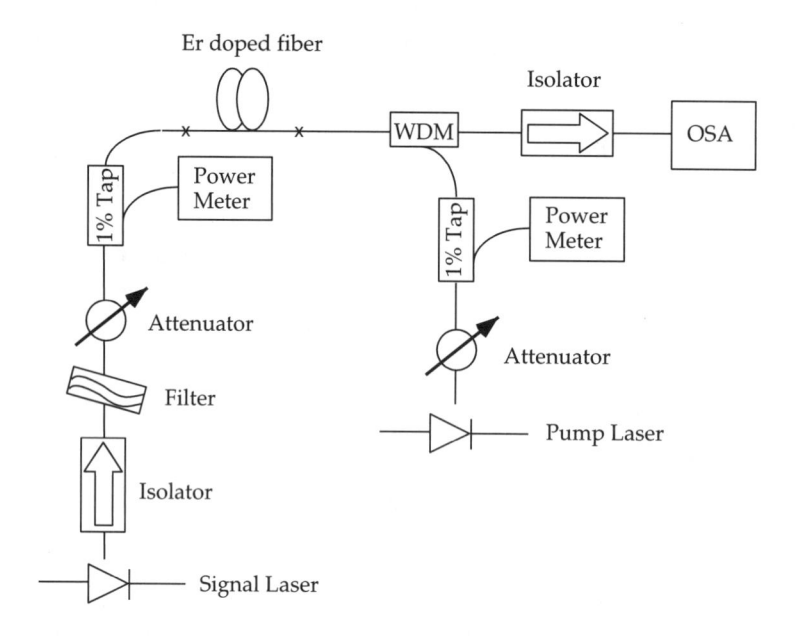

Figure 8.3: Typical experimental setup to measure signal gain in a length of erbium-doped fiber, with an optical spectrum analyzer (OSA).

An alternate method of discriminating signal from noise is to modulate the input signal and detect the signal component in the output by phase-sensitive detection.[1, 2] A schematic of this method is shown in Figure 8.4, where a mechanical chopper is used to modulate the input signal. The modulation frequency for this scheme should be greater than a few kilohertz to avoid the slow gain dynamics of the erbium ion, which would change the average inversion of the amplifier during the signal on time.[3] An automated demonstration of this method, modulating a tunable signal laser at a rate of 50 kHz and detecting the signal with an electrical spectrum analyzer has been described with a measurement accuracy of ±0.4 dB.[1] Another implementation of this method has been used to measure both polarization hole burning and spectral gain characteristics of an amplifier with a precision of 0.01 dB.[4, 5, 6]

Care must be taken with either method to avoid end reflections that would diminish the gain, increase the noise, and in the extreme case cause instability from lasing.[7, 8] To this end, isolators or angled fiber ends are usually placed at the amplifier input and output, reducing back-reflections to 50 dB below the forward-propagating signal.

Precise measurements of gain require consideration of the polarization-dependent sensitivity of the detector. For optical spectrum analyzers, this sensitivity is on the order of 0.05 dB, whereas for commercial optical power meters, this sensitivity can be as high as 0.1 dB. One method to diminish this sensitivity is to depolarize the signal by converting it to spontaneous emission.[9] This technique can diminish the polarization-dependent loss to an uncertainty of 0.001 dB. Alternatively, one can either select detectors for minimum polarization sensitivity or adjust the state of polarization for a maximum reading prior to each measurement. Another issue is that of detecting opti-

8.2. BASIC AMPLIFIER MEASUREMENT TECHNIQUES

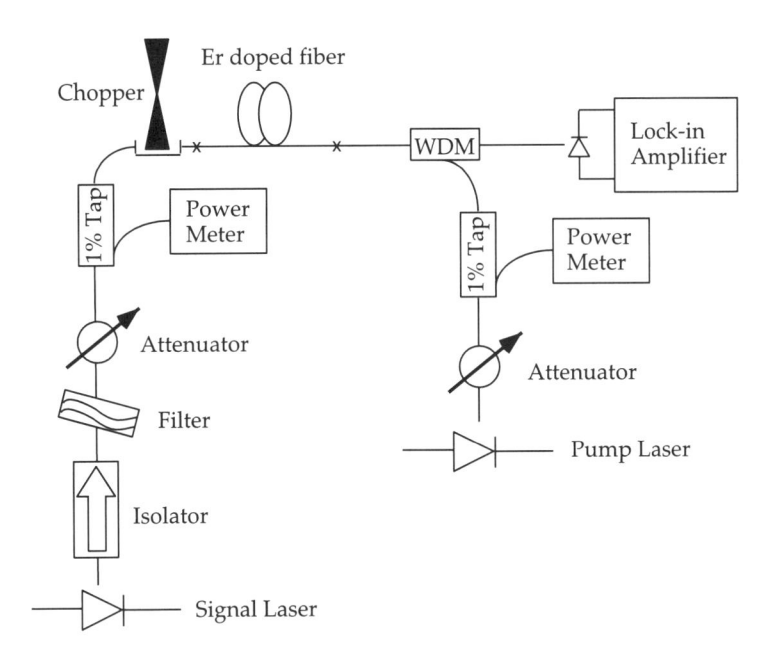

Figure 8.4: Typical experimental setup to measure signal gain in a length of erbium-doped fiber, with a modulated signal gain measurement setup (using phase densitive detection with a lock-in amplifier).

cal power from fibers with the high numerical apertures (NA) typical of erbium-doped fiber. Power meter manufacturers often provide collimating lenses to accommodate higher numerical aperture fibers. The preferred solution, however, is to splice the high NA (0.3) erbium-doped fibers to the lower NA (0.2) conventional communication fiber types. This solution defers the problem to one of accurately measuring the splice loss of such configurations and reproducing these splices. The combination of uncertainties typically encountered in these measurements leads to a precision in gain measurement of 0.50 dB.[1]

Measurement of the gain along the length of an erbium-doped fiber is occasionally of interest to verify its performance. Amplifying fibers a few meters long have been analyzed along their length by extracting power out of the side of the fiber by a small radius bend.[10] For low-gain or distributed amplifiers that do not use inline isolators, an optical time domain reflectometer (OTDR) may also be used to measure the signal level or gain along the span length.[11, 12, 13] Spans containing many amplifiers may also be observed by this technique as is also the case for remotely pumped amplifiers.[14] Typically, a 1 to 3 nm bandpass filter must be inserted between the amplifier and the input of the OTDR (which is quite sensitive) to reduce the backward-propagating ASE power. The setup and the experimental result are shown in Figure 8.5.

The OTDR method is best used as a system span integrity check or as a qualitative measure of the gain of an amplifier, given the limited control of the OTDR signal level and spectral width. Combining the OTDR signal with a saturating signal to eval-

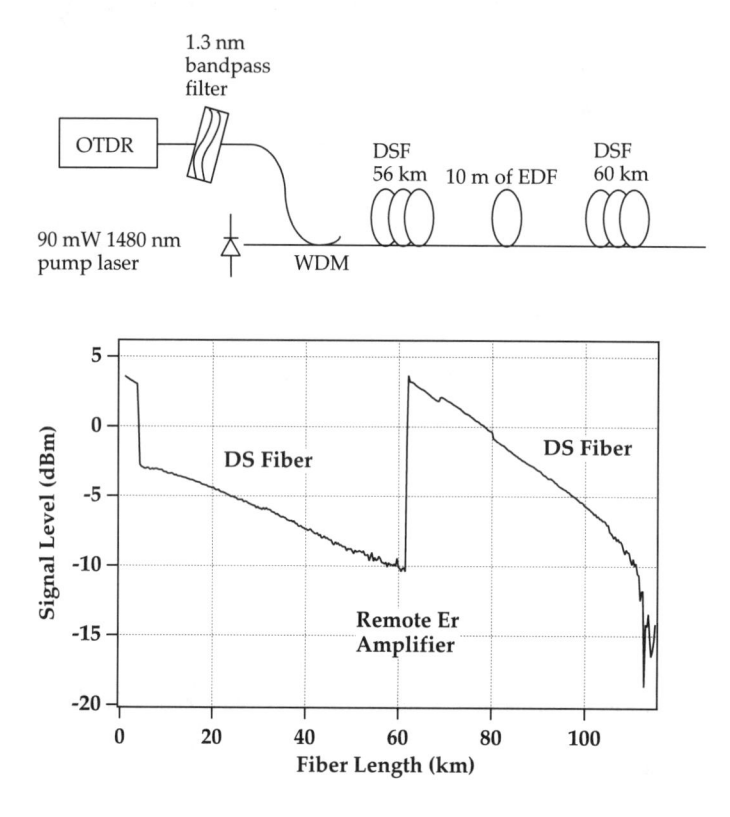

Figure 8.5: Top: setup for gain measurement by OTDR (DSF: dispersion-shifted fiber; EDF: erbium-doped fiber). Bottom: experimental gain measurement of the 10 m remotely pumped erbium-doped fiber using the OTDR setup above.

uate distributed amplifier performance has also been demonstrated and more closely simulates actual system signal conditions. [12, 15, 16, 17, 18]

Gain Measurements in Amplifiers for WDM

Transmission systems are now employing multiple wavelength signals to expand the capacity of amplified spans. The design and testing of amplifiers for this application require methods of distinguishing signals separated by small wavelength separations (less than 1 nm in some cases).

The measurement of amplifier gain for each signal is best accomplished by providing the collection of signal wavelengths to the amplifier input. The signals are then separated at the output with an optical spectrum analyzer, as shown in Figure 8.3. This allows the determination of the gain of an individual channel, with the available gain from the amplifier being shared among all channels. One can also simulate channel drop-out effects by turning selected signals on and off.

There are other ways of measuring the gain in a WDM amplifier. Multiple signal sources can be approximated by one signal with a specific wavelength and signal power. This is referred to as the reduced source method. It is effective because the amplifying

8.2. BASIC AMPLIFIER MEASUREMENT TECHNIQUES

transition for an erbium ion is predominantly homogeneously broadened.[19] In effect, one signal with the correct power establishes an inversion of the amplifier that is equal to that in the presence of multiple signals. The gain variation across the spectral region of interest may then be measured by using a low-power probe source simultaneously with the saturating signal. This broad source could be an edge-emitting LED, an ASE source, or a low-power tunable laser. One complication of this technique is the spectral hole burning surrounding the saturating signal.[20] The depth of the spectral hole increases with gain compression and the spectral hole width varies with wavelength (as discussed in Chapter 4, section 4.5.2). One observation for a alumino-germano-silica host material indicates an increase in hole depth with gain compression at a linear rate of 0.027 dB of gain decrease per 1 dB of gain compression at room temperature.[20] As a result of the hole burning, the probe signal will indicate a diminished gain in the spectral region of the saturating tone wavelength.

There are a number of pitfalls to be wary of in making gain measurements of an amplifier for WDM purposes. In contrast to the saturating tone and probe methods, a high-power saturating signal could be tuned across the spectral band, measuring the gain at each wavelength. This will not provide an accurate representation of the gain for a WDM system but will only indicate the spectral-dependent conversion efficiency. Similarly, using only a small signal probe (in the form of a low-power tunable laser or low-power broadband source) will not yield the WDM response of the amplifier. This method will not cause the amplifier to decrease the erbium ion inversion to the same degree as when it is subjected to the overall higher signal power of a collection of WDM signals.

Particularly applicable to the measurement of gain and noise figure in WDM systems is the time domain extinction (TDE) method.[21] In this method, the long recovery time of the ASE when the signal is turned off is used to differentiate the ASE from the signal. An input signal square wave modulation of 25 kHz (40 μs period) is used along with a gated optical spectrum analyzer to measure the output ASE value immediately following the extinction of the signal. Especially attractive in this method is the ability to measure the ASE at the wavelength of the signal without interpolating from either side of the signal. Rapid spectral-dependent gain and noise measurements can be made by combining a broadband source with a saturating tone, sampling noise and signal at the appropriate times.

8.2.2 Power Conversion Efficiency

When high output power is required (as in the case of power amplifiers used to boost the launched signal power), high conversion of pump power into output signal power is desired. The efficiency of this conversion is expressed as either the power conversion efficiency (PCE) or quantum conversion efficiency (QCE), as discussed in Chapter 6, section 6.4.6.[22]

$$\text{PCE} = \frac{(P_{\text{signal-out}} - P_{\text{signal-in}})}{P_{\text{pump}}} \tag{8.4}$$

$$\text{QCE} = \text{PCE} \times \left(\frac{\lambda_{\text{signal}}}{\lambda_{\text{pump}}}\right) \tag{8.5}$$

Amplifiers for this application are operated in compression to reach PCE values of 65% to 80%.[23] Measurement of conversion efficiency is identical to the gain measurements already described. However, the high output power typically encountered requires attenuation to bring the power level of the amplifier output down to the +10 dBm range of most optical spectrum analyzers and power meters.

Comparing the conversion efficiencies for a number of erbium-doped fibers can be tedious given that the optimum length for a given pump power would have to be determined for each fiber. A more rapid method of comparison has been described where the slope of the backward ASE as a function of pump power, for a long length of erbium-doped fiber, serves as an indicator of the relative conversion efficiency.[24] A long length of fiber is used since it gives the backward ASE enough distance over which to develop. Thus one obtains a measure of the fiber's inherent ability to stimulate gain, given the fiber's material properties. This is because the backward ASE travels over the input end of the fiber, a section that is highly inverted as it is the launch section for the pump. The backward ASE slope thus measured has been shown to be proportional to the slope efficiency (output signal power divided by launched pump power) of a length-optimized bidirectionally pumped amplifier.[24] For high signal input powers the latter slope efficiency is equal to the power conversion efficiency.

Optimizing the conversion efficiency of the erbium-doped fiber has resulted primarily from the proper choice of erbium concentration and host material. Thus, high concentrations of erbium, which lead to quenching effects, will act to reduce the conversion efficiency. As an example of the influence of host composition, in a GeO_2-SiO_2 host, the efficiency has been shown to drop from 60% to 5% for an increase in Er^{3+} concentration from 200 to 600 ppm, whereas an aluminum-based host allows concentrations near 1000 ppm and even a little higher without such a penalty.[25]

8.2.3 Noise Figure Measurements

The noise figure (NF) of an amplifier is a measure of the degradation of the signal to noise ratio for a signal passing through the amplifier (NF = SNR_{in}/SNR_{out}). Measurements of this property require accurate values for both the signal gain and noise power added by the amplifier. There are several methods to measure the noise figure, based on either optical or electrical methods.[26] They are the

- optical method — with spectral interpolation.

- optical method — with polarization nulling.

- optical method — with time domain extinction.

- electrical method — with RIN subtraction.

We will examine the optical methods first. As discussed in Chapter 7, section 7.3.3, the optical noise figure can be expressed as

$$\text{NF(dB)} = 10\log_{10}\left(\frac{P_{ASE}}{h\nu B_o G} + \frac{1}{G}\right) \tag{8.6}$$

8.2. BASIC AMPLIFIER MEASUREMENT TECHNIQUES

where we have assumed a shot-noise-limited input signal to noise ratio and have included both signal-spontaneous and spontaneous-spontaneous beat noise (see equation 7.76). The necessary quantities are as follows.

- P_{ASE} is the ASE power measured in the bandwidth B_o
- h = 6.626 × 10^{-34} J · s (Planck's constant)
- ν is the optical frequency (Hz)
- G is the gain of optical amplifier (linear units)
- B_o is the optical bandwidth in Hz (= $\frac{c}{\lambda}\left[\frac{\Delta\lambda}{\lambda}\right]$ = 6.2 × 10^{10} Hz for $\Delta\lambda = 0.5$ nm at 1555 nm)

The challenge of measuring noise figure then becomes one of accurately measuring the values of gain (G) and ASE power (P_{ASE}). If a continuous average signal power is sent through the amplifier, the ASE power at the signal wavelength may be estimated by fitting the ASE spectral curve on either side of the signal and extrapolating this curve under the signal peak (as in Figure 8.1).[27] Using the example of Figure 8.1, values of linear gain (G = 26.96) and spontaneous power ($P_{ASE} = 0.76 \times 10^{-6}$ W) are measured. These values then translate via equation 8.6 to a noise figure of 3.56 dB.

The two values with the least certainty are the spectrum analyzer optical bandwidth B_0 and the noise power P_{ASE}. For example, an error of 0.02 nm in bandwidth (eg., 0.48 nm instead of 0.5 nm) will result in a noise figure variation of 0.3 dB. The optical spectrum analyzer (OSA) may be calibrated for the signal wavelength region of interest, using a tunable laser source and wavemeter as a wavelength reference. Accurate OSA bandwidth and passband spectral shape measurements may be made by setting the OSA for a spectral span of zero, thereby running the OSA as a power meter with the spectral bandwidth of interest. The output of a tunable laser and wavemeter combination may then be scanned to the passband of the OSA. The passband spectral shape may then be measured noting the OSA indicated power at each wavelength setting.

Estimating the noise power under the output signal is difficult to do with precision, especially when the noise power becomes nearly equal to the signal power. It is also difficult to measure accurately the noise power when a filter is part of the amplifier, as in multistage amplifiers.

It is important to realize that equation 8.6 assumes that no noise is injected at the input to the amplifier (i.e., a shot-noise-limited input signal). In reality, noise accompanies all signal lasers and this will result in an error in the noise figure measurement. From Figure 8.6 it can be seen that a source-spontaneous emission of −50 dBm/nm results in a noise figure error of 0.5 dB.[28]

The second method to distinguish the signal from the noise is the polarization nulling technique. A typical setup to accomplish this is shown in Figure 8.7. The ASE power, which is randomly polarized, can be readily observed by substantially eliminating the polarized signal at the output of the amplifier with a polarizer.[29] This method requires a means for rotating the linear signal state of polarization exiting the amplifier relative to a polarizer to achieve signal extinction. Such rotation of the signal state of polarization can be achieved by selectively twisting a set of three inline loops

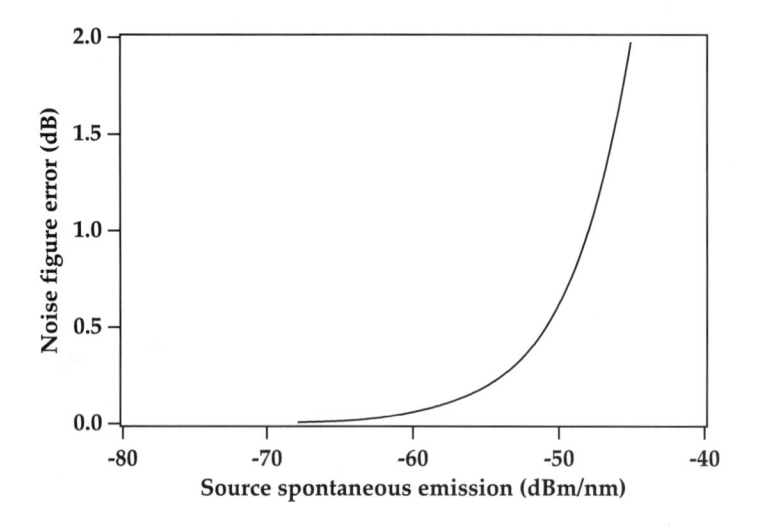

Figure 8.6: Increase in noise figure in an erbium-doped fiber amplifier due to spontaneous emission accompanying the signal at the input to the amplifier. From reference [28].

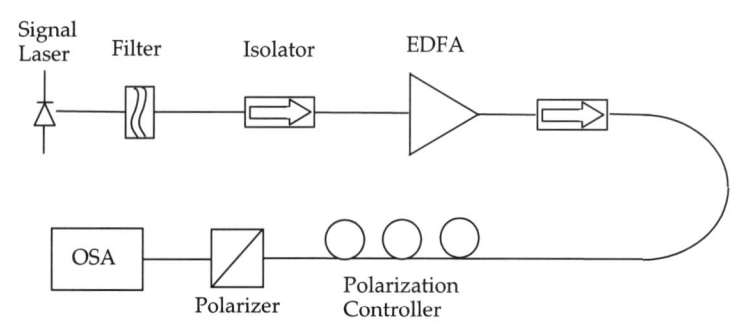

Figure 8.7: Typical setup used in making noise figure measurements using the polarization nulling technique. From reference [29].

of fiber.[30] In practice, the exiting signal polarization state changes with time as ambient conditions change the phase of the guided signal. With occasional adjustment of the fiber loops, extinction of the signal may be maintained long enough to complete the noise figure measurement. This process may of course be automated using a computer-adjusted polarization controller.

Alternatively, optical noise figure measurements may be made using a pulsed signal, taking advantage of the slow recovery period of the spontaneous emission to discriminate signal from noise.[1, 31, 32] This technique, the time domain extinction method, was described for gain measurements in WDM systems in Section 8.2.1.

Finally, the noise figure can also be measured using electrical methods. Comparisons between the optical and the electrical methods have been the subject of some

8.2. BASIC AMPLIFIER MEASUREMENT TECHNIQUES

debate.[33, 34, 35, 36] As discussed in Chapter 7, the noise added by an optical amplifier is the consequence of a number of interactions occurring both at the photodetector and through optical-optical heterodyne products. These interactions are the results of shot noise, signal-spontaneous, spontaneous-spontaneous, and multipath interference (mpi) components. Following Chapter 7, we express the noise figure as a sum of these contributions (with the addition of NF_{mpi}):

$$NF_{total} = NF_{shot} + NF_{s-sp} + NF_{sp-sp} + NF_{mpi} \qquad (8.7)$$

The multipath interference arises from reflections that cause time-delayed versions of the signal to interfere with itself. The cavity reflections may originate either within the amplifier, between the span and the amplifier, or entirely within the span.[37] This multipath interference creates noise that is especially detrimental in high signal to noise ratio systems such as analog systems (where 50 dB SNR is typically required). In digital systems, the signal to noise requirements are far less (eg., 20 dB) and the multipath interference much less of a factor.

The electrical noise figure methods measure the signal and noise powers in the electrical frequency domain instead of in the optical domain. This then measures noise in a manner nearly identical to that detected by optical receivers, which convert an optical signal into an electrical signal. However, the previously described optical methods of measuring noise figure do not determine NF_{mpi}.

Comparisons of the optical and electrical noise figure techniques have indicated a general agreement, with the exception of highly saturated amplifiers.[33, 34, 35, 36] Complications in the electrical methods may result from input signal noise and interpretation of low-frequency noise components. These latter typically result from phase to intensity noise conversion due to multipath interference.[38]

The most common method for of electrical noise figure measurements uses the relative intensity noise (RIN) subtraction method. The RIN of a signal is defined as

$$RIN = \frac{\Delta P^2}{P^2} \qquad (8.8)$$

expressed in units of dB/Hz. ΔP^2 is the mean square optical intensity fluctuations (in a 1 Hz bandwidth) at a specified frequency, and P^2 is the square of the optical power. The RIN is a measure of noise at a particular frequency, and electrical instruments that directly measure the RIN of a source are commercially available. To measure the RIN of the amplifier, the input signal RIN value (RIN_{sig}) is subtracted from the RIN of the signal at the amplifier output (RIN_{output}).

The noise figure may be calculated from the RIN as follows,

$$NF_{electrical} = \frac{1}{G} + \frac{P_{in}\left(\left(RIN_{output}\left(\frac{P_{out,total}}{P_{out,sig}}\right)^2\right) - RIN_{sig}\right)}{2h\nu} \qquad (8.9)$$

where G is the signal gain, ν is the optical signal frequency, P_{in} is the signal power into the amplifier, RIN_{output} is the RIN at the output of the amplifier (which is the RIN of combined signal and ASE power), $(P_{out,total}/P_{out,sig})^2$ is the power ratio correction

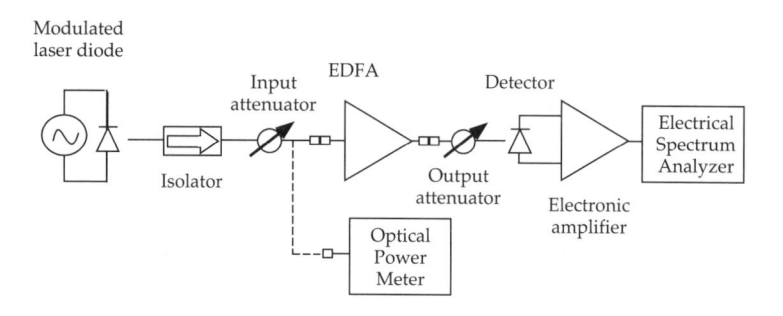

Figure 8.8: Electrical noise figure measurement setup (EDFA: erbium-doped fiber amplifier).

for RIN_{output}, $P_{out,total}$ is the total output power (signal and ASE), $P_{out,sig}$ is the signal power at amplifier output, and RIN_{sig} is the RIN of the input signal.[35] The first term, $1/G$, represents the signal shot noise term. The second term contains the contributions of signal-spontaneous, spontaneous-spontaneous, and multipath interference noise.

Electrical noise measurements require an isolated DFB source, attenuators for the input signal and amplifier output, an optical power meter, and a 1 MHz to 1 GHz electrical spectrum analyzer, as shown in Figure 8.8. The noise may be measured either with a continuous or modulated signal source. Using a commercial lightwave signal analyzer, one measures the RIN before and after the amplifier.[34] The two RINs, measured as a function of frequency in the electrical domain, can then be subtracted to obtain the noise figure.

The proper implementation of the RIN subtraction method requires a few experimental precautions. For an accurate measurement it is necessary that the amplifier RIN be larger than the signal source RIN and detector thermal RIN. It is also necessary that the signal source RIN be larger than the detector thermal RIN. It is also recommended that the input signal state of polarization be varied to search for the maximum value of RIN. This should yield the worst-case value, given that the multipath interference is highly polarization sensitive. The gain may be determined by subtracting the input power from the output power, measured electrically or optically. Given that RIN and the NF_{mpi} are RF frequency dependent, it is also suggested that the RIN values be measured at a number of frequencies to look for a stable value at a high frequency. Incoherent multipath interference effects will dominate at lower frequencies. Coherent multipath interference will be evident from an oscillating RIN with RF frequency. It is suggested that the source DFB for these measurements be selected with a width of greater than 10 MHz, to avoid multipath coherent reflections. It has also been suggested that the source optical power be kept near 0 dBm with a RIN less than -150 dB/Hz.[39] Source RIN values as low as -165 dB/Hz are believed to be necessary for minimum influence of source excess noise.[40] Modulating the source will increase its linewidth, which may influence the multipath component of the noise figure with neglible effect on the other noise sources. For a Lorentzian lineshape source and long

8.3. AMPLIFIER DESIGN ISSUES

Amplifier Type	High Gain	Low Noise Figure	High Power	Gain Flatness
Single-channel inline amplifier	L	M	M	L
Power amplifier	L	L	H	L, H
Preamplifier	H	H	L	L
Analog amplifier	M	H	H	H
WDM inline amplifier	M	M	H	H

Table 8.1: Characteristics required of erbium-doped fiber amplifiers for various applications, by importance (H: High; M: Medium; L: Low). Gain flatness requirements are low for single-channel digital power amplifiers and high for multichannel digital and analog power amplifiers.

delay paths ($2\pi \delta \nu \tau \geq 1$), this linewidth-dependent noise contribution is expressed as

$$\text{RIN}_{\text{mpi}} = \frac{4R^2}{\pi} \left[\frac{\Delta \nu}{f^2 + \Delta \nu^2} \right] \tag{8.10}$$

where R^2 is the roundtrip reflectivity of the interfering path, $\Delta \nu$ is the spectral linewidth of the source, τ is the round trip path delay time between the two reflecting elements, and f is the baseband frequency.[38, 41] Ignoring polarization fluctuations during the reflection, $R = \alpha \sqrt{R_1 R_2}$ where R_1 and R_2 are the two reflection coefficients from the two reflection points (e.g. connectors), and α is the single pass signal attenuation.

Electrical noise figure measurements are especially useful in evaluating amplifier performance for analog modulation applications where mpi due to fiber connections or Rayleigh backscattering may impact the noise power in the electrical spectrum below 20 MHz.[42]

8.3 AMPLIFIER DESIGN ISSUES

The main parameters in the design of an erbium-doped fiber amplifier include the fiber glass material, the waveguide characteristics of the fiber, the erbium concentration profile, the length of fiber used, the pump source, and any active and/or passive components. Active components include such items as automatic gain control electronics, while passive components include isolators and wavelength division multiplexers (WDM couplers). The design of the amplifier depends on the designer's intended application, and the parameters chosen need to be considered as a whole. The primary design goals—high gain, high output power, low noise figure, reliability, flatness of gain spectrum, etc.—also need to be considered. A preamplifier will have a different design than an inline amplifier or a power amplifier. The desired characteristics of the amplifier, depending on the intended application, are summarized in Table 8.1. Specific fiber parameters (length, core diameter, cutoff wavelength, etc.) can be obtained by numerically modeling the erbium-doped fiber amplifier performance for the required application and using constraints resulting from the manufacturing of the fiber. The models used are described in Chapter 6. In this section, we present an overview of

	Type 1	Type 2	Type 3	Type 4	Type 5
Peak absorption (dB/m):					
at 1530 nm	3.0-6.0	2.0-5.0	2.0-5.0	2.0-5.0	8.0-14.0
at 980 nm	2.5-6.0	1.4-4.5	1.4-4.5	1.4-4.5	5.6-14.0
Numerical aperture	≥ 0.33	0.29 ± 0.03	0.23 ± 0.03	0.17 ± 0.03	0.29 ± 0.04
Cutoff wavelength (nm)	800-950	800-950	800-950	800-950	800-950
Core radius (μm)	≤ 1.1	1.0-1.4	1.3-1.8	1.6-2.6	0.9-1.5
Mode field diameter (μm)	≤ 4.0	3.6-5.2	4.9-6.3	6.0-9.0	3.6-5.2
Bkgd loss 1550 nm (dB/km)	≤ 30	≤ 8	≤ 8	≤ 4	15
Typical efficiency:					
for 100 mW 980 nm pump	71%	78%	80%	80%	65%
for 500 mW 980 nm pump	44%	56%	64%	74%	45%
for 100 mW 1480 nm pump	79%	83%	82%	78%	65%
Typical Applications	Remote Amp	Preamp	Inline Amp	High Power	ASE Source
	Preamp	Inline Amp	2nd Stage	Final Stage	Short Amp
	First Stage	First Stage	Power Amp		

Table 8.2: Commercial data sheet describing erbium-doped fiber for amplifiers. From reference [43]. The mode field diameter is defined with the Petermann II convention at 1550 nm.[44, 45] The quantum conversion efficiency is computed with optimized length and a 1 mW signal at 1550 nm. First, second, and final stage refer to a multistage amplifier.

some of the key issues in the implementation of erbium-doped fibers for amplification, as well as some of the considerations raised in choosing and configuring the fiber and the pump sources.

Typical parameters for commercially available erbium-doped fiber for amplifier purposes are shown in Table 8.2.[43] The Type 1 fiber is suited for remote amplifier or preamplifier applications, i.e., ones where the signal input will be small and the gain desired high, while maintaining a low noise figure. This fiber thus has a fiber geometry such that the mode field diameters are small and the signal intensities high (achieved with a small core diameter and a high numerical aperture). This fiber also has a high gain efficiency and achieves high small signal gains at low pump power. Because this amplifier is not operated in a power mode, the lower conversion efficiency (resulting from a high local erbium ion concentration and also high background loss) is not a cause for concern. The higher background loss of the Type 1 fiber is a result of its high NA. Increasing the erbium concentration allows one to use a shorter length and thus minimize the effect of this loss. The Type 2 fiber provides a compromise between high small signal gain preamplifier applications and high conversion efficiency applications. The mode field diameters are slightly higher than the Type 1 fiber, yielding a slightly lower small signal gain coefficient. The background loss, however, is reduced compared to the Type 1 fiber (lower local erbium ion concentration due to a lesser confinement, and lower NA) and so the conversion efficiency at high input powers is better. The Type 3 and Type 4 fibers have a larger mode field diameter compared to Type 2 fiber. Thus the pump intensity is lower which limits the onset of nonlinear effects such

8.3. AMPLIFIER DESIGN ISSUES

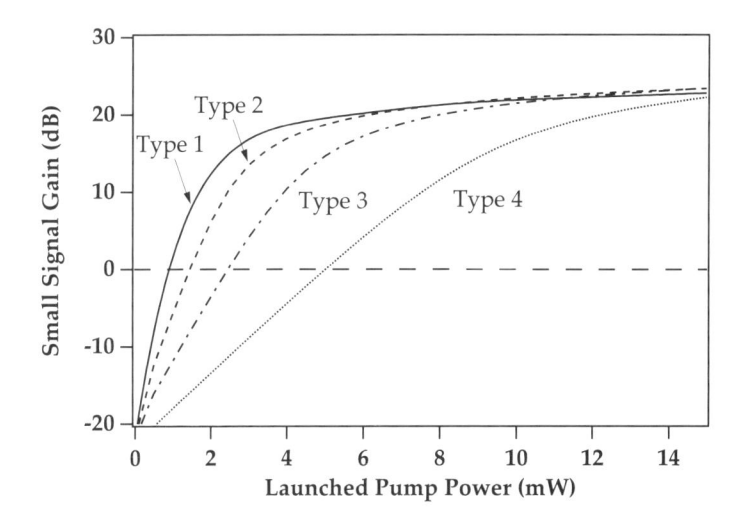

Figure 8.9: Small signal gain curves for the commercial fibers described in Table 8.2 (1550 nm signal, 980 nm pump). The fiber length is optimized for each pump power. From numerical simulations, courtesy P. Wysocki, Lucent Technologies, Murray Hill, NJ.

as pump excited state absorption, which can limit the efficiency of high power booster amplifiers. The improvement in efficiency of the high power 980 nm pumped amplifier in going from Type 2 to Type 3 to Type 4 fiber is a result of this. The small signal gain curves for these commercial fibers are shown in Figure 8.9. The gain thresholds increase with mode diameter, as expected. The selection of fiber lengths and geometries is discussed further in section 8.3.2.

Modern commercial amplifiers include a host of monitoring and electronic control components, such as automatic eye-safe power, multichannel input and output power monitors, add/drop control and equalization of output power per channel, and feedback for automatic gain control.

8.3.1 Copropagating and Counterpropagating Pumping Issues

Three different pump configurations are possible for pumping a length of erbium-doped fiber: copropagating pump and signal, counterpropagating pump and signal, and bidirectional pumping, as depicted in Figure 8.10. As far as small signal gain is concerned, copopagating and counterpropagating pumps yield the same gain and only the total amount of pump power matters. This is because the ASE patterns generated by the two pump patterns are mirror images of each other and so the average upper state population is the same in both cases. In the limit of a small signal the signal is too weak, at all points along the fiber, to influence or change the ASE pattern and hence the transfer of energy from pump to signal. Bidirectional pumping, in contrast, generates a different ASE pattern. When the fiber is sufficiently long, this can result in higher small signal gains at 1.5 μm, for equal amounts of pump power, then either the co or counterpropagating pumping patterns. In the presence of pump ESA the bidirectional

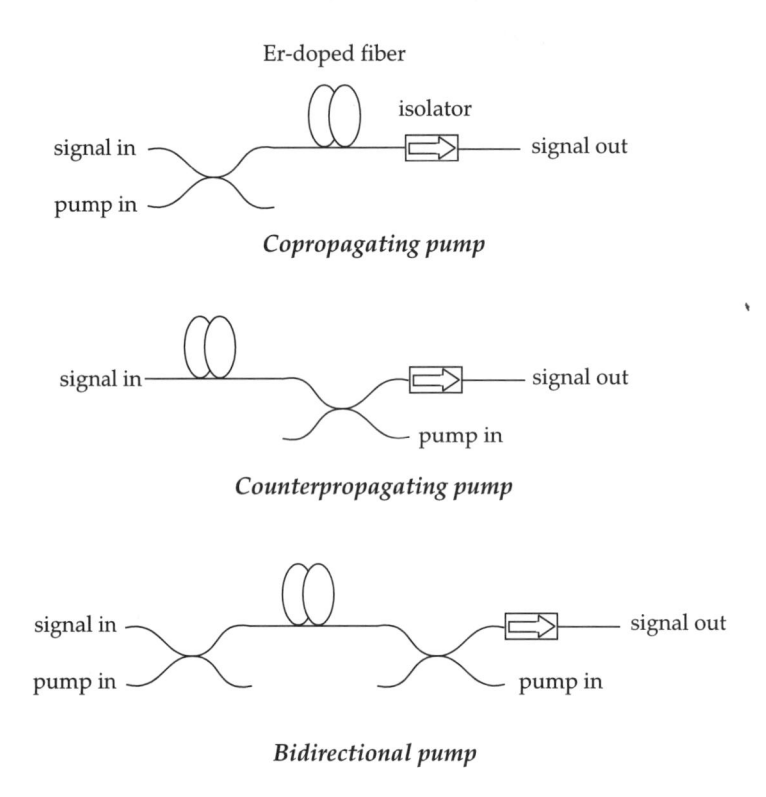

Figure 8.10: Pump configurations for a single-stage erbium-doped fiber amplifier.

configuration is also preferred for obtaining higher small signal gain. This is also the case when signal upconversion due to inter-ion energy exchange manifests itself. The reason for this is that bidirectional pumping results in a more uniform inversion, and avoids localized regions of high pump power where processes quadratic in the pump power have a higher probability of occuring. As the signal input level is raised, the choice of pumping direction between co and counterpropagating pump starts to matter in determining the signal gain of the amplifier. This was discussed in Chapter 6 in terms of the interplay between the gain experienced by the signal and that experienced by the forward- and backward-propagating ASE.

In the case of inline amplifiers, with signal input levels in the -10 dBm to -20 dBm range, both the bidirectional and the counterpropagating pumping schemes reveal themselves to be superior to the copropagating one for sufficiently long fibers where ASE effects start playing a significant role. Figure 8.11 shows the signal output level as a function of length for the specific case of the erbium-doped fiber A of Chapter 6 with a signal wavelength of 1550 nm. The input signal levels contrasted are -20 dBm and 0 dBm. In the -20 dBm case, the counterpropagating pump yields a higher output power than the copropagating pump at long fiber lengths due to the pump level evolution along the fiber being better tailored to the signal growth. The signal is strongest where the pump is strongest, allowing for an efficient transfer of energy between pump and

8.3. AMPLIFIER DESIGN ISSUES

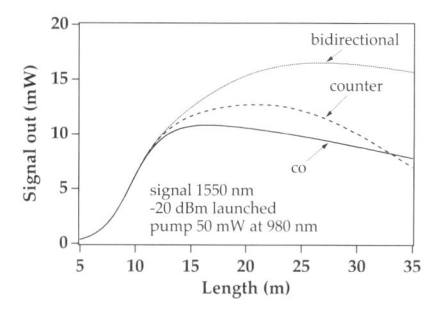

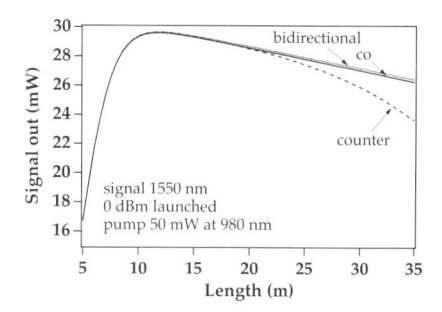

Figure 8.11: Signal output power (mW) at 1550 nm as a function of length, for erbium-doped fiber A of Chapter 6 pumped at 980 nm with 50 mW of pump power in either bidirectional, copropagating, or counterpropagating directions, from numerical simulations. The input signal power is −20 dBm (left) and 0 dBm (right).

signal without the ASE robbing too much of the pump energy. The counterpropagating pump is also well suited in the case when pump ESA is present as the signal can more effectively deplete the inversion and reduce the pump ESA. Figure 8.11 shows that the bidirectional pumping configuration yields the highest output power for a signal input of −20 dBm, at long fiber lengths. At the shorter fiber lengths where the ASE is low and the fiber well inverted the choice of pumping direction does not matter. Inline amplifiers for long haul applications are often operated at these short lengths so as to ensure a low-noise system and a saturated regime for power self-regulation. In the 0 dBm case, the signal is strong enough compared to the ASE—at all points along the fiber—that the choice of pumping direction does not really matter. Chapter 6 discusses in more detail this interplay between the signal and ASE along the fiber length. Note, finally, that at very long fiber lengths the counterpropagating pump does worse than the copropagating pump in terms of signal gain, as a growing portion of the pump then becomes consumed by the forward ASE.

Returning to the consideration of noise in small signal gain amplifiers, we contrast the two pumping schemes–co and counterpropagating. This is particularly applicable to preamplifier and inline amplifier design, in terms of obtaining the lowest noise figure possible. As shown in Figure 8.12, the copropagating pump configuration offers the lowest noise figure (considering only the erbium-doped fiber and ignoring any loss from passive components before or after the erbium-doped fiber). The intuitive reason for this is simple. In the copropagating configuration the portion of the fiber that the signal enters tends to be more inverted than the section by which the signal exits. Thus, the signal undergoes more gain per unit length at the beginning of the fiber than at the exit. In the counterpropagating configuration, the inverse situation is present, and the lower gain per unit length at the beginning of the fiber is equivalent to having some amount of loss for the signal before it enters the amplifier. Any loss that the signal experiences at the beginning of the fiber will degrade the noise figure. Thus, in the absence of any other effects, the copropagating pump configuration is preferred for obtaining a low noise figure. This argument is not as valid if the fiber is short enough; in that case, assuming that pump power is not limited, a strong enough pump will strongly invert

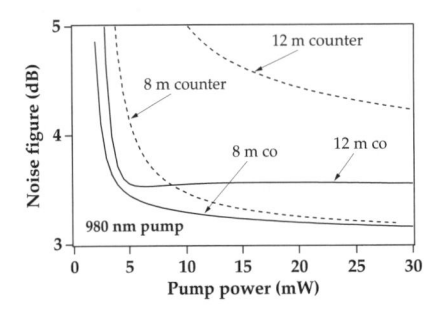

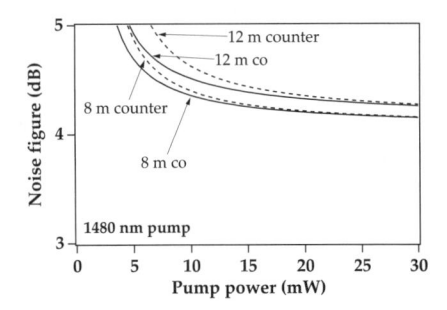

Figure 8.12: Noise figures for a copumped and counterpumped erbium-doped fiber amplifier (fiber A of Chapter 6) for 980 pumping (left) and 1480 pumping (right), from numerical simulations. Two lengths are contrasted, 8 m and 12 m.

the entire fiber and it will not matter whether the pump is copropagating or counterpropagating. Loss in the passive components used in the amplifier as a whole alters the picture somewhat. When choosing a copumping configuration, a WDM coupler that combines the pump and signal is placed immediately before the erbium-doped fiber. This WDM coupler has some loss at the signal wavelength and thus the noise figure is reduced by the loss (in dB) experienced by the signal through the WDM. This loss can be reduced to a few tenths of a dB by choosing a high quality WDM. In the undersea inline amplifier application of TAT-12, the amplifier is counterpumped, as shown in Figure 8.13. The amplifier in this case is a rather short piece of erbium-doped fiber that is operating in saturation, and for which the intrinsic noise figures of the erbium-doped fiber are approximately equal for the copumped and counterpumped configurations. By counterpumping, the loss of the WDM used in the copumped configuration is avoided in terms of its impact on the noise figure.

8.3.2 Choice of Fiber Lengths and Geometries for Various Applications

The discussion in this section affects considerations for the design of a low noise preamplifier or inline amplifier, with low to moderate input signal levels, as well as high output power amplifiers (discussed further in section 8.3.5) with large input signal levels.

Turning first to the small signal regime, one finds that, in a single-stage amplifier, it is difficult to simultaneously achieve both high gain and low noise figure. A long fiber will absorb all the pump light and offers the highest gain for a given pump power, assuming that the fiber is not so long that the end of the fiber is underpumped and the signal experiences loss. A shorter fiber will allow a significant amount of pump light to exit the fiber but will, on the other hand, be well inverted across its entire length, thus offering a good noise figure. Figure 8.14 shows, from numerical simulations of fiber A of Chapter 6, the gain and noise figure as a function of length for a 980 nm pumped fiber amplifier for a −40 dBm input signal at 1550 nm and a pump power of 25 mW. Also shown is the average upper state population across the fiber, where 1 denotes a fully inverted fiber and 0 a fiber where all the ions are in the ground state. As the inversion drops, the noise figure increases, as expected. The gain saturates and at some

8.3. AMPLIFIER DESIGN ISSUES 269

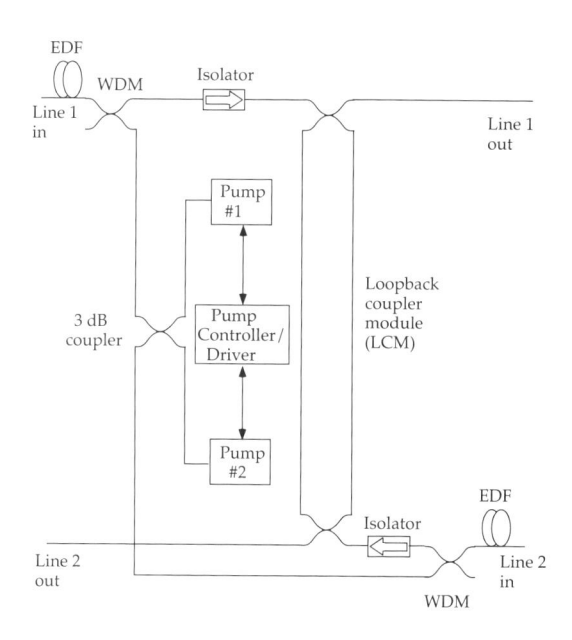

Figure 8.13: Amplifier pair architecture of AT&T design for transmission on a two-line path for TAT-12 undersea cable applications (EDF: erbium-doped fiber). From reference [46]. Copyright ©1995. *The AT&T Technical Journal*. All rights reserved. Reprinted with permission.

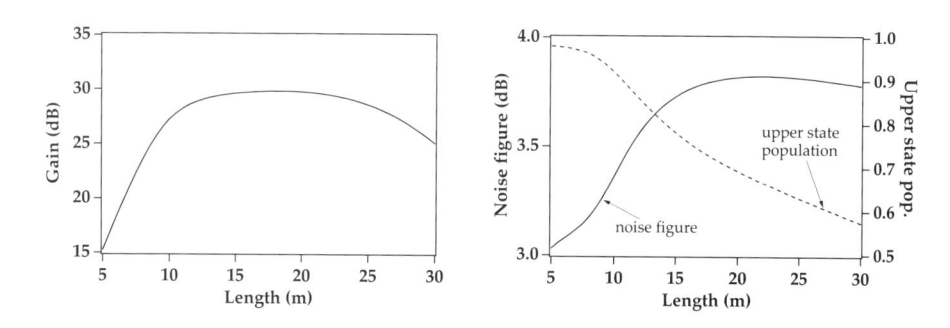

Figure 8.14: Gain (left) and noise figure and average upper state population (right) as a function of erbium-doped fiber length pumped with 25 mW of pump power at 980 nm, for a −40 dBm input signal at 1550 nm. From numerical simulations (fiber A of chapter 6).

point adding more fiber does not improve the gain much. As a result, there is a point beyond which it is not meaningful to increase the fiber length, as noted previously.[47] Note that at long lengths the noise figure varies slowly with length. This is due to the fact that the noise figure is determined primarily by the conditions at the input to the fiber, where the inversion is in turn determined by the backward ASE. Increasing the

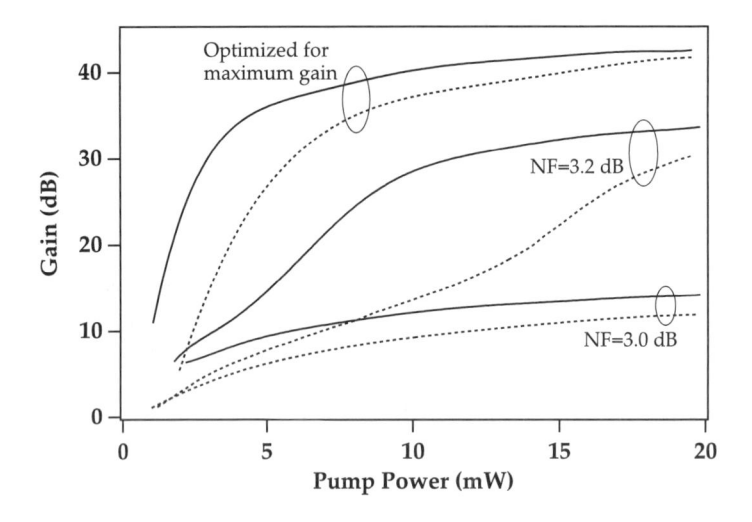

Figure 8.15: Maximum gain achievable for the indicated noise figures (NF), with a fiber optimized for length and cutoff wavelength at each pump power, for a −45 dBm signal at 1.53 μm, a 980 nm pump, and erbium confined to the center 70% of the core. The solid curves are for NA=0.3 and the dashed curves for NA=0.2. From numerical simulations presented in reference [48].

length of the fiber at low lengths increases the backwards ASE significantly, reducing the inversion level, and is thus the primary cause of the increase in the noise figure. At long lengths of fiber the input conditions stay very similar when increasing the fiber length. Added fiber at the end is loss faced by both the signal and the forward ASE and thus affects the noise figure less than the conditions at the beginning of the fiber.

In general, successively higher pump powers are required to obtain successively lower noise figures for a given gain, as shown in Figure 8.15. The lower noise figures typically require shorter lengths of fiber (to reduce the backwards ASE) and thus higher pump powers, to achieve the same gain. As the fiber NA is increased, reducing the mode field diameters, the pump power required to achieve a desired gain, for a specified noise figure, is decreased.[48]

As regards the transverse design of the fiber, the general rule is to use as much as possible a geometry which improves pump and signal intensities, their overlap with each other, their overlap with the erbium ion distribution, and the erbium ion inversion level. Thus, confining the erbium to the center of the core, the effect of which is shown in Figure 6.32, improves the noise figure, both at low pump powers and also at a given gain point. Reducing the mode field diameters (thus increasing pump and signal intensities) by increasing the numerical aperture of the fiber will improve the gain and noise performance at low pump power levels, by reducing the threshold for population inversion. As a note of caution, increasing the NA does not necessarily yield an improved noise figure at a given gain if the fiber length has been chosen to optimize the gain or the gain efficiency.[48] This is due again to the fact that optimization of

8.3. AMPLIFIER DESIGN ISSUES

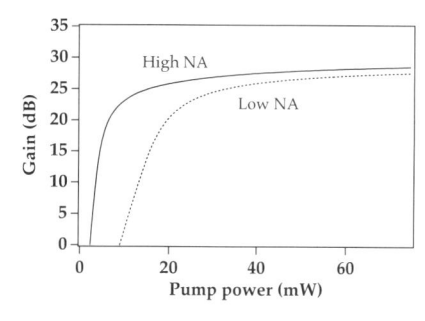

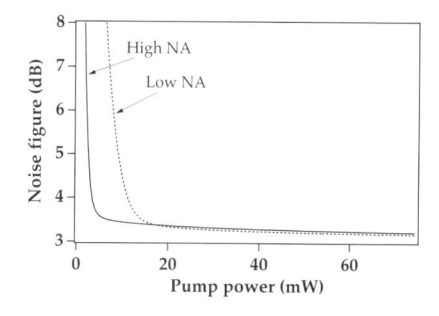

Figure 8.16: Gain (left) and noise figure (right) for a high NA fiber (NA=0.28) and a low NA (NA=0.15) fiber, as a function of pump power at 980 nm, of 12 m lengths and concentration such that the unpumped fibers have the same absorption coefficient at 1550 nm. The signal input is at 1550 nm with power −40 dBm). From numerical simulations.

gain alone often results in a higher noise figure, and is the more so true for a high NA fiber.

Figure 8.16 contrasts the small signal gain and noise performance at 1550 nm (for a 980 nm pump source) of two erbium-doped fibers of the same moderate length, corresponding, respectively, to a high NA (NA = 0.28, core radius 1.4 μm) and a low NA (NA = 0.15, core radius 2.0 μm) fiber, as a function of pump power. The erbium concentrations are such that the unpumped fibers have equal absorption at 1550 nm. At low pump powers, the advantage of the high NA fiber with its reduced core and mode field diameters is quite apparent. The advantage is removed at higher pump powers where both fibers are well inverted and the low NA fiber achieves high gains and low noise figures. At longer lengths the gains are improved and the noise figures deteriorated. The noise figure of the high NA fiber deteriorates more than that of the low NA fiber, due to the higher backward ASE level resulting from the more efficient operation at small signal levels. In the case of preamplifiers, very high gains are often required (i.e., greater than 30 dB) and high efficiency fibers are needed if the pump power is not to be too high. Very high NA (e.g., NA $\geq$ 0.3) fibers are indeed a good choice if the pump power is limited (e.g., $\leq$ 20 mW) as in the case of remote pumped amplifiers. Since these high NA fibers require lower pump powers to achieve a given gain and noise target, as examplified by Figure 8.15, they are the preferred choice for preamplifiers. Confining the erbium ions to the center of the core also aids in improving the gain efficiency and noise performance for such situations.

In the case of inline amplifiers, the desired performance characteristics are different than in the case of preamplifiers. Quite often only moderate gains and operation in saturation are needed. Pump powers are moderately high (e.g., on the order of 50 to 100 mW for long haul systems). For WDM amplifiers, moderate to large output powers are needed so that each channel has adequate power to maintain a good signal to noise ratio across the span. In this case, moderate NA fibers (0.2 to 0.3) are suitable for the application. As shown in Figure 8.17, which contrasts the high NA and low NA fibers of Figure 8.16, the higher NA fiber can achieve slightly higher output powers for

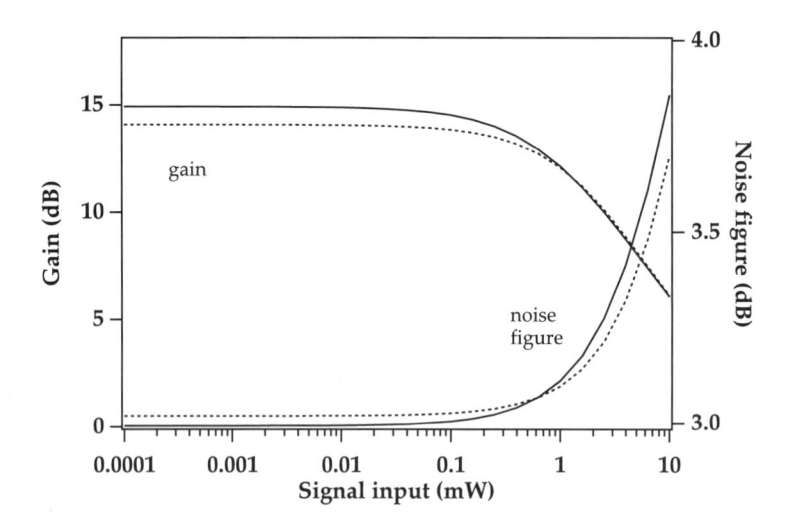

Figure 8.17: Gain and noise figure for a high NA fiber (NA=0.28; solid curves) and a low NA (NA=0.15; dashed curves) fiber, as a function of input signal power, of 6 m lengths and concentration such that the unpumped fibers have the same absorption coefficient at 1550 nm (pump power 60 mW at 980 nm). From numerical simulations.

the moderate signal input levels characteristic of inline repeaters (neglecting parasitic effects such as pump ESA). The fiber lengths are short and chosen for a typical long haul inline amplifier application. For the pump power shown, 60 mW at 980 nm, the difference between the high NA and low NA fibers is not that large. The ultra high NA fibers can suffer from higher background losses, as discussed in the context of Table 8.2, and this effect becomes more significant when higher signal output powers are needed. Whereas the small signal gain is minimally impacted by the background loss effect associated with high NA fibers, the conversion efficiency is significantly impacted.[49] The increased noise figure resulting from background losses is also of concern in systems with many inline amplifiers where the signal to noise ratio needs to not deteriorate too strongly during the signal's traversal of the span. Thus one would avoid very high NA (e.g., ≥ 0.3) fibers for inline amplifier applications. One would also tend to choose a lower NA fiber as the need for output power from an inline repeater increases and the pump power increases, since the lower NA fibers have higher pump efficiencies at high pump powers, as shown in Table 8.2.

Modeling results (without any pump ESA) lead to the conclusion that for power amplifiers an increased NA also leads to improved conversion efficiencies.[50] However, for very high output power amplifiers, low NA (on the order of 0.15) has been found to yield improved conversion efficiencies for high power applications. Higher NAs result in increased pump intensities and parasitic pump excited state absorption begins to occur with significant effects, especially when the pump is high (on the order of 500 mW or more).[51]

For a single stage amplifier, the requirements of high gain and gain efficiency are often in conflict with the objective of a low noise figure. This leads us to a discussion of multistage amplifiers, since the realization that backwards ASE, the main culprit in deteriorated noise figure situations, can be significantly reduced in multistage amplifiers.

8.3.3 Multistage Amplifiers

The key characteristic of multistage amplifiers is that they allow a significant portion of the ASE to be eliminated at some middle point along the fiber amplifier. This allows the gain to grow at the expense of the ASE and also reduces the noise. The ideal amplifier would contain "ASE eliminators" distributed along the fiber. To date, multistage amplifiers have been only of the two-stage or three-stage variety. As noted before, it is difficult to obtain simultaneously very large gains and very low noise figures in a single-stage amplifier. However, this can be achieved in a two-stage amplifier, as demonstrated in the first experiments with multistage amplifiers.[52, 53] By reducing the ASE, more gain is available for the signal and less noise is added to it from the spontaneous emission. Thus, it is not altogether surprising that the highest gain amplifiers have been achieved with a multistage design, all the while maintaining a low noise figure.

Various forms of multistage amplifiers have been reported, with more or less elaborate arrangements of passive elements to reduce the ASE and of multiplexers to deliver the pump power to the multiple sections of erbium-doped fiber.[54, 55] They all share the same basic characteristic that successive lengths of erbium-doped fiber are separated by optical elements that reduce the ASE preferentially over the signal. It is often advantageous to pump the first stage with 980 nm light and the second stage with 1480 nm light. The 980 nm pumping offers the benefit of low noise figure at the all-important input portion, while the 1480 nm pumping offers a much higher optical conversion efficiency for the power amplifier portion of the amplifier (see Figures 6.26 and 6.27).[56]

The most simple of the two-stage amplifiers contain two sections of erbium-doped fiber, separated by either an isolator or a spectral filter, as shown in Figure 8.18. Various pump configurations, as shown in the figure, can be employed. Counterpumping of the first stage can be employed if one wishes to avoid the signal loss from an input multiplexer, for a very low noise first stage. The positioning of the pump sources and any passive components required is important in extracting the highest possible gain and the lowest possible noise figure. One way to understand this is to view the first stage as a low-noise preamplifier and the second stage as a power amplifier. For each stage, one can follow the design principles valid for preamplifiers and power amplifiers. The isolator eliminates the backward-traveling ASE originating in the second portion of the fiber that would otherwise deplete the inversion in the beginning sections of the fiber. The first stage needs to be well inverted so that a moderate amount of gain is obtained with minimal noise added. The second stage then acts as a power amplifier for the signal. For a high gain and low noise preamplifier the NA of the first stage should be high. The NA of the second stage is better kept smaller than that of the first stage when the pump powers are very high (greater than $\simeq 100$ mW). Since the pump powers are higher for the second stage pump, excited state absorption can occur unless

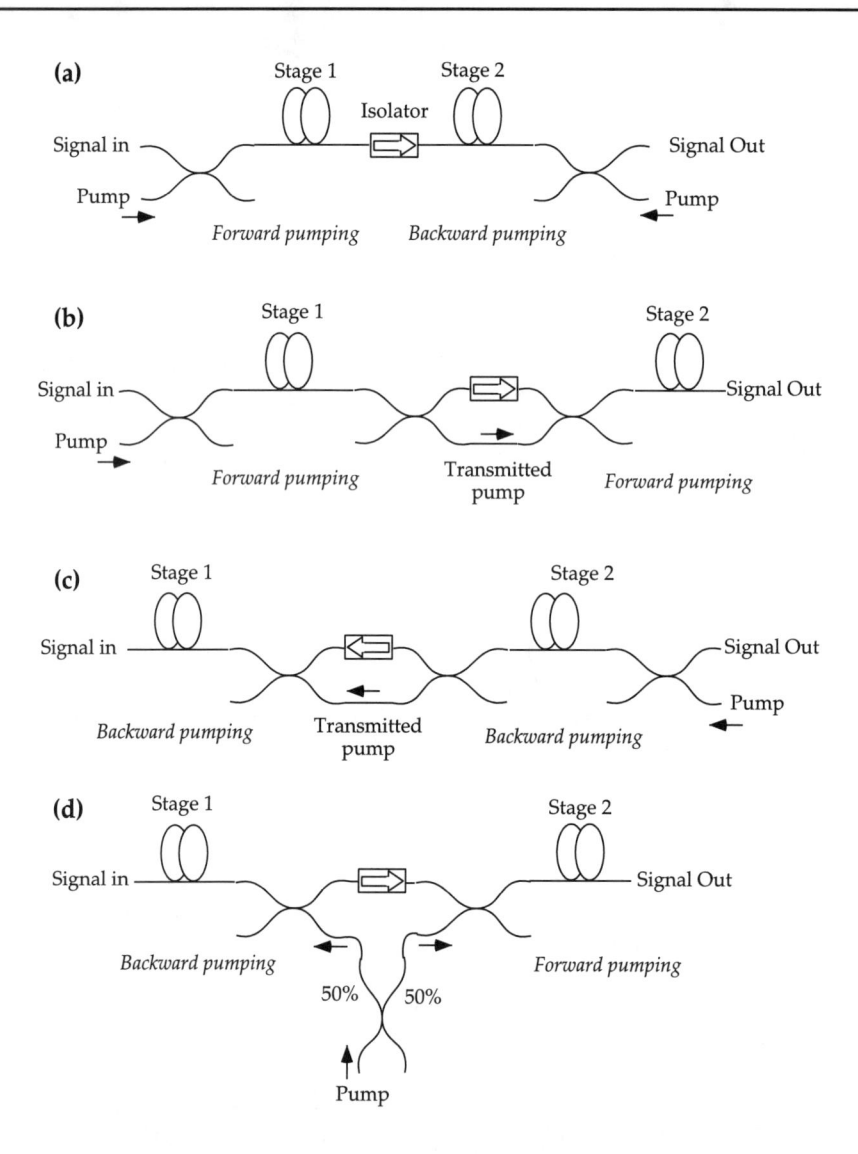

274   CHAPTER 8. AMPLIFIER CHARACTERIZATION AND DESIGN ISSUES

Figure 8.18: Experimental configurations of two-stage fiber amplifiers with a midway fiber isolator, with various possible pump configurations.

the pump intensity is kept lower by using a lower NA fiber. The power conversion efficiencies can still be high despite the lower NA.[57]

The experimental gain and noise figure obtained in a two-stage amplifier, with an isolator to separate the two stages, are shown in Figure 8.19. Such an amplifier has achieved a 54 dB gain while maintaining a 3.1 dB noise figure.[52] For a 980 nm pump, the pump light cannot traverse the isolator and thus in this case a multiplexer is used to allow the pump light exiting the first stage to bypass the isolator and enter the second stage. This kind of configuration is necessary in the case of a single pump that

8.3. AMPLIFIER DESIGN ISSUES   275

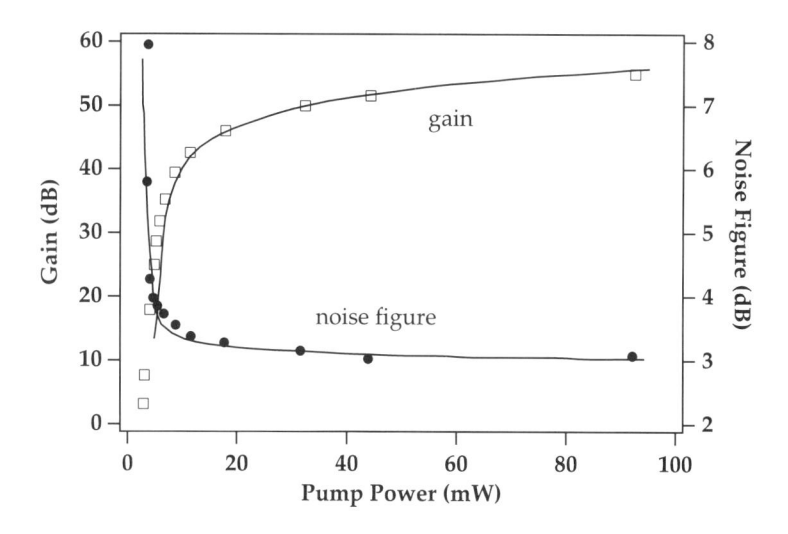

Figure 8.19: Experimental gain and noise figure as a function of pump power for a copumped two-stage amplifier, with a 25 m first stage and a 60 m second stage. The solid fit is from theoretical calculations. From reference [52] (©1992 IEEE).

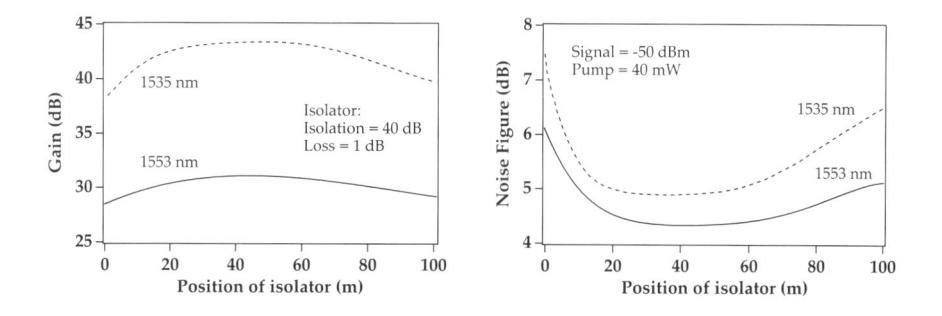

Figure 8.20: Gain (left) and noise figure (right) calculated as a function of the isolator position between two stages of erbium-doped fiber of total length 100 m, for the two signal wavelengths 1535 and 1553 nm and pumped with 40 mW at 1480 nm. From the numerical simulations presented in reference [53] (©1992 IEEE).

is co-propagating with the signal. Numerous other experiments have established the high performance and usefulness of such multistage amplifiers.[58, 59, 60, 61]

Simulations of a two-stage amplifier with an isolator placed between the two stages have shown that the optimum placement of the isolator is roughly at about one-quarter to one-half of the way from the beginning of the fiber.[53, 62] The optimum position of the isolator, as expected, depends on how the amplifier is pumped (copumped versus counterpumped versus double-pumped, etc.). Figure 8.20 shows the gain and noise figure as a function of the placement of the isolator.[53]

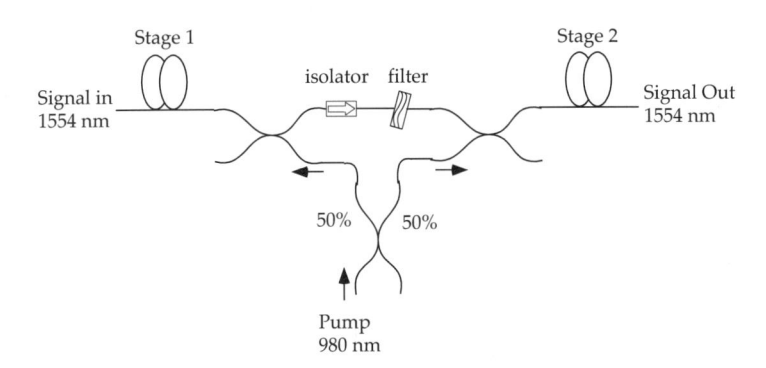

Figure 8.21: Configuration of a two-stage amplifier setup, with an interstage isolator and bandpass filter. From reference [66].

Another way to isolate the interaction between the stages of a complex amplifier is to use a filter.[60, 61, 63, 64, 65, 66] The filter can be a notch filter that rejects the 1530 nm ASE in the case of a 1550 nm region signal, which is the dominant ASE source in a relatively short piece of fiber. The filter can also be a bandpass filter that passes the wavelengths at the signal frequency. In contrast to the isolator, the filter rejects ASE traveling in both directions. For a small signal, a notch filter of width 20 nm has been shown to yield gain enhancements of about 4 dB while reducing the noise figure by 0.5 dB. For a bandpass filter of 40 nm at 1555 nm, a gain enhancement of 6 dB is obtained while reducing the noise figure by 0.5 dB.[61] Interstage filters can also be used to flatten the gain spectrum of the amplifier.[67, 68]

More complex amplifiers incorporate both an isolator and a filter, as shown in Figure 8.21 (which is similar to the configuration of Figure 8.18(d) with the combination of an isolator and a spectral filter between the two stages). The filter removes the 1530 nm ASE (which is the dominant source of ASE), while the isolator removes most of the backward-traveling ASE. The use of the isolator and filter combination provided a slight improvement over that of the filter alone, which itself was more effective than the isolator alone, for the application considered.[66]

Other passive elements can be added to multistage amplifiers. Circulators have been added, which allow the amplifier to operate in a reflective fashion where one or more lengths of erbium-doped fiber are double passed.[69, 70] This allows for the further incorporation of gain-flattening filters. Some passive elements that have been incorporated in multistage amplifiers, such as lengths of dispersion compensating fiber, can act to counterbalance other effects (i.e., dispersion).[71] Finally, multistage amplifiers in system applications have revealed themselves to be very useful. Two-stage amplifiers have been used to achieve record sensitivities of 76 photons per bit for detection at 1548 nm.[72] Multistage amplifiers have also been successfully used in high bit rate long distance transmission experiments.[73] Overall, multistage amplifiers have become a fruitful avenue of investigation to take amplifiers beyond the one-dimensional role of a high-gain single-channel amplifier.

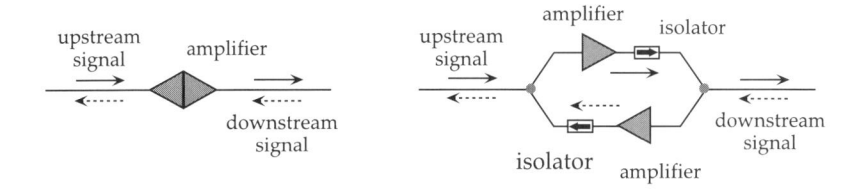

Figure 8.22: Basic amplifier configurations for a bidirectional transmission system using a single fiber link. In the pure bidirectional configuration (left) the same erbium-doped fiber is used for signals traveling in both directions, whereas in the separate amplifier bidirectional configuration (right) the signals are separated before being traversing different erbium-doped fibers each operating in unidirectional configuration.

8.3.4 Bidirectional Amplifiers

Erbium-doped fiber amplifiers are inherently bidirectional in nature, as exemplified by the presence of both forward and backward propagating ASE. The addition of unidirectional elements, such as isolators, removes the bidirectional amplification capability of a single erbium-doped fiber amplifier. These elements have traditionally served to protect the transmission system from the effect of unwanted reflections. Nevertheless, the significant success of long haul systems based on erbium-doped fiber amplifiers has led to extensive research on the potential for bidirectional transmission over a single fiber link. The advantages of such a scheme are several. The number of amplifiers can be reduced if the fiber amplifier is used as a bidirectional amplifier. In addition, by changing a system which previously transmitted N channels in a unidirectional configuration to a bidirectional system with N/2 channels in either direction, the potential for nonlinear effects such as four wave mixing can be reduced.[74] Bidirectional transmission systems where isolators are not used to restrict to transmission direction are in principle possible, at least from considerations of signal to noise issues.[75, 76]

Two main configurations for bidirectional transmission are possible, as shown in Figure 8.22. In the first configuration, the same erbium-doped fiber amplifier is used for signals transmitted in both directions. It is significant to note that this configuration includes both true bidirectional amplifiers, with signals propagating in both directions, as well as unidirectional amplifiers where the upstream and downstream signals have been arranged to copropagate through the amplifier. In the second configuration, the counterpropagating signals are separated at each amplifier node, and traverse different amplifiers arranged in a unidirectional configuration. The signals are then recombined after the repeater and transmitted over the same fiber span. The components which serve to separate the signals are passive components such as wavelength division multiplexers and/or circulators.

While the feasibility of the pure bidirectional configuration (counterpropagating signals in the same amplifier) has been demonstrated over short distances, it presents some very delicate implementation issues.[77] Reflections of any sort, whether discrete or continuous (Rayleigh scattering) will result in laser action by the amplifier which can significantly degrade the transmission capacity of a link consisting of an inline amplifier between two fiber spans. Even at reflection levels well below that needed

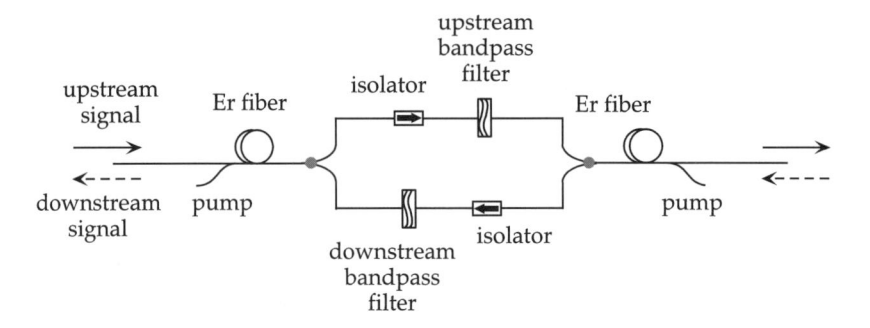

Figure 8.23: Repeater configuration for a bidirectional transmission system using a common erbium-doped fiber amplifier for both upstream and downstream signals, with reflections suppressed by a mid-point arrangement of isolators and bandpass filters. From reference [87] (©1996 IEEE).

for lasing, system degradation has been observed. This is due to the fact that Rayleigh scattering is a randomly fluctuating phenomenon and while the average value may be low, the instantaneous peak values are high. The relationship between the peak amplifier gain G and the reflection levels on either side of the amplifier, R_1 and R_2, is $G^2 R_1 R_2 \leq -15$ dB for a less than 0.5 dB power penalty (for digital transmission) from laser action of the amplifier.[78]

Another effect of reflections is that multiple reflections convert laser phase noise into intensity noise as reflections of the signal are interferometrically superimposed on the transmitted signal.[79] It has also been reported that given the Rayleigh scattering resulting from the transmission fiber (-31 dB to -34 dB of the launched signal power for long lengths of fiber) the amplifier gain for a single inline bidirectional amplifier (with counterpropagating signals in the same amplifier) needs to be restricted to less than roughly 20 dB for a less than 1 dB power penalty from multiple reflections in the transmission fiber.[80] With a link of cascaded amplifiers the Rayleigh scattering will accumulate and become even more of a problem.

The reflection problem is exacerbated in a bidirectional transmission system when signals in both direction are at close frequencies, such that a single reflection can overlap another signal.[81, 82, 83] By using different frequencies (1537 nm and 1545 nm) and short spans, pure bidirectional transmission has been demonstrated with one in-line amplifier over 95 km at 622 Mb/s.[77] Repeaterless bidirectional transmission has also been demonstrated using different transmitter frequencies, with both common and separate erbium-doped fiber amplifiers at each end.[74, 84, 85, 86] Both bandpass filters and fiber grating blocking filters are useful in separating the signal from unwanted Rayleigh backscattered light from the opposite direction signal.[86]

One way to combat the reflection problem is to separate the amplifier in two and place a reflection suppressing element in the middle, between two pieces of erbium-doped fiber. On can use, for example, a two-arm mid-point element consisting of isolators and bandpass filters which forces each signal through an isolator in the appropriate direction, as shown in Figure 8.23.[87, 88] This kind of configuration has been

8.3. AMPLIFIER DESIGN ISSUES

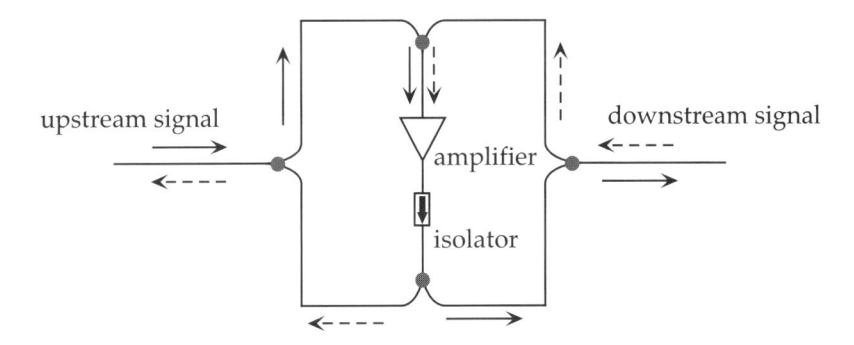

Figure 8.24: Repeater configuration for a bidirectional transmission system using a common erbium-doped fiber amplifier for both upstream and downstream signals, using a bidirectional bridge configuration. The signals are made to copropagate before traversing the common erbium-doped fiber operating in a unidirectional configuration. From reference [90].

used to demonstrate bidirectional transmission over 300 km at 2.5 Gb/s with one signal in each direction (1552 nm and 1545 nm).[87] Repeaterless transmission has also been demonstrated where a common erbium-doped fiber is used as both a preamplifier and postamplifier at both ends, and an arrangement of WDMs and isolators are used to separate the signals and reduce reflections.[89] A drawback of such architectures is that cross saturation of the erbium-doped fiber between the two counterpropagating signals may potentially occur. One signal will be strongest near its output end, which is the input end for the other signal. The strong signal will deplete the inversion at that point resulting a higher noise figure for the counter propagating signal. An alternative method consists of using a bridge circuit arrangement so as to ensure that signals copropagate in the amplifiers, as shown in Figure 8.24.[90] This permits arranging the erbium-doped fiber amplifier in a unidirectional configuration, thus reducing the effect of reflections. For the configurations shown in Figures 8.23 and 8.24, the upstream and downstream signals can be separated and recombined by circulators or wavelength selective couplers.

A number of schemes have been successfully investigated for bidirectional transmission where the amplifier nodes consist of a multiple arm arrangement where counterpropagating signals pass through separate amplifiers, as in the right hand side of Figure 8.22. Circulators and wavelength division multiplexers can be used to separate and recombine the signals before and after each amplifier node.[91, 92] Single reflections (discrete or Rayleigh) of signals propagating in a given direction will counterpropagate and be amplified by the amplifier assigned to the opposite direction signal. These spurious signals can be eliminated by narrow bandpass filters if the counterpropagating signals are of different wavelengths.[93, 94] Another technique to reject Rayleigh backscattered light from counterpropagating signals is to use combinations of circulators and fiber grating filters, which allowed bidirectional transmission at 10 Gb/s over 240 km of two channels separated by 1.5 nm.[91] Careful optimization of the passive elements used to separate the signals can lead to a system with a large number of

Year	Pump Input Power (mW)	Signal Output Power (mW)	Signal Input Power (mW)	Ref.
1989	25	6.3	0.15	[97]
1990	33	14	0.07	[98]
1990	50	35	0.09	[99]
1990	1100	530	0.8	[100]
1991	100	54	0.6	[101]
1991	363	145	0.25	[102]
1992	390	141	0.25	[103]
1992	900	288	2.5	[104]
1994	1900	700	1	[105]
1995	545	331	3	[106]
1996	12,600	4,004	2.6	[107]
1997	9,700	5,500	1.6	[108]

Table 8.3: Experimental demonstrations of erbium-doped fiber power amplifiers.

channels. One example of this is the transmission of seventy three channels (forty one upstream and thirty two downstream) over a repeaterless system length of 126 km.[95] As in the case of unidirectional WDM transmission, dispersion management has to be used to reduce four wave mixing in either direction when the number of channels becomes large and the channel spacing is reduced.

Assigning the collection of signals in each direction to non overlapping bands facilitates the architecture of systems.[94] The use of a gain flattened wide bandwidth amplifier allows the bidirectional transmission of a large number of channels. In particular, the use of the two arm gain flattened amplifiers described in section 8.3.6, where each arm corresponds to one of the erbium amplification bands, provides effective amplification as well as efficient crosstalk separation from Rayleigh backscattered light. Using such an amplifier architecture, bidirectional transmission of ten channels ranging from 1531 nm to 1607 nm at 2.5 Gb/s over 250 km has been demonstrated, with two inline amplifiers. Commercial applications of such bidirectional systems should increase as long distance transmission systems continue to improve in terms of efficient capacity utilization.

8.3.5 *Power Amplifiers*

The erbium-doped fiber amplifier has an output saturation power that can be increased with pump power, as discussed in section 5.2. Additionally, due to the long upper-state lifetime, pump to signal quantum conversion efficiencies approaching 100 % can in theory be obtained. The pump to signal energy conversion efficiency will be necessarily lower because the pump photon energy is larger than the signal photon energy. Thus, a high-power pump can transfer a significant fraction of its energy to a signal, resulting in signal outputs that can reach several watts when the signal input is large enough to extract the gain at the expense of the ASE. Table 8.3 lists the progress in power amplifier outputs over recent years.

8.3. AMPLIFIER DESIGN ISSUES

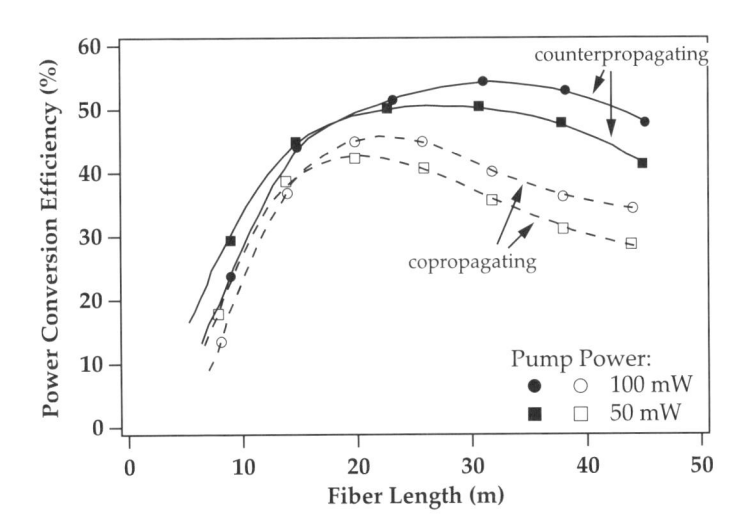

Figure 8.25: Experimental pump to signal power conversion efficiencies as a function of fiber length, for pump powers of 50 and 100 mW at 980 nm, in the copropagating (solid line) and counterpropagating (dashed line) configurations (signal input −3 dBm at 1533 nm). From reference [96] (©1991 IEEE).

Early experiments established quantum conversion efficiencies of 93 %, with a pump to signal energy conversion of 59 % for a 980 nm pump and a 1533 nm signal.[96] With an input signal power of −2 dBm to −3 dBm, output powers of up to 54 mW (17 dBm) were obtained for pump powers of 100 mW. This study established that the fiber length needs to be carefully chosen. When the fiber is too short, there are not enough erbium ions to absorb all the pump photons, and when the fiber is too long, the end section of fiber is underpumped. Figure 8.25 shows the pump to signal conversion efficiency as a function of fiber length. An optimum fiber length is clearly visible. In addition, the counterpropagating pump configuration clearly gives higher signal output powers than the copropagating one. We noted in sections 6.4.6 and 6.6.2, and in section 8.3.1, that this is true for intermediate signal input powers for amplifiers with and without pump ESA. Figure 8.11 confirms this as well from numerical simulations (note that the fiber of Figure 8.11 has a higher NA than that of Figure 8.25).

A power amplifier typically operates deep in saturation, with a gain significantly lower than the small signal gain. As a result, the gain is much flatter, as a function of signal wavelength, than that of an amplifier operating in the small signal mode. This is intuitively expected since, whatever the signal wavelength, almost all the energy stored in the erbium ions is extracted from the fiber. This energy is a constant that depends primarily on the total number of erbium ions for a strong enough pump (the signal is taking out nearly all the energy stored in the amplifier at the expense of the ASE, and this is a nearly wavelength-independent process). Figure 8.26 shows the signal gain as a function of wavelength for a 10 m length of fiber A (the characteristics of which were described in Chapter 6). The input power level was successively taken to be −40 dBm, −20 dBm, and 0 dBm. At high input power levels, with the amplifier operating as a

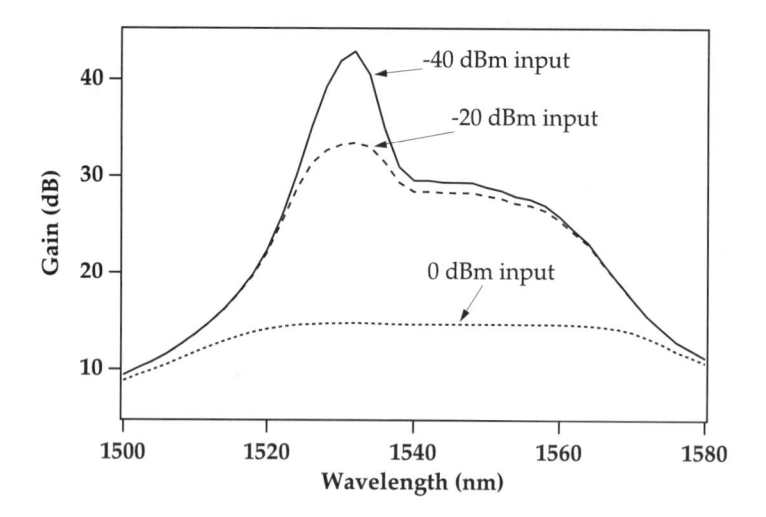

Figure 8.26: Spectrally resolved gain as a function of signal wavelength and signal input power, for a 10 m length of fiber A (described in Chapter 6) pumped with 50 mW at 980 nm. From numerical simulations.

power amplifier and running deep in saturation, the spectral gain is essentially flat over a wide range of wavelengths.

Commercially available laser diode pumped erbium-doped fiber power amplifiers produce typical output powers of approximately 15 dBm for input powers on the order of −5 dBm. These kinds of power amplifiers employ 1480 nm or 980 nm diode laser pumps. Higher power outputs been obtained with high-power 980 nm laser diode pumps.[105] The erbium-doped fiber power amplifier finds its fullest embodiment when the pump power is on the order of 1 W or more. These levels of pump power are typically obtained from diode array pumped solid state lasers or cladding pumped fiber lasers. Such pump sources were previously described in Chapter 3, section 3.13. The most common solid state laser is based on the Nd^{3+} laser operating near 1.06 μm. It can produce cw output powers in the multiwatt range and is thus suitable as a power amplifier pump. The 1.06 μm wavelength, however, is not in a pump range for Er^{3+}. To use this wavelength, it is necessary to transfer the energy via an intermediate ion. This intermediate ion, Yb^{3+}, has a strong absorption at 1.06 μm. Energy is then transferred from Yb^{3+} to Er^{3+}, as sketched in Figure 8.27. To ensure that transfer is mainly in the Yb^{3+} to Er^{3+} direction and not the reverse, it is important that the $^4I_{11/2}$ level of Er^{3+} deactivate as rapidly as possible to the $^4I_{13/2}$ level. The population of the $^4I_{11/2}$ level will then always be much smaller than that in the $^2F_{5/2}$ level of Yb^{3+}. This is accomplished by using a silica glass host that is strongly co-doped with phosphorus. The higher phonon energies in this glass increases the nonradiative deactivation of the Er^{3+} $^4I_{11/2}$ level. The absorption spectrum of Yb-Er fiber is compared to that of an Er-doped fiber in Figure 8.28.

The first experiments with a Yb-Er amplifying fiber were demonstrated by Grubb et al.[110] The pump source was an Nd:YLF laser pumped by AlGaAs diode arrays.

8.3. AMPLIFIER DESIGN ISSUES

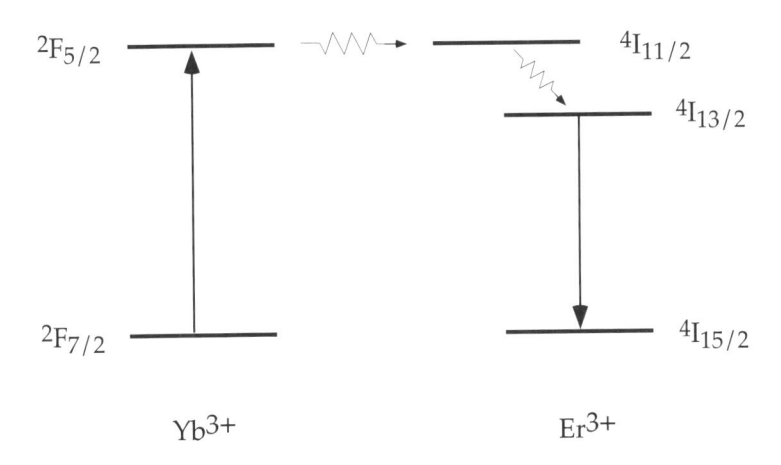

Figure 8.27: Yb^{3+} ion to Er^{3+} ion energy transfer process.

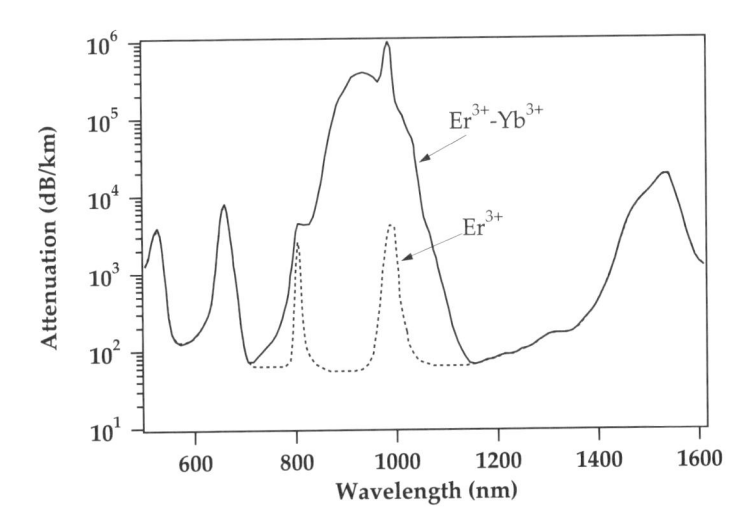

Figure 8.28: Yb^{3+}-Er^{3+} and Er^{3+} doped silica fiber absorption spectra. From reference [109].

The maximum fiber-coupled pump power obtained was 1 W. A maximum signal power of 24.6 dBm was obtained for an input signal power of 4 dBm at 1535 nm. By using two pump sources set up for bidirectional pumping of the Yb-Er fiber, an output power of 27 dBm was obtained. More recently, extremely high output powers, up to 4 W, have been obtained with solid state laser pumped Yb-Er amplifier fibers.[107] This power amplifier employed three Nd fiber lasers pumping three successive lengths of Yb-Er fiber separated by WDMs and isolators. Figure 8.29 shows the signal output power as a function of bidirectional pump power at 1 μm. At such high powers the signal output powers approach the limit for Brillouin scattering in single-mode fibers. Powers in

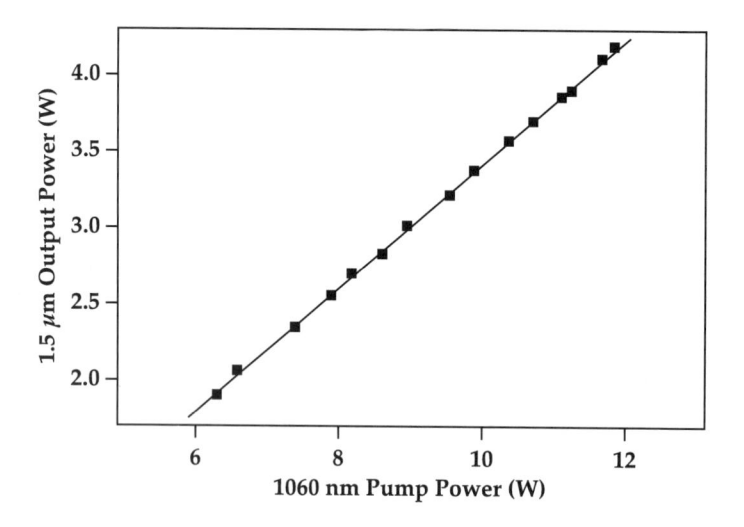

Figure 8.29: Signal output power as a function of pump power in a bidirectionally pumped Er-Yb-doped fiber power amplifier. From reference [107].

excess of 5 W at 1550 nm have been obtained by pumping with extremely high power pumps described in Chapter 3, section 3.13, such as cascaded Raman fiber lasers.[108]

8.3.6 WDM Amplifier Design Issues

Introduction

Even a cursory glance at the profile of the erbium ion gain spectrum reveals that it has several peaks and valleys and will not amplify with the same gain signals that are at different wavelengths. In a chain of cascaded amplifiers the signals near the peak of the gain will grow at the expense of the other signals. This can be a limitation if the receiver does not have a dynamic range sufficient to cover the spread of signal powers at the end of the system. One way to combat this effect is to keep all the wavelength channels in a very limited bandwidth region where the gain is more or less flat. This is restrictive, however, in terms of the available bandwidth and overall channel capacity, given a minimum channel spacing.

The shape of a single erbium-doped fiber amplifier gain profile may appear more flat when the amplifier is in saturation, from the point of view of a single wavelength being swept across the gain spectrum. This does not, however, imply that the amplifier gain is flat over that wavelength range with respect to multiple signal inputs. Figure 8.30 shows the example of the gain of a 0 dBm input signal swept in wavelength across the 1500-1580 nm band (see also Figure 8.26). The gain appears to be very flat over a wide spectral range. The average population in the upper state, also plotted in the figure, varies with each wavelength. A given amplifier will operate at a specific upper state population and cannot give rise to the flat gain spectrum of Figure 8.30. When saturated amplifiers are used in a cascade with multiple wavelength inputs there

8.3. AMPLIFIER DESIGN ISSUES   285

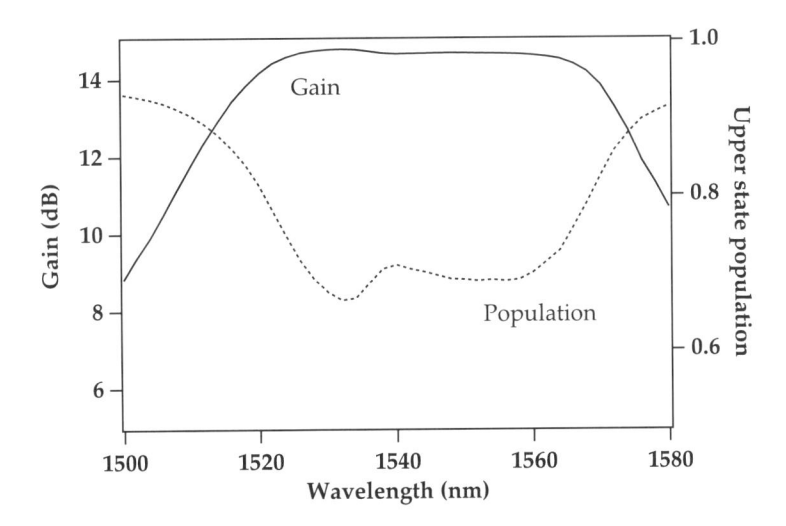

Figure 8.30: Gain spectrum of an erbium-doped fiber amplifier for an input signal with power 0 dBm swept in wavelength. The fiber is 10 m long and the pump power is 50 mW at 980 nm. Also shown is the average upper state population (relative to total population) present in the erbium ion population for operation at each wavelength. From numerical simulations of fiber A of Chapter 6.

is rapidly a large accumulation of interchannel power differences. The gain spectrum will then have one of the shapes discussed in Chapter 6, and will depend on both the pump, signal, and ASE powers present in the amplifier. Recall that the gain spectrum is proportional to the average erbium ion population inversion in the fiber, and the resulting shape will only be flat in limited spectral regions.

Gain flatness is a difficult goal to achieve in a system of cascaded erbium-doped fiber amplifiers, without the help of external gain-shaping and gain-flattening elements. Simulations of the behavior of cascaded amplifiers in a WDM system provide a guide for optimizing the transmission characteristics of the system, both in terms of the choice of the amplifier and possible external wavelength-shaping filters.[111] The various methods used over the years to achieve gain flatness in erbium-doped fiber amplifiers are listed in Figure 8.31. The glass composition category refers to intrinsic methods for single stage amplifiers that do not use external passive components to shape the gain. The equalizer category refers to methods that use external passive or active elements to flatten the gain and which are often placed at the midpoint in a two-stage amplifier. The hybrid methods refer to combinations of diverse single stage amplifiers, often in a cascade or split arm configuration, and possibly along with passive equalizers.

Gain flattening - "glass composition" methods

Astute choice of the pumping conditions as well as of the fiber length can contribute to increasing the gain flatness. 980 nm pumping and 1480 nm pumping have both been shown to give rise to flat spectral gain over a limited range of wavelengths.[112, 113]

Gain Flattening Techniques

- **Glass Composition / Single Stage Amplifier**
  - Standard silica - inversion shaping 1550 nm region 1580 nm region
  - High Al concentration co-doping
  - Standard silica - cooled to 77 K
  - Fluoride glass host
  - Tellurite glass host

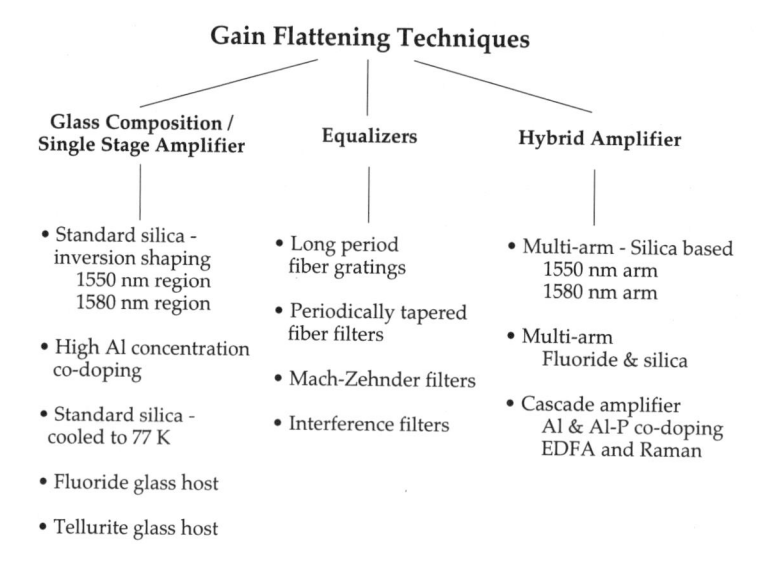

- **Equalizers**
  - Long period fiber gratings
  - Periodically tapered fiber filters
  - Mach-Zehnder filters
  - Interference filters

- **Hybrid Amplifier**
  - Multi-arm - Silica based 1550 nm arm 1580 nm arm
  - Multi-arm Fluoride & silica
  - Cascade amplifier Al & Al-P co-doping EDFA and Raman

Figure 8.31: Various methods used to achieve flat gain operation from erbium-doped fiber amplifiers. The methods are separated into three broad categories, described in the text.

Amplifier	Bandwidth (nm)
Single stage amplifier - Silica host 1550 nm band; no equalization	~ 12
Single stage amplifier - Silica host 1585 nm band; no equalization	~ 30
Single stage amplifier - Fluoride host 1550 nm band; no equalization	~ 25
Two-stage amplifier - Silica & fluoride hosts 1550 nm band; no equalization	~ 30
Two-stage amplifier - Silica host 1550 nm band; equalizing filter	~ 65
Two-stage amplifier - Silica host 1580 nm band; equalizing filter	~ 50
Two-stage amplifier - Tellurite host 1550 nm band; equalizing filter	~ 80
Dual-arm amplifier - Silica host Two-band operation; equalizing filters	~ 85

Table 8.4: Approximate bandwidths for "flat" gain operation for different types of erbium-doped fiber amplifiers described in the text.

8.3. AMPLIFIER DESIGN ISSUES

Since 980 nm pumping yields lower noise figures, it is the preferred pump source. The key here is to operate the amplifier with an inversion of about 0.7 between the upper state and lower state populations.[114, 115] This results in flat operation between 1546 nm and 1558 nm. A power spread of 1 dB over 60 dB of system loss has been demonstrated over this wavelength range.[115] In general, the optimal inversion depends on the desired bandwidth of operation. Once the average inversion level has been chosen, the noise figure and output power can be adjusted by an appropriate choice of fiber length, given the availability of the necessary pump power to achieve that inversion.

The underlying phenomenon is the interplay between the emission cross section and the absorption cross section. As discussed in Chapter 6, the gain spectrum for a section of fiber of length L, where the average populations of levels 1 and 2 are $\overline{N}_1$ and $\overline{N}_2$, is given by

$$G(dB) = 4.34 \times (\overline{N}_2 \sigma_s^{(e)} - \overline{N}_1 \sigma_s^{(a)}) \Gamma_s L \qquad (8.11)$$

where $\sigma_s^{(e)}$ and $\sigma_s^{(a)}$ are the emission and absorption cross sections, respectively, and Γ_s is the signal overlap factor. Under conditions of strong inversion where $\overline{N}_1 \simeq 0$, the amplifier is in a strongly pumped regime with no saturation. Under those conditions the gain spectrum is similar to the small signal gain spectrum of a well-pumped amplifier, which, to use the example of a silica host, exhibits a strong peak near 1530 nm and a relatively flat shoulder in the 1540–1560 nm region. Under conditions of moderate inversion, such as an amplifier with saturating inputs (signal or ASE), the 1530 nm signal will be more strongly attenuated than the longer wavelength signals and the region near 1550 nm has a much more pronounced curvature in the gain spectrum. The gain spectra, as represented by the net cross section (difference between emission and absorption cross sections, weighted by the respective populations in the upper and lower states) is shown in Figure 8.32. The situation is very similar to that of gain peaking in a cascade of typical single channel amplifiers. In the case of an ultra long distance single channel amplified system, the amplifiers are operating in compression. Thus the population inversion is moderate, and the gain peaks near 1560 nm while falling off sharply at short wavelengths. While this kind of operating condition is well suited to self-filtering in a single channel system, it will not provide the required flat gain for WDM operation.

Flat amplifiers for which the flatness is determined solely by the inversion level are not robust to system changes, since such changes will alter the operating conditions (e.g. signal power) and thus the inversion level. The link loss between amplifiers plays a role in determining the gain flatness of a cascaded system since it governs the strength of the input signals to the amplifiers and thus determines their level of saturation and of population inversion. Viewed in this light, unintended variations in link loss will affect these parameters and thus the flatness of the gain, impairing its robustness to changes in system conditions. The system robustness also becomes an issue when channels are added or dropped as the system is reconfigured.

Under conditions of high inversion, simulations have shown that a system with six cascaded highly inverted amplifiers would reduce the interchannel power spread by an order of magnitude, compared to an amplifier cascade with moderate inversion amplifiers.[114] This kind of system is more problematic for large numbers of amplifiers, since amplifiers operating at high inversion levels tend to favor a strong ASE peak

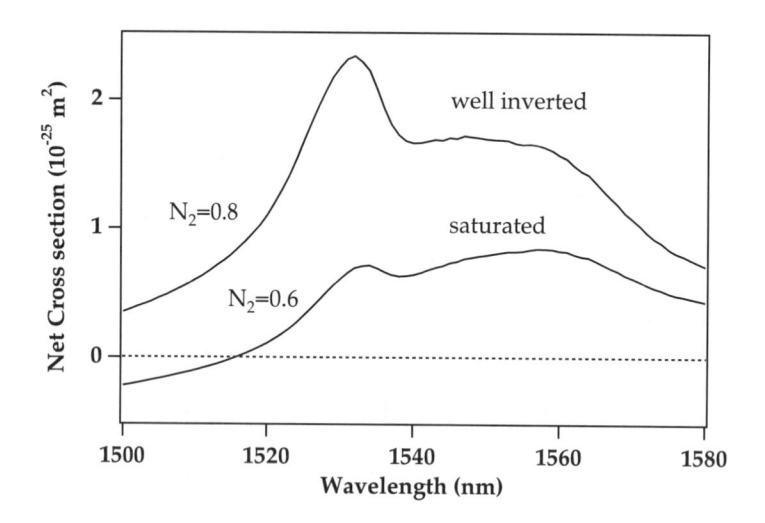

Figure 8.32: Effective cross section ($N_2\sigma_e - N_1\sigma_a$) for a well inverted and moderately inverted (saturated) erbium-doped fiber amplifier (Al-Ge fiber A of Chapter 6), where N_1 and N_2 are the populations in the lower and upper states, respectively, normalized to the total erbium ion population.

near 1530 nm, which robs the gain and reduces the SNR at the longer wavelengths. Filtering of this 1530 nm peak then becomes necessary. One proposal has been to use an alternating series of high inversion and moderate inversion amplifiers, so as to engineer an effectively flat gain spectrum in the 1550 nm region.[116]

Since the conditions for wavelength flat operation of a transmission system with unequalized single-stage silica-based erbium-doped fiber amplifiers comprise a quite narrow set of wavelengths, span losses, and amplifier parameters, it is not too surprising that other solutions have been investigated, as discussed in the following sections. Roughly speaking, operating erbium-doped silica host single stage amplifiers at an optimal inversion point results in a flat bandwidth of about 12 nm centered near 1550 nm, or 30 nm centered near 1585 nm.

Changes in the silica host glass composition can also produce a flatter gain spectrum in the longer wavelength regions of the erbium gain profile. A high Al co-doping concentration (on the order of 3 wt. %) has been shown to achieve this.[117, 118] The high Al concentration reduces the dip between the 1530 nm peak and the 1550 nm shoulder in the spectrum.[119] The spectrum is flattened from about 1544 nm to 1557 nm. System experiments including such amplifiers have demonstrated the transmission of sixteen channels spaced by 100 GHz as well as that of thirty two channels spaced by 62 GHz (this latter demonstration also included some gain equalizing filters).[118, 120]

The longer wavelength shoulder of the erbium gain profile, from approximately 1570 nm to 1600 nm, has also been demonstrated to offer the potential for flat gain operation.[123, 124, 125, 126, 127] Given the lower emission cross section in that region, longer fiber lengths have to be used to achieve reasonable gain. Referring to the inversion dependent gain spectra shown in Chapter 6 (see Figure 6.2), flat gain

8.3. AMPLIFIER DESIGN ISSUES

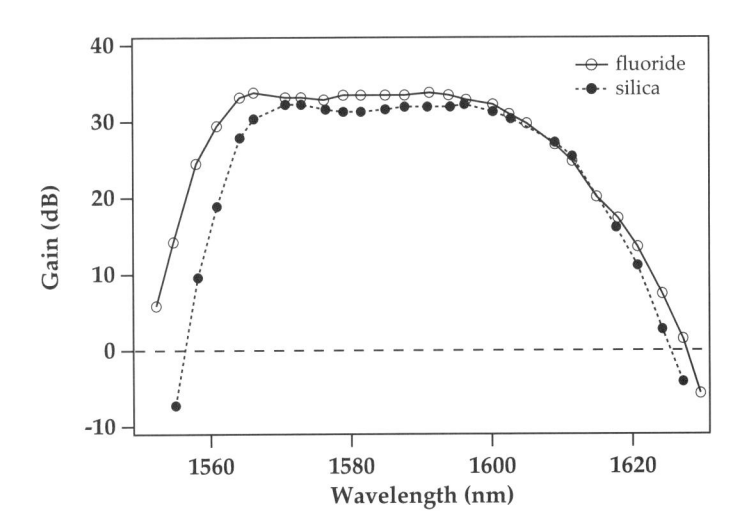

Figure 8.33: Gain spectra in the 1580 nm region of erbium-doped silica and fluoride fibers, with a signal input of −30 dBm and pumping at 1480 nm (forward 100 mW and and backward 13 mW for the silica fiber, forward 40 mW and backward 13 mW for the fluoride fiber). From reference [128].

operation is seen to be achieved for an inversion level in the range 0.3 to 0.4 (for the proportion of the population in the upper state). A gain excursion of only 0.9 dB (with an average gain of 29.5 dB) from 1573 nm to 1600 nm (for signal input powers of −30 dBm per channel) has been demonstrated in a silica host.[125] The noise figures in this region have been measured to be 4 dB to 6.5 dB, with a pump at 1480 nm. The gain spectrum is shown in Figure 8.33. Also shown is the 1580 nm region gain spectrum for an erbium-doped fluoride fiber which gives rise to a slightly larger flat bandwidth than silica, from 1570 nm to 1600 nm.[128] Amplifiers at 1580 nm have been used successfully for WDM transmission system demonstrations.[129, 130, 131]

Another alternative for achieving the requisite gain flatness for WDM transmission over long distances is offered by moving to a host fiber glass other than silica. Fluoride glass hosts yield an intrinsically flatter gain profile for the trivalent erbium ion in the 1500 nm region than do silica hosts. Amplifiers for WDM systems made with erbium-doped fluoride glass have been demonstrated.[132, 133] Erbium-doped fluoride glass amplifiers have a flat bandwidth of about 30 nm, from approximately 1530 nm to 1560 nm, where the gain variation is less than 1.5 dB (with an average gain of 26 dB).[134] The spectral region over which the gain varies by only 0.2 dB (with an average gain of 26 dB) is reported to be from 1534 nm to 1542 nm.[134] Figure 8.34 compares the output spectrum, after 440 km and four amplifiers, of sixteen channels at 2.5 Gb/s for both silica- and fluoride-based single stage amplifiers.[135] The fluoride amplifiers result in a significantly more equal distribution of powers of the various channels at the system output. This result has been extended to the transmission of sixteen 10 Gb/s channels over 531 km with seven inline single erbium-doped fluoride fiber amplifiers.[136] The drawback of the erbium-doped fluoride fiber amplifier, aside from the fact that the

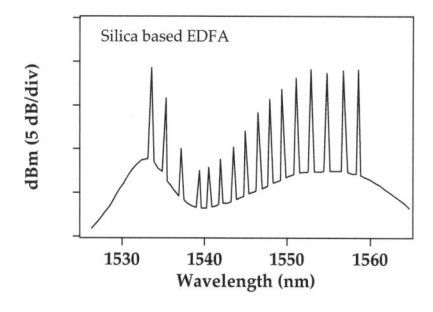

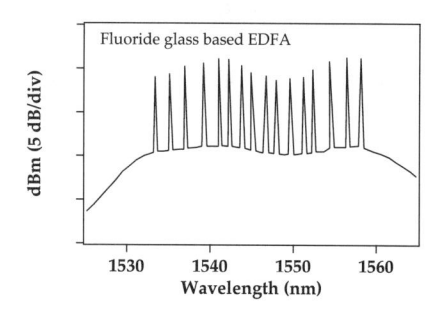

Figure 8.34: Spectral output of a 440 km 16 channel transmission system with four single-stage erbium-doped silica fiber inline amplifiers (left) and a similar system with four single-stage erbium-doped fluoride fiber inline amplifiers (right). From reference [135].

fiber itself is more problematic in terms of fabrication and handling, is that 1480 nm is the main pump wavelength. The 980 nm pump source is not a very a viable pumping option because the decay out of the $^4I_{11/2}$ pumping state is too slow to effectively populate the upper state of the amplifying transition. The 1480 nm pumping entails a higher noise figure than can be achieved with 980 nm pumped erbium-doped silica fiber amplifiers, and has only demonstrated noise figures of 6 to 7 dB.[134] Pumping at 970 nm, where there exists strong excited absorption to the $^4F_{7/2}$ level, followed by fast relaxation to the $^4I_{13/2}$ level, may offer a way to improve the noise figure at the expense of gain efficiency. Noise figures of 4.5 dB have been reported with 970 nm pumping, with reduced gain compared to a silica host.[137]

The use of tellurite as a host glass also represents a promising avenue. The spectrum of erbium doped in a tellurite host, shown in Figure 4.14, is wider than in a silica or fluoride host. Small signal gain in excess of 20 dB from 1530 nm to 1610 nm (with 200 mW of pump power) has been demonstrated in a tellurite host, with strong features that can be removed by an eqalizing filter.[121] With the use of an equalizing filter, one can achieve a 3 dB gain bandwidth (for a gain of 20 dB) of 83 nm from 1528 nm to 1611 nm, using a total pump power of 350 mW at 1480 nm for two stages of amplification.[122]

Some gain equalization methods are more apt to be robust with respect to gain cross saturation (the change in gain of one signal when another is turned on or off). These methods attempt to decouple the signal channels by essentially recreating conditions similar to inhomogeneous broadening. Twin core fibers have been used effectively in this respect with two signals at different wavelengths.[138] They are based on the principle that the two different signal wavelengths couple periodically between the two cores, both of which are doped with erbium, so that the two signals are amplified by different sets of erbium ions. The signal wavelengths are then effectively decoupled.

An effective way to decouple the various signal channels would be to employ an amplifier that is inhomogeneously broadened. In such a system, signals would be amplified by different subsets of ions and would propagate independently of one another, as is the case for the twin core fiber amplifier. Each channel would be able to take advantage of the self-regulation effect of a single-channel system. Thus, there would be

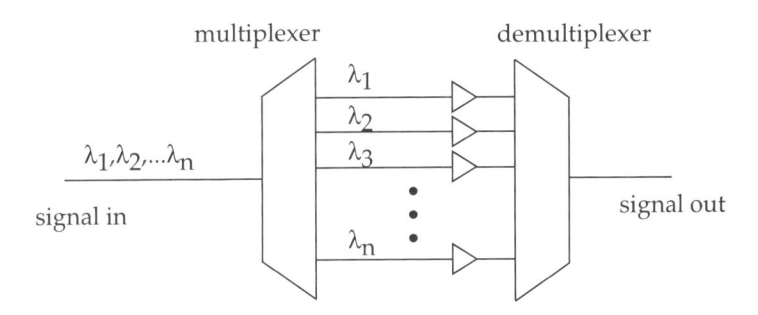

Figure 8.35: Signal channel decoupled amplification stage for WDM transmission, allowing self regulation on a channel by channel basis and flat gain operation. Adapted from reference [139].

no cross-saturation or overall saturation of the transition that would progressively shape the gain spectrum. One way to achieve this is to effectively decouple the signal channels at the end of the span, and provide a separate amplifier for each signal, as shown in Figure 8.35. This way the signals are decoupled and one can prevent inter-channel power differences from accumulating with each amplifier stage. This approach was demonstrated with a four channel system, where the pump source was shared among the four amplifiers.[139]

An elegant method to achieve decoupling of the signal channels has been to reduce the temperature of the silica host to 77 K, by cooling the erbium-doped fiber in liquid nitrogen, thereby creating an inhomogeneously broadened amplifier.[115, 140, 141, 142] In experiments conducted with cooled amplifiers, five channels in the 1546 to 1560 nm region, separated by 4 nm (to be compared to the 1 nm homogeneous linewidth of erbium ion 1550 nm transitions at 77 K), were propagated through a ten-amplifier chain separated by 10 dB loss fiber links. This system exhibited an interchannel power spread of 8 dB between the weakest and strongest channels, as compared to nearly 30 dB for a system where the amplifiers were kept at room temperature.[115] An inhomogeneously broadened system is also relatively immune to transient cross saturation from the addition or deletion of signal channels. The main drawback of this approach is the lack of a host with the requisite amplifier properties and in which inhomogeneous broadening is prevalent at room temperature.

Spectral flattening - passive element methods

A variety of methods to circumvent the gain flatness problem have been tried out by using passive or active equalizers. These equalizers are typically spectral filters of various kinds. The placement of such filters impacts the noise and gain characteristics of the amplifier. Given that the filter typically has a non negligible loss, placing the filter before the amplifier increases the noise figure, whereas placing it after the amplifier reduces the output power. A compromise consists in placing it midway in a two stage amplifier, where the noise figure is mostly determined by the first stage while the second stage acts as a power amplifier and also compensates for mid-point losses. Passive filters are usually designed for a specific operating point of the erbium-doped fiber amplifiers. Thus they are not necessarily that robust when system parameters vary and the spectral gain of the erbium-doped fiber they are compensating for changes. Note also that by allowing an increase in the loss that the filter provides can result in an increase in bandwidth at the expense of lower gain.

An example of the use of active equalization is the flattening achieved from 1540 to 1560 nm using an acousto-optic tunable filter placed after the amplifier.[143] The acousto-optic filters produce a wavelength dependent loss that can be tuned electrically. Transmission passbands can be set up to overlap the signal channels. In addition, feedback loops can be accomodated to provide variation in the filter response, based on the operating conditions.[144] In this way equalization can be provided which is robust with respect to changes in operating conditions. Acousto-optic tunable filters have also been used at the mid-point of a two-stage amplifier, resulting in about 35 nm of flat gain operation (to within 0.7 dB) over the 1530 nm to 1560 nm region.[145]

One of the first demonstrations of passive flattening of the gain profile was obtained by including a mechanical filter within the body of the amplifier. This can be achieved by using resonant coupling between the signal mode and a leaky cladding mode, by having the fiber proximate to a mechanical notch filter.[146] This filter serves to flatten the gain by reducing the 1530 nm peak to the same level as in the 1550 nm region when operating in the small signal amplification regime, and is more effective than if it had been placed at the output of the amplifier.

Another early example of a passive fiber filter has been the use of a D-shaped fiber with a silica based grating pressed against the flat surface of the D fiber.[147] This filter, placed after the erbium-doped fiber, flattened the gain spectrum by reducing the 1530 nm peak in a highly inverted erbium-doped fiber amplifier. Tunable Fabry-Perot etalon filters have been reported to be used effectively as gain equalizers in long distance transmission systems.[148] Periodically tapered filters also flatten the gain.[149] The principle of the periodically tapered fibers is that they couple the fundamental mode of the single mode fiber into cladding modes, i.e. loss. This is a function of wavelength, so that filters with given center wavelength, bandwidth, and loss, can be fabricated by adjusting the taper parameters. Another example of passive filters are Mach-Zehnder filters.[150] They can be tuned, in terms of the filter transmittance function, by means of a heater attached to one of the interferometric paths that makes up the Mach-Zehnder filter.

Gain-flattening filters made using long period gratings directly inscribed in a Ge-doped fiber are a very effective avenue for fabricating low-loss filters.[151, 152] These grating filters were described in Chapter 3, section 3.8. Ultraflat amplifiers can be fabricated by using these long period grating fiber filters. The most typical application includes the filter between the two stages of an erbium-doped fiber amplifier.[154, 155] The spectrum of an amplifier with gain flattened by a long period grating filter is shown in Figure 8.36, where the gain was measured with a signal input -20 dB lower in power than a saturating signal with input power ranging from -10 to -6 dBm. The shape of the filter is shown in Figure 3.16. The first stage was pumped at 980 nm with 75 mW of pump power, while the second was pumped at 1480 nm with 35 mW of pump power for the -10.2 dBm saturating input signal and with 74.5 mW of pump power for the -6.1 dBm saturating input signal. The average gain is about 22 dB with a variation of less than 1 dB from 1530 nm to 1570 nm, with a total output power of nearly 15 dBm.

8.3. AMPLIFIER DESIGN ISSUES

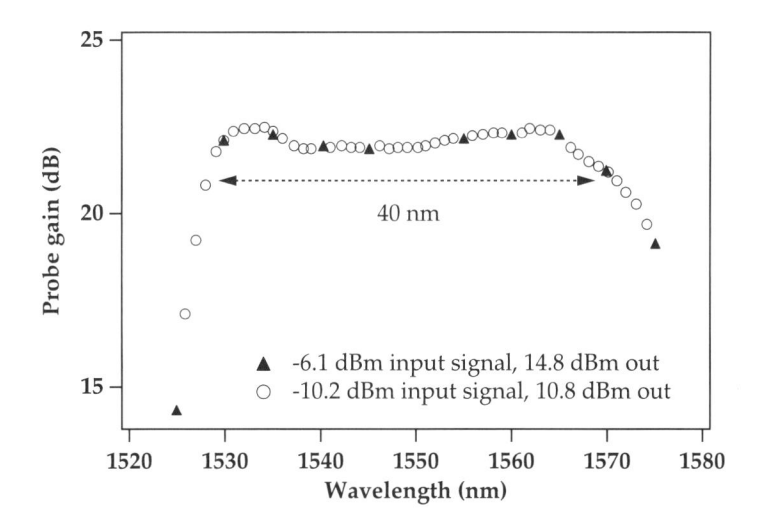

Figure 8.36: Gain spectrum of a gain-flattened two-stage erbium-doped fiber amplifier, with a saturating signal input levels of −10.2 dBm and −6.1 dBm and measured with a probe -20 dB lower in power than the saturating tone. From reference [155] (©1997 IEEE).

Over the 40 nm range of flat operation, the noise figure was maintained between 3.5 and 4.0 dB, approximately. An average signal gain of 17 dB has been achieved with gain flatness to within 3 dB over a bandwidth of 65 nm (noise figure less than 5 dB) by using a two-stage silica amplifier followed by a Mach-Zehnder type gain equalizing filter.[153] Here the total pump power required is on the order of 200 mW. Similarly, a two stage amplifier with a mid-point gain equalizer was used to flatten the gain over the range 1556 nm to 1608 nm to within 3 dB (for an average gain of 19 dB for a small signal probe with power −30 dBm and a saturating tone with an input power of −12 dBm).[154] The equalizer used in this experiment was a split-beam Fourier filter.[156]

Passive equalization of the gain has been used with success for transoceanic WDM transmission experiments with up to 32 channels operating at 5 Gb/s.[157, 158, 159, 160] For terrestrial applications as well, equalizers based on fiber grating filters have allowed up to 100 channels to be transmitted over several hundred km.[161]

One of the drawbacks of passive filters is the trade-off between bandwidth and loss. To make a filter with a wider bandwidth, more attenuation needs to be provided at the wavelengths of peak gain, thereby impacting either the noise figure or the output power and gain, depending on the placement of the filter and the pump power available for pumping the amplifier stages.

Spectral flattening - hybrid methods

Various methods used for producing flat gain amplifiers involve the use of multiple amplification stages which are either cascaded or positioned in different arms of a multi-arm amplifier. In particular, we will describe in this section hybrid amplifiers where the stages or arms of the amplifier use different kinds of host glasses.

Different hosts produce different shapes for the gain spectrum, both in its gross features as well as in its details. It has long been recognized that one way to flatten the gain is to superimpose the gain spectra arising from several different types of amplifiers. Two-component fiber amplifiers were early on demonstrated to be effective in equalizing the gain for specific signal wavelengths. For a two-wavelength system, an amplifier with two different composition fibers was demonstrated. The two stages were fabricated with Ge-Al silica and Al silica hosts.[162] Such multicomponent glass hosts have been used in amplifiers that are formed by connecting lengths of different types of erbium-doped fibers.[163, 164] Cascaded or multi-arm amplifier designs with large flat bandwidths include the following designs.

- A cascade design with a first stage made with an Al co-doped erbium-doped fiber and a second stage made with an P co-doped ytterbium-erbium-doped fiber. This produces a gain spectrum flat to within 0.5 dB over the range 1544 nm to 1561 nm. The first stage is pumped at 980 nm, determining the overall noise figure of 5 dB, while the second stage acts as a power amplifier and yields a total output power of 26 dBm.[119]

- A two stage amplifier cascade designed for 1580 nm band operation. The first stage is an erbium-doped silica host pumped at 980 nm for low noise, while the second stage is either a silica or fluoride host (erbium-doped) pumped at 1480 nm for high output power, with an isolator separating the two stages. The noise figure is about 5 dB, and the region of flat gain is from 1573 nm to 1600 nm (silica host) and 1567 nm to 1597 nm (fluoride host). In those regions the average gains are 33 dB (silica host) and 25 dB (fluoride host) with gain excursions under 1 dB.[165]

- A two stage amplifier with a silica first stage pumped at 980 nm and a fluoride second stage pumped at 1480 nm, for operation in the 1550 nm band. The gain excursion is under 1 dB in the range 1532 nm to 1560 nm (average signal gain 20 dB and a total WDM signal input power of -11 dBm), with a noise figure of 5 dB.[166]

- A cascade design which involves Raman amplification in the transmission fiber followed by a two-stage erbium-doped silica host fiber amplifier with a mid-point gain equalizing filter. The Raman gain has a slope opposite to that of the erbium-doped fiber amplifiers and the superposition is a flatter spectrum. The Raman fiber is pumped by 500 mW of light at 1535 nm. This design achieves a 3 dB gain bandwidth of 65 nm from 1549 to 1614 nm.[167]

- A two arm design with one arm optimized for operation in the 1550 nm band while the other arm is optimized for the 1580 nm band. With both silica-based and fluoride-based erbium-doped fiber, such a design first demonstrated a bandwidth of 54 nm (1530-1560 nm and 1576-1600nm) with a gain variation of less than 1.7 dB for an average signal gain of 30 dB (input signal strengths of -30 dBm).[168] Noise figures are under 6 dB for the 1550 nm band and under 7 dB for the 1580 nm band.

8.3. AMPLIFIER DESIGN ISSUES

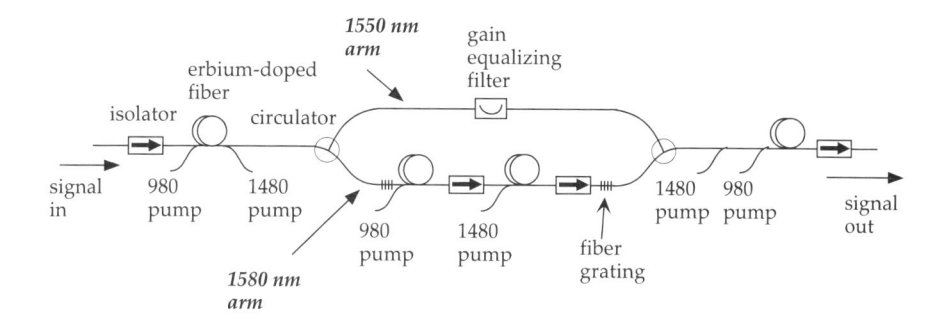

Figure 8.37: Architecture of a dual-arm gain-flattened erbium-doped fiber amplifier with more than 80 nm of bandwidth, with gain in both the 1550 nm and 1580 nm regions. From reference [170].

Two-arm amplifiers have, to date, produced the widest bandwidth of amplification. The advantage here is that each arm can be independently optimized for operation in a given wavelength band. A bandwidth of 80 nm has been demonstrated using silica host fibers in both arms of the amplifier.[169, 170] The idea here is to amplify in both the 1550 nm as well as the 1580 nm band. A demultiplexer consisting of a circulator combined with a fiber grating sends the 1550 nm band signals through the upper arm while the 1580 nm band signal travels through the lower arm, as shown in Figure 8.37. The signals are recombined after the two arms. The upper arm (for 1530-1560 nm amplification, referred to as the C band) contains a gain equalizing filter (and can contain amplification stages as well). The lower arm (for 1570-1600 nm amplification, referred to as the L band) is a two-stage amplifier with a midway isolator. The pump power required for high gain operation is on the order of 500 mW for the total of all laser diode output powers. This hybrid amplifier produces a gain of approximately 24 dB over a bandwidth of about 84 nm when pumped with 980 nm laser diodes.[161] The noise figure obtained is approximately 6 dB over the bandwidth of operation.[161] The spectrum of this amplifier is shown in Figure 8.38. This hybrid amplifier can support WDM transmission of 100 channels at 10 Gb/s, with experimental demonstrations achieving distances of 400 km.[161] Flat amplifiers thus open the way for massively large bandwidth transmission links. Their operation in the context of WDM transmission systems will be discussed further in Chapter 9, section 9.2.3.

8.3.7 Distributed Amplifiers

A key parameter determining the amplified system noise performance is the amplifier input power. Given that the output power of an amplifier is limited by the available pump power, it follows that low gain amplifiers can operate with higher input powers and thus contribute less overall system noise. Similarly in logic, the noise of an amplifier system is determined by the lowest signal level reached along the span. The ultimate low-gain amplifier might then be thought of as an attenuation-free fiber. An approximately loss free span could be constructed from a cascade of low-gain amplifiers. For a fixed span length, the accumulated noise would decrease as the number

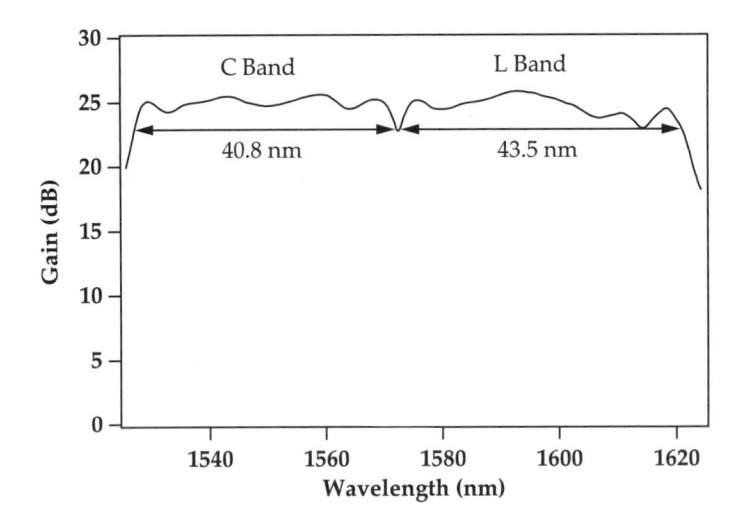

Figure 8.38: Gain spectrum of a dual-arm gain-flattened erbium-doped fiber amplifier with 84 nm of bandwidth, with an architecture as sketched in Figure 8.37. The C band is the conventional 1550 nm amplification band while the L band refers to the long wavelength band at 1580 nm. From reference [161].

of amplifiers increased (and the gain per amplifier decreased), as discussed in Chapter 7, section 7.3.4. This quasi-distributed amplifier span is then the routine practical solution for repeatered spans. The use of such lumped amplifier spans for either linear (NRZ) or nonlinear (soliton) systems then requires a compromise between the number of amplifiers used to compensate for the transmission loss and the amount of amplifier noise, as well as nonlinearities.[171] Such systems provide a signal excursion such as is sketched in the top portion of Figure 8.39. Tables 9.2 and 9.3 list the amplifier spacings for various system demonstrations.

An alternative method is to distribute the gain over the entire span using stimulated Raman amplification, a low Er dopant throughout the span, or a combination of both. These methods require significant pump power from each end and provide less signal excursion than quasi-distributed spans, as shown in the lower portion of Figure 8.39. They have the potential for lower span noise.[172] Distributed amplifiers have also been suggested as useful in multiple add/drop systems as well as soliton transmission spans.[173, 174]

The distributed Raman gain concept was first considered for soliton propagation where the distributed gain was believed to be necessary to retain both the soliton amplitude and width.[175] Soliton propagation requires that each pulse maintain a given power to sustain the nearly dispersion-free propagation and that the amplifier spacing should be less than the soliton period for acceptable pulse distortion.[171, 176] The feasibility of a distributed Raman gain optical communication link thousands of kilometers long was first demonstrated by optical soliton propagation over a 4000 km distance in a loop configuration.[177] This demonstration required a pump power of 300 milliwatts for a 40 km span. Subsequent high germanium content core fiber ($\Delta n = 0.03$)

8.3. AMPLIFIER DESIGN ISSUES

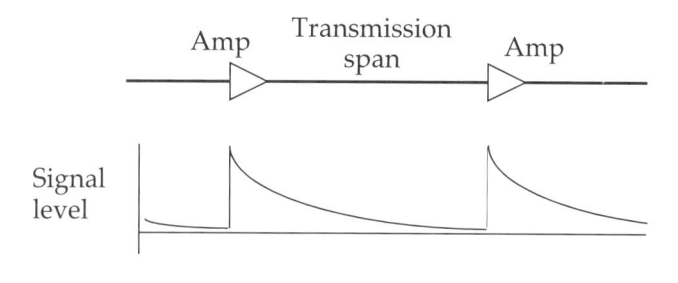

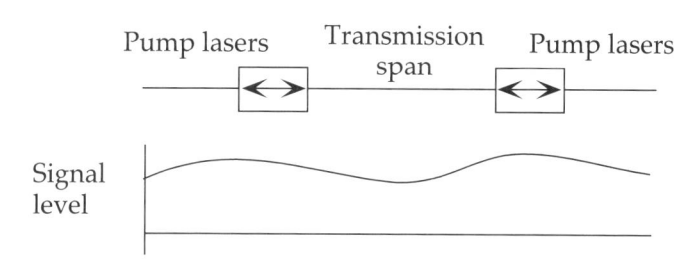

Figure 8.39: Variation of the signal power level along the length of a lumped gain amplified transmission system (top) and distributed gain amplified transmission system (bottom). The pump lasers are sets of copropagating and counterpropagating 1.48 μm lasers.

displayed higher efficiency requiring 200 mW of pump power for transparency in a 46 km length.[178]. This efficiency could, in theory, be doubled by aligning the pump and signal polarization states along the length using polarization maintaining fiber. However, the required pump power remains high with the dilemma that high Raman cross section fiber core compositions generally result in higher intrinsic losses.

On the other hand, the distributed erbium-doped fiber amplifier offers a potentially more efficient use of pump power. In a fashion similar to the distributed Raman gain approach, the pump power must propagate the length of a distributed erbium-doped fiber. A pump wavelength of 1.48 μm appears most suitable in that the silica fiber background loss is the lowest of all erbium pump wavelengths. The similarity between Raman gain and low levels of erbium doping has been recognized. Fabrication methods capable of producing fiber with the required low dopant levels retaining the purity level and reliability of undoped fiber appears feasible from a number of reports.[179, 180, 181, 182, 183, 184, 185]. A minimal background attenuation is especially desirable to prevent the loss of pump and signal power to absorption or scattering mechanisms.

The challenge is to provide a dopant level and low background loss such that span transparency and low signal noise will be achieved with the smallest signal excursion for the least amount of pump power. Details of the fabrication of these fibers were addressed in Chapter 2, section 2.3.3. Demonstrations of distributed erbium-doped fiber amplifiers began in 1990 showing that 10 km of such an active span could be made lossless for low signal levels and 20 mW of 1480 nm pump power.[181, 182, 186] Signal excursions along the length as small as 0.2 dB were demonstrated. Longer lengths (54 km) were used with launched signal powers of −2 dBm and a total bidirectional pump power of 110 mW.[187, 188] Amplifier models for the signal excursion, optimum dopant concentration, and pump power required for transparency indicate that longer wavelength signals (1550 nm) require less pump power and higher concentrations of erbium.[189]

Nonlinear optical processes must also be considered for distributed amplifiers, given that the signal levels may be maintained at higher levels over longer lengths than fiber spans with loss. Modeling has established that Raman pumping of the signal by the pump, generally adds favorably to both the gain and signal to noise ratio.[190, 191, 192] The influence of the Raman gain is, however, a complex function of core effective area, pump power as well as signal power, and wavelength. Stimulated Brillouin scattering (SBS) is generally a disadvantage, providing a mechanism for effectively reflecting either the signal or pump back to the source and degrading the performance.[193, 194]

Proof that distributed erbium-doped amplifier fiber could transmit high repetition rate pulses with little pulse spreading was demonstrated over a 28.5 km long distributed amplifier span using 18 ps wide pulses at a rate of 10 GHz.[195] Models of optimized distributed amplification fiber also indicate that with high numerical aperture fiber (NA = 0.2, core effective area of 28 μm^2), equal copropagating and counterpropagating pump power, and pump spacing between 10 and 100 km, error free transmission over 9000 km can be achieved at a data rate of 2.6 Gb/s.[196]

Simulations of the distributed amplifier performance may be made using extensions of the model developed for standard amplifiers.[173, 189, 197, 198, 199, 200] Let us consider, as an example, an erbium-doped distributed fiber made using the glass composition of fiber A of chapter 6 for the erbium-doped glass. We assume a background loss of 0.25 dB/km for both signal and copropagating pump, an input signal power of −20 dBm at 1550 nm and a span length of 50 km. The fiber core area is 35 μm^2, with the erbium dopant occupying a surface area of 5 μm^2 at the center of the core (we assume that the distributed fiber was made using the seed fiber method discussed in section 2.3.3). The pump and signal confinement factors are 0.13. A simulation for two erbium ion concentration levels, 8×10^{21} ions/m^3 and 4×10^{21} ions/m^3, respectively, is shown in Figure 8.40. The excursion in signal level drops from 7 to 1 dB as the erbium concentration is decreased by a factor of 2. The required pump power, however, more than doubles for the lower concentration, indicating the trade-off between allowed signal excursion and pump power. The computed noise figure for the 50 km span is is 8 dB for the higher concentration fiber, and 9 dB for the lower concentration fiber. This should be compared to the the overall noise figure of a span consisting of 50 km of lossy fiber and a discrete amplifier with a specific noise figure.

There is some concern that long amplifiers may have an additional background loss from impurities introduced along with the rare earth dopant. For the simulation shown in Figure 8.40 we have chosen a rather optimistic background loss of only five-hundredths of a dB/km higher than current state-of-the-art undoped fiber (0.2 dB/km). The effect of background losses from 0.2 to 0.4 dB/km are shown in Figure 8.41 where the same fiber characteristics as in Figure 8.40 were employed. Here, the case of a −20

8.3. AMPLIFIER DESIGN ISSUES

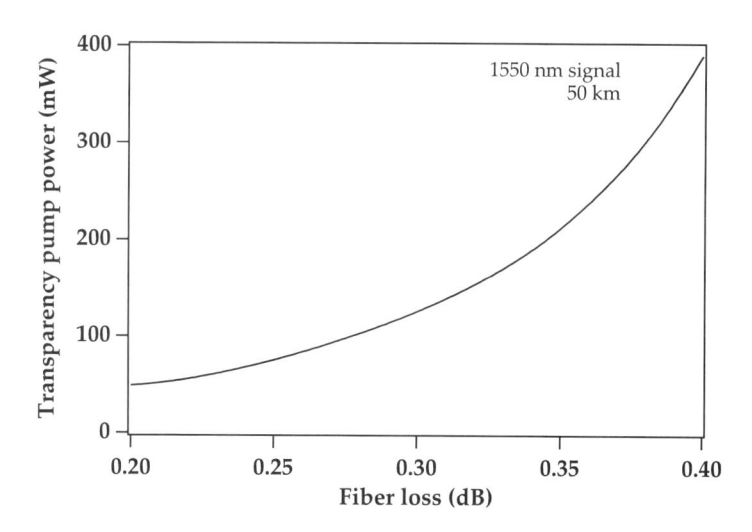

Figure 8.40: Distributed erbium amplifier concentration-dependent performance for a 50 km long fiber pumped at 1.48 μm and a signal at 1550 nm (the background loss is 0.25 dB/km). From numerical simulations with parameters described in the text.

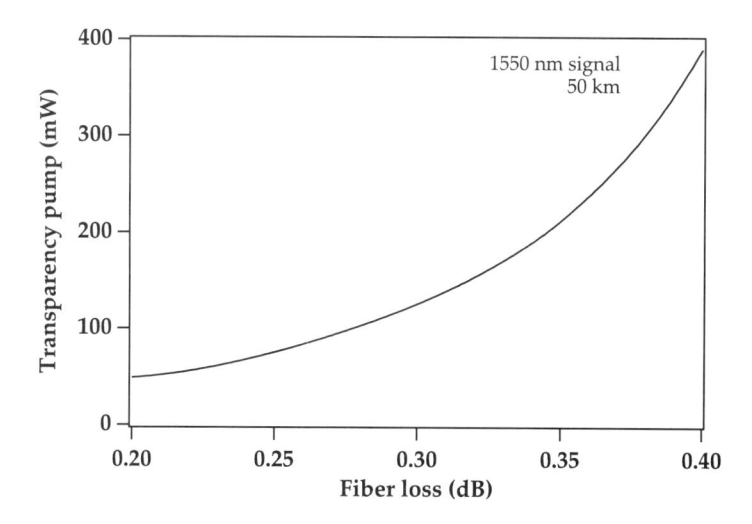

Figure 8.41: Distributed erbium-doped fiber amplifier background-loss-dependent performance: copropagating pump power required for transparency of a 50 km fiber pumped at 1.48 μm, with a 1550 nm signal. The signal input power is -20 dBm and the Er^{3+} concentration is 8×10^{21} ions/m^3. From numerical simulations (fiber parameters as in Figure 8.40).

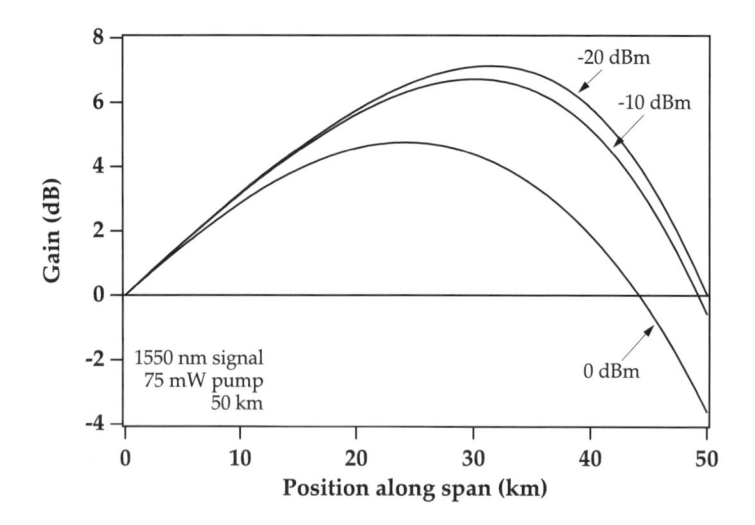

Figure 8.42: Distributed erbium-doped fiber amplifier saturation performance for a 50 km long fiber with 0.25 dB/km background loss at 1.5 μm, pumped at 1.48 μm with 75 mW, from numerical simulations. The different curves correspond to different signal input powers at 1550 nm.

dBm input signal at 1550 nm and 8×10^{21} ions/m^3 Er^{3+} dopant level indicates that the required pump power increases by a factor of 8 when background losses increase from 0.2 to 0.4 dB/km. This emphasizes the need for background losses comparable to those of undoped fiber for an efficient distributed amplifier.

A further practical issue for optical amplifiers is the reduced gain with higher input signal level (gain saturation). This effect was modeled for the distributed amplifier using the high concentration fiber of Figures 8.40 and 8.41, for a pump power of 75 mW, and is shown in Figure 8.42. While in the range -40 dBm to -10 dBm input the same approximate pump power will produce transparency, for 0 dBm input a significant increase in pump power is needed for transparency (from 75 mW for -20 dBm input to 100 mW for 0 dBm input).

The efficiency results obtained by various groups are shown in Figure 8.43. Distributed amplifier performance may be expressed as the pump power needed to render 1 km of fiber transparent (computed as the total pump power divided by the length of the span), for a small signal level and the span length used in the experiment. This is the figure of merit used to compare different results in Figure 8.43. The lower this number the better, and the lower the absorption at 1.53 μm the lesser the signal excursion across the span. Recently, signal level variation levels of less than 0.03 dB/km have been observed in a 68 km distributed erbium-doped fiber amplifier span.[201]

OTDR methods provide an excellent means for testing both the distribution of erbium and the distribution of gain along the length of fiber. The experimental setup used to measure the OTDR trace of an ultralong erbium-doped distributed fiber amplifier is shown in Figure 8.44. Figure 8.45 shows an experimentally obtained OTDR trace of a 22 km length of distributed amplifier fiber. The average OTDR signal power of -40

8.3. AMPLIFIER DESIGN ISSUES

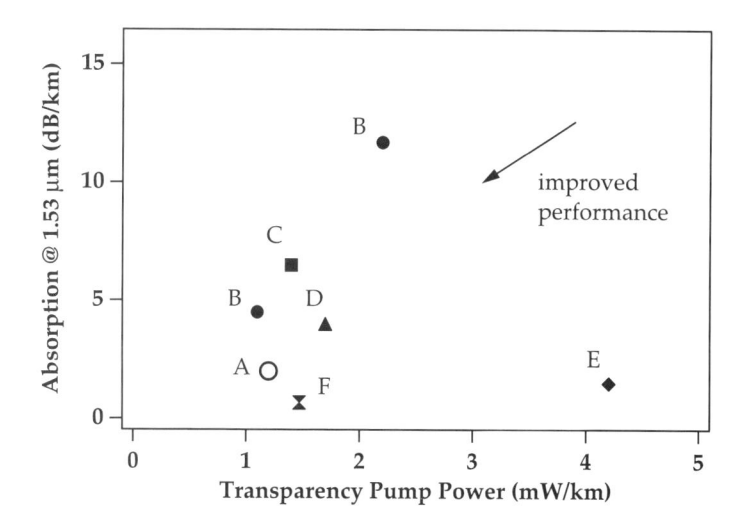

Figure 8.43: Experimental distributed erbium-doped fiber amplifier efficiencies reported by various groups. (the references are: A: [187]; B: [184]; C: [11]; D: [181]; E: [183]; F: [201]).

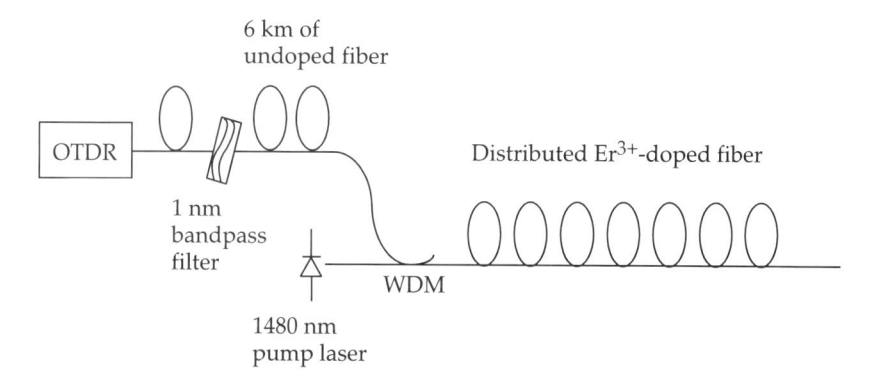

Figure 8.44: Experimental setup for measuring the OTDR trace of a distributed erbium-doped fiber amplifier.

dBm, measured at the launch end of the distributed erbium-doped amplifier fiber, ensures that the OTDR will probe the small signal gain characteristics of the fiber. Net transparency at a distance of 15 km into the distributed amplifier fiber is indicated for a pump power of 17.6 mW. Establishing the exact point of transparency from these OTDR traces is difficult given that the reflections and losses associated with the WDM coupler cannot be resolved. A net peak gain of 8 dB (including coupler losses) at a distance of 10 km for a pump power of 17.6 mW is indicated. The measured properties for the 22 km length are a pump attenuation $\alpha_p = 1.3$ dB/km at 1.48 μm, a signal attenuation $\alpha_s = 4.0$ dB/km at 1.53 μm and a background loss of 0.5 dB/km around 1.5

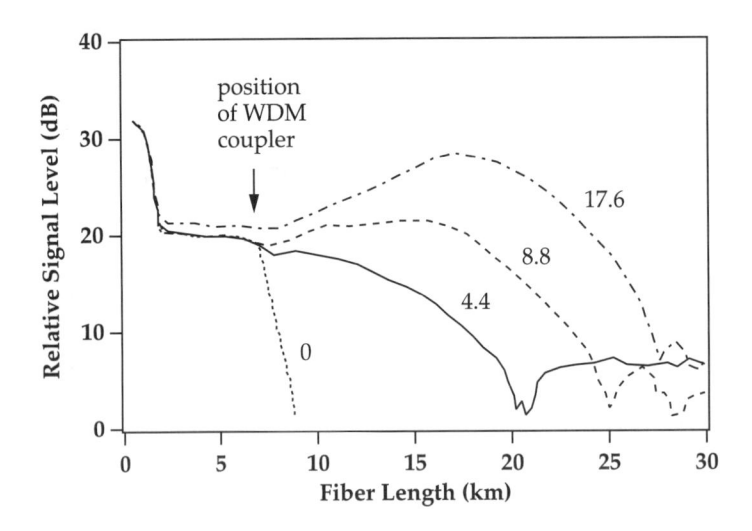

Figure 8.45: OTDR traces of a 22 km distributed erbium-doped fiber amplifier for a signal at 1.53 μm and 1.48 μm copropagating pump powers of 0, 4.4, 8.8, and 17.6 mW. From reference [200] (©1991 IEEE).

μm. Continued efforts to fabricate lower-erbium-concentration lower-loss fiber will no doubt improve the performance of these distributed optical fiber amplifiers, making them more viable for communications or device use.

8.3.8 Waveguide Amplifiers

It was long thought that waveguide amplifiers, where the amplifier lengths are on the order of a few cm to a few tens of cm, would be impractical due to concentration-quenching effects and other deleterious high-concentration effects. Waveguides, however, offer the advantage that host glass compositions can be employed that would otherwise be difficult to fabricate in single-mode fiber form. Waveguide amplifiers based on high-concentration erbium-doped glasses have been demonstrated.[202, 203, 204] Such waveguide amplifiers have been integrated with passive components, such as pump and signal WDM couplers.[205] It appears that silica-based compositions, similar to those used in erbium-doped silica fibers, are not the optimum choices for waveguide amplifiers due to clustering and other lifetime-reducing effects. Multicomponent glasses, for example, those from the soda-lime-silicate family, offer much more advantageous properties.[206]. The underlying processes that affect gain in such high-concentration glass hosts were discussed in Chapters 4 (section 4.6) and 6 (section 6.7).

Waveguide amplifiers doped with up to 5 mole % of Er^{3+} in the soda-lime-silicate glasses have shown gains of up to 15 dB in waveguides as short as 4.5 cm, as illustrated in Figure 8.46.[204, 207] This particular waveguide amplifier was made with waveguide number 2 of Figure 4.18. As seen from Figure 4.18, the effective lifetime of the $^4I_{13/2}$ level (defined by the 1/e decay point of the luminescence) is only 2 ms. In addition, the background loss of the waveguide is on the order of 0.75–1.0 dB/cm,

8.3. AMPLIFIER DESIGN ISSUES

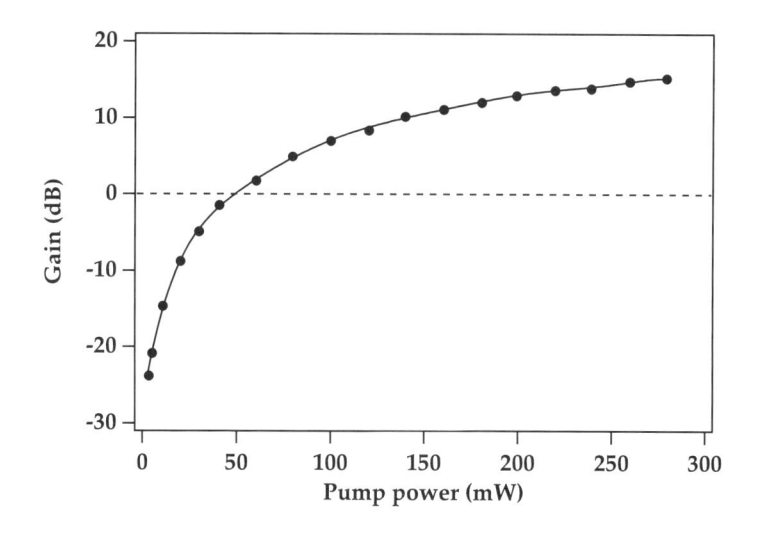

Figure 8.46: Small signal gain at 1537 nm versus pump power at 980 nm for an erbium-doped soda-lime-silicate glass waveguide amplifier. From reference [204] (©1993 IEEE).

so that it is not unreasonable to measure pump thresholds for 0 dB gain (transparency of the 4.5 cm long waveguide) that are roughly an order of magnitude higher than an efficient erbium-doped fiber amplifier. The peak gain of 15 dB corresponds to a gain of 3.3 dB per cm. The saturation output power of the amplifier is 3.5 dBm.

The performance of the waveguide amplifier of Figure 8.46 can be simulated by including the upconversion and background loss effects, as discussed in section 6.7. Modeling of waveguide amplifiers has also been performed where Er^{3+}-Er^{3+} ion pairs are taken into account.[208, 209] Comparisons of silica-based glasses and soda-lime-silicate glasses (from a study of fluorescence decay times) have determined that the latter have upconversion rates significantly lower then the former.[210] This is probably an important factor in the success of these glass hosts and suggests that the host composition will play a critical role in obtaining an efficient erbium-doped waveguide amplifier. Fabrication methods for the waveguide will also impact the upvonversion rate.[210] More recent fabrication advances in erbium-doped soda-lime-silicate glasses have brought the amplifier threshold down to 8 mW for a device 6 cm long.[211]

Further advances have been made with phosphate glass as a host material and with Yb^{3+} as a co-dopant, using ion exchange fabrication methods.[212] Phosphate glasses are also suitable hosts (from the point of view of maintaining high upper-state lifetimes) for incorporating erbium ions at high concentrations. The Yb^{3+} ion yields a higher level of absorption than Er^{3+} alone. These particular types of waveguides were first demonstrated with a low background loss (0.1 dB) and a high concentration of Er^{3+} (2.5 wt. %) and Yb^{3+} (3 wt. %) in a guide 4.1 cm long. By double passing the 1540 nm signal through the amplifier (with a Faraday rotator at one end and a circulator at the other), an internal gain of 34 dB was achieved.[212] Pumping was accomplished with 980 nm laser diodes. Such a waveguide amplifier was then pigtailed to fibers for ease of insertion into systems, with a resulting fiber to fiber gain of 16.5 dB at 1534 nm for 180 mW of pump power at 980 nm and with a saturation output power of 12.5 dBm.[213] The noise figure reached a minimum of 5.6 dB near 1540 nm. To prove the systems applicability of such an amplifier, it was used as a power amplifier in a 10 Gb/s transmission experiment over 72.5 km. The performance of these amplifiers has been improved by better design of the waveguide, demonstrating, for example, a gain of 27 dB at 1534 nm in an 8.6 cm long waveguide, for a pump power of 180 mW at 980 nm.[214] A first attempt at integrating the phosphate-glass-based Yb-Er waveguide amplifiers with passive components was performed by combining the amplifier with an integrated four-wavelength combiner. This allowed the demonstration of a lossless four-wavelength combiner, as the waveguide amplifier compensated for the losses associated with the wavelength combiner.[215]

The ultimate goal is to achieve the integration of waveguide amplifiers and the necessary passive components to obtain an "amplifier module on a chip." Continued progress in waveguide fabrication techniques and in host glasses for high concentration erbium doping should further that goal.

Bibliography

[1] D. M. Baney, C. Hentschel and J. Dupre, "Optical fiber amplifiers - measurement of gain and noise figure," *1993 Lightwave Symposium*, Hewlett-Packard Company (1993).

[2] D. Bonnedal, *IEEE Phot. Tech. Lett.* **5**, 1193 (1993).

[3] T. Kashiwada, M. Shigematsu, and M. Nishimura, "Accuracy of noise figure measurement for erbium-doped fiber amplifiers by the optical method," NIST Special Publication 839, *Technical Digest Symposium on Optical Fiber Measurements*, pp. 209–212 (1992).

[4] V. J. Mazurczyk and J. L. Zyskind, "Polarisation hole burning in erbium-doped fiber amplifiers," in *Conference on Lasers and Electro-Optics*, Vol. 11, 1993 OSA Technical Digest Series (Optical Society of America, Washington DC, 1993), pp. 720–721.

[5] V. J. Mazurczyk, "Spectral response of a single EDFA measured to an accuracy of 0.01 dB," in *Conference on Optical Fiber Communication*, Vol. 4, 1994 OSA Technical Digest Series (Optical Society of America, Washington, D.C., 1994), pp. 271–272.

[6] E. J. Greer, D. J. Lewis, and W. M. Macauley, *Elect. Lett.* **30**, 46 (1994).

[7] M. Kobayashi, T. Ishihara and M. Gotoh, *IEEE Phot. Tech. Lett.* **5**, 925 (1993).

[8] R. G. McKay, R. S. Vodhanel, R. E. Wagner, and R. I. Laming, "Influence of forward and backward traveling reflections on the gain and ASE spectrum of EDFAs," in *Conference on Optical Fiber Communication*, Vol. 5, 1992 OSA Technical Digest Series (Optical Society of America, Washington, D.C., 1992), pp. 176–177.

[9] B. Nyman and G. Wolter, *IEEE Phot. Tech. Lett.* **5**, 817 (1993).

[10] J. D. Evankow and R. M Jopson, *IEEE Phot. Tech. Lett.* **3**, 993 (1991).

[11] M. Nakazawa, Y. Kimura, and K. Suzuki, *Opt. Lett.* **15**, 1200 (1990).

[12] J. R. Simpson, H. T. Shang, L. F. Mollenauer, N. A. Olsson, P. C. Becker, K. S. Kranz, P. J. Lemaire, and M. J. Neubelt, *J. Light. Tech.* **9**, 228 (1991).

[13] S. T. Davey, D. L. Williams, and B. J. Ainslie, "Distributed erbium-doped fiber for lossless link applications," in *Conference on Optical Fiber Communication*, Vol. 4, 1991 OSA Technical Digest Series (Optical Society of America, Washington, D.C., 1991), p. 53.

[14] Y. Sato and K. Aoyama, *J. Light. Tech.* **10**, 78 (1992).

[15] D. M. Spirit and L. C. Blank, "10 Gb/s, 28.5-km distributed erbium fiber transmission with low-signal-power excursion," in *Conference on Optical Fiber Communication*, Vol. 4, 1991 OSA Technical Digest Series (Optical Society of America, Washington, D.C., 1991), p. 53.

[16] L. C. Blank and D. M. Spirit, *Elect. Lett.* **25**, 1693 (1989).

[17] J. D. Cox and L. C. Blank, "Optical time domain reflectometry on optical amplifier systems," in *Conference on Optical Fiber Communication*, Vol. 5, 1989 OSA Technical Digest Series (Optical Society of America, Washington, D.C., 1989), p. 98.

[18] L. C. Blank and J. D. Cox, *J. Light. Tech.* **7**, 1549 (1989).

[19] D. M. Baney and J. Stimple, *IEEE Phot. Tech. Lett.* **8**, 1615 (1996).

[20] A. K. Srivastava, J. L. Zyskind, J. W. Sulhoff, J. D. Evankow, Jr., and M. A. Mills, "Room temperature spectral hole-burning in erbium-doped fiber amplifiers," in *Conference on Optical Fiber Communication*, Vol. 2, 1996 OSA Technical Digest Series (Optical Society of America, Washington, D.C., 1996), pp. 33–34.

[21] W. Benger and J. Vobis, Hewlett Packard application notes (1996).

[22] B. Pedersen, M. L. Dakss, and W. J. Miniscalco, "Conversion efficiency and noise in erbium-doped fiber power amplifiers," in *Optical Fiber Amplifiers and Their Applications*, Vol. 13, 1991 OSA Technical Digest Series (Optical Society of America, Washington, D.C., 1991), pp. 170–176.

[23] J. F. Massicott, R. Wyatt, B. J. Ainslie and S. P. Craig-Ryan, *Elect. Lett.* **26**, 1038 (1990).

[24] D. Lee, D. J. DiGiovanni, J. R. Simpson, A. M. Vengsarkar, and K. L. Walker, "Backward amplified spontaneous emission from erbium-doped fibers as a measure of amplifier efficiency," in *Conference on Optical Fiber Communication*, Vol. 4, 1993 OSA Technical Digest Series (Optical Society of America, Washington, D.C., 1993), pp. 179–180.

[25] R. I. Laming, D. N. Payne, F. Meli, G. Grasso, and E. J. Tarbox, "Saturated erbium-doped-fibre power amplifiers," in *Optical Amplifiers and Their Applications*, Vol. 13, 1990 OSA Technical Digest Series (Optical Society of America, Washington, D.C., 1990), pp. 16–19.

[26] S. Poole, *Symposium on Optical Fiber Measurements*, Technical Digest, NIST Special Publication, pp. 1–6 (1994).

[27] J. F. Marcerou, H. Fevrier, J. Hervo, and J. Auge, "Noise characteristics of the EDFA in gain saturation regimes," in *Optical Amplifiers and Their Applications*, Vol. 13, 1991 OSA Technical Digest Series, (Optical Society of America, Washington, D.C., 1991), pp. 162–165.

[28] D. M. Baney, "Gain and noise characterization of EDFAs for WDM applications," in *Conference on Optical Fiber Communication*, Vol. 6, 1997 OSA Technical Digest Series (Optical Society of America, Washington, D.C., 1997), pp. 103–104.

[29] J. Aspell, J. F. Federici, B. M. Nyman, D. L. Wilson and D. S. Shenk, "Accurate noise figure measurements of erbium-doped fiber amplifiers in saturation conditions," in *Conference on Optical Fiber Communication*, Vol. 5, 1992 OSA Technical Digest Series (Optical Society of America, Washington, D.C., 1992), pp. 189–190.

[30] H. C. Lefevre, *Elect. Lett.* **16**, 778 (1980).

[31] D. M. Baney and J. Dupre, "Pulsed-source technique for optical amplifier noise figure measurement," in *18th European Conference on Optical Communication*, Proceedings Vol. 1, pp. 509–512 (1992).

[32] K. Bertilsson, P. A. Andrekson, and B. E. Olsson, *IEEE Phot. Tech. Lett.* **6**, 199 (1994).

[33] S. Nishi and M. Saruwatari, "Comparison of noise figure measurement methods for erbium doped fiber amplifiers," in *19th European Conference on Optical Communication*, Proceedings Vol. 2, pp. 177–180 (1993).

[34] F. W. Willems, J. C. van der Plaats, C. Hentschel and E. Leckel, "Optical amplifier noise figure determination by signal RIN subtraction," NIST Special Publication 864, *Technical Digest Symposium on Optical Fiber Measurements*, pp. 7–9 (1994).

[35] C. Hentschel, E. Muller and E. Lecker, "EDFA noise figure measurements — Comparison between optical and electrical technique," Hewlett-Packard company publication (1994).

[36] F. W. Willems and J. C. van der Plaats, "EDFA noise-figure reduction in the saturated operation regime," in *Conference on Optical Fiber Communication*, Vol. 8, 1995 OSA Technical Digest Series (Optical Society of America, Washington, D.C., 1995), pp. 44–45.

[37] P. F. Wysocki, G. Jacobovitz-Veselka, D. S. Gasper, S. Kosinski, J. Costelloe and S. W. Granlund, *IEEE Phot. Tech. Lett.* **7**, 1409 (1995).

[38] J. L. Gimlett and N. K. Cheung, *J. Light. Tech.* **7**, 888 (1989).

[39] IEC Publication 1290-3-1, "Basic specifications for optical fibre amplifier test methods, electrical spectrum analyzer test method," 1995.

[40] M. Artiglia, P. Bonanni, A. Cavaciuti, M. Potenza, M. Puleo, and B. Sordo, "Noise measurements of erbium-doped optical fibre amplifiers: influence of source excess noise," NIST Special Publication 839, *Technical Digest of the Symposium on Optical Fiber Measurements*, pp. 205–208 (1992).

[41] J. L. Gimlett, M. Z. Iqbal, N. K. Cheung, A. Righetti, F. Fontana, and G. Grasso, *IEEE Phot. Tech. Lett.* **2**, 211 (1990).

[42] F. W. Willems, J. C. van der Plaats, and D. J. DiGiovanni, *Elect. Lett.* **30**, 645 (1994).

[43] Commercial erbium-doped fiber product data sheets, Specialty Fiber Devices Group, Lucent Technologies, Franklin Township, NJ (1997).

[44] K. Petermann, *Elect. Lett.* **19**, 712 (1983).

[45] C. Pask, *Elect. Lett.* **20**, 144 (1984).

[46] R. L. Mortenson, B. Scott Jackson, S. Shapiro, and W. F. Sirocky, *AT&T Technical Journal* Vol. 74, Number 1, pp. 33-46 (January/February 1995).

[47] B. Clesca, P. Bousselet, J. Auge, J. P. Blondel, and H. Fevrier, *IEEE Phot. Tech. Lett.* **6**, 1318 (1994).

[48] M. N. Zervas, R. I. Laming, and D. N. Payne, in *Conference on Optical Fiber Communication*, Vol. 5, 1992 OSA Technical Digest Series (Optical Society of America, Washington, D.C., 1992), pp. 148–149.

[49] D. J. DiGiovanni, P. F. Wysocki, and J. R. Simpson, *Laser Focus World*, pp. 95-106 (September 1993).

[50] B. Pedersen, M. L. Dakss, B. A. Thompson, W. J. Miniscalco, T. Wei, and L. J. Andrews, *IEEE Phot. Tech. Lett.* **3**, 1085 (1991).

[51] J. C. Livas, S. R. Chinn, E. S. Kintzer, and D. J. DiGiovanni, "High power erbium-doped fiber amplifier pumped at 980 nm," in *Conference on Lasers and Electro Optics*, Vol. 15, 1995 OSA Technical Digest Series (Optical Society of America, Washington, D.C., 1995), pp. 521–522.

[52] R. I Laming, M. N. Zervas, and D. N. Payne, *IEEE Phot. Tech. Lett.* **4**, 1345 (1992).

[53] S. Yamashita and T. Okoshi, *IEEE Phot. Tech. Lett.* **4**, 1276 (1992).

[54] R. J. Nuyts, J-M. P. Delavaux, J. Nagel, and D.J. DiGiovanni, *Optical Fiber Tech.* **1**, 76 (1994).

[55] J-M.P Delavaux, D.M. Baney, D. J. DiGiovanni, and Y.K. Park, "BOA: balanced optical amplifier," in *Conference on Optical Fiber Communication*, Vol. 4, 1993 OSA Technical Digest Series (Optical Society of America, Washington, D.C., 1993), pp. 346–349.

[56] J.-M. P. Delavaux, C. F. Flores, R. E. Tench, T. C. Pleiss, T. W. Cline, D. J. DiGiovanni, J. Federici, C. R. Giles, H. Presby, J. S. Major, and W. J. Gignac, *Elect. Lett.* **28**, 1642 (1992).

[57] D. J. DiGiovanni, G. R. Jacobovitz-Veselka, P. F. Wysocki, US patent No. 5,633,964 (1997).

[58] V. Lauridsen, R. Tadayoni, A. Bjarklev, J. H. Povlsen, and B. Pedersen, *Elect. Lett.* **27**, 327 (1991).

[59] J. H. Povlsen, A. Bjarklev, O. Lumbolt, H. Vendeltorp-Pommer, K. Rottwit, and T. Rasmussen, "Optimizing gain and noise performance of EDFA's with insertion of a filter or an isolator," in *Fiber Lasers Sources and Amplifiers III, Proceedings SPIE* **1581**, pp. 107–113 (1991).

[60] S. Seikai, S. Shimokado, T. Fukuoka, and T. Tohi, *Proc. IEICE Trans. Comm.* **E77-B**, 454 (1994).

[61] J-M. P Delavaux and J. A. Nagel, *J. Light. Tech.* **13**, 703 (1995).

[62] O. Lumbolt, K. Schusler, A. Bjarklev, S. Dahl-Petersen, J. H. Povlsen, T. Rasmussen, and K. Rottwit, *IEEE Phot. Tech. Lett.* **4**, 568 (1992).

[63] H. Masuda and A. Takada, *Elect. Lett.* **26**, 661 (1990).

[64] H. Takara, A. Takada, and M. Saruwatari, *IEEE Phot. Tech. Lett.* **4**, 241 (1992).

[65] A. Yu, M. J. O'Mahony, and A. S. Siddiqui, *IEEE Phot. Tech. Lett.* **5**, 773 (1993).

[66] R. G. Smart, J. L. Zyskind, and D. J. DiGiovanni, *Elect. Lett.* **30**, 50 (1994).

[67] M. Tachibana, R. I. Laming, P. R. Morkel, and D. N. Payne, *IEEE Phot. Tech. Lett.* **3**, 118 (1991).

[68] G. Grasso, F. Fontana, A. Righetti, P. Scrivener, P. Turner, and P. Maton, "980-nm diode-pumped Er-doped fiber optical amplifiers with high gain-bandwidth product," in *Conference on Optical Fiber Communication*, Vol. 13, 1991 OSA Technical Digest Series (Optical Society of America, Washington, D.C., 1991), p. 195.

[69] Y. Sato and K. Aoyama, *IEEE Phot. Tech. Lett.* **3**, 1001 (1991).

[70] J-M. P. Delavaux, E. Chausse, J. A. Nagel, and D. J. DiGiovanni, "Multistage EDFA-circulator-based designs," in *Conference on Optical Fiber Communication*, Vol. 8, 1995 OSA Technical Digest Series (Optical Society of America, Washington, D.C., 1995), pp. 47–49.

[71] C. D. Chen, J-M. P. Delavaux, B. W. Hakki, O. Mizuhara, T. V. Nguyen, R. J. Nuyts, K. Ogawa, Y. K. Park, C. Skolnik, R. E. Tench, J. J. Thomas, L. D. Tzeng, and P. D. Yeates, "A field demonstration of 10 Gb/s - 360 km transmission through embedded standard (non-DSF) fiber cables," in *Conference on Optical Fiber Communication*, Vol. 4, 1994 OSA Technical Digest Series (Optical Society of America, Washington, D.C., 1994), pp. 442–445.

[72] J. C. Livas, "High sensitivity optically preamplified 10 Gb/s receivers," in *Conference on Optical Fiber Communication*, Vol. 2, 1996 OSA Technical Digest Series (Optical Society of America, Washington, D.C., 1996), pp. 343–345.

[73] K. Nakagawa, K. Hagimoto, S. Nishi, and K. Aoyama, "A bit-rate flexible transmission field trial over 300-km installed cables employing optical fiber amplifiers," in *Optical Fiber Amplifiers and Their Applications*, Vol. 13, 1991 OSA Technical Digest Series (Optical Society of America, Washington, D.C., 1991), pp. 341–344.

[74] C. R. Giles and A. McCormick, "Bidirectional transmission to reduce fiber FWM penalty in WDM lightwave systems," in *Optical Amplifiers and their Applications*, Vol. 18, 1995 OSA Technical Digest Series (Optical Society of America, Washington DC, 1995), pp. 76–79.

[75] L. Eskildsen and E. Goldstein, *IEEE Phot. Tech. Lett.* **4**, 55 (1992).

[76] J. Farré, E. Bødtker, G. Jacobsen, and K. E. Stubkjær, *IEEE Phot. Tech. Lett.* **4**, 425 (1993).

[77] J. Haugen, J. Freeman, and J. Conradi, *IEEE Phot. Tech. Lett.* **4**, 913 (1992).

[78] M. O. van Deventer and O. J. Koning, *IEEE Phot. Tech. Lett.* **7**, 1078 (1995).

[79] J. L. Gimlett and N. K. Cheung, *J. Light. Tech.* **7**, 888 (1989).

[80] J. L. Gimlett, M. Z. Iqbal, L. Curtis, N. K. Cheung, A. Righetti, F. Fontana, and G. Grasso, *Elect. Lett.* **25**, 1393 (1989).

[81] R. K. Staubli and P. Gysel, *J. Light. Tech.* **9**, 375 (1991).

[82] M. O. van Deventer, *IEEE Phot. Tech. Lett.* **5**, 851 (1993).

[83] M. O. van Deventer and O. J. Koning, *IEEE Phot. Tech. Lett.* **7**, 1372 (1995).

[84] K. Kannan and S. Frisken, *IEEE Phot. Tech. Lett.* **5**, 76 (1993).

[85] R. J. Orazi and M. N. McLandrich, *IEEE Phot. Tech. Lett.* **5**, 571 (1993).

[86] J.-M. P. Delavaux, O. Mizuhara, P. D. Yeates, and T. V. Nguyen, *IEEE Phot. Tech. Lett.* **7**, 1087 (1995).

[87] F. Khaleghi, J. Li, M. Kavehrad, and H. Kim, *IEEE Phot. Tech. Lett.* **8**, 1252 (1996).

[88] S. Seikai, K. Kusunoki, and S. Shimokado, *J. Light. Tech.* **12**, 849 (1994).

[89] S. Seikai, T. Tohi, and Y. Kanaoka, *Elect. Lett.* **30**, 1877 (1994).

[90] S. Seikai, S. Shimokado, and K. Kusunoki, *Elect. Lett.* **29**, 1268 (1993).

[91] J.-M. P. Delavaux, C. R. Giles, S. W. Granlund, and C. D. Chen, "Repeatered bi-directional 10-Gb/s 240-km fiber transmission experiment," in *Conference on Optical Fiber Communication*, Vol. 2, 1996 OSA Technical Digest Series (Optical Society of America, Washington, D.C., 1996), pp. 18–19.

[92] S.-K. Liaw, K.P. Ho, C. Lin, and S. Chi, *IEEE Phot. Tech. Lett.* **9**, 1664 (1997).

[93] C. W. Barnard, J. Chrostowski, and M. Kavehrad, *IEEE Phot. Tech. Lett.* **4**, 911 (1992).

[94] M. Yadlowsky, "Bidirectional optical amplifiers for high-performance WDM systems," in *OSA Trends in Optics and Photonics*, Vol. 16, Optical Amplifiers and Their Applications, M. N. Zervas, A. E. Willner, and S. Sasaki, Eds. (Optical Society of America, Washington, DC, 1997), pp. 307–310.

[95] J.M. P. Delavaux, T. Nuyen, O. Mizuhara, S. Granlund, P. Yeates, and A. Yeniay, "WDM repeaterless bi-directional transmission of 73 channels at 10 Gbit/s over 126 km of TrueWave™ fiber," in *23rd European Conference on Optical Communication*, Proceedings Vol. 5, pp. 21–24 (1997).

[96] R. I. Laming, J. E. Townsend, D. N. Payne, F. Meli, G. Grasso, and E. J. Tarbox, *IEEE Phot. Tech. Lett.* **3**, 253 (1991).

[97] M. Yamada, M. Shimizu, T. Takeshita, M. Okayasu, M. Horiguchi, S. Uehara, and E. Sugita, *IEEE Phot. Tech. Lett.* **1**, 422 (1989).

[98] A. Lidgard, J. R. Simpson, and P. C. Becker, *Appl. Phys. Lett.* **56**, 2607 (1990).

[99] Y. Kimura, M. Nakazawa, and K. Suzuki, *Appl. Phys. Lett.* **57**, 2635 (1990).

[100] J. F. Massicott, R. Wyatt, B. J. Ainslie, and S. P. Craig-Ryan, *Elect. Lett.* **26**, 1038 (1990).

[101] R. I. Laming, J. E. Townsend, D. N. Payne, F. Meli, G. Grasso, and E. J. Tarbox, *IEEE Phot. Tech. Lett.* **3**, 253 (1991).

[102] P. A. Leilabady, S. G. Grubb, K. L. Sweeney, R. Cannon, S. Vendetta, T. H. Windhorn, G. Tremblay, and G. Maurer, "Analog transmission characteristics of +18 dBm erbium amplifier pumped by diode-pumped Nd:YAG laser," in *17th European Conference on Optical Communication*, Proceedings Vol. 3, postdeadline paper, pp. 5–8 (1991).

[103] S. G. Grubb, W. F. Humer, R. Cannon, T. H. Windhorn, S. Vendetta, K. L. Sweeney, P. A. Leilabady, W. L. Barnes, K. P. Jedrzejewski, and J. E. Townsend, *IEEE Phot. Tech. Lett.* **4**, 553 (1992).

[104] S. G. Grubb, W. F. Humer, R. Cannon, S. Vendetta, K. L. Sweeney, P. A. Leilabady, M. R. Keur, J. G. Kwasegroch, T. C. Munks, and D. W. Anthon, *Elect. Lett.* **28**, 1275 (1992).

[105] J. C. Livas, S. R. Chinn, E. S. Kintzer, J. N. Walpole, C. A. Wang, and L. J. Missaggia, *Elect. Lett.* **30**, 1054 (1994).

[106] P. Bousselet, R. Meilleur, A. Coquelin, P. Garabedian, J. L. Beylat, "+25.2dBm output power from an Er-doped fiber amplifier with 1.48 μm SMQW laser-diode modules," in *Conference on Optical Fiber Communication*, Vol. 8, 1995 OSA Technical Digest Series (Optical Society of America, Washington, D.C., 1995), pp. 42–43.

[107] S. G. Grubb, D. J. DiGiovanni, J. R. Simpson, W. Y. Cheung, S. Sanders, D. F. Welch, and B. Rockney, "Ultrahigh power diode-pumped 1.5 μm fiber amplifiers," in *Conference on Optical Fiber Communication*, Vol. 2, 1996 OSA Technical Digest Series (Optical Society of America, Washington, D.C., 1996), pp. 30–31.

[108] G. R. Jacobovitz-Veselka, R. P. Espindola, C. Headley, A. J. Stentz, S. Kosinski, D. Iniss, D. Tipton, D. J. DiGiovannim M. Andrejco, J. DeMarco, C. Soccolich, S. Cabot, N. Conti, J. J. Veselka, T. Strasser, R. Pedrazzani, A. Hale, K. S. Kranz, R. G. Huff, G. Nykolak, P. Hansen, L. Gruner-Nielsen, D. Boggavarapu, X. He, D. Caffey, S. Gupta, S. Srinavasan, R. Pleak, K. McEven, and R. Patel, "A 5.5 W single-stage single-pumped erbium doped fiber amplifier at 1550 nm," in *OSA Trends in Optics and Photonics*, Vol. 16, Optical Amplifiers and Their Applications, M. N. Zervas, A. E. Willner, and S. Sasaki, Eds. (Optical Society of America, Washington, DC, 1997), pp. 148–151.

[109] S. B. Poole, "Fiber amplifiers," in Conference on Optical Fiber Communication, Tutorial WE, Tutorial Proceedings of OFC '94, (Optical Society of America, Washington, D.C., 1994), pp. 53–68.

[110] S. G. Grubb, P. A. Leilabady, K. L. Sweeney, W. H. Humer, R. S. Cannon, S. W. Vendetta, W. L. Barnes, and J. E. Townsend, "High output power Er^{3+}/Yb^{3+} co-doped optical amplifiers pumped by diode-pumped Nd^{3+} lasers," in *Optical Amplifiers and their Applications*, Vol. 17, 1992 OSA Technical Digest Series (Optical Society of America, Washington DC, 1992), pp. 202–205.

[111] A. E. Willner and S-M. Hwang, *J. Light. Tech.* **13**, 802 (1995).

[112] R. G. Smart, J. W. Sulhoff, J. L. Zyskind, J. A. Nagel, and D. J. DiGiovanni, *IEEE Phot. Tech. Lett.* **6**, 380 (1994).

[113] A. J. Lucero, J. W. Sulhoff, and J. L. Zyskind, "Comparison of 980- and 1480-nm pumped amplifiers in multiwavelength systems," in *Conference on Optical Fiber Communication*, Vol. 8, 1995 OSA Technical Digest Series (Optical Society of America, Washington, D.C., 1995), pp. 3–4.

[114] E. L. Goldstein, L. Eskildsen, C. Lin, and R. E. Tench, *IEEE Phot. Tech. Lett.* **6**, 266 (1994).

[115] E. L. Goldstein, L. Eskildsen, V. da Silva, M. Andrejco, and Y. Silberberg, *J. Light. Tech.* **13**, 782 (1995).

[116] J. Li, F. Khaleghi, and M. Kavehrad, *J. Light. Tech.* **13**, 2191 (1995).

[117] T. Kashiwada, K. Nakazato, M. Ohnishi, H. Kanamori, and M. Nishimura, "Spectral gain behavior of Er-doped fiber with extremely high concentration," in *Optical Amplifiers and their Applications*, Vol. 14, 1993 OSA Technical Digest Series (Optical Society of America, Washington DC, 1993), pp. 104–107.

[118] S. Yoshida, S. Kuwano, and K. Iwashita, *Elect. Lett.* **31**, 1765 (1995).

[119] P. F. Wysocki, N. Park, and D. DiGiovanni, *Opt. Lett.* **21**, 1744 (1996).

[120] N. Shimojoh, T. Naito, T. Terahara, H. Deguchi, K. Tagawa, M. Suyama, and T. Chikama, *Elect. Lett.* **33**, 877 (1997).

[121] A. Mori, Y. Ohishi, M. Yamada, H. Ono, Y. Nishida, K. Oikawa, and S. Sudo, "1.5 μm broadband amplification by tellurite-based EDFAs," in *Conference on Optical Fiber Communication*, Vol. 6, 1997 OSA Technical Digest Series (Optical Society of America, Washington, D.C., 1997), pp. 371–374.

[122] A. Mori, K. Kobayashi, M. Yamada, T. Kanamori, K. Oikawa, Y. Nishida, and Y. Ohishi, *Elect. Lett.* **34**, 887 (1998).

[123] J. F. Massicott, J. R. Armitage, R. Wyatt, B. J. Ainslie, and S. P. Craig-Ryan, *Elect. Lett.* **26**, 1645 (1990).

[124] J. F. Massicott, R. Wyatt, and B. J. Ainslie, *Elect. Lett.* **28**, 1924 (1992).

[125] H. Ono, M. Yamada, and Y. Ohishi, *IEEE Phot. Tech. Lett.* **9**, 596 (1997).

[126] H. Ono, M. Yamada, M. Shimizu, and Y. Ohishi, *Elect. Lett.* **34**, 1509 (1998).

[127] H. Ono, M. Yamada, M. Shimizu, and Y. Ohishi, *Elect. Lett.* **34**, 1513 (1998).

[128] H. Ono, M. Yamada, T. Kanamori, S. Sudo, and Y. Ohishi, *Elect. Lett.* **33**, 1471 (1997).

[129] M. Jinno, T. Sakamoto, J. Kani, S. Aisawa, K. Oda, M. Fukui, H. Ono, and K. Oguchi, *Elect. Lett.* **33**, 882 (1997).

[130] A, K, Srivasta, J. W. Sulhoff, L. Zhang, C. Wolf, Y. Sun, A. A¿ Abramov, T. A. Strasser, J. R. Pedrazzani, R. P. Espindola, and A. M. Vengsarkar, "L-band 64 × 10 Gb/s DWDM transmission over 500 km DSF with 50 GHz channel spacing," in *24th European Conference on Optical Communication*, Proceedings Vol. 3, pp. 73–75 (1998).

[131] B. Desthieux, Y. Robert, J. Hervo, and D. Bayart, "25-nm usable bandwidth for transoceanic WDM transmission systems using 1.58 μm erbium-doped fiber amplifiers," in *24th European Conference on Optical Communication*, Proceedings Vol. 3, pp. 133–135 (1998).

[132] D. Bayart, B. Clesca, L. Hamon, and J. L. Beylat, *IEEE Phot. Tech. Lett.* **6**, 613 (1994).

[133] B. Clesca, D. Ronarch, D. Bayart, Y. Sorel, L. Hamon, M. Guibert, J. L. Beylat, J. F. Kerdiles, and M. Semenkoff, *IEEE Phot. Tech. Lett.* **6**, 509 (1994).

[134] M. Yamada, T. Kanamori, Y. Terunuma, K. Oikawa, M. Shimizu, S. Sudo, and K. Sagawa, *IEEE Phot. Tech. Lett.* **8**, 882 (1996).

[135] B. Clesca, D. Bayart, L. Hamon, J. L. Beylat, C. Coeurjolly, and L. Berthelon, *Elect. Lett.* **30**, 586 (1994).

[136] S. Artigaud, M. Chbat, P. Nouchi, F. Chiquet, D. Bayart, L. Hamon, A. Pitel, F. Goudeseune, P. Bousselet, and J.-L. Beylat, "Transmission of 16×10 Gbit/s channels spanning 24 nm over 531 km of conventional single-mode fiber using 7 in-line fluoride-based EDFAs," in *Conference on Optical Fiber Communication*, Vol. 2, 1996 OSA Technical Digest Series (Optical Society of America, Washington, D.C., 1996), pp. 435–438.

[137] M. Yamada, Y. Ohishi, T. Kanamori, H. Ono, S. Sudo, and M. Shimizu *Elect. Lett.* **33**, 809 (1997).

[138] R. I. Laming, J. D. Minelly, L. Dong, and M. N. Zervas, *Elect. Lett.* **29**, 509 (1993).

[139] L. Eskildsen, E. L. Goldstein, G. K. Chang, M. Z. Iqbal, and C. Lin, *IEEE Phot. Tech. Lett.* **6**, 1321 (1994).

[140] E. L. Goldstein, V. da Silva, L. Eskildsen, M. Andrejco, and Y. Silberberg, *IEEE Phot. Tech. Lett.* **5**, 543 (1993).

[141] E. L. Goldstein, L. Eskildsen, V. da Silva, M. Andrejco, and Y. Silberberg, *IEEE Phot. Tech. Lett.* **5**, 937 (1993).

[142] L. Eskildsen, E. L. Goldstein, V. da Silva, M. Andrejco, and Y. Silberberg, *IEEE Phot. Tech. Lett.* **5**, 1188 (1993).

[143] S. F. Su, R. Olshansky, D. A. Smith, and J. E. Baran, *Elect. Lett.* **29**, 477 (1993).

[144] S. H. Huang, X. Y. Zou, A. E. Willner, Z. Bao, and D. A. Smith, *IEEE Phot. Tech. Lett.* **9**, 389 (1997).

[145] H. S. Kim, S. H. Yun, H. K. Kim, N. Park, and B. Y. Kim, *IEEE Phot. Tech. Lett.* **10**, 790 (1998).

[146] M. Tachibana, R. I. Laming, P. R. Morkel, and D. N. Payne, *IEEE Phot. Tech. Lett.* **3**, 118 (1991).

[147] M. Wilkinson, A. Bebbington, S. A. Cassidy, and P. McKee, *Elect. Lett.* **28**, 131 (1992).

[148] H. Taga, N. Takeda, K. Imai, S. Yamamoto, and S. Akiba, "110 Gbit/s (22×5 Gbit/s), 9500 k transmission experiment using 980 nm pump EDFA 1R repeater without forward error correction," in *OSA Trends in Optics and Photonics*, Vol. 5, Optical Amplifiers and Their Applications, 1996 OAA Program Committee, Eds. (Optical Society of America, Washington, DC, 1996), pp. 28–31.

[149] T. J. Cullen, N. E. Jolley, F. Davis, and J. Mun, "EDFA gain flattening using periodic tapered fibre filters," in *OSA Trends in Optics and Photonics*, Vol. 16, Optical Amplifiers and Their Applications, M. N. Zervas, A. E. Willner, and S. Sasaki, Eds. (Optical Society of America, Washington, DC, 1997), pp. 112–115.

[150] K. Inoue, T. Kominato, and H. Toba, *IEEE Phot. Tech. Lett.* **3**, 718 (1991).

[151] R. Kashyap, R. Wyatt, and R. J. Campbell, *Elect. Lett.* **29**, 154 (1993).

[152] J. F. Massicott, S. D. Wilson, R. Wyatt, J. R Armitage, R. Kashyap, D. Williams, and R. A. Lobbett, *Elect. Lett.* **30**, 963 (1994).

[153] M. Yamada, H. Ono, and Y. Ohishi, *Elect. Lett.* **34**, 1747 (1998).

[154] H. Masuda, S. Kawai, K.-I. Suzuki, and K. Aida, *Elect. Lett.* **33**, 1070 (1997).

[155] P. F. Wysocki, J. B. Judkins, R. P. Espindola, M. Andrejco, and A. Vengsarkar, *IEEE Phot. Tech. Lett* **9**, 1343 (1997).

[156] R. A. Betts, S. J. Frisken, and D. Wong, in "Split-beam Fourier filter and its application in a gain-flattened EDFA," in *Conference on Optical Fiber Communication*, Vol. 8, 1995 OSA Technical Digest Series (Optical Society of America, Washington, D.C., 1995), pp. 80–81.

[157] N. S. Bergano, C. R. Davidson, A. M. Vengsarkar, B. M. Nyman, S. G. Evangelides, J. M. Darcie, M. Ma, J. D. Evankow, P. C. Corbett, M. A. Mills, G. A. Ferguson, J. R. Pedrazzani, J. A. Nagel, J. L. Zyskind, J. W. Sulhoff, and A. J. Lucero, "100 Gb/s WDM transmission of twenty 5 Gb/s NRZ data channels over transoceanic distances using a gain flattened amplifier chain," in *21st European Conference on Optical Communication*, Proceedings Vol. 3, pp. 967–970 (1995).

[158] N. S. Bergano and C. R. Davidson, *J. Light. Tech.* **14**, 1299 (1996).

[159] N. S. Bergano, C. R. Davidson, M. A. Mills, P. C. Corbett, S. G. Evangelides, B. Pedersen, R. Menges, J. L. Zyskind, J. W. Sulhoff, A. K. Srivasta, C. Wolf, and J. Judkins, "Long-haul WDM transmission using optimum channel modulation: 160 Gb/s (32×5 Gb/s) 9,300 km demonstration," in *Conference on Optical Fiber Communication*, Vol. 6, 1997 OSA Technical Digest Series (Optical Society of America, Washington, D.C., 1997), pp. 432–435.

[160] N. S. Bergano, C. R. Davidson, M. Ma, A. Pilipetskii, S. G. Evangelides, H. D. Kidorf, J. M. Darcie, E. Golovchenko, K. Rottwitt, P. C. Corbett, R. Menges, M. A. Mills, B. Pedersen, D. Peckham, A. A. Abramov, and A. M. Vengsarkar. "320 Gb/s WDM transmission (64 × 5 Gb/s) over 7,200 km using large mode fiber spans and chirped return-to-zero signals," in *Conference on Optical Fiber Communication*, Vol. 2, 1998 OSA Technical Digest Series (Optical Society of America, Washington, D.C., 1998), pp. 472–475.

[161] A. K. Srivasta, Y. Sun, J. W. Sulhoff, C. Wolf, M. Zirngibl, R. Monnard, A. R. Chraplyvy, A. A. Abramov, R. P. Espindola, T. A. Strasser, J. R. Pedrazzani, A. M. Vengsarkar, J. L. Zyskind, J. Zhou, D. A. Ferrand, P. F. Wysocki, J. B. Judkins, and Y. P. Li, "1 Tb/s transmission of 100 WDM 10 Gb/s channels over 400 km of TrueWave™ fiber," in *Conference on Optical Fiber Communication*, Vol. 2, 1998 OSA Technical Digest Series (Optical Society of America, Washington, D.C., 1998), pp. 464–467.

[162] C. R. Giles and D. J. DiGiovanni, *IEEE Phot. Tech. Lett.* **2**, 866 (1990).

[163] M. Yamada, M. Shimizu, Y. Ohishi, M. Horiguchi, S. Sudo, and A. Shimizu, *Elect. Lett.* **30**, 1762 (1994).

[164] M. Semenkoff, M. Guibert, J. F. Kerdiles, and Y. Sorel, *Elect. Lett.* **30**, 1411 (1994).

[165] H. Ono, M. Yamada, T. Kanamori, and Y. Ohishi, *Elect. Lett.* **33**, 1477 (1997).

[166] M. Yamada, H. Ono, T. Kanamori, T. Sakamoto, Y. Ohishi, and S. Sudo, *IEEE Phot. Tech. Lett.* **8**, 620 (1996).

[167] H. Masuda, K.-I. Suzuki, S. kawai, and K. Aida, *Elect. Lett.* **33**, 753 (1997).

[168] M. Yamada, H. Ono, T. Kanamori, S. Sudo, and Y. Ohishi, *Elect. Lett.* **33**, 710 (1997).

[169] Y. Sun, J. W. Sulhoff, A. K. Srivasta, J. L. Zyskind, C. Wolf, T. A. Strasser, J. R. Pedrazzani, J. B. Judkins, R. P. Espindola, A. M. Vengsarkar, and J. Zhou, "Ultra wide band erbium-doped silica fiber amplifier with 80 nm of bandwidth," in *OSA Trends in Optics and Photonics*, Vol. 16, Optical Amplifiers and Their Applications, M. N. Zervas, A. E. Willner, and S. Sasaki, Eds. (Optical Society of America, Washington, DC, 1997), pp. 144–147.

[170] Y. Sun, J. W. Sulhoff, A. K. Srivasta, J. L. Zyskind, C. Wolf, T. A. Strasser, J. R. Pedrazzani, J. B. Judkins, R. P. Espindola, A. M. Vengsarkar, and J. Zhou, "An 80 nm ultra wide band EDFA with low noise figure and high output power," in *23rd European Conference on Optical Communication*, Proceedings Vol. 5, pp. 69–72 (1997).

[171] L. F. Mollenauer and S. G. Evangelides, "High bit rate, ultralong distance soliton transmission with erbium amplifiers," in *Conference on Lasers and Electro-Optics*, Vol. 7, 1990 OSA Technical Digest Series (Optical Society of America, Washington DC, 1990), pp. 512–514.

[172] A. Yariv, *Opt. Lett.* **15**, 1064 (1990).

[173] P. Urquhart and T. J. Whitley, *Appl. Opt.* **29**, 3503, (1990).

[174] M. N. Islam, L. Rahman, J. R. Simpson, *J. Light. Tech.* **12**, 1952 (1994).

[175] A. Hasagawa, *Opt. Lett.* **8**, 650 (1983).

[176] S. G. Evangelides and L. F. Mollenauer, "Comparison of solitons vs transmission at λ_0 for ultralong distance transmission using lumped amplifiers," in *Conference on Lasers and Electro-Optics*, Vol. 7, 1990 OSA Technical Digest Series (Optical Society of America, Washington DC, 1990), p. 514.

[177] L. F. Mollenauer and K. Smith, *Opt. Lett.* **13**, 675 (1988).

[178] S. T. Davey, D. L. Williams, D. M. Spirit, and B. J. Ainslie, "The fabrication of low loss high N.A. silica fibres for Raman amplification," in *Fiber Laser Sources and Amplifiers*, M. J. F. Digonnet, Ed., *Proc. SPIE* **1171**, pp. 181–191 (1989).

[179] M. Nakazawa, Y. Kimura, and K. Suzuki, "Soliton transmission in a distributed, dispersion-shifted erbium-doped fiber amplifier," in *Optical Amplifiers and Their Applications*, Vol. 13, 1990 OSA Technical Digest Series, (Optical Society of America, Washington, D.C., 1990), pp. 120–123.

[180] M. Nakazawa, Y. Kimura, K. Suzuki, and H. Kubota, *J. Appl. Phys.* **66**, 2803 (1989).

[181] J. R. Simpson, L. F. Mollenauer, K. S. Kranz, P. J. Lemaire, N. A. Olsson, H. T. Shang, and P. C. Becker, "A distributed erbium doped fiber amplifier", in *Conference on Optical Fiber Communication*, Vol. 13, 1990 OSA Technical Digest Series (Optical Society of America, Washington, D.C., 1990), pp. 313–316.

[182] S. P. Craig-Ryan, B. J. Ainslie and C. A. Millar, *Elect. Lett.* **26**, 185 (1990).

[183] D. Tanaka, A. Wada, T. Sakai, and R. Yamauchi, "Attenuation Free, Dispersion Shifted Fiber Doped with Distributed Erbium," in *Optical Amplifiers and Their Applications*, Vol. 13, 1990 OSA Technical Digest Series, (Optical Society of America, Washington, D.C., 1990), pp. 138–141.

[184] M. Morisawa, M. Yoshida, H. Ito, T. Gozen, H. Tanaka, and M. Yotsuya, "Er^{3+}-doped VAD fibers for lumped amplifiers and distributed active transmission lines," in *Optical Amplifiers and Their Applications*, Vol. 13, 1990 OSA Technical Digest Series, (Optical Society of America, Washington, D.C., 1990), pp. 142–145.

[185] T. J. Whitley, C. A. Milar, S. P. Craig-Ryan, P. Urquhart, "Demonstration of a Distributed Optical Fibre Amplifier BUS Network," in *Optical Amplifiers and Their Applications*, Vol. 13, 1990 OSA Technical Digest Series, (Optical Society of America, Washington, D.C., 1990), pp. 236–239.

[186] S. T. Davey, D. L. Williams, D. M. Spirit, and B. J. Ainslie, *Elect. Lett.* **26**, 1148 (1990).

[187] D. L. Williams, S. T. Davey, D. M. Spirit, and B. J. Ainslie, *Elect. Lett.* **26**, 1517 (1990).

[188] D. L. Williams, S. T. Davey, D. M. Spirit, G. R.Walker, and B. L. Ainslie, *Elect. Lett.* **27**, 812 (1991).

[189] E. Desurvire, *IEEE Phot. Tech. Lett.* **33**, 625 (1991).

[190] S. Wen and S. Chi, *IEEE Phot. Tech. Lett.* **4**, 189 (1992).

[191] S. Wen and S. Chi, *J. Light. Tech.* **10**, 1869 (1992).

[192] C. Lester, K. Rottwitt, J. H. Povlsen and A. Bjarklev, *Opt. Comm.* **106**, 183 (1994).

[193] S. L. Zhang and J. J. O'Reilly, *IEEE Phot. Tech. Lett.* **5**, 537 (1993).

[194] M. F. Ferreira, *Elect. Lett.* **30**, 40 (1994).

[195] D. M. Spirit, I. W. Marshall, P. D. Constantine, D. L. Williams, S. T. Davey, and B. J. Ainslie, *Elect. Lett.* **27**, 222 (1991).

[196] K. Rottwitt, J. Povlsen, A. Bjarklev, O. Lumholt, B. Pedersen, F. Pedersen, and T. Rasmussen, "Detailed analysis of distributed erbium doped fiber amplifiers," in *Optical Amplifiers and their Applications*, Vol. 17, 1992 OSA Technical Digest Series (Optical Society of America, Washington DC, 1992), pp. 178–181.

[197] D. N. Chen and E. Desurvire, *IEEE Phot. Tech. Lett.* **4**, 52 (1992).

[198] K. Rottwitt, A. Bjarklev, J. Povlsen, O. Lumholt and T. Rasmussen, *J. Light. Tech.* **10**, 1544 (1992).

[199] K. Rottwitt, J. H. Povlsen and A. Bjarklev, *J. Light. Tech.* **1**, 2105 (1993).

[200] J. R. Simpson, H. -T. Shang, L. F. Mollenauer, N. A. Olsson, P. C. Becker, K. S. Kranz, P. J. Lemaire, and M. J. Neubelt, *J. Light. Tech.* **LT-9**, 228 (1991).

[201] A. Altuncu, L. Noel, W. A. Pender, A. S. Siddiqui, T. Widdowson, A. D. Ellis, M. A. Newhouse, A. J. Antos, G. Kar, and P. W. Chu, *Elect. Lett.* **32**, 233 (1996).

[202] T. Kitagawa, K. Hattori, K. Shuto, M. Yasu, M. Kobayashi, and M. Horiguchi, *Elect. Lett.* **28**, 1818 (1992).

[203] K. Hattori, T. Kitagawa, M. Oguma, M. Wada, J. Temmyo, and M. Horiguchi, *Elect. Lett.* **29**, 357 (1992).

[204] G. Nykolak, M. Haner, P. C. Becker, J. Shmulovich, and Y. H. Wong, *IEEE Phot. Tech. Lett.* **5**, 1185 (1993).

[205] K. Hattori, T. Kitagawa, M. Oguma, Y. Ohmori, and M. Horiguchi, *Elect. Lett.* **30**, 856 (1994).

[206] J. Shmulovich, A. Wong, Y. H. Wong, P. C. Becker, A. J. Bruce, and R. Adar, *Elect. Lett.* **28**, 1181 (1992).

[207] J. Shmulovich, Y. H. Wong, G. Nykolak, P. C. Becker, R. Adar, A. J. Bruce, D. J. Muehlner, G. Adams, and M. Fishteyn, "15 dB net gain demonstration in Er^{3+} glass waveguide amplifier on silicon," in *Conference on Optical Fiber Communication*, Vol. 4, 1993 OSA Technical Digest Series (Optical Society of America, Washington, D.C., 1993), pp. 342–345.

[208] O. Lumholt, T. Rasmussen, and A. Bjarklev, *Elect. Lett.* **29**, 495 (1993).

[209] M. Federighi, I. Massarek, and P. F. Trwoga, *IEEE Phot. Tech. Lett.* **5**, 227 (1993).

[210] M. P. Hehlen, N. J. Cockcroft, T. R. Gosnell, A. J. Bruce, G. Nykolak, and J. Shmulovich, *Opt. Lett.* **22**, 772 (1997).

[211] R. N. Ghosh, J. Shmulovich, C. F. Kane, M. R. X. de Barros, G. Nykolak, A. J. Bruce, and P. C. Becker, *IEEE Phot. Tech. Lett.* **8**, 4 (1996)

[212] D. Barbier, J. M. Delavaux, A. Kevorkian, P. Gastaldo, J. M. Jouanno, "Yb/Er integrated optics amplifiers on phosphate glass in single and double pass configuration," in *Conference on Optical Fiber Communication*, Vol. 8, 1995 OSA Technical Digest Series (Optical Society of America, Washington, D.C., 1995), pp. 335–338.

[213] J-M. P. Delavaux, S. Granlund, O. Mizuhara, L. D. Tzeng, D. Barbier, M. Rattay, F. Saint Andre, and A. Kevorkian, *IEEE Phot. Tech. Lett.* **9**, 247 (1997).

[214] D. Barbier, P. Bruno, C. Cassagnettes, M. Trouillon, R. L. Hyde, A. Kevorkian, and J.-M. P. Delavaux, "Net gain of 27 dB with a 8.6-cm-long Er/Yb-doped glass-planar-amplifier," in *Conference on Optical Fiber Communication*, Vol. 2, 1998 OSA Technical Digest Series (Optical Society of America, Washington, D.C., 1998), pp. 45–46.

[215] D. Barbier, M. Rattay, F. Saint Andre, G. Clauss, M. Trouillon, A. Kevorkian, J-M. P. Delavaux, and E. Murphy, *IEEE Phot. Tech. Lett.* **9**, 315 (1997).

Chapter 9
System Implementations of Amplifiers

9.1 INTRODUCTION

In this chapter we will discuss the system implementations and applications of erbium-doped fiber amplifiers. To date, the main applications have been digital, although analog systems incorporating optical amplifiers are also of interest. In the digital domain, the advent of optical amplifiers has allowed systems engineers to reach a new and higher level of performance. The three dominant areas where the optical amplifier has proven invaluable in this way are preamplification, power amplification, and inline amplification for long-haul systems or high capacity WDM transmission links. Erbium-doped fiber amplifiers have also opened the way to transmission systems using soliton pulse propagation.

Considering first the preamplifier application, erbium-doped fiber amplifiers have contributed to a significant increase in the sensitivity of receivers, as compared to PIN or APD receivers used alone. The historical trend is shown in Figure 9.1 for various receiver families. The erbium-doped fiber experiments are based on intensity-modulated transmission formats. The results are for data rates ranging from 0.5 to 10 Gb/s and the optical amplifiers are typically used with PIN receivers. In the past, APD receivers were used due to their ability to provide gain; however, with the gain now being provided by the optical amplifier, PIN detectors become more attractive than APDs for signal to noise reasons. The quantum limited receiver sensitivity of 43 photons per bit is determined by the signal-spontaneous and spontaneous-spontaneous beat noises at the output of the optical amplifier (see section 7.3.3) and is an unavoidable consequence of the spontaneous emission from the amplifier. In contrast, an unamplified system has a shot noise detection quantum limit of 10 photons per bit.

Dramatic improvements have been obtained in power amplifier performance, where the maximum output power from an optical power amplifier has increased from about 60 mW in 1989 to 5.5 W in 1997, as shown in Figure 9.2. The high output powers now obtained have had a direct impact on remotely pumped transmission systems, where

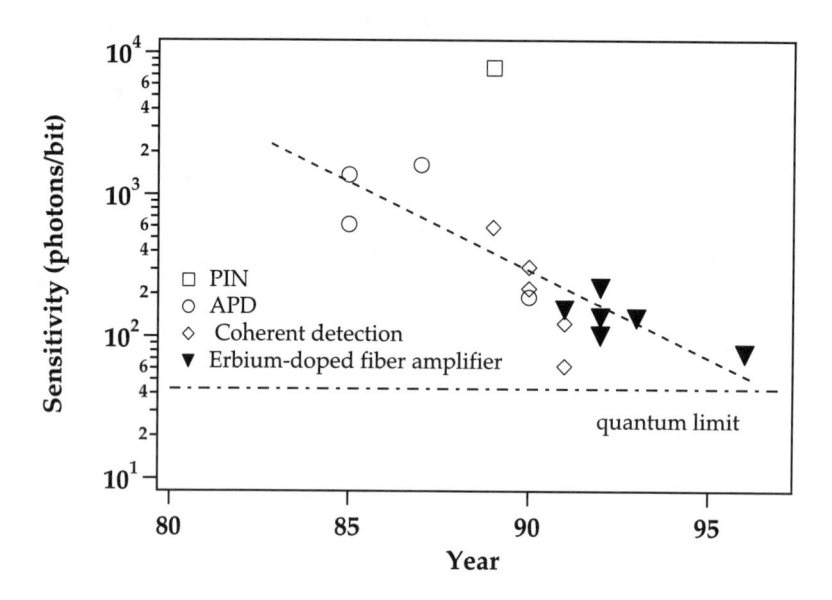

322 CHAPTER 9. SYSTEM IMPLEMENTATIONS OF AMPLIFIERS

Figure 9.1: Improvement over time in receiver sensitivity in photons per bit, from reported results. The limit of 43 photons per bit is the quantum limit for an optically preamplified direct-detection system.

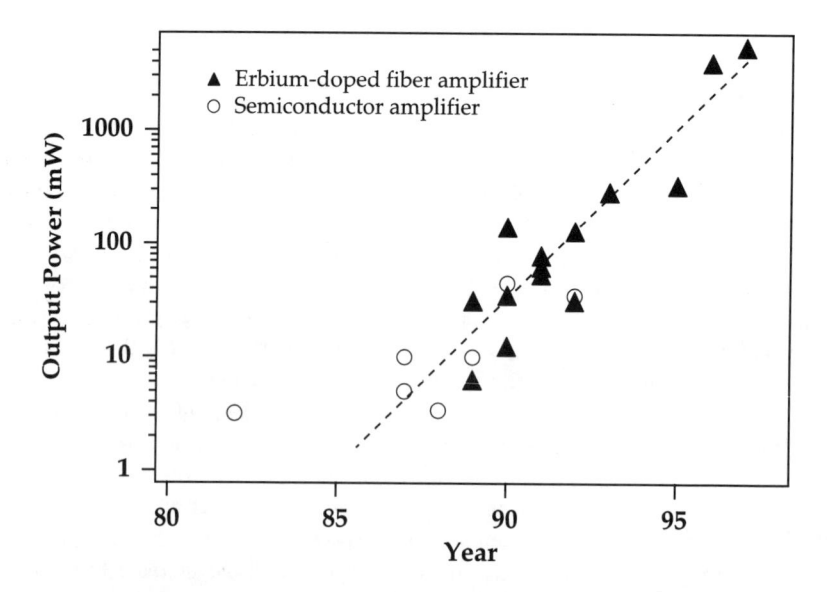

Figure 9.2: Improvement over time in optical power amplifier output, from reported results.

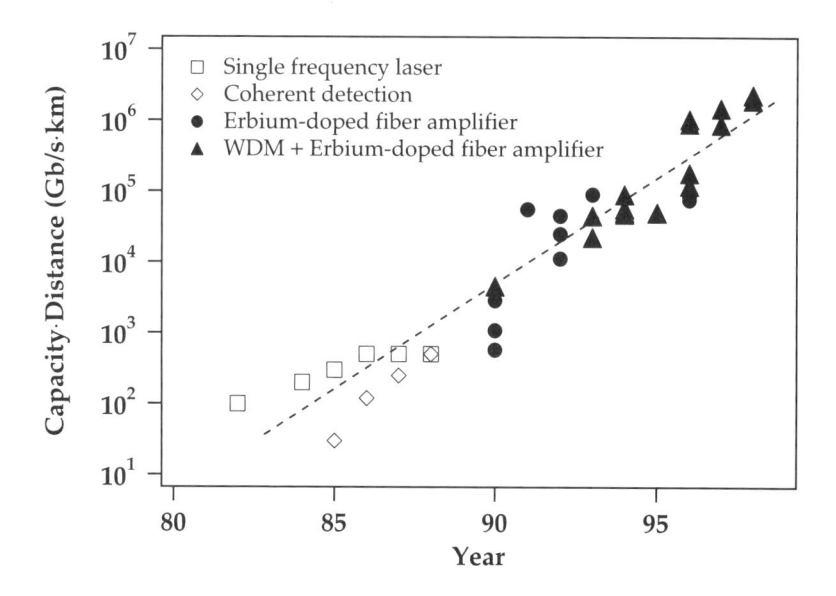

Figure 9.3: Improvement in long distance transmission during 1982 through 1998, from reported results.

the signal travels long distances (100 km to 500 km) before encountering an amplifier. The allowed length of such systems has significantly increased over the past few years.

Improvements in long distance transmission, as characterized by quantities such as capacity×distance achievements (in Gb/s · km), have probably been the most noteworthy results obtained with optical amplifiers in the past five years. These results are shown in Figure 9.3. A large part of the advance has been due to the utilization of WDM techniques. Figure 9.4 shows the capacities achieved in terms of the bit rate and distance product versus the number of channels used in the transmission system. The use of multiple channels has allowed a more than ten-fold increase in the capacity of long haul systems. The long distance application has been paramount in propelling the growth of research in the optical amplifier field, as it has led to the development of commercial undersea and terrestrial fiber transmission systems with vastly larger capacities than before.

9.2 SYSTEM DEMONSTRATIONS AND ISSUES

9.2.1 Preamplifiers

The key to high sensitivity receivers lies in low-noise amplification. This is the guiding theme of this section. The three choices for amplification are electronic amplification, inherent detector amplification (APD), and optical amplification.

Optical preamplifiers have a clear advantage over electrical preamplifiers as they introduce significantly less noise at high gains, especially at high (greater than 1 Gb/s)

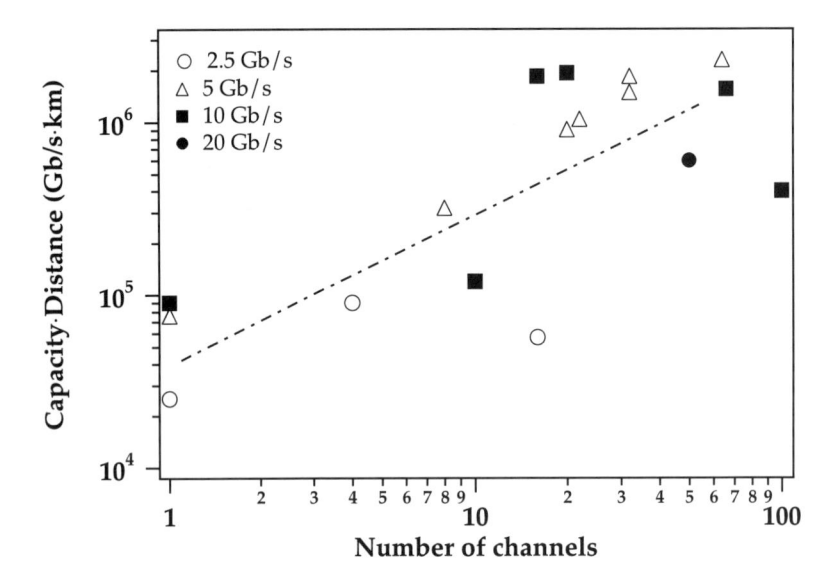

324   CHAPTER 9. SYSTEM IMPLEMENTATIONS OF AMPLIFIERS

Figure 9.4: Long distance WDM erbium-doped amplifier transmission systems capacities achieved experimentally as a function of the number of channels used in the transmission system, from published reports (up to 1998). Demonstrations at bit rates of 2.5, 5, 10, and 20 Gb/s have been included.

bit rates. The underlying reason is that the higher the bandwidth, the harder it is to obtain low-noise operation from electronic components.

There are two main categories of light detectors: PINs and APDs. PINs have no inherent gain, whereas APDs have gain from the avalanche multiplication process. Electrical amplifiers suffer from thermal noise, which impairs the performance of PIN-based receivers and to a lesser extent of APD-based receivers. The noise in most electronic amplifiers increases supralinearly with bandwidth. This is reflected in Figure 9.5, which shows the supralinear decrease in detector sensitivity as the bit rate (or bandwidth) increases. The limitation of PIN receivers is thus that of the electronic amplifiers that follow them.

APDs have low noise at low bit rates. Unfortunately, APDs have a fixed gain × bandwidth product. At high bandwidths the gain is lower, so that the receiver sensitivity is improved less. Figure 9.6 illustrates this fact with the supralinear decrease in receiver sensitivity as the bit rate increases. Erbium-doped amplifiers, on the other hand, have a low noise figure that is independent of the bit rate. Thus, optical preamplification is an attractive choice for amplifying the signal before its detection, as opposed to electrical amplification after detection.

For erbium-doped fiber based preamplifiers, a key decision is whether to use 980 nm or 1480 nm pump sources. It was shown in Chapter 7 (section 7.3.3) that the noise figure of a 980 nm pumped amplifier is lower, by 1 or more dB, than that of an amplifier pumped at 1480 nm. It is also important to have a narrow bandwidth optical filter at the output of the erbium-doped fiber to eliminate as much of the ASE that is exiting the

9.2. SYSTEM DEMONSTRATIONS AND ISSUES

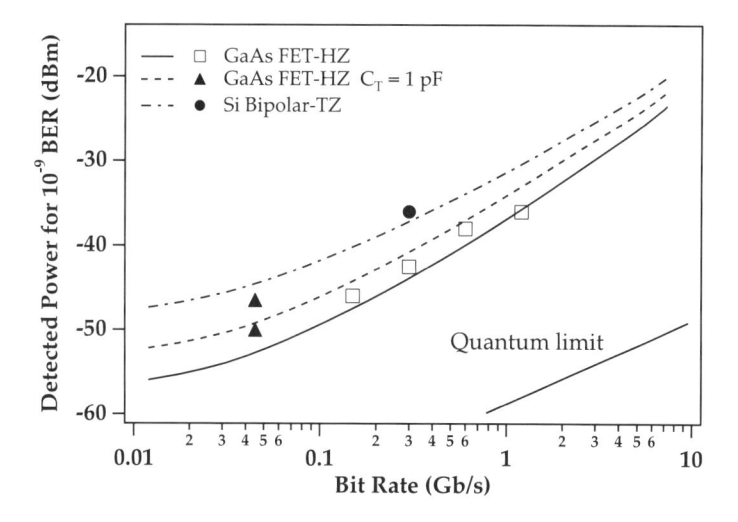

Figure 9.5: Receiver sensitivity for an electronically amplified PIN detector, as a function of signal transmission bit rate. From reference [1]. The three curves shown are for three receiver designs (HZ: high impedance design; TZ: transimpedance design). The quantum limit of 10 photons/bit refers to detector shot noise limit.

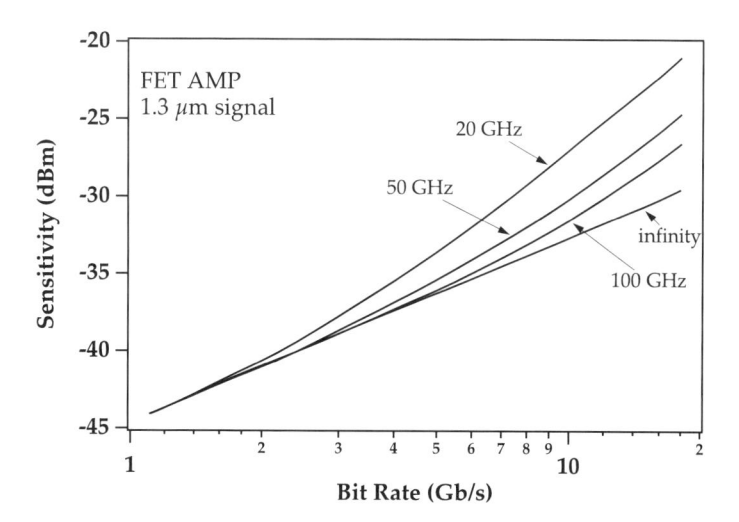

Figure 9.6: Receiver sensitivity for an APD detector for 1.3 μm signal wavelengths, as a function of signal transmission bit rate. From reference [2] (©1987 IEEE). The different curves are for gain × bandwidth products equal to 20, 50, and 100 GHz, and infinity.

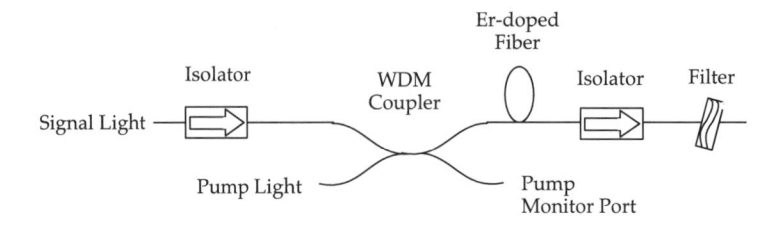

Figure 9.7: Schematic of a a possible configuration for a single-stage erbium-doped fiber preamplifier.

Year	Bit Rate (Gb/s)	Received Power (dBm)	Sensitivity (photons/bit)	Signal Wavelength (nm)	Reference
1989	1.8	-43	215	1531	[4]
1990	2.5	-49	152	1537	[9]
1991	0.6	-46.5	289	1548	[10]
1991	2.3	-40.7	186	1548	[10]
1991	10	-37.2	147	1536	[8]
1992	2.5	-41	247	1558	[11]
1992	2.5	-43.5	137	1530	[6]
1992	10	-36	193	1530	[6]
1992	10	-38.8	102	1536	[12]
1992	2.3	-43.3	92	1534	[13]
1993	5	-40.5	135	1534	[14]
1994	20	-31.6	270	1552	[15]
1996	10	-40.1	76	1547	[16]

Table 9.1: Experimental results for the performance of erbium-doped fiber preamplifiers. The receiver sensitivities and received powers are for 10^{-9} BER.

fiber as possible. We show in Figure 9.7 a sketch of one possible configuration erbium-doped fiber preamplifier. By reducing the input losses—including splices, connectors, isolators, and WDMs— the noise figure can be improved.

Early experiments established that the erbium-doped fiber amplifier was capable of supporting high bit rates (in the Gb/s range) and was effective as an optical preamplifier. The first such experiment, with an argon laser pumped erbium-doped fiber amplifier, showed a 6 dB enhanced receiver sensitivity for an amplifier with 17 dB gain at a data rate of 2.5 Gb/s and a signal wavelength of 1.53 μm.[4] Subsequent experiments demonstrated high-performance preamplifiers, pumped by large frame lasers at 980 nm and 1480 nm and by diode lasers operating at these wavelengths.[5, 6, 7, 8] The receiver enhancement has greatly improved with the advances in the amplifier performance. Table 9.1 shows results from recent years.

Considering first single-stage amplifiers, a sensitivity of 137 photons per bit (−43.5 dBm) was obtained at a bit rate of 2.5 Gb/s and a signal wavelength of 1.53 μm using a 980 nm pumped erbium-doped fiber preamplifier.[6] This sensitivity is defined at the input to the preamplifier. The difference between this sensitivity and the quantum limit of 43 photons per bit arises from the preamplifier input coupling loss, excess optical and electrical bandwidth, a nonideal transmitter extinction ratio, and receiver thermal noise.

More recently, preamplifiers have been demonstrated that combine very high gains with low noise figures.[14] These amplifiers are based on a two-stage design where two independently pumped erbium-doped fibers are separated by an isolator. In this way, the backward ASE originating from the second part of the fiber does not travel backward and saturate the gain of the first segment. As a result, much higher gains and lower noise figures are available at the signal wavelength. These multistage amplifiers and their design were discussed in section 8.3.3. As an example, a record sensitivity of 76 photons per bit (-40.1 dBm average received power) was obtained at a bit rate of 10 Gb/s using such a two-stage erbium-doped fiber preamplifier.[16] Each stage was pumped in a counterpropagating fashion by a 980 nm diode laser.

Figures 9.8 and 9.9 show the receiver characteristics for a 5 Gb/s experiment, where a two-stage erbium-doped fiber preamplifier was used.[17] Figure 9.10 shows the received optical power necessary for an adequate bit error rate as a function of optical bandwidth. In these figures, NF denotes the noise figure of the erbium-doped fiber preamplifier, η_{in} is the coupling loss into the optical preamplifier, B_o and B_e are the optical and electrical bandwidths, respectively, ϵ is the extinction ratio for the NRZ pulse pattern, I_c is the circuit noise spectral density, and G_{eff} is the gain of the erbium-doped fiber preamplifier minus optical loss of components and fiber between the input to the preamplifier and the PIN receiver. G_{eff} is varied in this experiment by means of an optical attenuator placed before the receiver. The signal wavelength is 1534 nm. The experimentally obtained points fit the theory outlined in Chapter 7 quite well.

9.2.2 Inline Amplifiers - Single Channel Transmission

Single Channel Transmission Issues

Inline amplifiers are typically used in long distance transmission links. In such transmission links the most straightforward means for data encoding is intensity modulation with direct detection by the receiver (IM/DD) format. The intensity modulation can be, in its simplest forms, patterns of ones and zeros for digital transmission. The most common transmission format is NRZ (Non Return to Zero) where a long succession of sequential 1s or 0s, in a random pattern, corresponds to a constant average optical power level.

A number of issues arise in single channel transmission that influence strongly the design of the transmission link; the type of transmission fiber used, the design of the amplifiers, and the passive components used throughout the link. Many of the performance degradation issues are related to the onset of fiber nonlinearities. Nonlinearities have been discussed in Chapter 7 (section 7.3.6). While erbium-doped fiber amplifiers offer many advantages over traditional electronic repeaters, one drawback (except in the case of soliton transmission) is that they offer a fertile ground for nonlinearities to develop. This is due to the long interaction length that different optical pulses have in

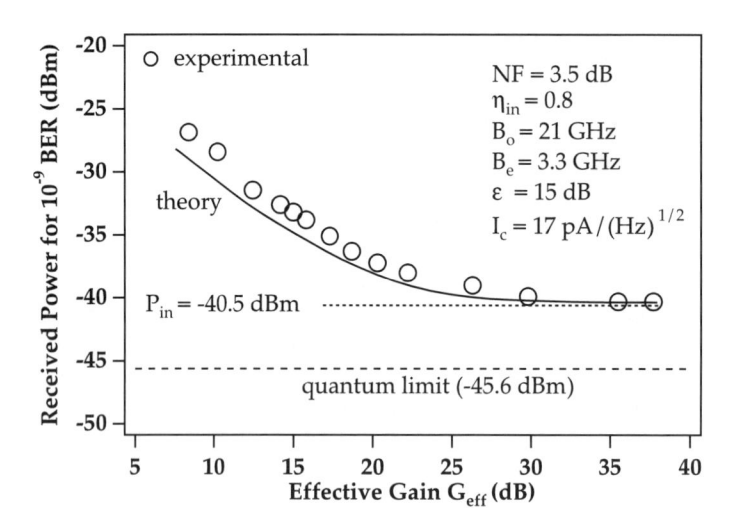

Figure 9.8: Received optical power at 1534 nm for 10^{-9} bit error rate as a function of amplifier gain for a 5 Gb/s optical preamplifier and receiver (parameters as defined in the text). From reference [17] The preamplifier quantum limit corresponds to 43 photons per bit.

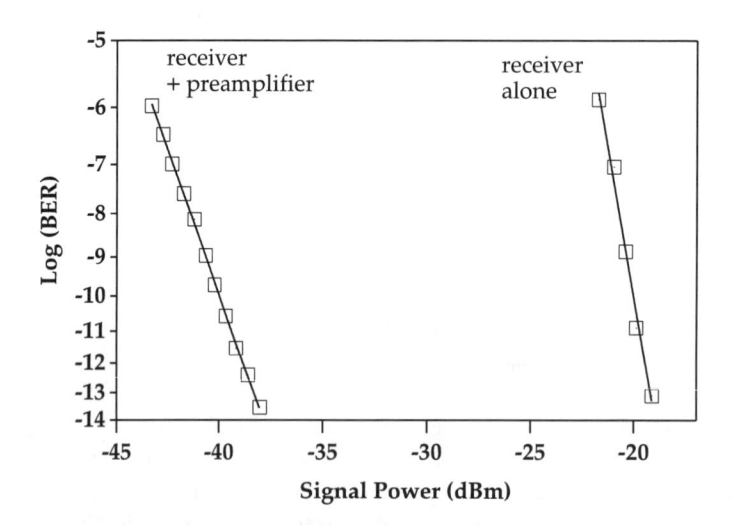

Figure 9.9: Bit error rate as a function of received optical power for a 5 Gb/s optical preamplifier and receiver, and for a receiver alone. The preamplifier offers a performance improvement of 18 dB. The parameters are as in Figure 9.8. From reference [17].

9.2. SYSTEM DEMONSTRATIONS AND ISSUES 329

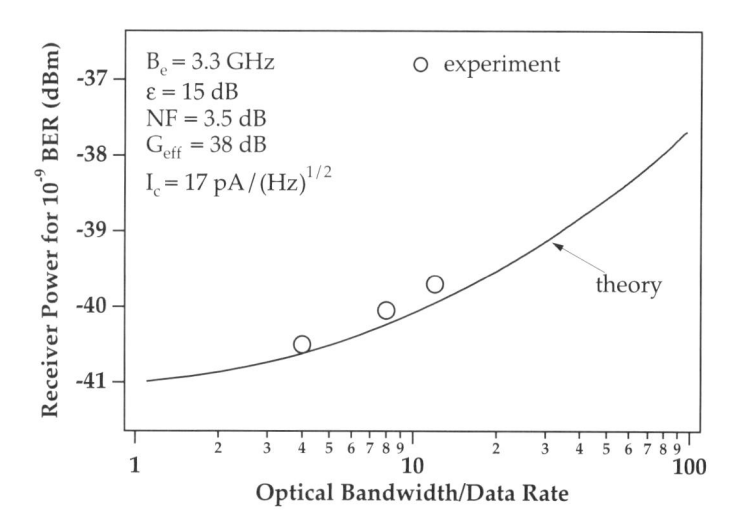

Figure 9.10: Received optical power for a 10^{-9} bit error rate as a function of optical bandwidth for a receiver with optical preamplification. Parameters as defined in the text. From reference [17].

an optically amplified link as well as the higher optical powers. It is a logical thought to design the transmission system so that the zero dispersion of the fiber coincides with the signal wavelength, since in that case the pulse will suffer no temporal distortion from GVD (group velocity distortion). However, this results in a tremendous increase in the strength of the nonlinear interactions. As a result, a major part of the experimental work in demonstrating high capacity transmission has consisted of developing techniques for transmitting at a wavelength other than that of zero dispersion of the link. Figure 9.11 shows the various nonlinearities and system degradation causes which occur both in single channel and WDM transmission. A WDM system will face all the issues of a single channel system as well as WDM specific problems. While it is desirable to have a high signal strength so as to ensure a good SNR at the receiver, once the power is too high, nonlinearities begin setting in, degrading system performance. In practice one uses as high a power as possible while staying below the onset of any nonlinearities generated by the choice of fiber and transmission method.

The first major problem which can arise in an amplified link is four-wave mixing between the optical pulse and the ASE from the amplifiers.[18, 19] When the two fields propagate together, as is the case when the signal wavelength is at the zero dispersion of the transmission fiber, then significant spectral broadening and distortion of the pulse occurs. Effectively, most of the signal pulse energy gets absorbed in optical filters used in the system as the edges of the broadened signal spectrum get clipped. Even without filters, the erbium-doped fiber amplifier bandwidth acts as a filter. Increasing the filter bandwidth of the optical filters in the system only serves to increase the amount of ASE accepted by the receiver and hence to increase the noise. The simplest way to eliminate the nonlinear coupling between the signal and the ASE is to operate the link away from zero dispersion, by at least a few tenths of ps/km·nm.[18] Both an

```
Single Channel & WDM                          WDM Specific

ASE noise
Chromatic Dispersion
                                              FWM
signal SPM+GVD      Transmission Degradation
                       Spectral Distortion    SRS
FWM: signal+ASE        Temporal Distortion
                            Crosstalk
SBS                                           XPM

Polarization effects
```

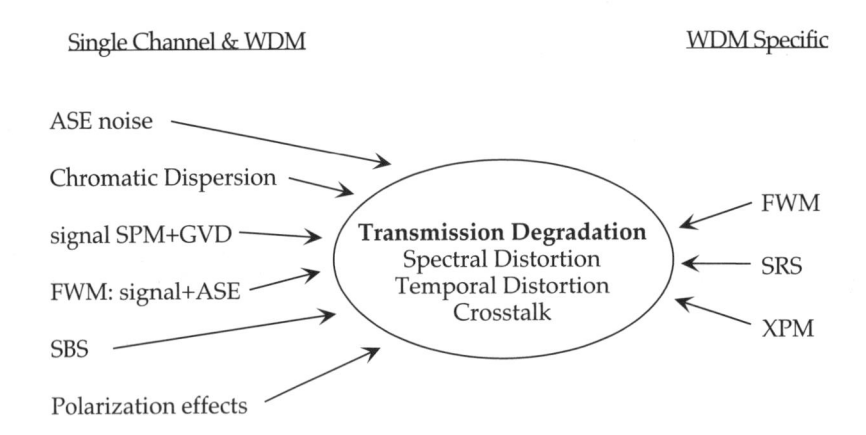

Figure 9.11: Performance degradation problems faced by high capacity single channel and WDM transmission systems (ASE: amplified spontaneous emission; SPM: self phase modulation; GVD: group velocity dispersion; FWM: four wave mixing; SBS:stimulated Brillouin scattering; SRS: stimulated Raman scattering; XPM: cross phase modulation). These problems arise from the presence of nonlinearities in optically amplified links as well as noise from amplifier ASE.

alternating dispersion map as well as a sawtooth dispersion map (long lengths of fiber with slight dispersion followed by a short length of high dispersion fiber) are effective in accomplishing this. These dispersion maps are sketched in Figure 9.12. We will find later that these dispersion maps are also very useful in reducing four wave mixing between channels.

SPM (discussed in section 7.3.6) will cause pulse broadening if the optical power of the pulse is sufficiently high. Coupled with GVD this will create temporal pulse distortion if a large amount of dispersion is accumulated over the entire system length.[18, 20, 21] By ensuring that the total dispersion of the transmission link is near zero, this problem can be significantly reduced as long as the spectrum of the pulse has not been broadened so much that the edges are clipped by optical filters present in the link (e.g. from the amplifier bandwidth itself). For this reason, it is important in long haul systems to ensure that any left-over dispersion for the signal channel after it has traversed the entire system is either pre-compensated or post-compensated by a length of dispersive fiber in the appropriate amount. A careful choice of the transmission fiber used mitigates many of the nonlinear problems which can arise in single channel optically amplified systems.[22]

Since 1989 many experiments have been conducted that have established the very significant advantage of inline optical amplifiers for long haul transmission systems. These experiments continue to push back the frontiers in terms of transmission distances achieved, amplifier span lengths, bit rate, and number of channels transmitted. The experiments are conducted in either of two formats: loop experiments or straight line experiments. Table 9.2 lists some of the many single-channel long distance transmission experiments that have been performed to date with erbium-doped fiber amplifiers. By 1996, the bulk of the effort to construct high capacity long haul transmission

9.2. SYSTEM DEMONSTRATIONS AND ISSUES 331

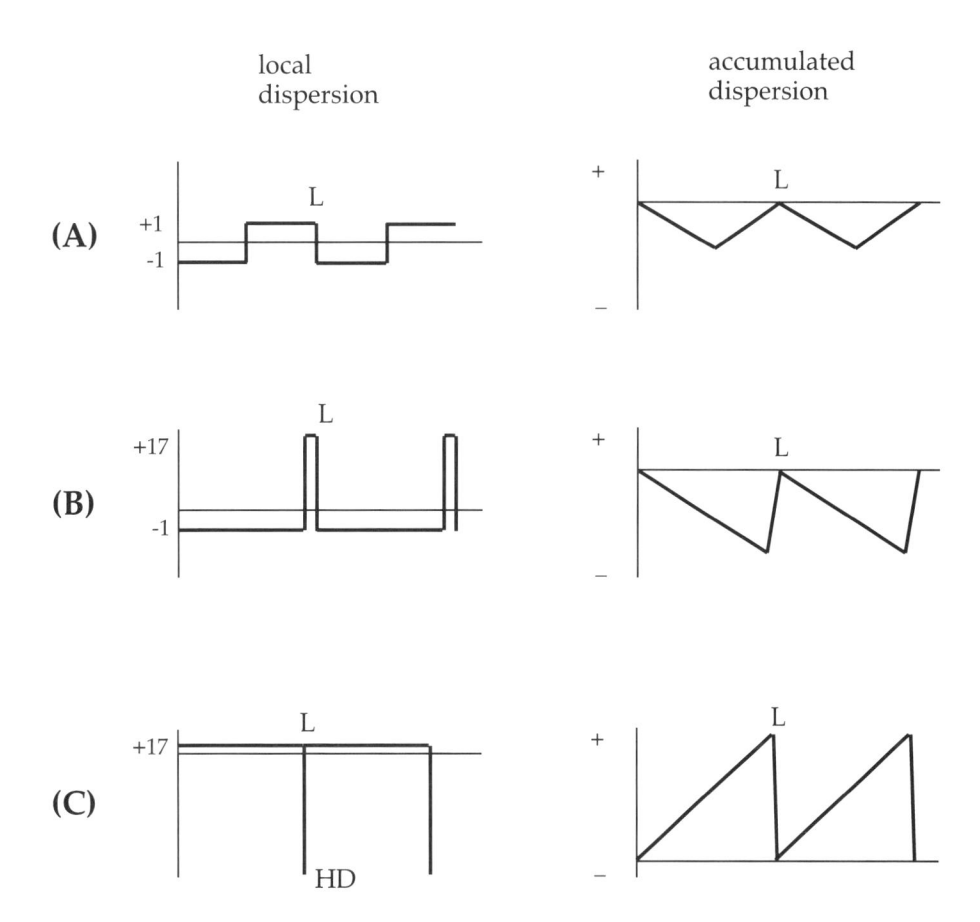

Figure 9.12: Examples of typical transmission fiber dispersion maps for high capacity long haul transmission links that strongly reduce nonlinear coupling between signal and amplifier ASE, as well as between channels in a WDM transmission system. The link is constituted of multiple spans of length L. Map (A) is an alternating dispersion map which was first used in single channel transoceanic transmission. With higher values of dispersion (as with TrueWave™fiber) this map is also suited for WDM transmission. The alternating dispersion map has also been used with positive dispersion fiber preceding the negative dispersion fiber. Map (B) is known as the sawtooth dispersion map and uses primarily one type of non zero dispersion shifted fiber with a short length of conventional fiber to reduce to zero the accumulated dispersion. For WDM systems the non zero dispersion shifted fiber usually has higher absolute values of dispersion. Map (C) uses conventional single mode fiber with periodic compensation by a high dispersion (HD) element such as dispersion compensating fiber or fiber gratings.

Year	Bit Rate Gb/s	System Length (km)	Capacity (Gb/s · km)	Signal (nm)	Num. Amps.	Amp. Spacing (km)	Ref.
1989	1.2	267	320	1536	2	77	[23]
1989	11	151	1,660	1536	1	70/81	[24]
1990	2.4	710	1,704	1536	9	50-78	[25]
1990	0.56	1,028	576	1536	10	72-103	[26]
1990	10	505	5,050	1552	4	90-115	[27]
1991	2.5	2,223	5,557	1554	25	80	[28]
1991	5	14,300	71,500	1550	357	40	[29]
1991	2.4	21,000	50,400	1550	525	40	[30]
1992	2.5	10,073	25,180	1552	199	50	[31]
1992	10	4,550	45,500	1558	138	33	[32]
1992	5	9,000	45,000	1558	274	33	[33]
1992	2.5	4,520	11,300	1550	48	80-100	[34]
1993	10	9,040	90,400	1559	274	33	[35]
1994	10	360	3,600	1552	2	120	[36]
1995	2.5	9,720	24,300	1552	108	90	[37]
1995	10	6,480	64,800	1552	72	90	[37]
1995	40	140	5,600	1552	1	70/70	[38]
1995	100	500	50,000	1553	11	40-55	[39]
1996	5	15,200	76,000	1559	336	45	[40]
1996	10	8,460	84,600	1559	124	68	[41]

Table 9.2: Selected single wavelength transmission system experiments involving erbium-doped fiber amplifiers (Signal (nm): signal wavelength (nm); Num. Amps.: number of amplifiers; Amp. spacing: amplifier spacing).

systems had shifted in focus to WDM systems.

Loop Experiments

Loop experiments are a convenient way of simulating a long distance transmission experiment. The signal is injected into a loop of length L and after it has completed R round-trips it is switched out from the loop; thus it travels an equivalent distance RL. This is often cheaper and easier than purchasing and setting up a straight line experiment with RL length of fiber and a commensurate number of inline amplifiers. However, one must be cautious about loop experiments. Effects might appear that would not do so in an equivalent straight line experiment, since effects (due to dispersion, for example) are averaged over a much shorter length of fiber and a lesser number of amplifiers. In a loop, one does not have the natural statistics of the fiber parameters that one would have in a real system. Hence a loop experiment will tend to emphasize either the very good or the very bad attributes of the fiber used in the loop. Fiber loop experiments were first used by Mollenauer and coworkers to study the long distance transmission of Raman amplified solitons.[42]

9.2. SYSTEM DEMONSTRATIONS AND ISSUES

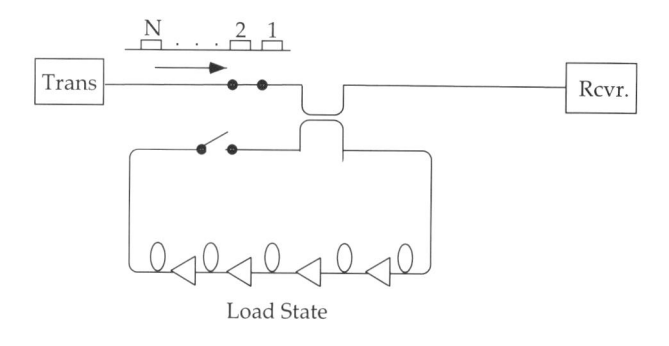

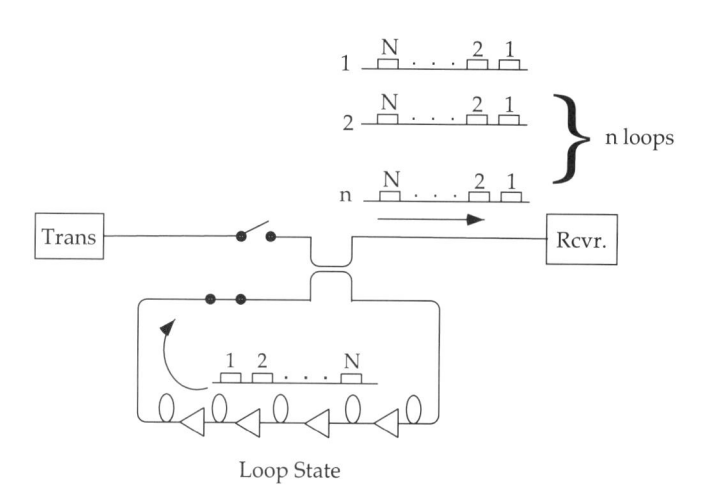

Figure 9.13: Experimental setup of a loop experiment. The top figure depicts the loading of the loop by a sequence of pulses. The lower figure depicts the loop in operation and producing as output the sequence of pulses after each successive traversal of the loop.

Loop experiments simplify the optics aspect of the transmission experiment; however, the electronics become commensurately more complex. The main issue in loop experiments is the timing synchronization, especially when measuring bit error rates. A typical loop setup is shown in Figure 9.13. One of the main requirements in a loop experiment is that the output of the loop needs to be synchronized with the bit error rate test equipment. When closed, the loop switch, typically a lithium niobate switch, loads the initial data train into the loop. A 3 dB coupler allows for data patterns to be loaded in and also lets them exit the loop after each round-trip. The switching in and out of the data trains needs to be synchronized both with the loop transit time and the bit error rate test set clock. The loop round-trip time needs to be an integer multiple of the length of the pattern of bits loaded, to within a fraction of a bit period. The bit error rate after transmission of varying distances can be measured by using the data patterns exiting

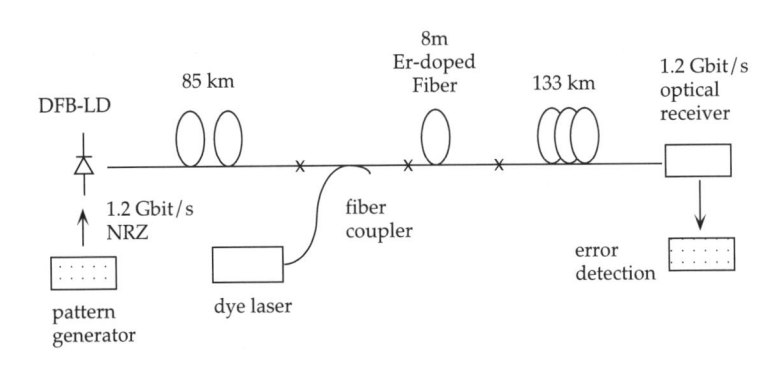

Figure 9.14: Experimental setup of an early straight line transmission experiment involving inline erbium-doped fiber amplifiers at 1.2 Gb/s over 218 km. From reference [44].

the loop after the desired number of round-trips, so that any transmission degradation with distance can be observed.[43]

High-Capacity Single Channel Transmission Results

The first demonstration of the effective use of an inline optical amplifier for a long haul transmission system was done by Edagawa and coworkers in both a laboratory setting and an actual submarine field trial.[23, 44] The first laboratory demonstration was done at bit rate of 1.2 Gb/s, over a distance of 218 km, with a single erbium-doped fiber amplifier repeater, as shown in Figure 9.14.[44] The amplifier provided a gain of 20 dB, thus allowing the transmission distance to be extended by more than 100 km. The bit error rate measurements demonstrated that the amplifier provided no penalty; the error performance was the same at both 0 km and 218 km distances. This first paper on long distance transmission applications of erbium-doped fiber amplifiers also noted the self-filtering effect of these amplifiers, a fact which several years later was found to be an important characteristic of transoceanic transmission systems. At the same time, an actual submarine field trial conducted over 94 km of fiber optic cable, with an amplifier pumped by a 1.47 μm diode laser, established for the first time the practicality of fiber amplified systems and spurred an interest in these systems that has continued unabated to date.

Further extensions of the straight line experiment soon brought transmission distances to near 900 km, using twelve inline optical amplifiers. These early experiments established that to overcome the ASE noise from amplifiers, that accumulates in long distance systems, it is necessary to keep the signal level high and not let it fall too low (i.e., less than -20 dBm) after traversing each span. The amplifiers are usually operating in saturation, with output powers on the order of 0 to 10 dBm. Higher output powers can lead to nonlinear effects. The signal power needs to be maintained in a narrow range to be strong enough to overcome ASE noise yet not strong enough to generate nonlinearities.[34]

Loop experiments accomplished by Bergano and coworkers in 1991 established the feasibility of transmission over transoceanic distances using erbium-doped fiber

9.2. SYSTEM DEMONSTRATIONS AND ISSUES 335

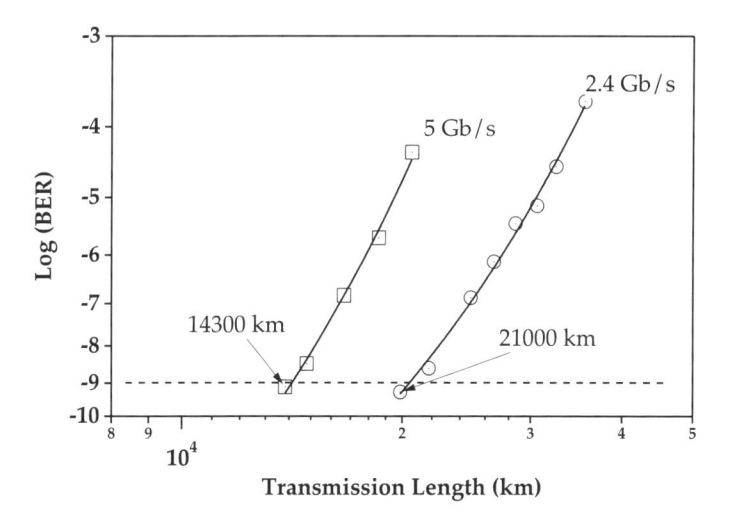

Figure 9.15: Measured bit error rate versus transmission distance in a loop experiment. From reference [30]. The dashed line indicates the 10^{-9} BER limit.

amplifiers.[30] The experiment demonstrated a 9000 km and 5 Gb/s system as well as a 21,000 km 2.5 Gb/s system, using standard NRZ pulses, for a bit error rate of 10^{-9}. In this experiment the amplifier spans were 40 km, and the amplifiers were counterpumped at 1480 nm. Figure 9.15 shows the bit error rate achieved with this loop experiment as a function of transmission length and bit rate. Straight line experiments soon confirmed the loop experiment results and established that multi-Gb/s transmission over transoceanic distances with inline erbium-doped fiber amplifiers was feasible. The first reported experiment, a collaboration between AT&T and KDD, used 274 erbium-doped fiber amplifiers operating at 5 Gb/s and accomplished error-free operation over 9000 km.[33] The amplifier spacing was 33 km. Each amplifier was counterpumped with 20 mW of 1480 nm diode power and provided a net saturated gain of 7.5 dB with a noise figure of 6 dB. The signal power after each amplifier was 2 mW. The signal wavelength was 1558.2 nm, which was the peak of the gain spectrum of the cascaded system of amplifiers and fiber spans. 10 Gb/s experiments were also reported at the same time, over a distance of 4500 km.[32]

Recent experiments have continued to push back the technical frontiers in terms of bit rates, distances, and number of channels involved in the transmission experiments, while using the standard linear NRZ format for pulse information encoding. Overall capacity improvements have been achieved mainly by increasing the number of channels transmitted, using the WDM technique.[45, 46]

9.2.3 Inline Amplifiers - WDM Transmission

WDM Transmission Issues

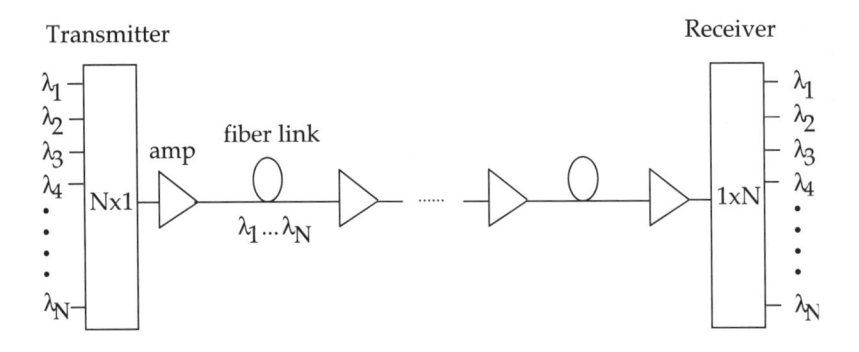

Figure 9.16: Schematic of a generic WDM (wavelength division multiplexed) system.

Wavelength division multiplexing (WDM), used to increase the capacity of an optical communication system, has led to large increases in the transmission capacity of fiber optic transmission systems. Several wavelength channels, each operating at some specific bit rate, are collected together and transmitted along the same optical fiber, as shown in Figure 9.16.

The single channel capacity can be limited, for example, by available amplifier output power since signal to noise considerations will govern the maximum bit rate attainable for a given transmitted power. For an amplifier output power limited system, WDM does not add capacity since the output power then becomes distributed over several channels rather than a single channel, so that the bit rate needs to be reduced to maintain a specific signal to noise ratio. If sufficient power exists, then adding more wavelength channels becomes a simple way to increase transmission capacity. WDM is a fruitful avenue to pursue if dispersion effects limit the possibility of increasing the bit rate of a single channel system or if one desires to keep the same electronics as those of a current single-channel system operating at a given bit rate. WDM has recently become a subject of great interest as long haul systems–first established for a single channel–migrate to WDM for greater capacity. New problems arise and need to be dealt with in the case of WDM systems. For example, nonlinear interactions between channels will degrade their individual bit error rates. These nonlinearities were discussed from a theoretical standpoint in Chapter 7 (section 7.3.6).

Transmission systems specifically designed for single channel operation are usually not easily upgradeable to WDM operation. Several of the first optically amplified undersea links installed for single channel operation have been demonstrated to be upgradeable to two or perhaps three channel operation.[47] TAT-12 has already been upgraded to two-channel operation (as will be TPC-5) and will be upgraded to three-channel operation. Many of the features which are attractive for single channel operation, for example self regulation of the amplifiers or a fiber dispersion as low as possible, do not have as significant an advantage in the case of a WDM system. Power self regulation of the amplifiers when operating in compression operates on the total optical power in the amplifier and cannot regulate the power on a channel by channel basis in the case of WDM systems. In addition, amplifiers operating in compression tend to exhibit strong gain peaking. While this feature is attractive in filtering out ASE

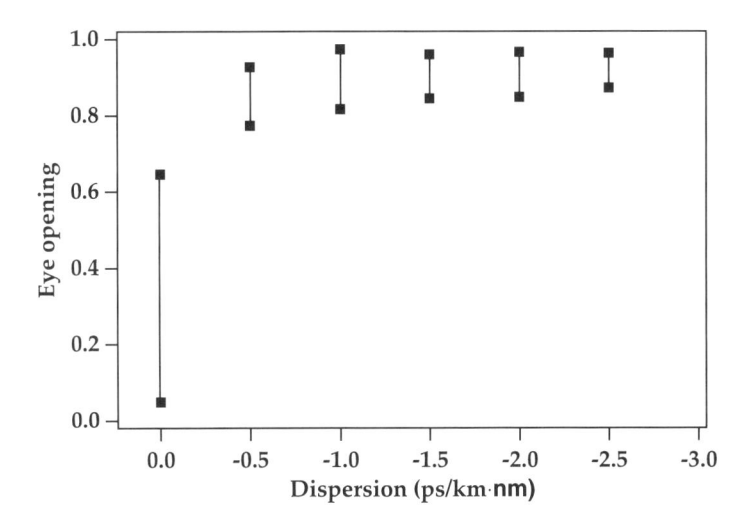

Figure 9.17: Range of fractional eye openings in an eight-channel WDM system operating at 10 Gb/s with a constant channel spacing of 200 GHz and an average power of +6 dBm per channel, as a function of the dispersion of the transmission fiber (at the center of the signal band). The transmission fiber is periodically compensated by lengths of conventional fiber with positive dispersion placed before each amplifier. From reference [48] (©1995 IEEE).

in the case of single channel systems, it prevents the kind of flat amplifier operation which needed for multiple channel WDM. To eliminate this problem, passsive equalizing filters can be used within the amplifier, allowing operation in saturation. Finally for WDM systems with a large number of channels and a small channel spacing (less than 200 GHz) a larger dispersion (at least ±1 to ±2 ps/km · nm) is needed to reduce the effects of four wave mixing between channels.

A key to achieving error-free WDM transmission is to find an effective way to combat the four-wave mixing process. For single-channel experiments, one tends to use dispersion-shifted fiber to minimize pulse-broadening effects. This has a deleterious effect for WDM systems as it allows the different channels to travel together, with minimum walk-off from each other, thus reinforcing any nonlinear interaction between them. The solution for WDM systems is to use fibers with different dispersions so that locally the dispersion is rather large, while overall the dispersion is near zero. By offsetting the zero dispersion point of the fiber relative to the wavelengths of the signal channels, the four-wave mixing can be reduced by the increased walk-off between channels. Numerical solutions have shown that dispersions in the range 1 to 2 ps/km·nm in absolute value are suitable for channel spacings of 200 GHz with average powers per channel of a few dBm.[48] Figure 9.17 shows the range of fractional eye openings for all the channels obtained when the channels are operated at 10 Gb/s, as a function of the dispersion of a transmission fiber with negative dispersion, from numerical simulations. The link length was 360 km with amplifiers spaced by 120 km. Prior to each amplifier, a length of conventional fiber with large positive dispersion reduces

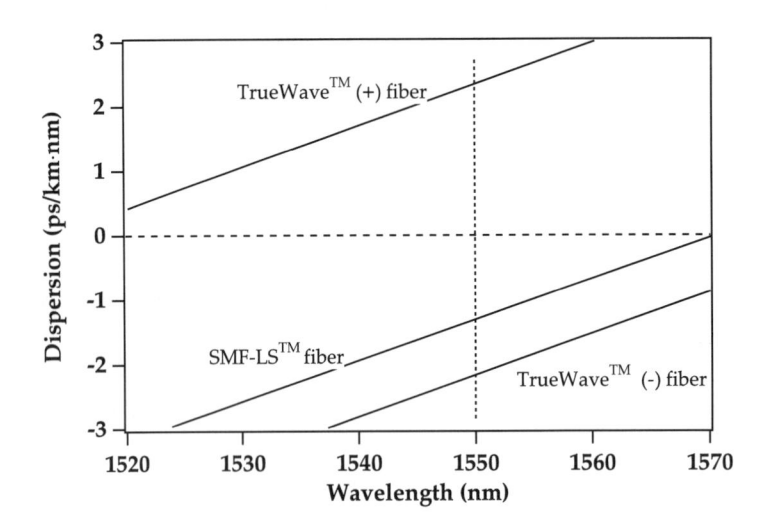

Figure 9.18: Dispersion of several commercial non-zero dispersion shifted fibers. These fibers offer a moderate amount of dispersion near 1550 nm, so as to combat the four wave mixing effect in WDM systems. TrueWave™ fiber is fabricated by Lucent Technologies and SMF-LS™ by Corning, Inc.

the overall span dispersion to zero. The degradation in performance at zero dispersion is rapidly reduced as the dispersion of the fiber is increased to the 1 ps/km · nm range. Transmission fibers have been developed which have small values of either positive or negative dispersion near 1550 nm, so as to reduce the four wave mixing effect. Such fibers are sometimes referred to as non-zero dispersion shifted fibers (NZ DSF). These fibers consitute a third class of transmission fiber, beside the conventional single mode fibers with dispersion zero near 1300 nm, and the dispersion shifted fibers with dispersion zero at 1550 nm (sometimes referred to as Z DSF). Figure 9.18 shows the dispersion parameters of commercially available fibers for use as non zero dispersion shifted fibers. Such transmission fibers are widely used for WDM system applications. An alternative method for dispersion management is to use strongly dispersive fiber, such as conventional single mode fiber, and periodically use dispersion compensating fiber.

Since different wavelength channels will experience a different total accumulated dispersion after traversing the entire system length, at most one channel will end up with an overall accumulated dispersion of zero after traversing a dispersion managed system. Channels well removed in wavelength from this central channel will have significant accumulated dispersion. For long haul systems, this implies potential temporal pulse distortion from the combination of SPM and GVD. Thus, there needs to be appropriate dispersion compensation on a channel by channel basis. Such dispersion compensation can be accomplished with highly dispersive fiber, such as conventional fiber with dispersion zero near 1300 nm or dispersion compensating fiber. Both precompensation and postcompensation are used to zero out the accumulated dispersion on a channel by channel basis.[49]

9.2. SYSTEM DEMONSTRATIONS AND ISSUES

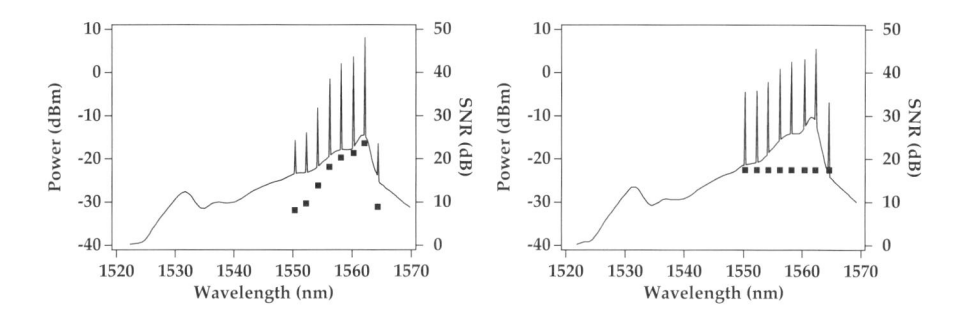

Figure 9.19: Calculated output powers (solid curve) and SNRs (squares) for an eight-channel (2 nm spacing) WDM system after transmission through an 840 km system with 70 km fiber spans separated by erbium-doped fiber amplifiers, for unequalized (left) and equalized (right) launch conditions. From reference [50] (©1992 IEEE).

Another issue with WDM networks is that they are subject to being constantly reconfigured as the nature of the signal traffic in the system changes—for example, in the case of a local loop system. When channels are added or dropped, the resulting increase or decrease in the power spectrum will alter the gain profile of the erbium-doped fiber amplifier. This happens because the erbium energy level populations change as a result of the change in the stimulated transition rates arising from the channels being reconfigured, and the gain spectrum depends on these populations in a very sensitive way. This effect is accentuated by the fact that the erbium-doped silica fiber amplifier is predominantly a homogeneously broadened amplifier, so that the presence or absence of a given wavelength signal will affect the gain of a signal at another wavelength.

Amplifier flatness is also a critical issue for WDM systems. This was discussed in Chapter 8 in the context of amplifier architectures for specific system applications. In an ideal world, optical amplifiers along the signal path need to amplify each channel equally; that is, the gain profile needs to be flat. When one channel is preferentially amplified at the expense of another and this effect is cumulated along many amplifier spans, degradation of the WDM system's information-carrying capacity will result.

For erbium-doped silica fibers, the flattest gain profile observed is that of the alumino-silica type in the 1550 to 1560 nm gain region. Nevertheless, after a cascaded series of amplifiers, large differences in transmitted output power appear among the different signals. Figure 9.19 shows the transmitted spectrum and SNR of eight channels traversing a cascade of thirteen erbium-doped alumino-silica fiber amplifiers for two cases: first where the launched powers are all equal and second where they are adjusted to make the SNRs equal at the output (this is known as the pre-emphasis method). The adjustment of the input powers equalizes the SNRs at the output of the amplifier cascade.[50] The pre-emphasis method presupposes that the system characteristics, such as signal wavelengths and incoming signal powers, either are fixed ahead of time or are monitored by a servo loop. This can be a limitation if the system is to be used in the wider context of a local loop type architecture where the set of signals can be changing due to varying traffic patterns. The most optimum situation for the

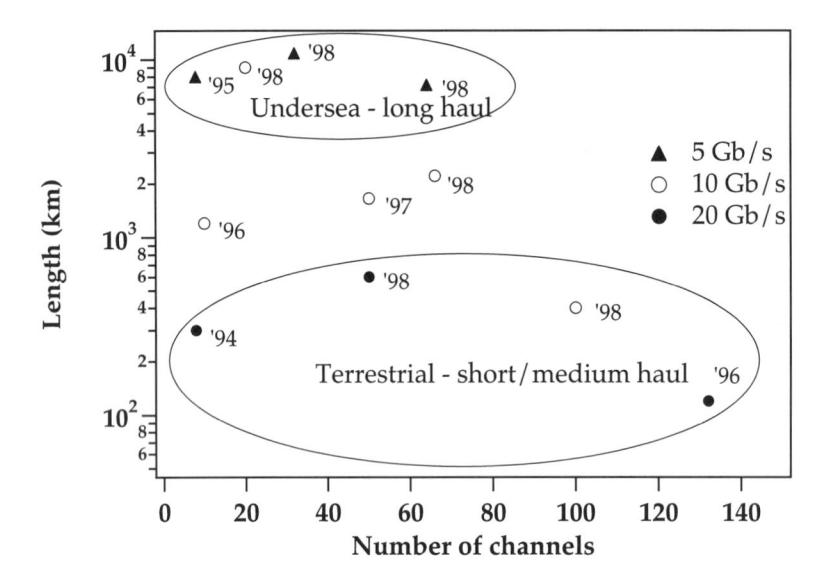

340 CHAPTER 9. SYSTEM IMPLEMENTATIONS OF AMPLIFIERS

Figure 9.20: Selected WDM transmission system demonstrations involving erbium-doped fiber amplifiers. The year of each demonstration is indicated next to the point and the individual channels are modulated at 5, 10, and 20 Gb/s as indicated. The two regimes of system applications are undersea long-haul and terrestrial short and medium haul.

pre-emphasis method is that of a point-to-point system where the number of channels is kept constant.

Advanced amplifier architectures are a more elegant way of solving the problem addressed by the pre-emphasis method. In recent years, fiber component technology progress has led to much flatter amplifiers (see section 8.3.6), which have in turn led to dramatic increases in the tranmission capacity of WDM systems.

WDM Transmission Results

In this section, we will discuss some of the system results that have been achieved with WDM systems, and the tactical decisions made by the system designers to achieve these results. The increase in capacity of transmission links using optical amplifiers has been achieved primarily by adding wavelength channels. In the relatively short time span of a few years, transmission achievements with optically amplified links have gone from single channel to over 100 channels transmitted. Figure 9.20 shows examples of transmission distances achieved using systems with multiple channels. Two applications are of interest for WDM systems, and are outlined on the figure. These are long-haul systems suitable for transoceanic transmission, and short to medium haul transmission systems for terrestrial applications. The higher bit-rate and higher channel capacity systems are understandably easier to implement in shorter length systems, for signal to noise reasons, as observed from Figure 9.20.

9.2. SYSTEM DEMONSTRATIONS AND ISSUES 341

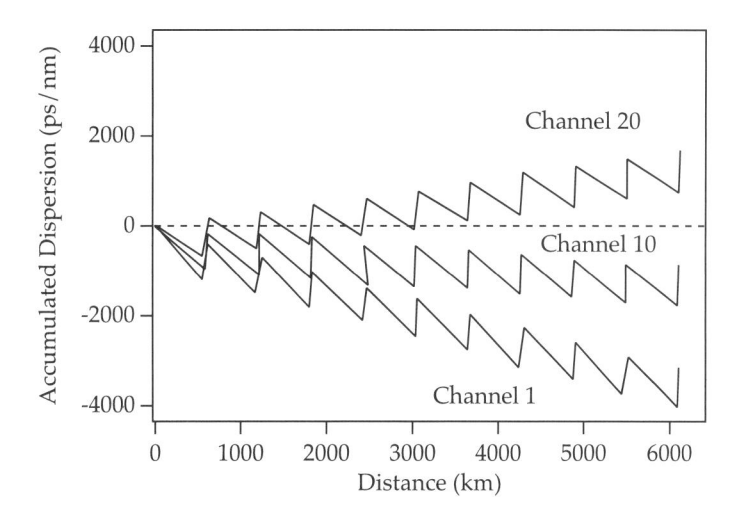

Figure 9.21: Accumulated chromatic dispersion versus transmission distance for three channels of a twenty-channel WDM transmission system. From reference [51] (©1995 IEEE). The nonzero local dispersion serves to minimize four-wave mixing between channels while the overall dispersion management prevents excessive dispersion over the entire fiber span.

Table 9.3 lists selected multiple wavelength channel experiments performed with inline erbium-doped fiber amplifiers. The progress made in a relatively short time frame is quite significant, both in terms of distances achieved as well as number of channels.

Early experiments demonstrated 40 Gb/s over 8000 km by transmitting 8 × 5 Gb/s channels, with channels spaced by 0.5 nm.[62] Dispersion management was used in this experiment as both negative dispersion (−2 ps/km·nm) and positive dispersion (+17 ps/km·nm) fiber was used and judiciously placed in the system to negate the interchannel nonlinearities. The average amplifier separation was 45 km, and the total launched power for all eight channels was +2.3 dBm. The average power in each channel was thus only 90 μW.

This result was soon extended to 20 channels operating at 5 Gb/s. The 20 channels were spaced by 0.6 nm from 1550 to 1562 nm, and were transmitted over a distance of 6300 km. This result was obtained by the use of gain-equalizing filters, fabricated with long period fiber grating filters, placed after every three or four amplifiers.[51] As shown in Figure 9.21, alternating spans of fiber–900 km with λ_0 = 1585 nm and 100 km with λ_0 = 1310 nm–were used for the dispersion management. Since the fibers have different dispersion for each wavelength the dispersion map is different for each channel. The equalizing filters increased the bandwidth of the transmission span from 3.5 nm to 11 nm, as shown in Figure 9.22. These filters allowed the transmitted spectrum to stay relatively constant after 6300 km. The transmission length in this experiment was later extended to 9100 km by using a forward error correction encoding protocol.[67] These latter experiments were carried out with 1480 nm pumped erbium-

Year	Num. of Ch.	Total Bit Rate Gb/s	System Length (km)	Total Capacity (Gb/s · km)	Channel Spacing (Å)	Num. Amps.	Amp. Spacing (km)	Ref.
1990	4	9.6	459	4,406	20	5	47-85	[52]
1992	5	5	1,111	11,110	31	21	53	[53]
1993	2	10	4,550	45,500	20	138	33	[54]
1993	8	80	280	22,400	10-20	4	55	[55]
1994	16	40	1420	56,800	8	12	95-123	[56]
1994	8	160	300	48,000	9	5	48	[57]
1994	4	10	9,000	90,000	10	195	46	[58]
1995	17	340	150	51,000	8	2	48-55	[59]
1995	10	100	500	50,000	5-8	10	40	[60]
1995	8	20	275	5,500	15	1	125-150	[61]
1995	8	40	8,000	320,000	5	200	40	[62]
1995	16	160	1,000	160,000	10	27	20-40	[63]
1996	10	100	1,200	120,000	8-24	11	100	[64]
1996	55	1100	150	165,000	6	2	50	[65]
1996	8	40	4,500	180,000	8	69	68	[66]
1996	20	100	9,100	910,000	5.5	198	46	[67]
1996	50	1000	55	55,000	8	1*	-	[68]
1996	10	1000	40	40,000	32	1*	-	[69]
1996	132	2600	120	312,000	2.6	1*	-	[70]
1997	22	110	9,500	1,045,000	4	237	40	[71]
1997	32	160	9,300	1,488,000	4	199	45/64	[72]
1997	32	320	640	204,800	8	8	80	[73]
1997	8	320	240	76,800	8	2	80	[74]
1997	7	1400	50	70,000	47	1*	-	[75]
1997	16	160	6,000	960,000	8	133	45	[76]
1997	24	256	2,526	646,656	8	50	50	[77]
1997	50	533	1,655	882,115	3	33	50	[78]
1997	32	320	500	160,000	8	3	125	[79]
1998	100	1000	400	400,000	4,8	3	100	[80]
1998	50	1000	600	600,000	8	9	60	[81]
1998	32	170	10,850	1,844,500	4	210	52	[82]
1998	20	213	9,000	1,930,632	8	180	50	[83]
1998	64	320	7,200	2,304,000	3	144	50	[84]

Table 9.3: Selected multiwavelength transmission system experiments involving erbium-doped fiber amplifiers (Num. of Ch.: number of channels; Num. amps.: number of amplifiers; Amp. spacing: amplifier spacing; *: power amplifier only). Both long haul systems (total length > 1000 km) suited for undersea applications as well as medium haul systems (total length < 1000 km) suited for terrestrial applications are included.

9.2. SYSTEM DEMONSTRATIONS AND ISSUES 343

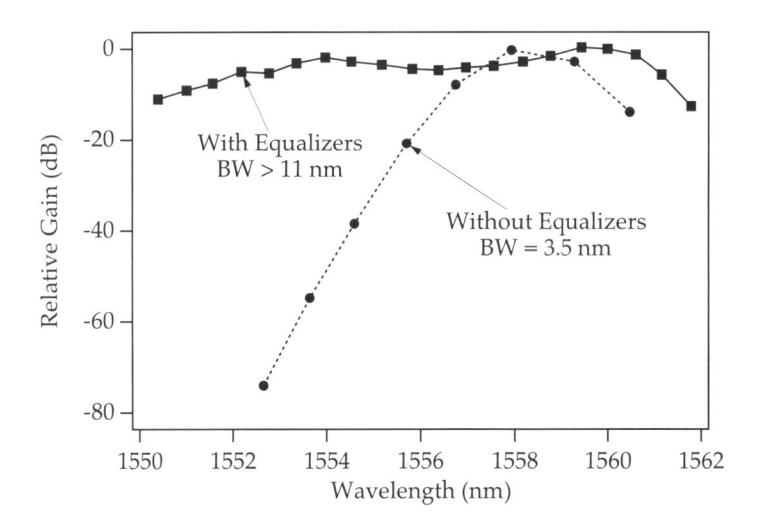

Figure 9.22: Relative gain as a function of transmission wavelength, for a 6300 km amplifier chain, with and without passive gain equalization. From reference [51] (©1995 IEEE).

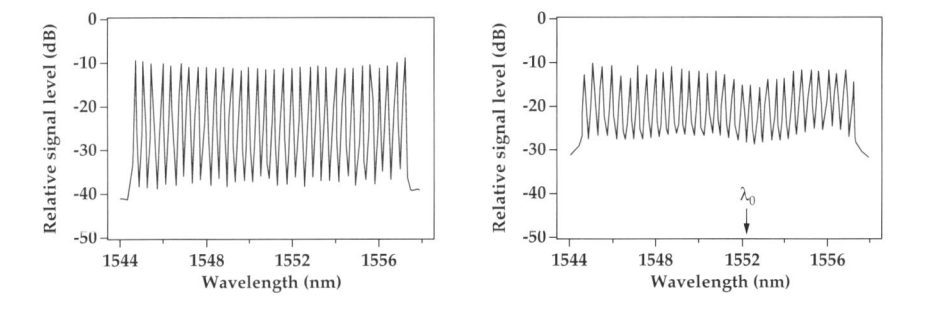

Figure 9.23: Transmitted (left) and received (right) spectra for a 32 channel system transmitted over a 9300 km amplifier chain with passive gain equalization and dual stage amplifiers pumped at 980 nm. From reference [72]. λ_0 indicates the average dispersion-zero of the entire fiber span.

doped fiber amplifiers. Making use of 980 nm pumped amplifiers, transmission of 22 channels at 5 GB/s over 9500 km, without forward error correction, was achieved.[71]

Making use of more sophisticated broadband amplifiers the transmission of 32 channels at 5 Gb/s at distances greater than 9000 km has been demonstrated.[72, 82, 85] One can, for example, make use of long period grating filters and tunable etalon filters to reduce the gain variation to 0.4 dB over the 13 nm of bandwidth used.[72] The received spectrum for the 32 channels transmitted is remarkably well preserved compared to the transmitted spectrum, as shown in Figure 9.23. Further improvements can be achieved by continuing to reduce the importance of nonlinearities. An effective way to accomplish this is through the use of large core transmission fibers (e.g. ≥ 75 μm^2 effective area) where the light intensity is significantly reduced compared to standard dispersion shifted transmission fiber (effective area $\simeq 50$ μm^2).[82, 84] This allows the use of a close channel spacing. In this way, 64 channels at 5 Gb/s have been packed into a 19 nm bandwidth and transmitted over 7,200 km.[84]

Even higher bit rates can be achieved over shorter distances. The 1 Tb/s barrier was broken in 1996.[65, 68, 69] A 1.1 Tb/s transmission system over 150 km was achieved by using 55 wavelength channels spaced by 0.6 nm from 1531.7 nm to 1564.1 nm at 20 Gb/s.[65] The amplifier spacing in this experiment was 50 km with a total launched signal power of 13 dBm. Standard fiber with high dispersion was used, thus mitigating nonlinear mixing effects. The chromatic dispersion was corrected at the receiver end with a single dispersion compensating fiber. In a second experiment, 25 wavelength channels were chosen from 1542 nm to 1561 nm with 0.8 nm spacing, and 50 channels were obtained by using two orthogonal polarizations for each wavelength.[68] Subsequent experiments have made use of even greater numbers of channels to increase the capacity of the system, with 132 channels being reported.[70]

For terrestrial systems with transmission lengths less than 1000 km, DWDM (dense WDM) with a large number of channels has proved itself to be a very effective way to increase the capacity of a fiber link. The system will perform well, i.e., without impairments due to four wave mixing and other nonlinearities, if the fiber is sufficiently dispersive for the band of wavelengths used in the transmission. Hence, conventional fiber links with dispersion zero near 1300 nm will be appropriate for DWDM transmission. Particularly well suited to this application as well are the non zero dispersion shifted fibers with moderate values of positive or negative dispersion (e.g., ± 2ps/km $\cdot$ nm). Densely packed channels can be transmitted, even at very high bit rates.[86] Amplifier spacings can be larger than in undersea long haul systems, given the reduced overall system length. Coupled with the use of gain flattened amplifiers, a large number of channels can be transmitted. As an example, it has been demonstrated that thirty two channels with spacings of 0.8 nm can be transmitted over a distance of 640 km with an amplifier spacing of 80 km.[73] With conventional fiber (highly dispersive at 1550 nm), the use of dispersion compensating fiber or dispersion compensating fiber gratings allows for high capacity WDM transmission.[87, 88] Conventional fiber can in turn be used as a dispersion compensating fiber when non-zero dispersion shifted fiber is used as a transmission fiber.[89] The dispersion compensating elements null the overall dispersion of the link while the local dispersion of the fiber prevents the four-wave mixing from occuring. Thus, one can achieve, as an example, the transmission of 32 channels at 10 Gb/s over 500 km of conventional fiber (dispersion zero at 1300 nm) with an amplifier spacing of 125 km.[87]

Another option for terrestrial systems is to use the channels in the 1580 nm signal band since this wavelength band has finite dispersion when used with zero dispersion shifted fiber ($\lambda_0 \simeq 1550$ nm), high capacity WDM transmission at 1580 nm can be demonstrated with traditional (zero) dispersion shifted fiber.[90] WDM transmission at 1580 nm has been demonstrated in terrestrial systems (64 channels at 10 Gb/s over 500 km) as well as transoceanic (8 channels over 6600 km).[91, 92] In particular, the reduced four-wave mixing allows the use of closer channel spacings, i.e., 50 GHz has been demonstrated.[91] This despite the fact that dispersion shifted fiber implies

9.2. SYSTEM DEMONSTRATIONS AND ISSUES

smaller mode field diameters for the signals and thus higher powers (increasing the role of nonlinearities).

The use of wideband amplifiers allows the use of even more channels over a large wavelength range. Systems operating in both the 1550 nm and 1580 nm bands with a capacity of 1 Tb/s (100 channels at 10 Gb/s or 50 channels at 20 Gb/s) have been transmitted over distances in the 400 km to 600 km range, when used with appropriate dispersion management.[80, 81] These systems allow for tremendous capacity increase in terrestrial short haul and medium haul links.

The long distance inline amplifier experiments have established the technical feasibility of multi-Gb/s bit rate long distance (greater than 10,000 km) WDM transmission systems. Such results will be put to use in the undersea transmission systems currently being developed. The high-density WDM experiments at lesser distances have opened the way to ultra-high bit rate digital transmission capable of handling the high capacity requirements of data transmission systems.

9.2.4 Repeaterless Systems

Advancing the technology that allows the longest distance between optical transmitter and receiver is of great interest to both undersea and terrestrial communication providers. Common to both is the desire to reduce the amount of hardware along the route. A number of repeaterless demonstrations have been reported; Table 9.4 indicates the practical trade-offs in span distances achievable for data rates from 0.565 to 100 Gb/s. These results indicate the progress in this field from 1989 to 1996 during advances in achievable transmitted power, in receiver sensitivity, in control of span dispersion, and in control of nonlinear effects from the transmission fiber.

Limits to the maximum power that can be launched into a span are determined by nonlinear optical effects. These effects are dominated by stimulated Brillouin scattering, which is in turn controlled by the modulation bit rate, type and modulation index.[111] A launched signal power as high as +25 dBm with a spectral width of 2.2 nm has been used in demonstrations of repeaterless systems.[109]

Receiver sensitivities are frequently enhanced by either low-noise erbium-doped fiber preamplifiers or by using coherent detection. Sensitivities as high as 65 photons per bit are reported using continuous phase frequency shift keying (CPFSK).[98] Receiver sensitivities in Table 9.4 are expressed in photons per bit. Coherent systems have a fundamental advantage in providing the most sensitive receivers; however, the increased complexity of these methods has kept them from widespread commercial application.

Span losses have been reported as high as 68.5 dB (transmission distance of 365 km) incorporating a power amplifier providing +16.2 dBm launched signal power and a 980 nm pumped preamplifier/receiver combination providing a 60 photons per bit receiver sensitivity at a data rate of 0.622 Gb/s.[99] Also significant in this demonstration was the use of forward error correction (FEC), which adds a 14% redundancy to the data rate and improves the receiver sensitivity by 5 dB.

High data rate demonstrations have required dispersion management of the span along with receiver and transmitter improvements. For applications that upgrade existing spans, the dispersion must be adjusted at the terminal ends. Ignoring nonlinear

Year	Bit Rate (Gb/s) Format	Length (km)	Span Loss (dB)	Launched Signal Power (dBm)	Receiver Sensitivity (ph/bit)	Ref.
1989	12 NRZ	84	20	+1.7	644	[93]
1989	11 NRZ	151	36	—	9835	[94]
1990	2.4 FSK	250	33	-0.5	1823	[95]
1990	0.5 DPSK	218.5	62.9	+12.2	116	[96]
1990	17 NRZ	150	32.5	+9.8	1500	[97]
1990	2.5 CPFSK	364	66.4	+19.1	65	[98]
1992	0.6 NRZ/FEC	365	68.5	+16.2	72	[99]
1992	2.5 NRZ/FEC	341	62.5	+15.4	60	[99]
1992	0.6 NRZ	304	59.8	+14.8	500	[100]
1992	1.2 CPFSK	195	51.9	+13	414	[101]
1993	100 RZ	50	24.8	+1.8	391	[102]
1993	10 NRZ	215	47.8	+15	409	[103]
1993	10 NRZ	180	41.7	+13.5	1179	[103]
1993	10 NRZ	151	—	+13.5	2874	[104]
1993	10 NRZ	252	52.5	+17	219	[105]
1994	4×10 NRZ	80	15.8	—	8360	[106]
1994	8×10 NRZ	137	32.9	+8 total	—	[107]
1994	10 RZ	300	36	—	—	[108]
1996	40 NRZ	198	—	+25	381	[109]

Table 9.4: Selected repeaterless system experiments (FEC: forward error correction; FSK: frequency shift keying; CPFSK: continuous phase frequency shift keying; DPSK: differential phase shift keying; NRZ: non return to zero; RZ: return to zero; —: not reported). The receiver sensitivities and received powers are for 10^{-9} BER.

effects, the signal to noise ratio improvement for a dispersion compensated span is largest when the compensator is located following the receiver preamplifier.[112] Dispersion compensating fiber can be used to upgrade preexisting dispersion-shifted fiber plant. In one such example, a 138 km long submarine cable was upgraded from its originally installed data rate of 10 Gb/s to 40 Gb/s, using dispersion compensating fiber at the receiver end.[109]

9.2.5 Remote Pumping

The repeaterless approach to increasing span distance is limited at the transmitter end by fiber nonlinearity to launched signals of near 20 dBm (for a single channel). It is also limited at the receiver end by the ultimate receiver sensitivity. Further increases in span length require amplification several tens of kms from each end. This could be achieved with amplifiers located in standard repeater housings, powered by electrically active cables. However, unpowered cables and smaller device housings are frequently more desirable. Remote erbium-doped fiber amplifiers, in the form of lengths of erbium-doped fiber located along the span, powered optically from the ends, then serve this

9.2. SYSTEM DEMONSTRATIONS AND ISSUES

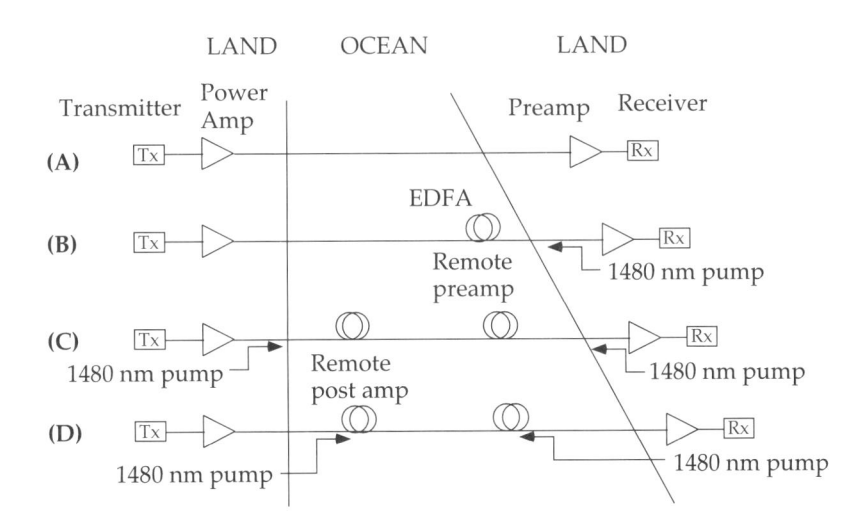

Figure 9.24: Progression of repeaterless systems from a simple preamplifier and power amplifier combination to remote pumping from both transmitter and receiver ends. In (D) the pump power is directly delivered to the remote amplifiers by means of a separate fiber. From reference [113] (©1995 IEEE).

need. The fiber amplifiers required for such systems should ideally be of very high efficiency, since the pump powers are usually weak by the time they reach the amplifier. The historical evolution of repeaterless systems is shown in Figure 9.24, showing how remote pumping is the most powerful embodiment of repeaterless systems, allowing the span length to be extended.

This method was first demonstrated by overcoming a 19.7 dB portion of the 62 dB span loss in a 310 km experiment at 1.866 Gb/s. Remote pumping from the receiver end was used as shown in diagram (B) of Figure 9.24.[110] This remote amplifier with 9.9 dB of gain was constructed from a 15 meter long erbium-doped fiber, pumped by launching 192 mW of 1480 nm power from the receiver end, 34.6 kilometers distant. An additional gain of 5.5 dB was obtained from Raman amplification in the span of fiber between the receiver and the remote amplifier. Raman gain of 4.3 dB from the transmitter end was also realized using 1480 nm copropagating pumping. The fortuitous additional Raman gain is the result of the proper wavelength separation between the 1550 nm signal and a copropagating or counterpropagating 1480 nm pump, closely coincident to the stimulated Raman shift frequency for silica (440 cm^{-1}).

To appreciate the effect of remote pumping from the receiver end, it is appropriate to define the configuration as an "extended" receiver that is, in practice, a combination of the remotely pumped amplifier, Raman amplifier, and a low-noise preamplifier. The signal power budget improvement ΔB may then be defined as a bit rate independent relationship,

$$\Delta B = P_R - (P_{REM} - \alpha L) \qquad (9.1)$$

where P_{REM} is the sensitivity of the extended receiver, P_R is the sensitivity of the optically preamplified receiver, α is the attenuation of the fiber at the signal wavelength,

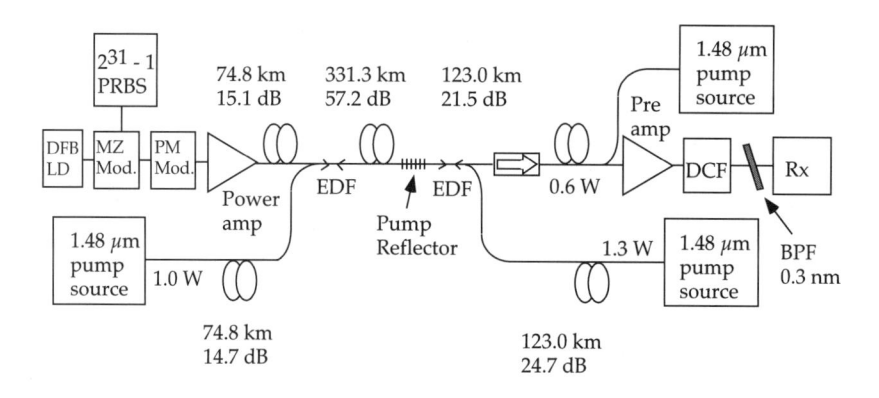

Figure 9.25: Configuration of a 2.488 Gb/s unrepeatered transmission experiment over 529 km using remotely pumped postamplifiers and preamplifiers, forward error correction and dispersion compensation. From reference [116] (MZ Mod.: Mach–Zehnder modulator; PRBS: pseudo-random bit generator; PM Mod.: phase modulator; DCF: dispersion-compensating fiber; BPF: band pass filter).

and L is the length of the transmission fiber between the remote amplifier and the receiver.[113]

In practice, the largest change in power budget can be made by increasing the gain of the remotely pumped erbium-doped fiber amplifier by increasing the launched pump power. For the case of a single span fiber carrying pump to the remote amplifier and signal to the receiver, numerical modeling predicts nearly 0.85 dB of power budget improvement for every 1 dB of additional pump power.[113] In the preceding statement, it has been assumed that the position of the erbium-doped fiber, as well as its length, has been optimized for maximum power budget improvement. As the pump power is increased, the erbium-doped fiber length is moved farther from the pump source for maximum span budget improvement. Typically there is a trade-off here between Raman gain and pump attenuation, with the smaller effective area (35 μm^2), higher Raman gain fiber experiencing 0.09 dB/km higher attenuation at the 1480 nm pump wavelength than its higher effective area (82 μm^2), standard fiber counterpart.[114] Although this difference may seem small, for a typical remote receiver span of 100 kilometers this represents a 9 dB difference in pump power delivered to the remote amplifier.

The highest remote amplifier gain of 18.2 dB has been reported using a combination of two fiber pump paths, with the remote erbium-doped span placed near the receiver end (see Figure 9.25). One low-loss, 123 kilometer long fiber was dedicated to delivering pump (1.3 W of 1480 nm power launched) in a counterpropagating direction. The main fiber span of 123 kilometer carried both the signal and a counterpropagating pump (0.6 W of pump launched) to the remote amplifier. The large pump power available in this demonstration was provided by a Raman laser pumped by a cladding pumped fiber laser.[115] In addition, an isolator follows the amplifier and a fiber grating serves to reflect residual pump power back through the amplifier in a copropagating direction (as shown in Figure 9.25).[116, 117, 118] A similar demonstration with a co-

9.2. SYSTEM DEMONSTRATIONS AND ISSUES

propagating pump power at the amplifier of 2.3 mW and counterpropagating power of 4.2 mW realized 16.5 dB of signal gain.[119] When the gain of these remote amplifiers is less than about 15 dB, no isolator before or after the amplifier appears necessary. Modeling of these systems has indicated less than a 1 dB influence on the budget from an isolator added before the remote amplifier to reduce the Rayleigh scattering-ASE interaction.[120, 121]

Remote pumping may be applied to the transmitter end of the span and simultaneously at the receiver end, as shown in diagram (C) of Figure 9.24. One preferred configuration for this remote postamplifier is an erbium-doped fiber amplifier remotely pumped by a copropagating pump delivered through a low-loss fiber, separate from the signal-carrying fiber.[122] One such configuration, using a 1 W 1480 nm pump and 24 dBm launched signal, resulted in a 6 dB budget improvement.[123] A gain of 7 dB has also been reported using a combination of copropagating and counterpropagating pump sources.[119] A power conversion efficiency of 64% for this demonstration is consistent with other power amplifier performance demonstrations.

The signal power launched is limited by stimulated Brillouin scattering from the entry span fiber. The threshold power for Brillouin scattering may be increased by a few dB through fiber processing or dissimilar fiber concatenation. To maintain a low-loss fiber with the required dispersion properties, however, increasing the Brillouin threshold by broadening the signal bandwidth is generally preferred. This may be accomplished by adding a few sinusoidal modulations (typically frequencies in the tens of kHz range are used) to the signal laser bias. This creates a broadened signal because of the chirp of the laser (the change in laser wavelength with drive current), which is of the order of $\simeq 100$ MHz/mA for a DFB laser. The line broadening is on the order of a few GHz with this method. Alternatively, the signal spectral width may be increased through a separate multitone phase modulation.[116, 124, 125, 126]

In addition to remote amplification at both ends of the span, a transmitted pattern encoding method such as forward error correction (FEC) can be applied to these systems. This provides a budget improvement of near 4 dB for a BER of 10^{-4} with an increase in bit rate of 7% to 14% for bit redundancy.[121, 127]

As the span lengths are pushed to near 500 kilometers the total dispersion (near 8000 ps/nm), as well as the local dispersion along the span, must also be managed. The total dispersion may be brought to near zero by providing negative dispersion at the end of the span in the form of dispersion compensating fiber (DCF).[119, 128, 129] The loss associated with this dispersion compensation then requires an amplifier after the DCF to make up for the loss. The dispersion of the fiber near the transmitter end may also be selected to be high to prevent modulation instability from self-phase modulation.[129]

Over a period of 6 years, unrepeatered span lengths have been extended from 300 kilometers to over 530 kilometers (corresponding to 62 dB to 93.8 dB span loss). This progress is outlined in Table 9.5, which indicates the practical trade-offs in span distances achievable for various data rates from 0.622 Gb/s to 80 Gb/s. Relatively simple remote amplification from the receiver end has brought the span lengths to 425 kilometers for 2.5 Gb/s bit rates.[134] Remote amplification has also been applied to WDM systems. Eight channels at 10 Gb/s were transmitted over a 352 km span as shown in Figure 9.26.[129] Progressing from 300 to 500 kilometers has required advances in

Year	Bit Rate (Gb/s)	Length (km)	Span Loss (dB)	Launched Signal Power (dBm)	Receiver Sensitivity (ph/bit)	Reference
1989	1.8 NRZ	310	62.2	+14.4	936	[110]
1992	2.5 FEC	357	68	+15.4	60	[127]
1992	0.6 FEC	401	73	+16.2	52.4	[127]
1994	2.5	410	74.5	+19.5	125	[128]
1995	2.5 FEC	529	93.8	+26.5	135	[116]
1995	2.5 FEC	511	88.8	+15.5	58.9	[119]
1995	0.6 FEC	531	92.2	+16.5	118	[119]
1995	8 × 10	352	65.3	+11.25	497	[129]
1995	5.0 FEC	405	83.3	+24.3	238	[130]
1996	16 × 2.5 FEC	427	74.7	+19.2 total	444	[131]
1996	8 × 2.5 FEC	450	78.8	+21.0 total	444	[131]
1996	4 × 2.5 FEC	475	83.1	+21.0 total	444	[131]
1996	10	442.5	80.6	+20.0	336	[132]
1996	4 × 10 FEC	254	61	+36.0 total	343	[133]

Table 9.5: Selected remotely pumped amplifier system experiments. The receiver sensitivities and received powers are for 10^{-9} BER (FEC: forward error correction). Several experiments are WDM (e.g., 16×2.5 is 16 channels at 2.5 Gb/s each) and the launched signal power is that of the channels combined.

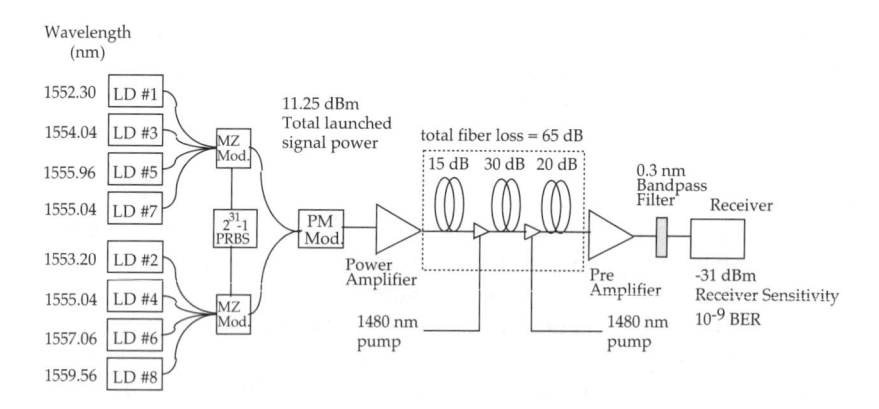

Figure 9.26: High capacity WDM remote pumping experiment, with 1 W pump power supplied from the transmitter end and 1.3 W from the receiver end. From reference [129] (MZ Mod.: Mach-Zehnder modulator; PRBS: pseudo-random bit generator). The pump light is carried to the amplifiers by separate fibers.

available pump power, lower threshold erbium-doped fiber, and dispersion engineering and modeling to optimize the designs of these systems.

9.2.6 Analog Applications

Analog Transmission Results

Analog systems are important for television signal broadcasting. The main application for erbium-doped fiber amplifiers in analog transmission systems is for compensating the link losses between transmitters and receivers. The increased link loss that can thus be tolerated can be used for longer transmission distances in trunk systems or for more subscribers in distribution systems. Because analog systems require very high signal to noise ratios, about 50 dB compared to about 20 dB in digital systems, the signal power levels are kept high. Therefore, the main application for amplifiers are as power amplifiers and high-power inline amplifiers. Preamplification at the receiver is not used because the high signal levels result in shot-noise or near-shot noise limited operation and preamplification is thus not necessary. Some of the main issues associated with analog transmission using erbium-doped fiber amplifiers were discussed in Chapter 7, section 7.3.7. A summary of selected analog transmission results, all directed at video transmission, is shown in Table 9.6.

One of the requirements for analog transmission is to maintain a suitable signal transmission quality, while broadcasting to the maximum number of users with the minimum number of expensive analog grade transmitter lasers. The conditions on signal quality derive from the necessity to transmit a large number of channels (typically 40 to 120) with the least amount of interference between channels (as measured by composite second order (CSO) and composite triple beat distortion (CTB) and good signal to noise (measured by CNR). Typical numbers for a 70 channel system are a CNR of at least 48 dB, a CSO of at least −53 dBc, and a CTB of at least −56 dBc.

Higher power booster amplifiers allow for an increase in the optical power budget. The first analog transmission experiments using an erbium-doped fiber amplifier were reported in 1989.[135] Nineteen AM-modulated video CATV channels superimposed on one laser transmitter were transmitted through, and amplified in, an 11.5 dB gain erbium-doped fiber amplifier used as a power amplifier. This experiment was expanded upon by the use of 16 lasers (transmitting a total of 100 analog video channels) amplified in an erbium-doped fiber amplifier situated between a series of 1 by N splitters and a 9 km section of fiber, simulating delivery to 4096 users.[136] The amplifier was clearly shown to allow greater splitting and transmission capacity for the system, without introducing significant additional noise or distortion. Further experiments shortly thereafter demonstrated the effective use of amplifiers for systems with larger numbers of channels per laser. For example, 35 channels were transmitted with the help of an erbium-doped fiber power amplifier with an output power of 9.6 dBm in a system demonstration with a power budget of 16 dB.[140] The power budget can be directly translated into fiber span loss or splitting among subscribers. A solid state laser pumped erbium-doped fiber amplifier yielding an output power of 19 dBm resulted in a power budget of 18 dB for a 42 channel experiment.[150]

352 CHAPTER 9. SYSTEM IMPLEMENTATIONS OF AMPLIFIERS

Year	Optical Channels	Analog Media Channels per Optical Channel	Total Head End Power (dBm)	Fiber Length Used (km)	Fiber λ_0 (μm)	Number of Amplifiers	Ref.
1989	1	19	8	A	A	1(b)	[135]
1990	16	10	1.8	9	1.3	1(b)	[136]
1990	16	10	8	9	1.3	1(i)	[137]
1990	1	13	-	15	1.3	1(i)	[138]
1990	1	47	0	40	1.5	4(i)	[139]
1991	1	35	9.6	A	A	1(b)	[140]
1991	1	20	13	32	1.3	1(b)	[141]
1991	1	35	-	20	1.5	1(b)+1(i)	[142]
1991	1	40	13.6	A	A	1(b)	[143]
1992	1	32	15	24	1.3	1(b)+2(i)	[144]
1992	1	35	3.3	2	1.3	4(i)	[145]
1992	1	60	13	10	1.3	1(b)	[146]
1992	2	40	-	A	A	1(b)	[147]
1992	1	30	15	A	A	1(b)+2(i)	[148]
1993	1	61	14	A	A	1(b)+3(i)	[149]
1993	1	42	18.2	40	1.5	1(b)	[150]
1994	1	50	17.3	A	A	1(b)+3(i)	[151]
1995	1	50	14	30	1.3	1(b)+2(i)	[152]
1995	2	60	15	A	A	1(b)	[153]
1996	1	80	17	100	1.3	1(b)+1(i)	[154]
1996	1	50	14.3	38	1.3	3(i)	[155]

Table 9.6: Analog transmission experiments involving erbium-doped fiber amplifiers. "A" denotes that only attenuators were used to simulate the presence of splitters or fiber span loss (in some of the experiments above where fiber was used, attenuators were employed as well). Amplifiers are employed either as boosters (b) or used as inline amplifiers (i) to increase transmission length. In most of the experiments 1 by N power splitters were used to simulate the necessity of transmitting to many subscribers.

By cascading several optical amplifiers, systems can be constructed to serve a large number of end users, by the use of 1 by N splitters placed between the amplifiers. For example, a three-amplifier system was demonstrated with a 73 dB power budget that could potentially serve more than half a million subscribers.[156]. The setup for this experiment is shown in Figure 9.27. An actual system to serve all the subscribers would need far more than the three amplifiers used to demonstrate the feasibility of the scheme. Cascading a fourth amplifier increases the power budget to 89 dB.[145]

The first demonstration of erbium-doped fiber amplifiers in CATV analog systems did not exhibit any system degradations from the amplifiers themselves. Any system degradations caused by amplifiers were in general masked by the poor performance of other system components, notably the laser transmitter nonlinearities. However, as

9.2. SYSTEM DEMONSTRATIONS AND ISSUES

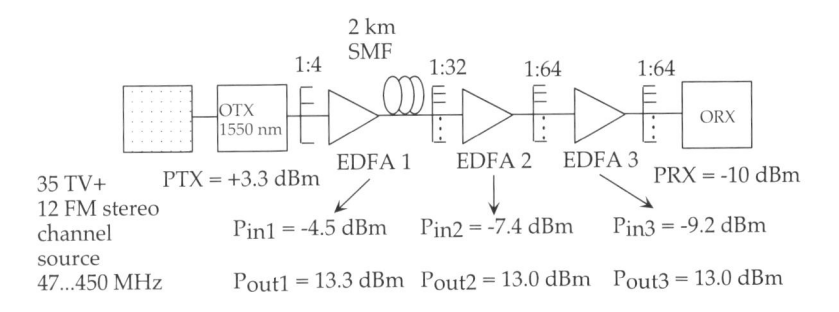

Figure 9.27: Experimental setup for an amplified AM-VSB transmission system with three cascaded amplifiers, with potential service to more than half a million subscribers. EDFA: erbium-doped fiber amplifier. From reference [156] (©1992 IEEE).

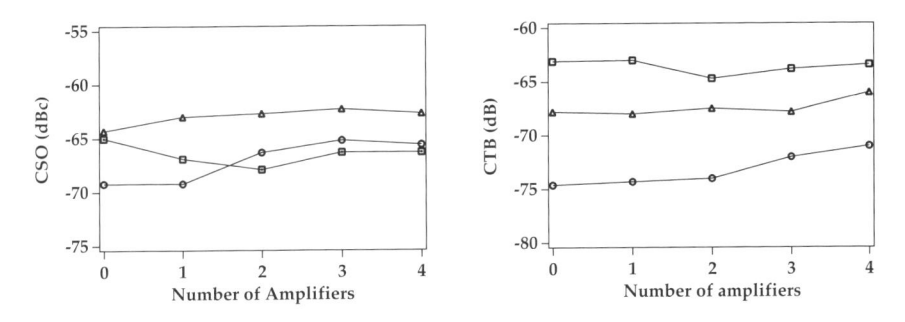

Figure 9.28: Dependence of system performance parameters on number of amplifiers in an amplified AM-VSB transmission system (circles: channel 1, 91.25 MHz; triangles: channel 24, 259.25 MHz; squares: channel 50, 445.25 MHz). From reference [151].

the overall system performance was improved and the number of analog channels increased, it was discovered that the amplifiers do generate system nonlinearities. As discussed in Chapter 7 (section 7.3.7), a major source of degradation originates in amplifier gain tilt (i.e., wavelength-dependent gain) when used in conjunction with directly modulated lasers. As these limitations were uncovered, compensation techniques were developed. In one experiment, the original 61.2 dB CSO level was degraded to 54.8 dB by the amplifier, but by deploying a gain tilt equalizer the CSO was restored to 61.2 dB.[147] The use of composite materials or filters can be instrumental in maintaining the gain flatness, which is key to obtaining linearity in the system.[153] Four such amplifiers were cascaded to show the increase in power budget that can be obtained without a commensurate increase in signal distortion, as shown in Figure 9.28.

It is the interaction between the chirp in directly modulated lasers and the gain tilt in the amplifier that causes the degradation in linearity. Therefore, the use of external modulation, which results in very low chirp, almost entirely removes the gain flatness issue. Figure 9.29 shows the CNR obtained with an externally modulated DFB laser

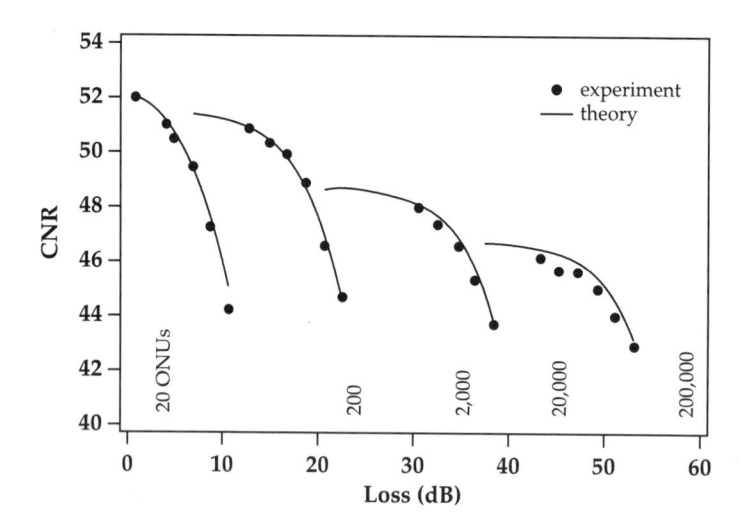

Figure 9.29: CNR as a function of optical loss for a system of three cascaded erbium-doped fiber amplifiers, with an externally modulated transmitter. From reference [152] (©1995 IEEE). The number of subscribers (ONUs) that can be served for a given optical loss are indicated assuming a 3.5 dB loss per 1:2 split.

transmitted through three cascaded erbium-doped fiber amplifiers as a function of total optical link loss.[152] The CNR suffers some degradation with each amplifier, due to ASE-generated noise, however, the power budget is strongly enhanced by the use of the amplifiers. The number of subscribers or optical network units (ONU) potentially served is shown, assuming a 3.5 dB loss per 1:2 split. By careful system engineering and by using high performance components, very high-performance systems can be built. For example, an 80-channel AM CATV signal was transported over 100 km while maintaining a carrier to noise ratio of 52.5 dB and a CSO and CTB distortion of -70 dB and -65 dB, respectively.[154]

The gain flatness issue can be circumvented by a careful design of the amplifier. A two-stage amplifier specifically designed for analog application has been demonstrated, which incorporates a fiber grating filter between the two stages for gain flattening.[157] This amplifier has a gain tilt of less than 0.2 dB/nm over an 11 nm wavelength range and 4 dB input power range. It was used to demonstrate power amplification of a 77-channel NTSC video signal without significant distortion. As the above results clearly demonstrate, erbium-doped fiber amplifiers can play an important role in analog video distribution networks.

9.2.7 Gain Peaking and Self-Filtering

Optical communications systems based on a cascade of inline fiber amplifiers require an understanding of signal gain and noise accumulation along the link. Control of the spectral-dependent gain and signal to noise is critical for single wavelength long haul systems with tens to hundreds of amplifiers typical of undersea trunk routes, as well as

9.2. SYSTEM DEMONSTRATIONS AND ISSUES 355

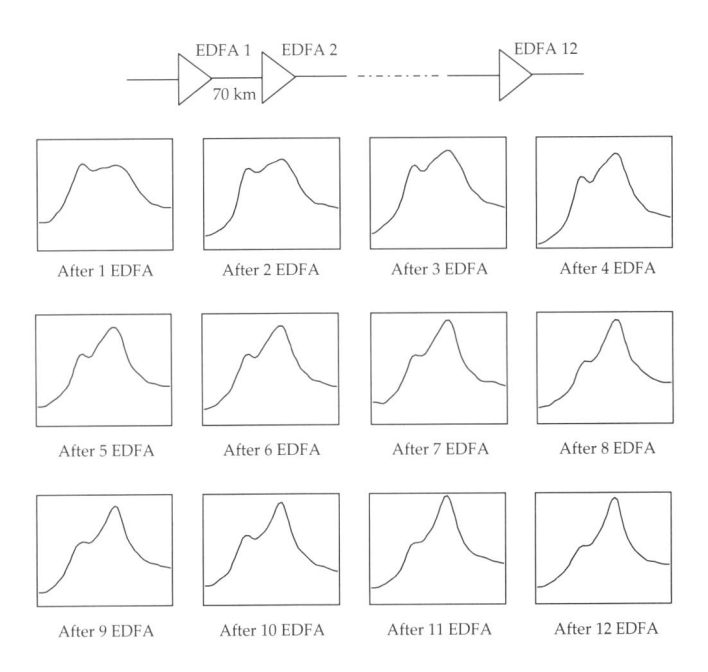

Figure 9.30: Gain peaking in a cascade of twelve inline erbium-doped fiber amplifiers (EDFA) spaced by 70 km. Each graph shows the ASE spectrum (1 nm bandwidth) after a specific number of amplifiers. The horizontal axis spans 1500 nm to 1600 nm and the vertical axis spans 80 dB. From reference [158] with additional data provided courtesy R. Tench, Lucent Technologies, Breinigsville, PA.

wavelength division multiplexed (WDM) systems used in both undersea and terrestrial links. The evolution of the spontaneous emission spectra from cascaded single stage unequalized inline amplifiers is shown in Figure 9.30, indicating that for a series of twelve cascaded amplifiers the gain peak is in the longer wavelength region (typically 1.558 μm). The gain peak also narrows with each stage of amplification.

We discuss here the situation for single stage amplifiers without spectral equalization components, typical for single channel systems. An intuitive understanding of the long wavelength gain peak can be obtained by a simple explanation involving the gain spectrum of the amplifier under saturated conditions.[159] We revisit for this purpose the simple model discussed in Chapter 5. For a hypothetical two-level amplifier model with constant populations N_1 and N_2 in the lower and upper levels, respectively, the gain will be proportional to $N_2 \sigma_e(\lambda) - N_1 \sigma_a(\lambda)$, where $\sigma_e(\lambda)$ and $\sigma_a(\lambda)$ are the emission and absorption cross sections, respectively. Assume units such that $N_1 + N_2 = 1$. For a well pumped amplifier operating in small signal conditions ($N_1 = 0$ and $N_2 = 1$), the gain spectrum has the same shape as that of the emission cross section $\sigma_e(\lambda)$. For an amplifier operating deep in saturation we have $N_1 = 1/2$ and $N_2 = 1/2$, in which case the gain spectrum is proportional to the difference between the emission and absorption spectra $\sigma_e(\lambda) - \sigma_a(\lambda)$. The situation is summarized in Figure 9.31. The gain peak is near 1.558 μm once the amplifier has reached saturation. In most amplifier chains

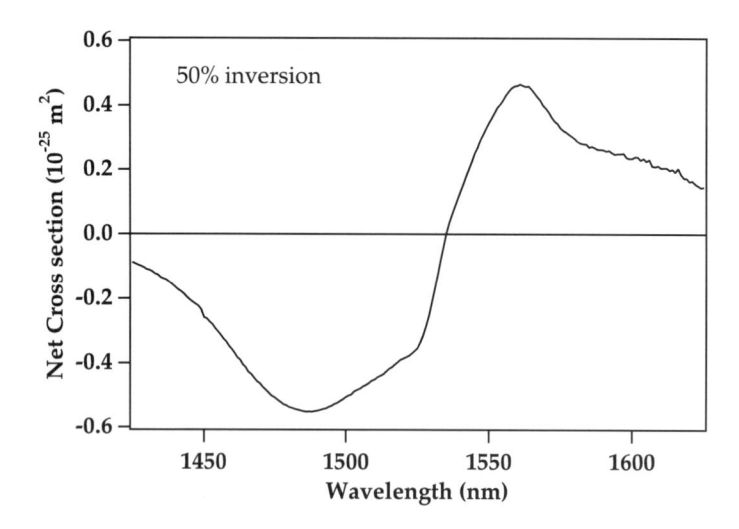

Figure 9.31: Effective per-ion gain spectrum for a 50% inverted amplifier, computed from the emission and absorption cross sections for an erbium-doped Al-Ge fiber (Figure 6.1).

the level of saturation builds up along the chain as the ASE and signal power build up, therefore the individual gain spectra vary with position along an amplifier chain. This behavior depends on the launched signal power and the specific pump power levels for each amplifier. Once the amplifiers have reached saturation, at some distance from the input to the system, the total power is maintained constant and the signal power drops at the expense of ASE (as described in Chapter 7, section 7.3.4). The gain peak is near 1558 nm in a typical single channel long haul saturated amplifier system.

It is critical in the single-channel NRZ transmission systems that the gain peak wavelength (GPW), determined by the self-filtering or gain peaking function of cascaded amplifiers, be close to the zero dispersion wavelength λ_0 of the entire span.[160, 161] The proximity of these wavelengths is most crucial when the width of the spectral gain profile is extremely narrow about the peak. Such narrow windows occur in long undersea systems (as long as 10,000 km) where nearly 300 amplifiers may be connected inline. The autofiltering function of this narrow spectral gain profile fortuitously diminishes the build up of ASE or noise power along the span. Spectral filters could be inserted along the span to control the ASE and force the gain to a specified wavelength. However, a simpler and more practical solution is to allow the gain spectrum of the amplifier to determine the peak gain wavelength and gain bandwidth.[160] Amplifier chains in single-channel long undersea systems, where self-filtering has been used, have been designed to operate near 1558 nm with the gain compressed from the small signal level by a few dB for self-regulated gain control along the chain.

The amplifier GPW has been shown to be determined by the average inversion of the erbium-doped fiber amplifier and the peak gain per unit length.[162] A representative dependence of these two properties is shown in Figure 9.32 for a GeO_2-Al_2O_3-SiO_2 core composition erbium-doped fiber amplifier and a 1480 nm pump. Note also

9.2. SYSTEM DEMONSTRATIONS AND ISSUES

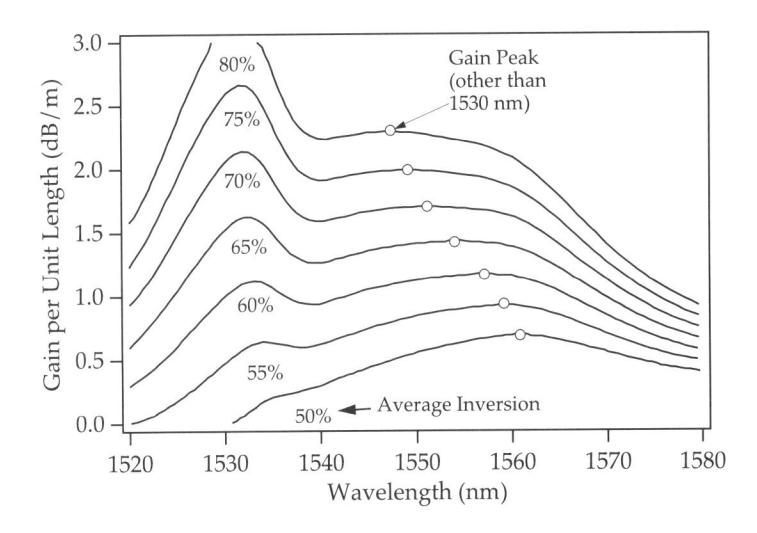

Figure 9.32: Inversion-dependent gain peaking in a Ge-Al erbium-doped fiber amplifier pumped at 1480 nm. From reference [162] (©1994 IEEE).

that the fiber loss in the spans between the amplifiers is wavelength dependent and is higher at 1530 nm than at 1550 nm (due to the $1/\lambda^4$ dependence of the Rayleigh scattering component of the loss). Thus, strictly speaking, the spectral variation of the transmitted signal and ASE is a convolution of the fiber spectral loss and amplifier spectral properties. Using phenomenological erbium-doped fiber amplifier modeling parameters, as discussed in Chapter 6 (section 6.4.7), the gain per unit length ($G(\lambda)/L$) for an amplifier with an average fractional population density in the upper state $\overline{N_2}/N$ may be expressed as

$$\frac{G(\lambda)}{L} = \left\{ [g^*(\lambda) + \alpha(\lambda)] \frac{\overline{N_2}}{N} - \alpha(\lambda) - l(\lambda) \right\} \quad (9.2)$$

where $g^*(\lambda)$ is the gain per unit length (dB/m) for the fiber with the erbium ions completely inverted, $\alpha(\lambda)$ is the absorption per unit length (dB/m) when the fiber erbium ions are not inverted, $l(\lambda)$ is the fiber background loss (dB/m), $\overline{N_2}$ is the average upper state population density (along the length of fiber) and N is the total erbium ion density.[162] Equation 9.2 implies that the GPW is independent of pump wavelength, pump power, signal power, amount of gain compression, or amplifier configuration for a given operating gain and fiber length. Measured peak wavelengths versus gain per unit length for various lengths and pump powers are shown in Figure 9.33. Different fiber compositions do, however, result in different GPW versus gain per unit length dependences as well as gain peak bandwidths.[159, 162] For example, higher aluminum levels, from 1% to 6%, have been shown to influence this dependence.[162]

A practical issue from the point of view of the systems designer is how to predict and control the GPW to the desired tolerance of less than 1 nm. In theory, $g^*(\lambda)$, $\alpha(\lambda)$, and $l(\lambda)$ could be measured and computer modeling could predict the GPW.

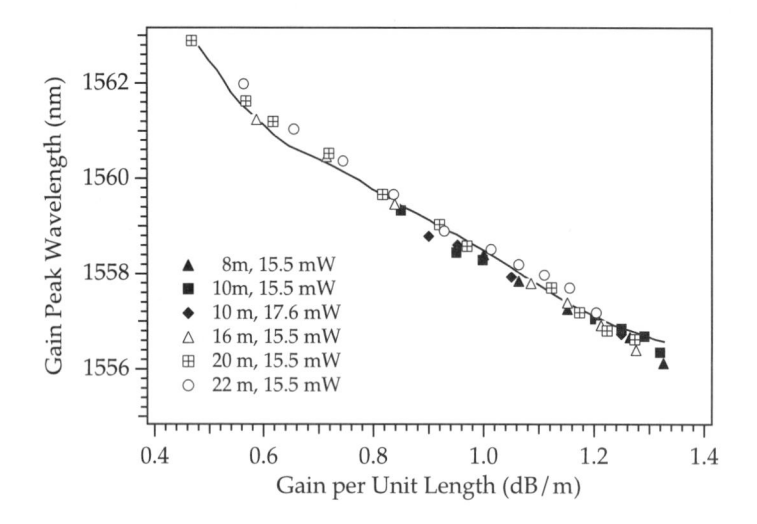

358   CHAPTER 9. SYSTEM IMPLEMENTATIONS OF AMPLIFIERS

Figure 9.33: Gain peak versus gain per unit length. From reference [162] (©1994 IEEE).

In practice, this cannot provide the required accuracy. Options for measuring GPW include: constructing a full length test bed; assembling a partial test bed with a few amplifiers; characterizing one amplifier; or using a loop technique. Practicality and accuracy have resulted in two dominant experimental approaches, one measuring the GPW versus gain per unit length in a recirculating loop configuration and the other measuring the GPW for a single amplifier.[163, 164]

Recirculating loop configurations have been used with some success to measure the GPW.[160, 165, 166] The gain spectrum of an amplifier chain results from the superposition of the gain spectra of the individual amplifiers. Even if the individual amplifier gain spectrum is broad, the region near the gain peak is amplified more than other regions, resulting in spectral narrowing after many stages. Equivalently, the emitted spectrum of a multipass loop simulates the concatenation and spectral narrowing of such an amplifier chain. The gain peak of multiple spans or multiple passes through a loop is the same as the gain peak of the amplifier chain. Loop configurations may, however, be plagued with instability in the form of polarization dependent multiple peaks. A combination of polarization hole burning (PHB), residual spectral hole burning (SHB), and polarization-dependent loss (PDL) can account for this behavior.[167] Stabilization of these recirculating loops has been demonstrated by the introduction of a polarizer.[168] However, such operation does not guarantee emission at the peak gain wavelength of the amplifiers. An alternate loop configuration incorporating polarization scrambling has been shown to be particularly effective.[163, 169] Wavelength-dependent loss can also be a critical parameter in determining the gain peak wavelength.[162]

In WDM systems, the gain peaking is actually a hindrance since it favors single-channel operation, not multi-channel operation. As a result, self-filtering is specifically avoided in large bandwidth WDM systems. This is accomplished by flattening the amplifier gain as described in Chapter 8, section 8.3.6.

9.2. SYSTEM DEMONSTRATIONS AND ISSUES

9.2.8 Polarization Issues

Shortly after the first successful long haul transmission experiments involving erbium-doped fiber amplifiers, polarization-dependent effects—some of which are of subtle origin—were uncovered. It was long thought that erbium-doped fiber amplifiers had no polarization dependence, because the medium is isotropic and the erbium ions were presumed to be randomly oriented in the disordered glass environment. This is a particular advantage over semiconductor amplifiers, with which it is very difficult to obtain polarization-independent gain. However, erbium-doped fiber amplifiers do exhibit a very weak polarization-dependent gain that becomes significant when many amplifiers are cascaded, as is the case in long haul systems that are thousands of km in length. There are three main polarization effects:

- polarization-dependent loss.
- polarization-dependent gain.
- polarization mode dispersion effects.

Polarization-Dependent Loss

Polarization-dependent loss (PDL) arises mainly from the passive components used in the transmission line. The loss for one polarization axis can be different than for the orthogonal axis. The signal light is polarized and its polarization direction rotates randomly as it traverses the transmission fiber. A situation might arise where, by chance, the polarization state is predominantly aligned along the preferential directions of the various components. Another situation would be for the polarization state to be aligned along the more lossy axes. The actual situation will fluctuate between these two extremes. Over time, the signal strength at the receiver end will thus fluctuate. The ASE, however, is unpolarized and will have a constant power. Thus, the SNR will fluctuate over time and the bit error rate performance of the system, as measured by the Q parameter, will fluctuate as well.

Polarization-Dependent Gain

Polarization-dependent gain (PDG), also referred to as polarization hole burning (PHB), has been the object of intensive study.[170] The effect can be summarized as follows. The signal, which traverses the amplifier in a given polarization state, usually experiences gain in the saturated regime. ASE polarized in the direction orthogonal to the signal experiences small signal gain, which is higher than that experienced by the signal (or ASE polarized parallel to the signal). The total ASE noise thus grows faster than would be expected from the standard theories that neglect any polarization dependence of the gain. Since the PDG is believed to originate in the amplifiers, the effect is cumulative with the number of amplifiers and the system impairment increases with total system length.[170]

An understanding of this effect can be obtained by considering the erbium ion environment on a microscopic level. The local sites surrounding the erbium ions are non-isotropic with a randomly oriented symmetry axis. The cross section for a transition between two levels will depend on the state of polarization of the light and whether it is

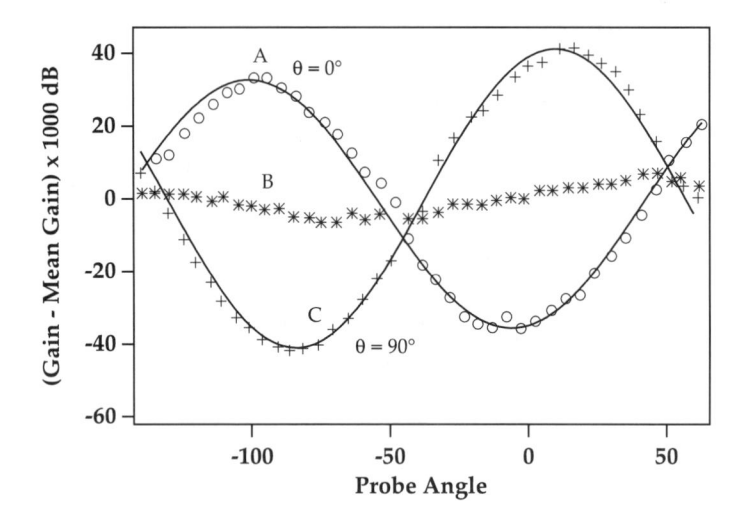

Figure 9.34: Gain variation, for an erbium-doped fiber amplifier, of a weak probe in the presence of an unpolarized pump, as a function of probe polarization angle relative to that of the signal (for two orthogonal polarizations of the signal), and for a depolarized signal. For curves A and C the angle θ is that between the linearly polarized signal and an arbitrarily chosen axis. For curve B, the signal is unpolarized. From reference [172] (©1994 IEEE).

parallel or orthogonal to the symmetry axis. Thus a polarized signal will preferentially saturate a subset of the erbium ions in the amplifier. A probe polarized orthogonal to the signal will interact preferentially with the subset of ions oriented orthogonally to the signal subset. It will see more gain than the signal when the probe is weak because it will be in the small signal regime whereas the signal is in the saturation regime. This is the polarization hole burning effect, which was first observed in the context of Nd^{3+} glass lasers.[171]

For a single amplifier the PDG is a very small effect, on the order of 0.1 dB. Over many amplifiers, however, this effect cumulates to several dB. This effect has been measured using a system with an experimental gain resolution of 0.01 dB.[172] A weak polarized probe measures the gain for a single stage amplifier in the presence of a strong signal and a pump, each of which is polarized with a given state and degree of polarization. Figure 9.34 shows the gain variation obtained by rotating the polarization of the probe relative to that of the signal, in the presence of an unpolarized pump.

There also exists a polarization effect due to the pump polarization. A polarized pump will preferentially excite a subset of the erbium ions. Figure 9.35 shows the differential gain in the presence of a polarized pump. When the signal polarization is aligned along an axis defined presumably by the pump polarization, the differential gain variation is very large. When the signal polarization is orthogonal to the pump-defined axis, the differential gain variation is negligible. This results from the fact that the important angle is now that between the pump polarization and the probe polarization. In a long haul system, the polarization effects due to the pump are averaged over

9.2. SYSTEM DEMONSTRATIONS AND ISSUES

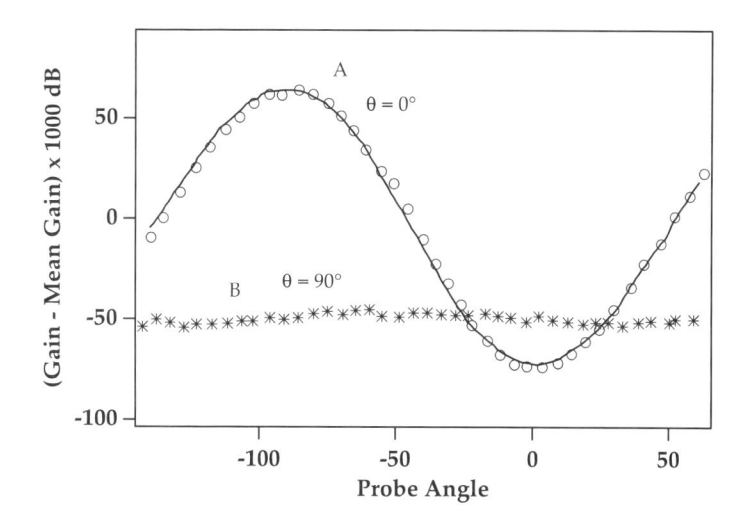

Figure 9.35: Gain variation, for an erbium-doped fiber amplifier, of a weak probe as a function of probe polarization angle relative to that of the signal, for a polarized pump, for two orthogonal polarization states of the signal relative to a chosen axis (θ is the angle between the signal polarization axis and the chosen axis). From reference [172] (©1994 IEEE).

many amplifiers since the pump and the signal have uncorrelated polarizations. The relative polarization of pump and probe can change with time so this effect will give rise to a penalty similar to polarization-dependent loss. Its effect can be eliminated by depolarizing the pumps for all the amplifiers. The differential gain is smaller than in the case of the polarized pump. When the signal is, in turn, depolarized, the differential gain is reduced to near zero.

The magnitude of PDG increases with saturating signal power, starting from zero for a weak enough signal.[173, 174] Figure 9.36 shows the variation of PDG with the signal power (and thus the amount of gain compression), for an amplifier with a small signal gain of about 28 dB.[173] This effect has been confirmed by computer modeling.[175] The PDG also increases with pump power.

PDG can be virtually eliminated by polarization scrambling of the signal.[176, 177, 178, 179] In an 8800 km loop experiment at 5 Gb/s, using 192 amplifiers with a saturated gain of 10 dB, this improved the Q factor by 2 dB. Figure 9.37 shows the Q factor measured every 5 minutes in the 8800 km loop experiment, with the scrambler on and off. In this experiment, the scrambling is performed by using a piezoelectric transducer driven at 600 kHz attached to a single-mode fiber through which the signal passes, thus producing a time-varying birefringence in the fiber. The scrambling can also be performed with lithium niobate modulator. Since the time constant for polarization hole burning has been measured to be 130 μs, any polarization modulation has to be carried out at a frequency larger than 7 kHz.[180] Subsequent experiments have suggested that the optimum scrambling frequency is the bit rate of the transmission.[181]

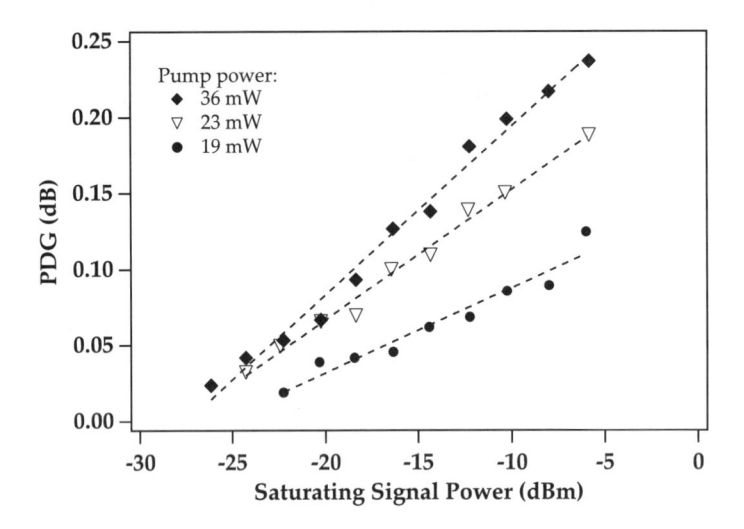

Figure 9.36: PDG of an erbium-doped fiber amplifier as a function of the input signal power. From reference [173].

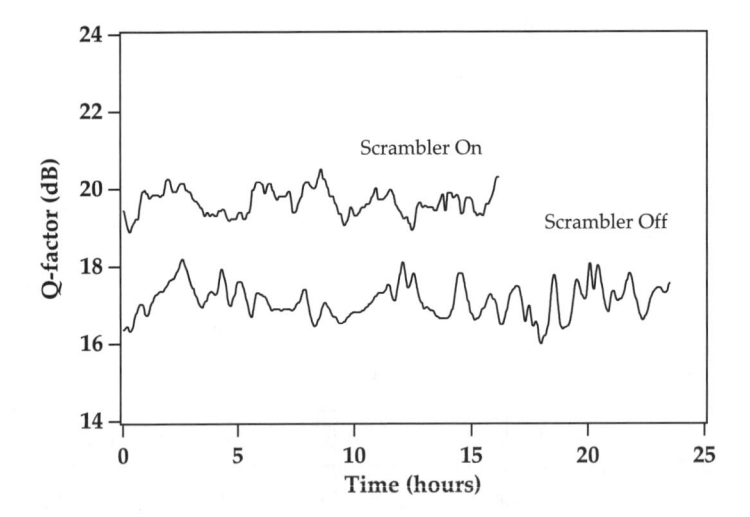

Figure 9.37: Q factor versus time for a single channel 5 Gb/s 8800 km loop transmission experiment, with and without polarization scrambling of the signal. From reference [176].

Polarization Mode Dispersion Effects

Polarization mode dispersion (PMD) denotes the effect whereby the two polarization states of a fiber can have different velocities. Hence the optical pulse shape can change as the two polarization components get separated as they travel along the fiber. This is a time-varying process because the polarization of the signal varies randomly. This will cause penalties in the bit error rate performance of the system. In addition, the PMD of the fiber, which has a Maxwellian distribution, can vary with time. It is thus necessary to know both the mean and the standard deviation of the PMD along the fiber optic cable used for the transmission system. It has been calculated that a maximum mean PMD of about 10 to 15 ps is allowed for a 5 Gb/s system to limit the probability of the Q penalty exceeding 1 to 2 dB.[182] For an actual link 6200 km long with a PMD of $0.1\,\text{ps}/\sqrt{\text{km}}$, it was found that the effects of PMD on system performance were negligible.[183] The most effective way to combat the effects of PMD is to use low PMD fiber. Modern fiber manufacturing techniques now allow this to be accomplished.

9.2.9 Transient Effects

The time dynamics of the gain process in an erbium-doped fiber amplifier give rise to transient effects. These effects can manifest themselves when a non-steady-state signal is amplified. The classic examples of such a situation are

- a single-channel signal is modulated in time.

- one or more channels are added to a collection of WDM channels.

- one or more channels are removed from a collection of WDM channels.

In digital signal transmission with intensity modulation the signal is modulated at the bit rate. In most cases the bit rate is much higher than the frequency with which the erbium ion gain dynamics can respond, so that the signal appears to be essentially steady state. The slow gain dynamics arise from the long upper-state lifetime of the erbium ion as well as from the rather long stimulated lifetimes (which correspond to the stimulated emission rates). This latter lifetime is long (several tens of μs or more) for the typical power levels of the pump and signal in the erbium-doped fiber. When the modulation frequency is low enough—on the order of the characteristic response time of the erbium gain dynamics—transient effects begin to appear.[184]

A characteristic measure of the strength of the stimulated emission rate is the ratio of the power of the light field to its intrinsic saturation value. Since near 1550 nm this saturation power is on the order of a few hundred μW for signal powers on the order of 1 to 10 mW this ratio can be on the order of 10 to 30, for example. For strong signals the transient response time is given by the ratio of the upper-state lifetime and the signal power measured in units of the saturation power.[184] The transient response is then reduced below the 10 ms time by one or two orders of magnitude.

An early study of the gain dynamics investigated how modulation in the pump is transferred over to the signal. Laming et al. measured the relative modulation in the amplified signal as a function of the modulation frequency of the pump.[185] At low frequencies (tens of Hz), there is significant modulation in the amplified signal. As the frequency of the pump modulation is increased, however, the modulation in the signal rolls off. The roll off point is 100 Hz (the inverse of the upper-state lifetime) for low signal inputs and increases to a few hundred Hz for strong signal inputs. This increase in frequency in the latter case corresponds to a decrease in the characteristic response time of the erbium ion gain dynamics, which happens because now the

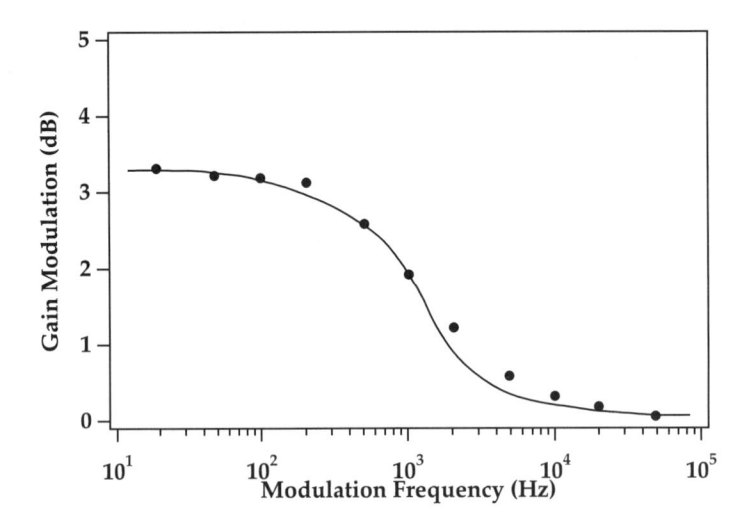

Figure 9.38: Modulation in dB of a weak probe at 1537 nm amplified in an erbium-doped fiber amplifier as a function of the modulation frequency of a copropagating square wave modulated strong signal at 1531 nm. From reference [186].

response time is dominated not by the upper-state lifetime but by the signal stimulated emission rate.[185] A similar effect is observed when a strong saturating signal is simultaneously amplified along with a weak probe, when the strong signal is square wave modulated. Figure 9.38 shows the modulation impressed on the probe signal as a result of the presence of the modulated saturating signal, as a function of the frequency of that modulation.[186] The crosstalk starts rolling off at 250 Hz, consistent with the strong signal regime.

A picture of the gain dynamics can be obtained by following the temporal behavior of the amplified strong signal and weak probe over the period of modulation of the signal. Figure 9.39 shows the dynamics of the gain during and after the passage of the strong signal.[186, 187] At the leading edge of the strong signal there is a gain overshoot as the inversion drops down to a steady-state value, corresponding to a balance between the signal stimulated emission rate and the pumping rate. The front of the pulse initially has more gain as it sees a higher inversion. The gain overshoot can be as high as 7 dB over a saturated gain of 23 dB for strong signal inputs (i.e., 25 μW).[186] Such effects were witnessed early on in studies of pulse amplification in single-pass amplifiers.[188] At the same time, the gain of the weak probe drops with the same time constant as that of the strong square pulse. This time constant decreases with signal power (from 285 μs at an input of 0.75 μW to 155 μs at an input of 25 μW), as expected from the increase in the stimulated emission rate. Increases in the pump power also decrease the overshoot decay time constant. In contrast, the recovery time constant of the weak probe, 300 μs in Figure 9.39, is determined solely by the pump rate as the probe stimulated emission rate is too low to be a factor. For low pump powers, the gain recovery times can stretch to more than a ms.[187]

9.2. SYSTEM DEMONSTRATIONS AND ISSUES

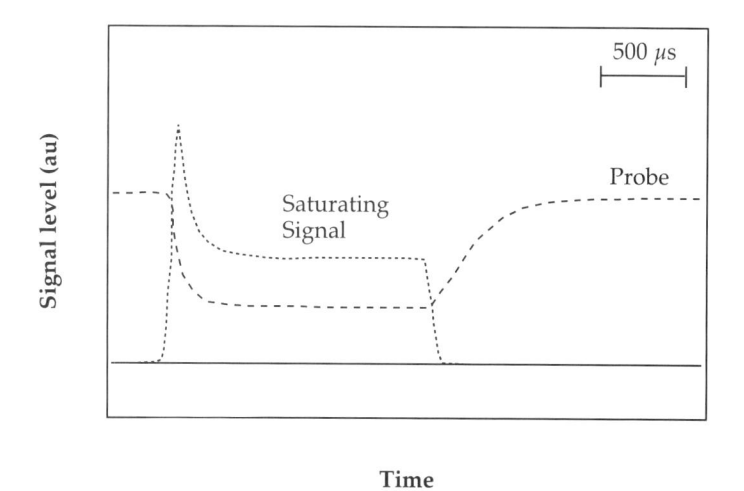

Figure 9.39: Oscilloscope trace of the signal level of a weak probe at 1537 nm (3.5 μW) and a square wave modulated strong saturating signal at 1531 nm (34 μW), simultaneously amplified in an erbium-doped fiber amplifier. From reference [186].

The gain dynamics prevalent when one or more channels is dropped from a WDM system have been modeled by means of an analytical solution.[189, 190] This approach builds on the analytical model of Saleh et al (which was described in Chapter 5, section 5.5) and is valid when ASE is not a significant factor in the signal gain (as would be satisfied when the small signal gain is low or in the strong signal case). Using this model, a two-stage amplifier was simulated in terms of the power excursion of the remaining channels in an eight channel WDM system, when one or more of the channels are dropped (as shown in Figure 9.40). Each channel had an input of −2 dBm to the amplifier and a gain of 9 dB. The transient response times are 29, 34, and 52 μs when dropping one, four, and seven channels, respectively, given the pump and signal characteristics of this particular amplifier operated well in saturation with high pump powers.[190] The reason for the increase in the response lifetime is that the power of the remaining channels, and hence the stimulated response time, is increasingly reduced for a larger number of dropped channels. The fastest response time is for a small number of dropped channels.

Faster transients, as well as oscillatory features, have been observed in the case of a WDM system multiamplifier chain subjected to channel dropping.[191] The transient response time is markedly reduced from the single amplifier value, to values on the order of a few μs at the output of a ten-amplifier chain.[191]

Methods to reduce the effect of transients include feedback on the pump power via the drive current applied to the pump laser and the addition of a "dummy" signal channel to keep the saturation level of the amplifier constant.[192] Motoshima et al showed that pump feedback to maintain the output of a weak probe signal constant resulted in a significant decrease of transient overshoots in the output waveform of a modulated strong signal pulse.[193] A recent result on pump feedback to stabilize an eight-channel

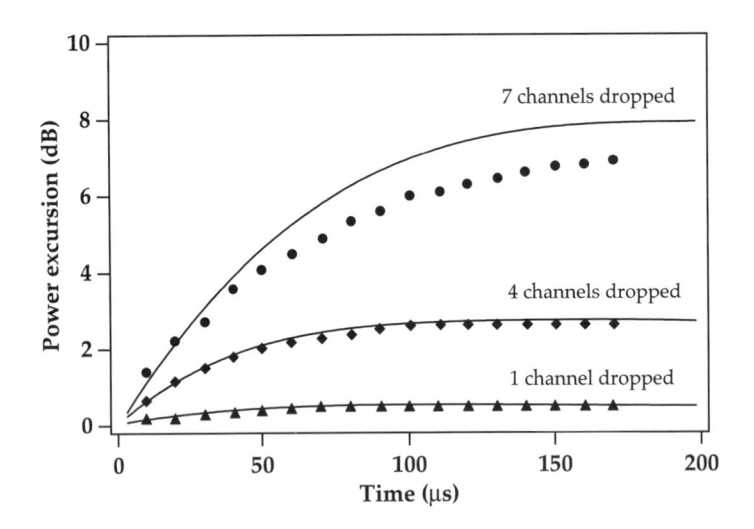

Figure 9.40: Power excursion as a function of time of the remaining channels in an eight channel WDM system when one, four, and seven channels are suddenly removed from the signal (markers: experimental measurement; smooth curves: analytic model). From reference [189]

WDM system where channels were added and dropped was the reduction of signal excursions from 6 dB to less than 0.5 dB. This was accomplished with feedback response times on the order of 10 μs by means of feedback on the pump power determined by the signal level, after the signal had traversed a single amplifier.[194]

The "dummy" signal method has been used to maintain the total signal input power to the amplifier, and as a consequence, the level of saturation of the amplifier, at a constant level.[195] The advantage of this technique is that it avoids feedback on the pump power level that may change the operating characteristics of the pump laser. The compensating signal is at a wavelength different than any of the signals, so that it can be filtered out before detection. The technique has shown a reduction in power penalty from 5.6 dB to 1.0 dB in a 622 Mb/s transmission experiment with 16 channels, where several channels are turned on and off periodically with turn-on and turn-off times of 0.5 ms.[195] The compensation signal method was used in a seven-channel WDM system where an eighth channel was added before the first amplifier of a eight-amplifier long haul span.[196] This control channel was modulated, based on the adding and dropping of signal channels, to maintain the total signal power constant. Signal ouput power transients were thus significantly reduced.[196]

As erbium-doped fiber amplifiers become a more pervasive component of multi-wavelength networks for signal distribution or access networks, the compensation of transient gain dynamics in response to changing network traffic will become an important tool.

9.3. SOLITON SYSTEMS

Limiting Factor	Effect NRZ	Remedy NRZ	Effect Soliton	Remedy Soliton
Dispersion	Pulse distortion.	Dispersion by management. Regeneration. Phase conjugation.	Canceled by SPM. Pulse collision in WDM.	No remedy needed.
PMD	Pulse distortion. (negligible effect for current fibers)	Polarization scrambling.	Pulse distortion. (negligible effect for current fibers)	
SPM	Pulse distortion. Spectral distortion.	Spectral broadening.	Canceled by dispersion.	No remedy needed.
FWM/XPM	WDM crosstalk.	Dispersion management.	Pulse collision in WDM.	Sliding frequency filter.
Amplifier noise accumulation.	SNR degradation.	Power management. Regeneration.	Timing jitter.	Sliding frequency filter. Synchronous modulation.

Table 9.7: Limiting factors for soliton and NRZ transmission systems.

9.3 SOLITON SYSTEMS

Soliton systems have the potential to lead the way to ultrahigh bit rate transmission systems. At very high data rates, nonlinearities start playing an important role, as discussed in Chapter 7. Solitons are able to capture these nonlinearities and make use of them, rather than combatting them. Soliton transmission systems are thus inherently nonlinear. NRZ systems are linear and try to eliminate or reduce the nonlinearities by a judicious choice of system parameters. Table 9.7 compares soliton and NRZ systems based on their ability to overcome or subjugate limiting factors. One can note that historically all transmission systems have been linear in nature, starting with the telegraph. Solitons would represent the first nonlinear transmission system employed to date.

In this section we will briefly review some important principles of solitons, in particular soliton pulses in optical fibers. We will also discuss the characteristics of solitons and the limiting factors of soliton transmission systems, followed by a description of systems demonstration experiments involving solitons in optical amplifier based transmission systems.

9.3.1 Principles

When short pulses of light with a certain frequency bandwidth travel through a glass medium such as a silica fiber, the dispersion of the fiber (material and waveguide) broadens the pulse. For long enough pulses this is a negligible effect. However, as the bit rate increases and the bit widths become shorter in time, the frequency bandwidth becomes larger and the dispersion broadens the pulse by a significant amount. Two

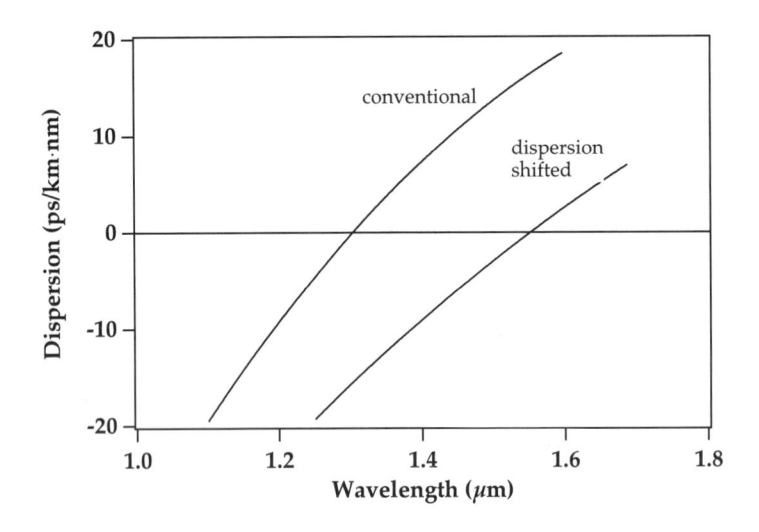

Figure 9.41: Dispersion parameters as a function of wavelength of conventional (dispersion zero at 1.3 µm) and dispersion shifted single mode silica glass fiber.

adjacent bits can then overlap in time and the information-transmitting capacity of the system is lost.

The dispersion parameter of a silica single-mode fiber is shown in Figure 9.41. For wavelengths above 1.3 µm the group velocity dispersion is negative (the dispersion parameter of a fiber is opposite in sign to the group velocity dispersion). Thus, higher frequencies travel faster than shorter ones (blue colors will lead the red colors). This is known as the anomalous dispersion regime. In the normal dispersion regime, below 1.3 µm, the shorter frequencies travel faster than the longer frequencies (i.e., red leads blue).

The index of refraction of a fiber also depends on the optical intensity, an effect known as the Kerr effect. The Kerr effect results in the index of refraction increasing with increasing light intensity. Since a pulse of light has a higher intensity at its peak than at the wings, this effect will vary along the profile of a pulse. The intensity-dependent refractive index shift leads to a intensity dependent phase shift across the pulse. The derivative of the nonlinear phase with respect to time, added to the optical carrier frequency, is the instantaneous frequency. The frequency shift relative to the optical carrier frequency is thus greatest where the optical power envelope of the pulse is changing the fastest, i.e., at the leading and trailing edges of the pulse. This effect is also known as self-phase modulation (see section 7.3.6). The leading edge of the pulse has an increasing intensity and an increasing nonlinear refractive index, and, as a result, experiences a negative nonlinear frequency shift. The leading edge of the pulse is thus red shifted by the Kerr effect. Similarly, the trailing edge is blue shifted. In a medium with positive group velocity dispersion, this effect will broaden the pulse temporally. The leading edge of the pulse will travel faster than the trailing edge of the pulse since the former has longer wavelength components than the latter. In a medium with negative group velocity dispersion, however, the Kerr effect will work in a

9.3. SOLITON SYSTEMS

direction opposite to the dispersion. The dispersion will slow down the red frequencies in the front of the pulse and advance the blue pulses at the trailing edge of the pulse.

The two effects, dispersion and Kerr effect, work against each other, and when certain conditions are met the pulse becomes stable, does not change shape or spectrum over long distance, and becomes what is known as a soliton. Solitons can be extremely advantageous for telecommunications as they can be propagated over extremely long distances without changing form and thus preserve the information content that is being transmitted. In contrast, nonsoliton pulses can change shape, causing adjacent pulses to soon overlap and the information content of each pulse to be lost.

A pulse needs to have a certain peak power and pulse shape for it to be a soliton pulse, given the fiber characteristics (geometry and glass medium, notably). The peak power for what is known as a fundamental soliton, which has a pulse shape with a sech^2 intensity profile, is

$$P_0 = \frac{A_{eff}\lambda}{4n_2 Z_0} \tag{9.3}$$

where A_{eff} is the effective cross sectional area of the fiber core, λ is the wavelength in vacuum of the optical carrier wave, n_2 is the nonlinear coefficient for the Kerr effect such that the dependence of the refractive index n on the intensity I of the pulse is given by

$$n(\omega, I) = n_0(\omega) + n_2 I \tag{9.4}$$

and Z_0 is the soliton period

$$Z_0 = 0.322c \left(\frac{\pi}{\lambda}\right)^2 \frac{\tau^2}{D} \tag{9.5}$$

where c is the speed of light in vacuum, τ is the width of the pulse (full width at half maximum, and D is the fiber dispersion parameter in ps/km · nm. For silica glass, a typical value of n_2 is 3×10^{-16} cm^2/W. The fiber dispersion parameter can be related to the group velocity dispersion parameter β_2 (which is merely the derivative of the group velocity with respect to frequency) by the equation

$$D = -\frac{2\pi c}{\lambda^2} \beta_2 \tag{9.6}$$

The fundamental soliton shape remains stable as the soliton travels along the fiber so long as the energy of the pulse is maintained relatively close to the soliton energy. For soliton transmission systems with periodic amplification, it has been shown that it is the "path average" soliton energy that needs to be close to the soliton energy. Indeed, with periodic amplification to compensate the loss in the interamplifier span, the pulse energy may deviate significantly from the soliton energy. What is important here to maintain soliton propagation is that the span length between amplifiers be short compared to the soliton period.[197, 198] Note also that it is the "path average" dispersion that matters in a soliton transmission system and that the local dispersion may vary from point to point along the fiber span.

A key feature of solitons is their restorative properties: If the pulse is perturbed not too far from the soliton conditions, it will tend to recover the characteristics of the soliton pulse. For example, a soliton pulse that has acquired some extra bandwidth will shed that bandwith as it propagates. A pulse that has the energy required for a soliton pulse will naturally evolve toward a soliton shape. This result originates from the fact that solitons are stable solutions of a certain class of equation, the nonlinear Schrodinger equation. The forces that create a soliton combine to push a soliton back to its original state if it deviates from it. This is an exceedingly valuable result for optical communications as it renders the system robust and resistant to small perturbations.

As for the polarization mode dispersion of the fiber, it is worth noting that if the PMD is small enough the soliton will not be significantly perturbed. The soliton will be resistant to the PMD of the fiber and will preserve its temporal and spectral characteristics, if the randomly varying birefringence index difference is "weaker" than the nonlinear index change. This will happen if the PMD is lower than a threshold value. Specifically, Mollenauer et al have shown that this condition can be written as PMD $\leq 0.3\sqrt{D}$, where PMD is the polarization mode dispersion of the fiber, in ps/$\sqrt{km}$, and D is the dispersion of the fiber in ps/nm · km.[199]

The noise characteristics of a soliton system have more features than does a linear system such as an NRZ-based transmission system. The soliton system, like the linear one, is still subject to the limit provided by ASE noise as embodied in the signal-spontaneous and spontaneous-spontaneous beat noises. In addition, soliton systems are subject to a noise limit known as the Gordon-Haus limit, which arises from a jitter in the arrival times of the solitons at the receiving end. In extreme cases, such jitter will result in temporal overlap between adjacent pulses and thus information loss. This jitter comes from the interaction between the ASE generated by the amplifiers and the soliton pulse. As described by Gordon and Haus, this interaction will produce a random frequency shift in the soliton's central frequency.[200] The soliton, because of dispersion, will then travel the remaining portion of the fiber at a different velocity. There will thus be, overall, a spread in the velocities of the different solitons that represent each bit and jitter in pulse arrival times relative to each another.

Figure 9.42 shows the experimentally obtained soliton pulse widths and standard deviation of pulse arrival times, compared to behavior expected from the Gordon-Haus theory. It can be shown that the variance of the jitter scales proportionally to the soliton energy. Hence, the greater the soliton energy, the greater the jitter. This effect runs contrary to the spontaneous-spontaneous beat noise effect, which is reduced in importance as the pulse energy is increased.

The combination of the two types of noise leads to a narrow window of optimal pulse energies for a given transmission distance. For long transmission systems, the limiting noise factor is the Gordon-Haus effect. Figure 9.43 shows the distance limits arising from the Gordon-Haus effect and the ASE noise (and resulting window) for a soliton system operating at 5 Gb/s.[202]

It was discovered in 1991–1992 that the Gordon-Haus effect could be effectively reduced by means of frequency filters inserted along the fiber.[203, 204] The idea is based on the restorative conditions under which solitons operate. The peaks of the filters are set at the original soliton center frequency. For example, consider a soliton that has been subject to a frequency shift: its center frequency is no longer aligned with that of the filter. Those frequencies that are near the filter peak will be favored over those farther from it. New frequency components at the filter peak can be generated by

9.3. SOLITON SYSTEMS 371

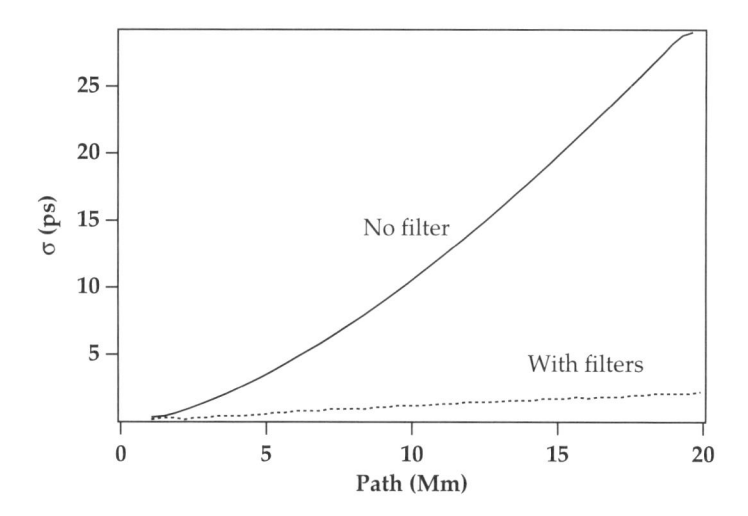

Figure 9.42: Soliton pulse width (σ) and standard deviation of pulse arrival times (τ_{eff}) as a function of transmission distance and calculated from the Gordon-Haus theory (dashed lines). From reference [201].

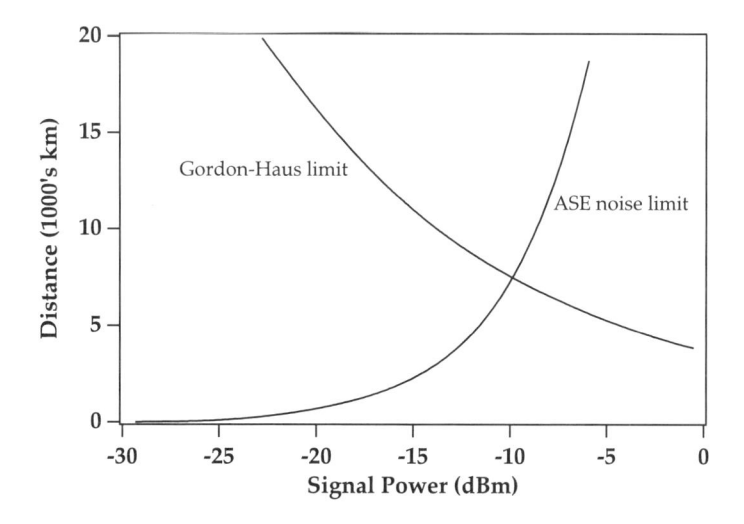

Figure 9.43: Maximum transmission distances (10^{-9} bit error rate), as limited by the Gordon-Haus effect and ASE noise, as a function of the soliton input power. From the numerical simulations of reference [202].

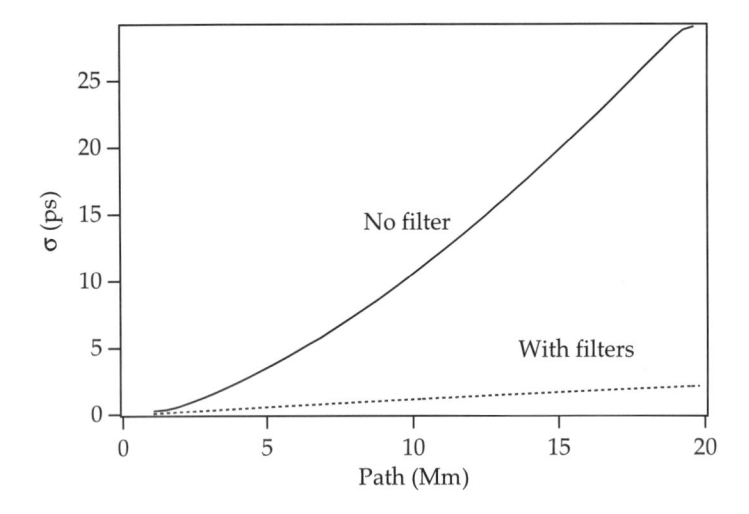

Figure 9.44: Calculated standard deviation in soliton pulse arrival times, with and without sliding guiding filters, as a function of transmission distance. From reference [205].

the soliton pulse via self-phase modulation. As a result, the central frequency of the soliton is returned to its original position and the jitter in arrival times arising from the ASE-noise-induced frequency shift is strongly reduced. These filters also act to reduce amplitude variation in the soliton power.

One drawback of using filters that are all tuned to an identical fixed central wavelength is that ASE noise centered at the filter peak can build up exponentially over the transmission span. This significantly limits the usefulness of the fixed center frequency filter technique. A solution was proposed by Mollenauer et al in the form of filters that have differing center frequencies as one moves along the transmission span.[205] This technique is known as sliding guiding filtering. Its applicability stems again from the robustness of solitons, which allows them to adjust their central frequency to that of the filters. This occurs when the center frequencies of the filters are gradually translated along the transmission span. The change in center frequency essentially ensures that one dominant ASE frequency cannot grow exponentially with distance. In essence, the use of the filter technique moves the Gordon-Haus limit curve of Figure 9.43 much farther to the right-hand side. The sliding frequency technique has yielded impressive results for ultra long soliton transmission at high bit rates.[206, 207, 208] Figure 9.44 shows the calculated standard deviation in soliton arrival times, as a function of transmission distance, for transmission both with and without filters. The filters produce a dramatic reduction in pulse arrival time jitter.

Other schemes have been explored for combating the Gordon-Haus effect. One notable technique is that of synchronous modulation.[209] This technique involves placing optical modulators in soliton transmission systems for retiming and reshaping of the pulses. The clock signal of the solitons is extracted from the data stream and then applied to a $LiNbO_3$ or Mach-Zehnder modulator, which synchronously reshapes

9.3. SOLITON SYSTEMS

the solitons.[210, 211] The solitons are thus retimed and their jitter eliminated. ASE noise is also significantly reduced. In one experiment, the modulators were placed every 500 km and, by reshaping and retiming the solitons, they helped to reduce the effects of the Gordon-Haus limit, the soliton interaction effects, and the accumulated ASE noise. 10 Gb/s solitons were thus propagated over 1,000,000 km of fiber.[212] A WDM version of this single-channel experiment transmitted 4 channels at 10 Gb/s over 10,000 km.[213] Other techniques reported to combat the Gordon-Haus effect are the use of alternating amplitude solitons (5% amplitude variations) and periodic dispersion compensation.[214, 215]

Polarization division multiplexing also combats the soliton interaction effects. It has been shown that solitons conserve their respective polarization states over distances up to 10,000 km. Solitons that enter the transmission system with perpendicular polarization states, for example, will remain so throughout the entire system.[216] It has also been shown that the interaction between perpendicularly polarized solitons is significantly reduced, as compared to that between parallel polarized solitons.[216, 217]. Thus alternating the polarization of the soliton pulses in a transmission system allows for a higher bit rate than would be achievable with parallel polarized solitons.

Solitons can also form the basis for a WDM system. They have an advantage over linear systems in this respect, because in the latter the pulses are centered near the wavelength of zero dispersion of the fiber to minimize pulse broadening via dispersion. A WDM system then has an issue in that some of the channels will be removed in wavelength from the zero dispersion point and will suffer some dispersive broadening which needs to be compensated for. For the soliton system the dispersion is compensated for by the nonlinear effects. In soliton WDM systems, however, the main worry is the temporal displacement of the pulses resulting from soliton collisions. In the case of a fiber with uniform loss, the soliton collisions leave the solitons unaltered: any frequency shifts or energy transfer suffered by the solitons in the first half of the collision are reversed in the second half. The presence of amplifiers removes the symmetry of the collision, since a portion of the collision can now take place at an amplifier.[218, 219] It has been shown that when the collision length is long compared to the spacing between amplifiers, the interaction caused by cross-phase modulation is minimal and the situation is essentially the same as that of the uniform fiber.[220] As an example, the collision of two soliton pulses of pulse width 50 ps with wavelength separation $\Delta\lambda = 1$ nm, in a fiber of dispersion $D = 1$ ps/nm $\cdot$ km, lasts for a distance $2\tau\, D\Delta\lambda = 100$ km (measured at the overlap of the half power points). Sliding frequency filters also act to reduce the collision effect.[221] A technique to correct any remaining nonrandom jitter (other causes include interaction between solitons and acoustic phonons and polarization scattering during collisions) has been proposed by Mollenauer.[222]

In addition to the cross-phase modulation that occurs during a soliton collision, there can also occur the creation of four-wave mixing components during the collision of two solitons with different wavelengths. This effect was discovered to have a significant negative impact on the ability of a WDM soliton system to operate with high capacity (even two-channel operation is hindered).[223] The four-wave mixing effect is increased by the presence of the pseudo phase matching created by the periodic variation in pulse intensity engendered by the span loss and amplifiers. When the amplifier

Year	Num. of ch.	Total bit rate (Gb/s)	System length (km)	Total capacity (Gb/s·km)	Pulse width (ps)	Amp. spacing (km)	Num. of inline amps.	Ref.
1990	2	1	106	106	70	36-70	1	[229]
1990	1	5	100	500	20	25	3	[230]
1990	1	20	200	4,000	20	25	7	[231]
1990	1	5	250	1,250	20	25	9	[232]
1990	1	5	250	1,250	20	25	9	[233]
1991	1	10	1,000	10,000	45	50	19	[234]
1992	1	32	90	2,880	16	30	2	[235]
1992	1	5	3,000	15,000	35	33	91	[236]
1992	1	10	1,200	12,000	30	50	23	[237]
1993	1	20	1,850	37,000	12	50	36	[238]
1993	1	40	750	30,000	12	50	36	[238]
1993	1	20	2,000	40,000	12	50	39	[239]
1993	1	40	1,000	40,000	12	50	39	[239]
1994	1	4	310	1,240	55	38-59	6	[240]
1994	1	20	3,000	60,000	11	50	60	[210]
1995	1	9	416	3,744	24	48-57	7	[241]
1995	1	160	200	32,000	1.5	25	9	[242]
1995	1	10	4000	40,000	19	40	100	[243]
1995	1	20	8100	162,000	10	33-36	240	[244]
1996	1	40	500	20,000	10	100	4	[245]
1996	1	40	4000	160,000	6-10	50	79	[246]

Table 9.8: Soliton transmission system straight line experiments with erbium-doped fiber amplifiers as inline amplifiers (Num. of ch.: number of wavelength channels; Amp. spacing: amplifier spacing; Num. of inline amps.: number of inline amplifiers).

separation allows for the phase-matching condition between two wavelength channels, the effect leads to both amplitude and timing jitter for the soliton pulses. The solution to this problem proposed by Mamyshev and Mollenauer is to use fiber with decreasing dispersion in the span between two amplifiers. In practice, this continuous downwards taper of the fiber dispersion can be replaced by a stepwise drop in dispersion with as little as two steps.[223] This advance has opened the way to high capacity WDM transmission with solitons.[224]

9.3.2 System Results and Milestones

Very encouraging results have been obtained with ultra long distance transmission using soliton pulses. The first experimental demonstration of soliton transmission in optical fibers occurred in 1980.[225] In the experiment performed by Mollenauer et al, a train of pulses with pulsewith 7 ps was launched into a 700 m long fiber. The pulse shape evolved periodically with propagation, exhibiting the signatures of high-order solitons. In 1988, Mollenauer and Smith built the first soliton transmission system for

9.3. SOLITON SYSTEMS 375

Year	Num. of ch.	Total bit rate (Gb/s)	System length (km)	Total capacity (Gb/s·km)	Pulse width (ps)	Amp. spacing (km)	Ref.
1990	1	4	10,000	40,000	50	25	[247]
1991	2	4	9,000	18,000	60	30	[248]
1991	1	2.5	14,000	38,000	50	25	[249]
1991	1	10	1,000,000	10,000,000	35-45	50	[212]
1992	2	10	11,250	75,000	40	27	[250]
1993	1	8.2	4,200	34,440	20	25-28	[251]
1993	1	8.2	4,200	34,440	20	25	[252]
1993	1	5	13,100	65,000	24	30	[253]
1993	1	10	20,000	200,000	18	26	[254]
1993	2	20	13,000	260,000	18	26	[254]
1994	1	5	11,000	55,000	23	33	[255]
1994	1	10	7,200	72,000	15	50	[256]
1994	1	20	11,500	230,000	12	30	[257]
1994	1	20	18,000	360,000	23	33	[258]
1994	1	15	25,000	375,000	16	26	[259]
1995	1	20	9,000	180,000	10	30-36	[260]
1995	8	20	10,000	200,000	60	45	[261]
1995	1	10	2,700	27,000	20	90	[262]
1995	1	10	19,000	190,000	15	63	[263]
1995	1	10	30,000	300,000	15	30	[207]
1996	4	40	10,000	400,000	22-25	50	[213]
1996	3	60	10,000	600,000	10-12	50	[264]
1996	7	70	9,400	658,000	20	33	[208]
1996	8	80	9,500	760,000	20	33	[224]
1996	3	60	7,000	420,000	10-15	35	[265]

Table 9.9: Recirculating loop soliton transmission experiments with erbium-doped fiber amplifiers as inline amplifiers (Num. of ch.: number of wavelength channels; Amp. spacing: amplifier spacing).

transoceanic distances.[226] In this experiment, a propagation distance of 4000 km was simulated by using a fiber loop, similar to the one shown in Figure 9.13 and described in Section 9.2.2. The soliton pulse duration was 55 ps.

In the initial soliton transmission experiments, the propagation loss was compensated for by using distributed Raman gain.[226, 227, 228] In each one of these experiments, the Raman gain was obtained by pumping the fiber using a color center laser or a diode laser. Since solitons involve continual balance of dispersion and nonlinear effects along the fiber, the prevalent notion was that only distributed amplifiers could be used to compensate for the loss accumulated in long distance transmission.

In 1991, Mollenauer et al proved analytically and by numerical simulations that soliton energy loss can be compensated for by using lumped amplifiers.[198] The im-

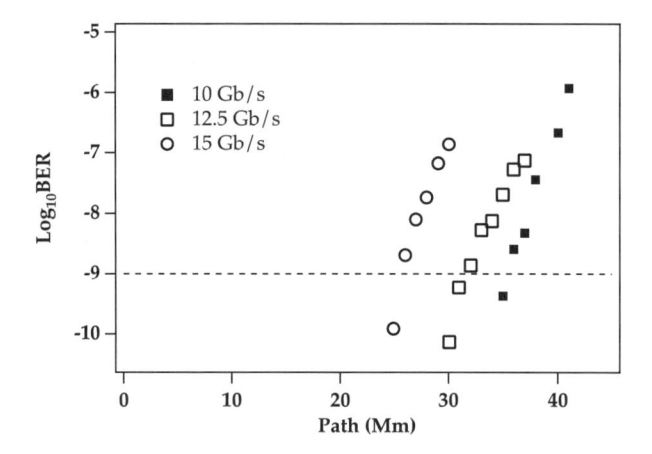

Figure 9.45: Experimental bit error rates measured in a single wavelength channel soliton loop transmission experiment, as a function of distance. From reference [259].

portant condition obtained in their calculations is that the separation between the amplifiers in the system has to be short compared to the soliton period to ensure that the nonlinear and dispersion terms in the propagation equation do not produce significant broadening in the time or frequency domains.

Soliton transmission results using erbium-doped fiber amplifiers are shown in Tables 9.8 and 9.9. The first table has results of straight line experiments. The second table contains results of soliton transmission using fiber loops.

The first demonstration of soliton transmission using erbium-doped fiber amplifiers succeeded in transmitting two wavelengths simultaneously over a 30 km length at a 100 MHz repetition rate.[266] The first high bit rate error-free soliton transmission experiment using erbium-doped fiber amplifiers was a straight line experiment performed by Nakazawa et al. The data rate was 5 Gb/s and the signal was transmitted over a distance of 100 km of fiber, with the help of three erbium-doped fiber amplifiers spaced every 25 km.[230] The amplitudes of the solitons at launch were set slightly above the energy required for $N = 1$ solitons to precompensate for energy loss over the span.

The first soliton transmission experiment over transoceanic distances using erbium-doped amplifiers involved a fiber loop with four amplifiers. Solitons at 4 Gb/s were transmitted for a distance of 10,000 km over dispersion-shifted fiber.[201] Subsequent experiments demonstrated that solitons could be transmitted error-free at high bit rates over distances exceeding 20,000 km, as shown in Figure 9.45 from a 1994 experiment by Mollenauer et al.[259] The synchronous modulation and reshaping technique described earlier allowed soliton recirculation in a loop for a total distance of 1 million km.[212] Straight line experiments have reached distances over 8000 km at 20 Gb/s.[244]

Soliton transmission over non dispersion-shifted single-mode fiber has also been demonstrated.[240] 55 ps pulses at 4 Gb/s were propagated over a distance of 310 km of standard single-mode fiber. Since the fiber dispersion is high, the power needed to generate solitons at 1.5 μm is higher than that needed when using dispersion-shifted

9.3. SOLITON SYSTEMS

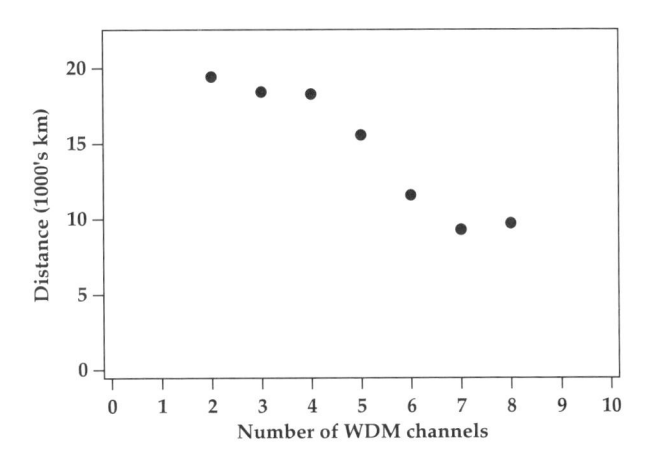

Figure 9.46: Error-free distances (10^{-9} bit error rate), as a function of the number of soliton channels at 10 Gb/s. For a single channel (N = 1), the error-free distance is 35,000 km. For N = 2 through N = 7 the erbium-doped fiber amplifiers are pumped at 1480 nm, and for N = 8 the amplifiers are pumped at 980 nm. From reference [224].

fiber. Soliton transmission at 10 Gb/s has been demonstrated over an installed undersea cable of 2700 km length.[262] The cable was made of dispersion-shifted fiber and the amplifier spacing was 90 km. This system was designed to be used with NRZ pulses. A terrestrial network in the Tokyo metropolitan area, designed for a 2.4 Gb/s intensity-modulated NRZ transmission system, was also used to demonstrate the transmission of solitons at 10 Gb/s over 2000 km.[267]

Significant results have been obtained in soliton transmission at multiple wavelengths. Eight soliton channels at 2.5 Gb/s were transmitted over a distance of 10,000 km.[261] This experiment used a 2400 km long fiber loop, with 54 erbium-doped fiber amplifiers. The separation between channels was 0.2 nm. This channel spacing is significantly reduced compared to NRZ systems, where four-wave mixing between adjacent channels can be a serious problem. In 1996, Mollenauer et al reported an eight-channel (each at 10 Gb/s) soliton transmission experiment over more than 10,000 km.[224] This experiment used dispersion tapering in each span of 33.3 km length between successive amplifiers. The dispersion tapering was accomplished in three to four steps. The channel wavelengths ranged from 1554 nm to 1558 nm, with channel separations of 1 nm. A summary of the transmission distance attainable, as a function of the number of channels, is shown in Figure 9.46. Soliton based transmission systems have started stirring commercial interest and have been tested in a real network environment, with a demonstration of a four channel WDM system over 450 km over an existing terrestrial link.[268]

Bibliography

[1] B. L. Kasper, in *Optical Fiber Telecommunications II*, Vol. II, Chapter 18, S. E. Miller and I. P. Kaminow, Eds., (Academic Press, San Diego, 1988).

[2] B. L. Kasper and J. C. Campbell, *J. Light. Tech.* **LT-5**, 1351 (1987).

[3] T. Saito, Y. Aoki, K. Fukagai, S. Ishikawa, and S. Fujita, "High receiver sensitivity in multi-Gbit/s region with 0.98-μm backward-pumped Er-doped fiber preamplifier," in *Conference on Optical Fiber Communication*, Vol. 5, 1992 OSA Technical Digest Series (Optical Society of America, Washington, D.C., 1992), pp. 206–207.

[4] C. R. Giles, E. Desurvire, J. R. Talman, J. R. Simpson, and P. C. Becker, *J. Light. Tech.* **4**, 651 (1989).

[5] C. R. Giles, E. Desurvire, J. L. Zyskind, and J. R. Simpson, *IEEE Phot. Tech. Lett.* **1**, 367 (1989).

[6] A. H. Gnauck and C. R. Giles, *IEEE Phot. Tech. Lett.* **4**, 80 (1992).

[7] K. Hagimoto, K. Iwatsuki, A. Takada, M. Nakazawa, M. Saruwatari, K. Aida, K. Nakagawa, and M. Horiguchi, "A 212 km non-repeatered transmission experiment at 1.8 Gb/s using ld pumped Er^{3+}-doped fiber amplifiers in an IM/direct-detection repeaterless system," in *Conference on Optical Fiber Communication*, Vol. 5, 1989 OSA Technical Digest Series (Optical Society of America, Washington, D.C., 1989), pp. 253–256.

[8] T. Saito, Y. Sunohara, K. Fukagai, S. Ishikawa, N. Hanmi, S. Fujita, and Y. Aoki, "High receiver sensitivity at 10 Gb/s using an Er-doped fiber preamplifier pumped with a 0.98 μm laser-diode," in *Conference on Optical Fiber Communication*, Vol. 4, 1991 OSA Technical Digest Series (Optical Society of America, Washington, D.C., 1991), pp. 292–295.

[9] P. P. Smyth, R. Wyatt, A. Fidler, P. Eardley, A. Sayles, and S. Craig-Ryan, *Elect. Lett.* **26**, 1604 (1990).

[10] P. M. Gabla, V. Lemaire, H. Krimmel, J. Otterbach, J. Auge, and A. Dursin, *IEEE Phot. Tech. Lett.* **3**, 727 (1991).

[11] Y. K. Park, S. W. Granlund, T. W. Cline, L. D. Tzeng, J. S. French, J.-M. P. Delavaux, R. E. Tench, K. Korotky, J. J. Veselka, and D. J. DiGiovanni, *IEEE Phot. Tech. Lett.* **4**, 179 (1992).

[12] R. I. Laming, A. H. Gnauck, C. R. Giles, M. N. Zervas, and D. N. Payne, *IEEE Phot. Tech. Lett.* **4**, 1348 (1992).

[13] P. M. Gabla, E. Leclerc, J. F. Marcerou, and J. Hervo, "Ninety-two photons/bit sensitivity using an optically preamplified direct-detection receiver," in *Conference on Optical Fiber Communication*, Vol. 5, 1992 OSA Technical Digest Series (Optical Society of America, Washington, D.C., 1992), p. 245.

[14] Y. K. Park, J.-M. P. Delavaux, O. Mizuhara, L. D. Tzeng, T. V. Nguyen, M. L. Kao, P. D. Yeastes, S. W. Granlund, and J. Stone, "5-Gbit/s optical preamplifier receiver with 135-photons/bit usable receiver sensitivity," in *Conference on Optical Fiber Communication*, Vol. 4, 1993 OSA Technical Digest Series (Optical Society of America, Washington, D.C., 1993), pp. 16–17.

[15] T. Kataoka, Y. Miyamoto, K. Hagimoto, and K. Noguchi, *Elect. Lett.* **30**, 715 (1994).

[16] J. C. Livas, "High sensitivity optically preamplified 10 Gb/s receivers," in *Conference on Optical Fiber Communication*, Vol. 2, 1996 OSA Technical Digest Series (Optical Society of America, Washington, D.C., 1996), pp. 343–345.

[17] Y. K. Park and S. W. Granlund, *Opt. Fiber Tech.* **1**, 59 (1994).

[18] D. Marcuse, *J. Light. Tech.* **9**, 356 (1991).

[19] E. Lichtman and S. Evangelides, *Elect. Lett.* **30**, 346 (1994).

[20] A. Naka and S. Saito, *Elect. Lett.* **28**, 2221 (1992).

[21] A. Naka and S. Saito, *J. Light. Tech.* **12**, 280 (1994).

[22] M. Murakami, T. Takahashi, M. Aoyama, T. Imai, M. Amemiya, M. Sumida, and M. Aiki, *J. Light. Tech.* **14**, 2657 (1996).

[23] N. Edagawa, K. Mochizuki, and H. Wakabayashi, "First sea trial of an optical amplifier submarine cable system," in *Seventh International Conference on Integrated Optics and Optical Fiber Communications*, Proceedings Vol 5, pp. 66–67.

[24] M. Z. Iqbal, J. L. Gimlett, M. M. Choy, A. Yi-Yan, M. J. Andrejco, L. Curtis, M. A. Saifi, C. Lin, and N. K. Cheung, *IEEE Phot. Tech. Lett.* **1**, 334 (1989).

[25] N. Edagawa, Y. Yoshida, H. Taga, S. Yamamoto, and H. Wakabayashi, *IEEE Phot. Elect. Lett.* **2**, 274 (1990).

[26] S. Ryu, N. Edagawa, Y. Yoshida, and H. Wakabayashi, *IEEE Phot. Tech. Lett.* **2**, 428 (1990).

[27] K. Hagimoto, S. Nishi, and K. Nakagawa, *J. Light. Tech.* **8**, 1387 (1990).

[28] S. Saito, T. Imai, and T. Ito, *J. Light. Tech.* **9**, 161 (1991).

[29] N. S. Bergano, J. Aspell, C. R. Davidson, P. R. Trischitta, B. M. Nyman, and F. W. Kerfoot, *Elect. Lett.* **27**, 1889 (1991).

[30] N. S. Bergano, J. Aspell, C. R. Davidson, P. R. Trischitta, B. M. Nyman, and F. W. Kerfoot, "A 9000 km 5 Gb/s and 21,000 km 2.4 Gb/s feasibility demonstration of transoceanic EDFA systems using a circulating loop," in *Conference on Optical Fiber Communication*, Vol. 4, 1991 OSA Technical Digest Series (Optical Society of America, Washington, D.C., 1991), pp. 288–291.

[31] T. Imai, M. Murakami, Y. Fukada, M. Aiki, and T. Ito, "Over 10,000 km straight line transmission system experiment at 2.5 Gb/s using in-line optical amplifiers," in *Optical Amplifiers and Their Applications*, Vol. 17, 1992 OSA Technical Digest Series (Optical Society of America, Washington, D.C., 1992), pp. 280–283.

[32] H. Taga, N. Edagawa, Y. Yoshida, S. Yamamoto, M. Suzuki, and H. Wakabayashi, "10 Gbit/s, 4500 km transmission experiment using 138 cascaded Er-doped fiber amplifiers," in *Conference on Optical Fiber Communication*, Vol. 5, 1992 OSA Technical Digest Series (Optical Society of America, Washington, D.C., 1992), pp. 355–358.

[33] N. S. Bergano, C. R. Davidson, G. M. Homsey, D. J. Kalmus, P. R. Trischitta, J. Aspell, D. A. Gray, R. L. Maybach, S. Yamamoto, H. Taga, N. Edagawa, Y. Yoshida, Y. Horiuchi, T. Kawazawa, Y. Namihira, and S. Akiba, "9,000 km, 5 Gb/s NRZ transmission experiment using 274 erbium-doped fiber-amplifiers," in *Optical Amplifiers and Their Applications*, Vol. 17, 1992 OSA Technical Digest Series (Optical Society of America, Washington, D.C., 1992), pp. 276–279.

[34] S. Saito, M. Murakami, A. Naka, Y. Fukada, T. Imai, M. Aiki, and T. Ito, *IEEE J. Light. Tech.* **10**, 1117 (1992).

[35] H. Taga, N. Edagawa, H. Tanaka, M. Suzuki, S. Yamamoto, H. Wakabayashi, N. S. Bergano, C. R. Davidson, G. M. Homsey, D. J. Kalmus, P. R. Trischitta, D. A. Gray, and R. L. Maybach, "10 Gbit/s, 9,000 km IM-DD transmission experiments using 274 Er-doped fiber amplifier reepeaters," in *Conference on Optical Fiber Communication*, Vol. 4, 1993 OSA Technical Digest Series (Optical Society of America, Washington, D.C., 1993), pp. 275–278.

[36] C. D. Chen, J-M. P. Delavaux, B. W. Hakki, O. Mizuhara, T. V. Nguyen, R. J. Nuyts, K. Ogawa, Y. K. Park, C. S. Skolnick, R. E. Tench, J. Thomas, L. D. Tzeng, and P. D. Yeates, "A field demonstration of 10 Gb/s - 360 km transmission through embedded standard (non-DSF) fiber cables," in *Conference on Optical Fiber Communication*, Vol. 4, 1994 OSA Technical Digest Series (Optical Society of America, Washington, D.C., 1994), pp. 442–445.

[37] M. Murakami, T. Takahashi, M. Aoyama, M. Amemiya, M. Sumida, N. Ohkawa, Y. Fukada, T. Imai, and M. Aiki, *Elect. Lett.* **31**, 814 (1995).

[38] W. S. Lee and A. Hadjifotiou, "Optical transmission over 140 km at 40 Gbit/s by OTDM," in *Conference on Optical Fiber Communication*, Vol. 8, 1995 OSA Technical Digest Series (Optical Society of America, Washington, D.C., 1995), pp. 286–287.

[39] S. Kawanishi, H. Takara, O. Kamatani, and T. Morioka, "100-Gbit/s 500-km optical transmission experiment," in *Conference on Optical Fiber Communication*, Vol. 8, 1995 OSA Technical Digest Series (Optical Society of America, Washington, D.C., 1995), pp. 287–288.

[40] D. Simeonidou, K. P. Jones, A. Rey, M. S. Chaudhry, P. A. Farrugia, G. G. Windus, N. H. Taylor, and P. R. Markel, "5 Gbit/s transmission over 15,200 km on a subsegment of the TAT-12 transatlantic submarine cable system," in *Conference on Optical Fiber Communication*, Vol. 2, 1996 OSA Technical Digest Series (Optical Society of America, Washington, D.C., 1996), pp. 71–72.

[41] Y. K. Park, T. V. Nguyen, O. Mizuhara, C. D. Chen, L. D. Tzeng, P. D. Yeates, F. Heismann, Y. C. Chen, D. G. Ehrenberg, and J. C. Feggeler, "Field demonstration of 10 Gbit/s line-rate transmission on an installed transoceanic submarine lightwave cable," in *Conference on Optical Fiber Communication*, Vol. 2, 1996 OSA Technical Digest Series (Optical Society of America, Washington, D.C., 1996), pp. 75–76.

[42] L. F. Mollenauer and K. Smith, *Opt. Lett.* **13**, 675 (1988).

[43] N. S. Bergano and C. R. Davidson, *J. Light. Tech.* **13**, 879 (1995).

[44] N. Edagawa, K. Mochizuki, and H. Wakabayashi, *Elect. Lett.* **25**, 363 (1989).

[45] N. S. Bergano and C. R. Davidson, *J. Light. Tech.* **14**, 1299 (1996).

[46] H. Taga, *J. Light. Tech.* **14**, 1287 (1996).

[47] J. C. Feggeler, C.-C. Chen, Y. C. Chen, J.-M. P. Delavaux, F. L. Heisman, G. M. Homsey, H. D. Kidorf, T. M. Kissel, J. L. Miller, Jr., J. A. Nagel, R. Nuyts, Y.-K. Park, and W. W. Patterson, *Elect. Lett.* **32**, 1314 (1996).

[48] R. W. Tkach, A. R. Chraplyvy, F. Forghieri, A. H. Gnauck, and R. M. Derosier, J. Light. Tech. **13**, 841 (1995).

[49] T. Terahara, T. Naito, N. Shimojoh, T. Chikama, and M. Suyama, *Elect. Lett.* **33**, 603 (1997).

[50] A. R. Chraplyvy, J. A. Nagel, and R. W. Tkach, *IEEE Phot. Tech. Lett.* **4**, 920 (1992).

[51] N. S. Bergano, C. R. Davidson, A. M. Vengsarkar, B. M. Nyman, S. G. Evangelides, J. M. Darcie, M. Ma, J. D. Evankow, P. C. Corbett, M. A. Mills, G. A. Ferguson, J. R. Pedrazzani, J. A. Nagel, J. L. Zyskind, J. W. Sulhoff, and A. J. Lucero, "100 Gb/s WDM transmission of twenty 5 Gb/s NRZ data channels over transoceanic distances using a gain flattened amplifier chain," *Proceedings of the 21st European Conference on Optical Communication*, Vol. 3, pp. 967–970 (1995).

[52] H. Taga, Y. Yoshida, N. Edagawa, S. Yamamoto, and H. Wakabayashi, "459 km, 2.4 Gbit/s 4 wavelength multiplexing optical fiber transmission experiment using 6 Er-doped fiber amplifiers," in *Conference on Optical Fiber Communication*, Vol. 13, 1990 OSA Technical Digest Series (Optical Society of America, Washington, D.C., 1990), pp. 273–276.

[53] P. M. Gabla, J. O. Frorud, E. Leclerc, S. Gauchard, and V. Havard, *IEEE Phot. Tech. Lett.* **4**, 717 (1992).

[54] H. Taga, N. Edagawa, S. Yamamoto, Y. Yoshida, Y. Horiuchi, T. Kawazawa, and H. Wakabayashi, "Over 4,500 km IM-DD 2-channel WDM transmission experiments at 5 Gbit/s using 138 in-line Er-doped fiber amplifiers," in *Conference on Optical Fiber Communication*, Vol. 4, 1993 OSA Technical Digest Series (Optical Society of America, Washington, D.C., 1993), pp. 287–290.

[55] A. P. Chraplyvy, A. H. Gnauck, R. W. Tkach, and R. M. Derosier, *IEEE Phot. Tech. Lett.* **5**, 1233 (1993).

[56] A. R. Chraplyvy, J.-M. Delavaux, R. M. Derosier, G. A. Ferguson, D. A. Fishman, C. R. Giles, J. A. Nagel, B. M. Nyman, J. W. Sulhoff, R. E. Tench, R. W. Tkach, and J. L. Zyskind, *IEEE Phot. Tech. Lett.* **6**, 1371 (1994).

[57] A. P. Chraplyvy, A. H. Gnauck, R. W. Tkach, and R. M. Derosier, "160-Gb/s (8 × 20 Gb/s WDM) 300-km transmission with 50-km amplifier spacing and span-by-span dispersion reversal," in *Conference on Optical Fiber Communication*, Vol. 4, 1994 OSA Technical Digest Series (Optical Society of America, Washington, D.C., 1994), pp. 410–413.

[58] N. S. Bergano and C. R. Davidson, "Four-channel WDM transmission experiment over transoceanic distances," in *Optical Amplifiers and Their Applications*, Vol. 14, 1994 OSA Technical Digest Series (Optical Society of America, Washington, D.C., 1994), pp. 202–206.

[59] A. R. Chraplyvy, A. H. Gnauck, R. W. Tkach, R. M. Derosier, C. R. Giles, B. M. Nyman, G. A. Ferguson, J. W. Sulhoff, and J. L. Zyskind, *IEEE Phot. Tech. Lett.* **7**, 98 (1995).

[60] K. Oda, M. Fukutoku, M. Fukui, T. Kitoh, and H. Toba, "10-channel × 10-Gbit/s optical FDM transmission over a 500-km dispersion-shifted fiber employing unequal channel spacing and amplifier-gain equalization," in *Conference on Optical Fiber Communication*, Vol. 8, 1995 OSA Technical Digest Series (Optical Society of America, Washington, D.C., 1995), pp. 27–29.

[61] K. Motoshima, K. Takano, J. Nakagawa, and T. Katayama, "Eight-channel 2.5-Gbit/s WDM transmission over 275 km using directly modulated 1.55-μm MQW DFB-LDs," in *Conference on Optical Fiber Communication*, Vol. 8, 1995 OSA Technical Digest Series (Optical Society of America, Washington, D.C., 1995), pp. 30–32.

[62] N. S. Bergano, C. R. Davidson, B. M. Nyman, S. G. Evangelides, J. M. Darcie, J. D. Evankow, P. C. Corbett, M. A. Mills, G. A. Ferguson, J. A. Nagel, J. L. Zyskind, J. W. Sulhoff, A. J. Lucero, and A. A. Klein, "40 Gb/s WDM transmission of eight 5 Gb/s data channels over transoceanic distances using the conventional NRZ modulation format," in *Conference on Optical Fiber Communication*, Vol. 8, 1995 OSA Technical Digest Series (Optical Society of America, Washington, D.C., 1995), pp. 397–400.

[63] K. Oda, M. Fukutoku, M. Fukui, T. Kitoh, and H. Toba, "16-channel × 10-Gbit/s optical FDM transmission over a 1000 km conventional single-mode fiber employing dispersion-compensating fiber and gain equalization," in *Conference on Optical Fiber Communication*, Vol. 8, 1995 OSA Technical Digest Series (Optical Society of America, Washington, D.C., 1995), pp. 408–411.

[64] S. Yoshida, S. Kuwano, N. Takachio, and K. Iwashita, "10 Gbit/s × 10-channel WDM transmission experiment over 1200 km with repeater spacing of 100 km without gain equalization or preemphasis," in *Conference on Optical Fiber Communication*, Vol. 2, 1996 OSA Technical Digest Series (Optical Society of America, Washington, D.C., 1996), pp. 19–21.

[65] H. Onaka, H. Miyata, G. Ishikawa, K. Otsuka, H. Ooi, Y. Kai, S. Kinoshita, M. Seino, H. Nishimoto, and T. Chikama, "1.1 Tb/s WDM transmission over a 150 km 1.3 μm zero-dispersion single-mode fiber," in *Conference on Optical Fiber Communication*, Vol. 2, 1996 OSA Technical Digest Series (Optical Society of America, Washington, D.C., 1996), pp. 403–406.

[66] M. Suyama, H. Iwata, S. Harasawa, and T. Naito, "Improvement of WDM transmission performance by non-soliton RZ coding - a demonstration using 5 Gb/s 8-channel 4500 km straight line test bed," in *Conference on Optical Fiber Communication*, Vol. 2, 1996 OSA Technical Digest Series (Optical Society of America, Washington, D.C., 1996), pp. 431–434.

[67] N. S. Bergano, C. R. Davidson, D. L. Wilson, F. W. Kerfoot, M. D. Tremblay, M. D. Levonas, J. P. Morreale, J. D. Evankow, P. C. Corbett, M. A. Mills, G. A. Ferguson, A. M. Vengsarkar, J. R. Pedrazzani, J. A. Nagel, J. L. Zyskind, and J. W. Sulhoff, "100 Gb/s erro free transmission over 9100 km using twenty 5 Gb/s WDM channels," in *Conference on Optical Fiber Communication*, Vol. 2, 1996 OSA Technical Digest Series (Optical Society of America, Washington, D.C., 1996), pp. 419–422.

[68] A. H. Gnauck, F. Forghieri, R. M. Derosier, A. R. McCormick, A. R. Chraplyvy, J. L Zyskind, J. W. Sulhoff, A. J. Lucero, Y. Sun, R. M. Jopson, and C. Wolf, "One terabit/s transmission experiment," in *Conference on Optical Fiber Communication*, Vol. 2, 1996 OSA Technical Digest Series (Optical Society of America, Washington, D.C., 1996), pp. 407–410.

[69] T. Morioka, H. Takara, S. Kawanishi, O. Kamatani, K. Takiguchi, K. Uchiyama, M. Saruwatari, H. Takahashi, M. Yamada, T. Kanamori, and H. Ono, "100 Gbit/s

× 10 channel OTDM/WDM transmission using a single supercontinuum WDM source," in *Conference on Optical Fiber Communication*, Vol. 2, 1996 OSA Technical Digest Series (Optical Society of America, Washington, D.C., 1996), pp. 411–414.

[70] Y. Yano, T. Ono, K. Fukuchi, T. Ito, H. Yamazaki, M. Yamaguchi, and K. Emura, "2.6 Terabit/s WDM transmission experiment using optical duobinary encoding," *Proceedings of the 22nd European Conference on Optical Communication*, Vol. 5, pp. 5.3–5.6 (1996).

[71] H. Taga, K. Imai, N. Takeda, S. Yamamoto, and S. Akiba, *Elect. Lett.* **33**, 700 (1997).

[72] N. S. Bergano, C. R. Davidson, M. A. Mills, P. C. Corbett, S. G. Evangelides, B. Pedersen, R. Menges, J. L. Zyskind, J. W. Sulhoff, A. K. Srivasta, C. Wolf, and J. Judkins, "Long-haul WDM transmission using optimum channel modulation: a 160 Gb/s (32×5 Gb/s) 9,300 km demonstration," in *Conference on Optical Fiber Communication*, Vol. 6, 1997 OSA Technical Digest Series (Optical Society of America, Washington, D.C., 1997), pp. 432–435.

[73] Y. Sun, J. B. Judkins, A. K. Srivasta, L. Garrett, J. L. Zyskind, J. W. Sulhoff, C. Wolf, R. M. Derosier, A. H. Gnauck, R. W. Tkach, J. Zhou, R. P. Espindola, A. M. Vengsarkar, and A. R. Chraplyvy, *IEEE Phot. Tech. Lett.* **9**, 1652 (1997).

[74] D. Garthe, R. A. Saunders, W. S. Lee, and A. Hadjifotiou, "Simultaneous transmission of eight 40 Gbit/s channels over standard single mode fiber," in *Conference on Optical Fiber Communication*, Vol. 6, 1997 OSA Technical Digest Series (Optical Society of America, Washington, D.C., 1997), pp. 447–450.

[75] S. Kawanishi, H. Takara, K. Uchiyama, I. Shake, O. Kamatani, and H. Takahashi, *Elect. Lett.* **33**, 1716 (1997).

[76] N. S. Bergano, C. R. Davidson, M. A. Mills, P. C. Corbett, R. Menges, J. L Zyskind, J. W. Sulhoff, A. K. Srivasta, and C. Wolf, "Long-haul WDM transmission using 10 Gb/s channels: a 160 Gb/s (16×10 Gb/s) 6,000 km demonstration," in *OSA Trends in Optics and Photonics*, Vol. 16, Optical Amplifiers and Their Applications, M. N. Zervas, A. E. Willner, and S. Sasaki, Eds. (Optical Society of America, Washington, DC, 1997), pp. 402–405.

[77] T. Terahara, T. Naito, N. Shimojoh, H. Iwata, S. Kinoshita, and T. Chikama, "A quarter Tb/s WDM transmission experimentsover 2,526 km of twenty-four, 10.663-Gb/s data channels using RZ modulation format," in *OSA Trends in Optics and Photonics*, Vol. 16, Optical Amplifiers and Their Applications, M. N. Zervas, A. E. Willner, and S. Sasaki, Eds. (Optical Society of America, Washington, DC, 1997), pp. 410–413.

[78] H. Taga, T. Miyakawa, K. Murashige, N. Edagawa, M. Suzuki, H. Tanaka, K. Goto, and S. Yamamoto, "A half Tbit/s (50×10.66 Gbit/s) over 1600 km transmission experiment using widely gain-flattened EDFA chain," *Proceedings of*

the 23rd European Conference on Optical Communication, Vol. 5, pp. 13–16 (1997).

[79] S. Bigo, A. Bertaina, M. Chbat, S. Gurib, J. Da Loura, J.-C. Jacquinot, J. Hervo, P. Bousselet, S. Borne, D. Bayart, L. Gasca, and J.-L. Beylat, "320 Gbit/s WDM transmission over 500 km of conventional single-mode fiber with 125-km amplifier spacing," *Proceedings of the 23rd European Conference on Optical Communication*, Vol. 5, pp. 17–20 (1997).

[80] A. K. Srivasta, Y. Sun, J. W. Sulhoff, C. Wolf, M. Zirngibl, R. Monnard, A. R. Chraplyvy, A. A. Abramov, R. P. Espindola, T. A. Strasser, J. R. Pedrazzani, A. M. Vengsarkar, J. L. Zyskind, J. Zhou, D. A. Ferrand, P. F. Wysocki, J. B. Judkins, and Y. P. Li, "1 Tb/s transmission of 100 WDM 10 Gb/s channels over 400 km of TrueWave™fiber," in *Conference on Optical Fiber Communication*, Vol. 2, 1998 OSA Technical Digest Series (Optical Society of America, Washington, D.C., 1998), pp. 464–467.

[81] S. Aisawa, T. Sakamoto, M. Fukui, J. Kani, M. Jinno, and K. Oguchi, "Ultrawide band, long distance WDM transmission demonstration: 1 Tb/s (50 × 20 Gb/s) 600 km transmission using 1550 and 1580 nm wavelength bands," in *Conference on Optical Fiber Communication*, Vol. 2, 1998 OSA Technical Digest Series (Optical Society of America, Washington, D.C., 1998), pp. 468–471.

[82] M. Suzuki, H. Kidorf, N. Edagawa, H. Taga, N. Takeda, K. Imai, I. Morita, S. Yamamoto, E. Shibano, T. Miyakawa, E. Nazuka, M. Ma, F. kerfoot, R. Maybach, H. Adelmann, V. Arya, C. Chen, S. Evangelides, D. Gray, B. Pedersen, and A. Puc, "170 G/s transmission over 10,850 km using large core transmission fiber," in *Conference on Optical Fiber Communication*, Vol. 2, 1998 OSA Technical Digest Series (Optical Society of America, Washington, D.C., 1998), pp. 491–494.

[83] H. Taga, N. Edagawa, M. Suzuki, N. Takeda, K. Imai, S. Yamamoto, and S. Akiba, "213 Gbit/s (20 × 10.66 Gbit/s), over 9000 km transmission experiment using dispersion slope compensator," in *Conference on Optical Fiber Communication*, Vol. 2, 1998 OSA Technical Digest Series (Optical Society of America, Washington, D.C., 1998), pp. 476–479.

[84] N. S. Bergano, C. R. Davidson, M. Ma, A. Pilipetskii, S. G. Evangelides, H. D. Kidorf, J. M. Darcie, E. Golovchenko, K. Rotwitt, P. C. Corbett, R. Menges, M. A. Mills, B. Pedersen, D. Peckham, A. A. Abramov, and A. M. Vengsarkar, *Proceedings of the Optical Fiber Communication Conference*, postdeadline paper PD12 (1998). "320 Gb/s WDM transmission (64 × 5 Gb/s) over 7,200 km using large mode fiber spans and chirped return-to-zero signals," in *Conference on Optical Fiber Communication*, Vol. 2, 1998 OSA Technical Digest Series (Optical Society of America, Washington, D.C., 1998), pp. 472–475.

[85] N. Shimojoh, T. Naito, T. Terahara, H. Deguchi, K. Takagawa, M. Suyama, and T. Chikama, *Elect. Lett.* **33**, 877 (1997).

[86] C. D. Chen, I. Kim, O. Mizuhara, T. V. Nguyen, K. Ogawa, R. E. Tench, L. D. Tzeng, and P. D. Yeates, *Elect. Lett.* **34**, 1002 (1998).

[87] S. Bigo, A. Bertaina, M. W. Chbat, S. Gurib, J. Da Loura, J.-C. Jacquinot, J. Hervo, P. Bousselet, S. Borne, D. Bayart, L. Gasca, and J.-L. Beylat, IEEE Phot. Tech. Lett. **10**, 1045 (1998).

[88] A. H. Gnauck, L. D. Garrett, F. Forghieri, V. Gusmeroli, and D. Scarano, IEEE Phot. Tech. Lett. **10**, 1495 (1998).

[89] T. Terahara, T. Naito, N. Shimojoh, T. Tanaka, T. Chikama, and M. Suyama, *Elect. Lett.* **34**, 1001 (1998).

[90] M. Jinno, T. Sakamoto, J. Kani, S. Aisawa, K. Osa, M. Fukui, H. Ono, and K. Oguchi, *Elect. Lett.* **33**, 882 (1997).

[91] A, K, Srivasta, J. W. Sulhoff, L. Zhang, C. Wolf, Y. Sun, A. A¿ Abramov, T. A. Strasser, J. R. Pedrazzani, R. P. Espindola, and A. M. Vengsarkar, "L-band 64 × 10 Gb/s DWDM transmission over 500 km DSF with 50 GHz channel spacing," in *24th European Conference on Optical Communication*, Proceedings Vol. 3, pp. 73–75 (1998).

[92] B. Desthieux, Y. Robert, J. Hervo, and D. Bayart, "25-nm usable bandwidth for transoceanic WDM transmission systems using 1.58 μm erbium-doped fiber amplifiers," in *24th European Conference on Optical Communication*, Proceedings Vol. 3, pp. 133–135 (1998).

[93] H. Nishimoto, I. Yokota, M. Suyama, T. Okiyama, M. Seino, T. Horimatsu, H. Kuwahara, and T. Touge, "Transmission of 12 Gb/s over 100 km using an LD-pumped erbium-doped fiber amplifier and a Ti:LiNbO3 Mach-Zehnder modulator," in *Seventh International Conference on Integrated Optics and Optical Fiber Communications*, Proceedings Vol 5, pp. 26–27.

[94] M. Z. Iqbal, J. L. Gimlett, M. M. Choy, A. Yi-Yan, M. J. Andrejco, C. Curtis, L. Curtis, M. A. Saifi, C. Lin, and N. K. Cheung, "An 11 Gb/s, 151 km transmission experiment employing a 1480 nm pumped erbium-doped in-line fiber amplifier," in *Seventh International Conference on Integrated Optics and Optical Fiber Communications*, Proceedings Vol 5, pp. 24–25.

[95] E. G. Bryant, S. F. Carter, A. D. Ellis, W. A. Stallard, J. V. Wright, and R. Wyatt, *Elect. Lett.* **26**, 528 (1990).

[96] B. Clesca, J. Auge, B. Biotteau, P. Bousselet, A. Dursin, C. Clergeaud, P. Kretzmeyer, V. Lemaire, O. Gautheron, G. Grandpierre, E. Leclerc, and P. Gabla, *Elect. Lett.* **26**, 1426 (1990).

[97] K. Hagimoto, Y. Miyamoto, T. Kataoka, K. Kawano, and M. Ohhata, "17 Gb/s long-span fiber transmission experiment using a low-noise broadband receiver with optical amplification and equalization," in *Optical Amplifiers and Their Applications*, Vol. 13, 1990 OSA Technical Digest Series (Optical Society of America, Washington, D.C., 1990), pp. 100–103.

[98] T. Sugie, N. Ohkawa, T. Imai, and T. Ito, *J. of Light. Tech.* **9**, 1178 (1991).

[99] P. M. Gabla, J. L. Pamart, R. Uhel, E. Leclerc, J. O. Frorud, F. X. Ollivier, and S. Borderieux, "401 km, 622 Mbit/s and 357 km, 2.488 Gbit/s IM/DD repeaterless transmission experiments using erbium-doped fiber amplifiers and error correcting code," in *Optical Amplifiers and Their Applications*, Vol. 17, 1992 OSA Technical Digest Series (Optical Society of America, Washington, D.C., 1992), pp. 292–295.

[100] J.-M. P. Delavaux, C. F. Flores, R. E. Tench, T. C. Pleiss, T. W. Cline, D. J. DiGiovanni, J. Federici, C. R. Giles, H. Presby, J. S. Major, and W. J. Gignac, *Elect. Lett.* **28**, 1642 (1992).

[101] S. Ryu, T. Miyazaki, T. Kawazawa, Y. Namihira, and H. Wakabayashi, *Elect. Lett.* **28**, 1965 (1992).

[102] S. Kawanishi, H. Takara, K. Uchiyama, T. Kitoh, and M. Saruwatari, "100 Gbit/s 50 km optical transmission employing all-optical multi/demultiplexing and PLL timing extraction," in *Conference on Optical Fiber Communication*, Vol. 4, 1993 OSA Technical Digest Series (Optical Society of America, Washington, D.C., 1993), pp. 279–282.

[103] B. L. Patel, E. M. Kimber, N. E. Jolley, and A. Hadjifotiou, "Repeaterless transmission at 10 Gb/s over 215 km of dispersion shifted fibre, and 180 km of standard fibre," in *Conference on Optical Fiber Communication*, Vol. 4, 1993 OSA Technical Digest Series (Optical Society of America, Washington, D.C., 1993), pp. 295–298.

[104] B. Wedding and B. Franz, *Elect. Lett.* **29**, 402 (1993).

[105] G. Grandpierre, O. Gautheron L. Pierre, J.-P. Thiery, and P. Kretzmeyer, *IEEE Phot. Tech. Lett.* **5**, 531 (1993).

[106] A. D. Ellis and D. M. Spirit, *Elect. Lett.* **30**, 72 (1994).

[107] F. Forghieri, A. H. Gnauk, R. W. Tkach, A. R. Chraplyvy, and R. M. Derosier, "Repeaterless transmission of 8 10-Gb/s channels over 137 km (11 Tb/s-km) of dispersion-shifted fiber," in *Conference on Optical Fiber Communication*, Vol. 4, 1994 OSA Technical Digest Series (Optical Society of America, Washington, D.C., 1994), pp. 438–441.

[108] A. Sano, Y. Miyamoto, T. Kataoka, H. Kawakami, and K. Hagimoto, *Elect. Lett.* **30**, 1695 (1994).

[109] A. Sano, T. Kataoka, H. Tsuda, A. Hirano, K. Murata, H. Kawakami, Y. Tada, K. Hagimoto, K, Sato, K. Wakita, K. Kato, and Y. Miyamoto, *Elect. Lett.* **32**, 1218 (1996).

[110] K. Aida, S. Nishi, Y. Sato, K. Hagimoto, and K. Nakagawa, "1.8 Gb/s 310 km fiber transmission without repeater equipment using a remotely pumped in-line Er-doped fiber amplifier in an IM/direct-detection system," *Proceedings of the 15th European Conference on Optical Communication*, Vol. 3, pp. 29–32 (1989).

[111] T. Sugie, *Opt. and Quant. Elect.* **27**, 643 (1995).

[112] Y. H. Cheng, *J. Light. Tech.* **11**, 1495 (1993).

[113] V. DaSilva, D. L. Wilson, G. Nykolak, J. R. Simpson, P. F. Wysocki, P. B. Hansen, D. J. DiGiovanni, P. C. Becker, and S. G. Kosinski, *IEEE Phot. Tech. Lett.* **7**, 1081 (1995).

[114] V. L. DaSilva and J. R. Simpson, "Comparison of Raman efficiencies in optical fibers," in *Conference on Optical Fiber Communication*, Vol. 4, 1994 OSA Technical Digest Series (Optical Society of America, Washington, D.C., 1994), pp. 136–137.

[115] S. G. Grubb, "High power diode-pumped fiber lasers and amplifiers," in *Conference on Optical Fiber Communication*, Vol. 8, 1995 OSA Technical Digest Series (Optical Society of America, Washington, D.C., 1995), pp. 41–42.

[116] P. B. Hansen, L. Eskildsen, S. G. Grubb, A. M. Vengsarkar, S. Korotky, T. A. Strasser, J. E. J. Alphonsus, J. J. Veselka, D. J. DiGiovanni, D. W. Peckham, E. C. Beck, D. Truxal, W. Y. Cheung, S. G. Kosinski, D. Gasper, P. F. Wysocki, V. L. daSilva, and J. R. Simpson, "2.488-Gb/s unrepeatered transmission over 529 km using remotely pumped post- and pre- amplifiers, forward error correction, and dispersion compensation," in *Conference on Optical Fiber Communication*, Vol. 8, 1995 OSA Technical Digest Series (Optical Society of America, Washington, D.C., 1995), pp. 420–423.

[117] C. E. Soccolich, V. Mizrahi, T. Erdogan, P. J. LeMaire, and P. Wysocki, "Gain enhancement in EDFAs by using fiber-grating pump reflectors," in *Conference on Optical Fiber Communication*, Vol. 4, 1994 OSA Technical Digest Series (Optical Society of America, Washington, D.C., 1994), pp. 277–278.

[118] V. L. DaSilva, P. B. Hansen, L. Eskildsen, D. L. Wilson, S. G. Grubb, V. Mizrahi, W. Y. Cheung, T. Erdogan, T. A. Strasser, J. E. J. Alphonsus, J. R. Simpson, and D. J. DiGiovanni, "15.3 dB power-budget improvement by remotely pumping an EDFA with a fiber-grating pump reflector," in *Conference on Optical Fiber Communication*, Vol. 8, 1995 OSA Technical Digest Series (Optical Society of America, Washington, D.C., 1995), pp. 146–147.

[119] S. S. Sian, O. Gautheron, M. S. Chaudhry, C. D. Stark, S. M. Webb, K. M. Guild, M. Mesic, J. M. Dryland, J. R. Chapman, A. R. Docker, E. Brandon, T. Barbier, P. Garabedian, and P. Bousselet, "511 km at 2.5 Gbit/s and 531 km at 622 Mbit/s - unrepeatered transmission with remote pumped amplifiers, forward error correction and dispersion compensation," in *Conference on Optical Fiber*

Communication, Vol. 8, 1995 OSA Technical Digest Series (Optical Society of America, Washington, D.C., 1995), pp. 424–427.

[120] J. P. Blondel, F. Misk, and G. Gabla, *IEEE Phot. Tech. Lett.* **5**, 1433 (1993).

[121] P. M. Gabla, J.-P. Blondel, and O. Gautheron, "Progress and perspectives of repeaterless submarine systems," *Proceedings of the 19th European Conference on Optical Communication*, Vol. 1, pp. 107–113 (1993).

[122] L. Eskildsen, P. B. Hansen, V. L. DaSilva, and S. G. Grubb, "Remotely pumped postamplifiers for repeaterless lightwave systems," in *Conference on Optical Fiber Communication*, Vol. 8, 1995 OSA Technical Digest Series (Optical Society of America, Washington, D.C., 1995), pp. 147–148.

[123] L. Eskildsen, P. B. Hansen, S. G. Grubb, and V. L. Da Silva, *Elect. Lett.* **31**, 1163 (1995).

[124] L. D. Pedersen, B. Velschow, C. F. Pedersen, and F. Ebskamp, "Uncoded NRZ 2.488 Gbit/s transmission over 347 km standard fiber 67 dB span loss, using a remotely pumped amplifier, SBS suppression, SPM optimization and Raman gain," in *Conference on Optical Fiber Communication*, Vol. 4, 1994 OSA Technical Digest Series (Optical Society of America, Washington, D.C., 1994), pp. 446–449.

[125] J. E. J. Alphonsus, P. B. Hansen, V. L. DaSilva, D. A. Truxal, D. W. Wilson, D. J. DiGiovanni, and J. R. Simpson. "5-Gbit/s repeaterless transmission over 339 km by using an Er-doped fiber post, preamplifiers, and remote amplifiers," in *Conference on Optical Fiber Communication*, Vol. 8, 1995 OSA Technical Digest Series (Optical Society of America, Washington, D.C., 1995), pp. 121–122.

[126] S. K. Korotky, P. B. Hansen, L. Eskildsen, and J. J. Veselka, "Efficient phase modulation scheme for suppressing stimulated Brillouin scattering," *Tenth International Conference on Integrated Optics and Optical Fibre Communication, IOOC-95*, Technical Digest Vol. 2, pp. 110–111 (1995).

[127] P. M. Gabla, J. L. Pamart, R. Uhel, E. Leclerc, J. O. Frorud, F. X. Ollivier, and S. Borderieux, *IEEE Phot. Tech. Lett.* **4**, 1148 (1992).

[128] M. S. Chaudhry, S. S. Sian, K. Guild, P. R. Morkel, and C. D. Stark, *Elect. Lett.* **30**, 2061 (1994).

[129] P. B. Hansen, L. Eskildsen, S. G. Grubb, A. M. Vengsarkar, S. K. Korotky, T. A. Strasser, J. E. J. Alphonsus, J. Veselka, D. J. DiGiovanni, D. W. Peckham, D. Truxal, W. Y. Cheung, S. G. Kosinski, and P. F. Wysocki, "8×10 Gb/s WDM repeaterless transmission experiment over 352 km," *Tenth International Conference on Integrated Optics and Optical Fibre Communication, IOOC-95*, Technical Digest Vol. 5, pp. 27–28 (1995).

[130] J. E. J. Alphonsus, P. B. Hansen, L. Eskildsen, D. A. Truxal, S. G. Grubb, D. J. DiGiovanni, T. A. Strasser, and E. C. Beck, *IEEE Phot. Tech. Lett.* **7**, 1495 (1992).

[131] S. Sian, S. M. Webb, K. M. Guild, and D. R. Terrence, *Elect. Lett.* **32**, 50 (1996).

[132] P. B. Hansen, L. Eskildsen, S. G. Grubb, A. M. Vengsarkar, S. K. Korotky, T. A. Strasser, J. E. J. Alphonsus, J. J. Veselka, D. J. DiGiovanni, D. W. Peckham, and D. Truxal, *Elect. Lett.* **32**, 1018 (1996).

[133] C. Cremer, U. Gaubatz, and P. Krummrich, *Elect. Lett.* **32**, 1116 (1996).

[134] P. B. Hansen, V. L. DaSilva, L. Eskildsen, S. G. Grubb, V. Mizrahi, W. Y. Cheung, T. Erdogan, T. A. Strasser, J. E. J. Alphonsus, G. Nykolak, D. L. Wilson, D. J. DiGiovanni, D. Truxal, A. M. Vengsarkar, S. G. Kosinski, P. F. Wysocki, J. R. Simpson, and J. D. Evankow, "423-km repeaterless transmission at 2.488 Gb/s using remotely pumped post- and pre-amplifiers," *Proceedings of the 20th European Conference on Optical Communication*, Vol. 4, pp. 57–60 (1994).

[135] W. I. Way, M. M. Choy, A. Yi-Yan, M. J. Andrejco, M. Saifi, and C. Lin, *IEEE Phot. Tech. Lett.* **1**, 343 (1989); W. I. Way, M. M. Choy, A. Yi-Yan, M. J. Andrejco, M. Saifi, and C. Lin, "Multi-channel AM-VSB television signal transmission using an erbium-doped optical fiber power amplifier," in *Seventh International Conference on Integrated Optics and Optical Fiber Communications*, Proceedings Vol 5, pp. 30–31.

[136] W. I. Way, M. W. Maeda, A. Yi-Yan, M. J. Andrejco, M. M. Choi, M. Saifi, and C. Lin, *Elect. Lett.* **26**, 139 (1990).

[137] W. I. Way, S. S. Wagner, M. M. Choi, C. Lin, R. C. Menedez, A. Tohme, A. Yi-Yan, A. C. Von Lehman, R. E. Spicer, M. J. Andrejco, M. Saifi, and H. L. Lemberg, "Distribution of 100 FM-TC channels and six 22 Mb/s channels to 4096 terminals using high-density WDM and a broadband in-line erbium-doped fiber amplifier," in *Conference on Optical Fiber Communication*, Vol. 1, 1990 OSA Technical Digest Series (Optical Society of America, Washington, D.C., 1990), pp. 321–324.

[138] H. E. Tohme, C. N. Lo, and M. A. Saifi, *Elect. Lett.* **26**, 1280 (1990).

[139] K. Kikushima, E. Yoneda, K. Suto, and H. Yoshinaga, "Simultaneous distribution of AM/FM FDM TV signals to 65,536 subscribers using 4 stage cascade EDFAs," in *Optical Amplifiers and Their Applications*, Vol. 13, 1990 OSA Technical Digest Series (Optical Society of America, Washington, D.C., 1990), pp. 232–235.

[140] P. M. Gabla, V. Lemaire, H. Krimmel, J. Otterbach, J. Auge, and A. Dursin, *IEEE Phot. Tech. Lett.* **3**, 56 (1991).

[141] D. R. Huber and Y. S. Trisno, "20 channel VSB-AM CATV link utilizing an external modulator, erbium laser and a high power erbium amplifier," in *Conference on Optical Fiber Communication*, Vol. 4, 1991 OSA Technical Digest Series (Optical Society of America, Washington, D.C., 1991), pp. 301–304.

[142] R. Heidemann, B. Jungiger, H. G. Krimmel, J. Otterbach, D. Schlump, and B. Wedding, "Simultaneous distribution of analogue AM-TV and multigigabit HDTV with optical amplifier," in *Optical Amplifiers and Their Applications*, Vol. 13, 1991 OSA Technical Digest Series (Optical Society of America, Washington, D.C., 1991), pp. 210–213.

[143] G. R. Joyce, R. Olshansky, R. Childs, and T. Wei, "A 40 channel AM-VSB distribution system with a 21 dB link budget," in *Optical Amplifiers and Their Applications*, Vol. 13, 1991 OSA Technical Digest Series (Optical Society of America, Washington, D.C., 1991), pp. 218–221.

[144] B. Clesca, J.-P. Blondel, P. Bousselet, J.-P. Herbert, F. Brillouet, L. Sniadower, and Y. Cretin, "32 channel, 48 dB CNR and 46 dB budget AM-VSB transmission experiment with field-ready post-amplifiers," in *Optical Amplifiers and Their Applications*, Vol. 17, 1992 OSA Technical Digest Series (Optical Society of America, Washington, D.C., 1992), pp. 91–94.

[145] H. Bülow, R. Fritschi, R. Heidemann, B. Jungiger, H. G. Krimmel, and J. Otterbach, *Elect. Lett.* **28**, 1836 (1992).

[146] M. R. Phillips, A. H. Gnauck, T. E. Darcie, N. J. Frigo, G. E. Bodeep, and E. A. Pitman, "112 channel WDM split-band CATV system," in *Conference on Optical Fiber Communication*, Vol. 5, 1992 OSA Technical Digest Series (Optical Society of America, Washington, D.C., 1992), pp. 332–335.

[147] K. Kikushima, *J. Light. Tech.* **10**, 1443 (1992).

[148] P. M. Gabla, C. Bastide, Y. Cretin, P. Bousselet, A. Pitel, and J. P. Blondel, *IEEE Phot. Tech. Lett.* **4**, 510 (1992).

[149] E. Yoneda, K. Suto, K. Kikushima, and H. Yoshinaga, *J. Light. Tech.* **11**, 128 (1993).

[150] S. G. Grubb, P. A. Leilabady, and D. E. Frymyer, *J. Light. Tech.* **11**, 27 (1993).

[151] M. Shigematsu, H. Go, M. Kakui, and M. Nishimura, "Distortion-free optical fiber amplifier for analog transmission based on hybrid erbium-doped fiber configuration," in *Optical Amplifiers and Their Applications*, Vol. 14, 1994 OSA Technical Digest Series (Optical Society of America, Washington, D.C., 1994), pp. 162–164.

[152] W. Muys, J. C. van der Plaats, F. W. Willems, H. J. van Dijk, J. S. Leong, and A. M. Koonen, *IEEE Phot. Tech. Lett.* **7**, 691 (1995).

[153] M. Shigematsu, M. Kakui, M. Onishi, and M. Nishimura, *Elect. Lett.* **31**, 1077 (1995).

[154] C. Y. Kuo, D. Piehler, C. Gall, A. Nilsson, and L. Middleton, "High-performance optically amplified 1550-nm lightwave AM-VSB CATV transport system," in *Conference on Optical Fiber Communication*, Vol. 2, 1996 OSA Technical Digest Series (Optical Society of America, Washington, D.C., 1996), pp. 196–197.

[155] W. Muys, J. C. van der Plaats, F. W. Willems, A. M. Vengsarkar, C. E. Soccolich, M. J. Andrejco, D. J. DiGiovanni, D. W. Peckham, S. G. Kosinski, and P. F. Wysocki, "Directly modulated AM-VSB lightwave video transmission system using dispersion-compensating fiber and three cascaded EDFAs, providing 50-dB power budget over 38 km of standard single-mode fiber," in *Conference on Optical Fiber Communication*, Vol. 2, 1996 OSA Technical Digest Series (Optical Society of America, Washington, D.C., 1996), pp. 198–199.

[156] H. Bülow, R. Fritschi, R. Heidemann, B. Jungiger, H. G. Krimmel, and J. Otterbach, *IEEE Phot. Tech. Lett.* **4**, 1287 (1992).

[157] N. Park, T. Nielsen, J. Simpson, P. Wysocki, R. Pedrazzani, D. DiGiovanni, S. Grubb, and K. Walker, *Elect. Lett.* **32**, 913 (1996).

[158] N. S. Bergano, "System applications of optical amplifiers," in Conference on Optical Fiber Communication, Tutorial ThN, Tutorial Proceedings of OFC '93, (Optical Society of America, Washington, D.C., 1993), pp. 227–280.

[159] B. M. Desthieux, M. Suyama, and T. Chikama, *J. Light. Tech.* **12**, 1405 (1994).

[160] P. Blondel, J. F. Marcerou, J. Auge, H. Fevrier, P. Bousselet, and A. Dursin, "Erbium-doped fiber amplifier spectral behavior in tranoceanic links," in *Optical Amplifiers and Their Applications*, Vol. 13, 1991 OSA Technical Digest Series (Optical Society of America, Washington, D.C., 1991), pp. 82–85.

[161] N. Bergano, "Undersea lightwave transmission systems using erbium-doped fiber amplifiers," *Optics and Photonics News* **4**, pp. 6–14 (1993).

[162] P. F. Wysocki, J. R. Simpson and D. Lee, *IEEE Phot. Tech. Lett.* **6**, 1098 (1994).

[163] D. Lee, P. F. Wysocki, J. R. Simpson. D. J. DiGiovanni, K. L. Walker, and D. Gasper, *IEEE Phot. Tech. Lett.* **6**, 1094 (1994).

[164] V. J. Mazurczyk, "Spectral response of a single EDFA measured to an accuracy of 0.01 dB," in *Conference on Optical Fiber Communication*, Vol. 4, 1994 OSA Technical Digest Series (Optical Society of America, Washington, D.C., 1994), pp. 271–272.

[165] J. P. Blondel, A. Pitel, and J. F. Marcerou, "Gain-filtering stability in ultralong-distance links," in *Conference on Optical Fiber Communication*, Vol. 4, 1993 OSA Technical Digest Series (Optical Society of America, Washington, D.C., 1993), pp. 38–39.

[166] B. M. Desthieux, M. Suyama, and T. Chikama, "Self-filtering characteristics of concatenated erbium-doped fiber amplifiers," in *Optical Amplifiers and Their Applications*, Vol. 14, 1993 OSA Technical Digest Series (Optical Society of America, Washington, D.C., 1993), pp. 100–103.

[167] D. W. Hall, R. A. Haas, W. F. Krupke, and M. J. Weber, *IEEE J. Quant. Elect.* **QE-19**, 1704 (1983).

[168] J. T. Lin and W. A. Gambling, "Polarization effets in fibre lasers: Phenomena, theory and applications," in *Fiber Laser Sources and Amplifiers II*, M. J. F. Digonnet, Ed., *Proc. SPIE* **1373**, pp. 42–53 (1990).

[169] W. K. Burns and A. D. Kersey, *J. Light. Tech.* **10**, 992 (1992).

[170] M. G. Taylor, *IEEE Phot. Tech. Lett.* **5**, 1244 (1993).

[171] D. W. Hall, M. J. Weber, and R. T. Brundage, *J. Appl. Phys.* **55**, 2642 (1984).

[172] V. L. Mazurczyk and J. L. Zyskind, *IEEE Phot. Tech. Lett.* **6**, 616 (1994).

[173] E. J. Greer, D. J. Lewis, and W. M. Macauley, *Elect. Lett.* **30**, 46 (1994).

[174] F. Bruyere, *Elect. Lett.* **31**, 401 (1995).

[175] P. F. Wysocki, "Computer modeling of polarization hole burning in EDFAs," in *Conference on Optical Fiber Communication*, Vol. 4, 1994 OSA Technical Digest Series (Optical Society of America, Washington, D.C., 1994), pp. 307–308.

[176] N. S. Bergano, V. J. Mazurczyk, and C. R. Davidson, "Polarization scrambling improves SNR performance in a chain of EDFAs," in *Conference on Optical Fiber Communication*, Vol. 4, 1994 OSA Technical Digest Series (Optical Society of America, Washington, D.C., 1994), pp. 255–256.

[177] M. G. Taylor and S. J. Penticost, *Elect. Lett.* **30**, 805 (1994).

[178] Y. Fukada, T. Imai, and A. Mamoru, *Elect. Lett.* **30**, 432 (1994).

[179] F. Bruyere, O. Audouin, V. Letellier, G. Bassier, and P. Marnier, *IEEE Phot. Tech. Lett.* **6**, 1153 (1994).

[180] N. S. Bergano, "Time dynamics of polarization hole burning in an EDFA," in *Conference on Optical Fiber Communication*, Vol. 4, 1994 OSA Technical Digest Series (Optical Society of America, Washington, D.C., 1994), pp. 305–306.

[181] N. S. Bergano, C. R. Davidson, and F. Heismann, *Elect. Lett.* **32**, 52 (1996).

[182] P. R. Morkel, V. Syngal, D. J. Butler, and R. Newman, *Elect. Lett.* **30**, 806 (1994).

[183] F. Bruyere and O. Audouin, *IEEE Phot. Tech. Lett.* **6**, 443 (1994).

[184] E. Desurvire, *IEEE Phot. Tech. Lett.* **1**, 19 (1989).

[185] R. I. Laming, L. Reekie, P. R. Morkel, and D. N. Payne, *Elect. Lett.* **25**, 455 (1989).

[186] C. R. Giles, E. Desurvire, and J. R. Simpson, *Opt. Lett.* **14**, 880 (1989).

[187] E. Desurvire, C. R. Giles, and J. R. Simpson, *J. Light. Tech.* **7**, 2095 (1989).

[188] L. M. Frantz and J. S. Nodvik, *J. Appl. Phys.* **34**, 2346 (1963).

[189] Y. Sun, G. Luo, J. L. Zyskind, A. A. M. Saleh, A. K. Srivasta, and J. W. Sulhoff, *Elect. Lett.* **32**, 1490 (1996).

[190] A. K. Srivasta, Y. Sun, J. L. Zyskind, and J. W. Sulhoff, *IEEE Phot. Tech. Lett.* **9**, 386 (1997).

[191] J. L. Zyskind, Y. Sun, A. K. Srivasta, J. W. Sulhoff, A. J. Lucero, C. Wolf, and R. W. Tkach, "Fast power transients in optically amplified multiwavelength optical networks," in *Conference on Optical Fiber Communication*, Vol. 2, 1996 OSA Technical Digest Series (Optical Society of America, Washington, D.C., 1996), pp. 451–454.

[192] E. Desurvire, M. Zirngibl, H. M. Presby, and D. DiGiovanni, *IEEE Phot. Tech. Lett.* **3**, 453 (1991).

[193] K. Motoshima, L. M. Leba, D. N. Chen, M. M. Downs, T. Li, and E. Desurvire, *IEEE Phot. Tech. Lett.* **5**, 1424 (1993).

[194] A. K. Srivasta, Y. Sun, J. L. Zyskind, J. W. Sulhoff, C. Wolf, and R. W. Tkach, "Fast gain control in an erbium-doped fiber amplifier," in *OSA Trends in Optics and Photonics*, Vol. 5, Optical Amplifiers and Their Applications, 1996 OAA Program Committee, Eds. (Optical Society of America, Washington, DC, 1996), pp. 24–27.

[195] K. Motoshima, K. Shimizu, K. Takano, T. Mizuochi, and T. Kitayama, "EDFA with dynamic gain compensation for multiwavelength transmission systems," in *Conference on Optical Fiber Communication*, Vol. 4, 1994 OSA Technical Digest Series (Optical Society of America, Washington, D.C., 1994), pp. 191–192.

[196] J. L. Zyskind, A. K. Srivasta, Y. Sun, J. C. Ellson, G. W. Newsome, R. W. Tkach, A. R. Chraplyvy, J. W. Sulhoff, T. A. Strasser, J. R. Pedrazzani, and C. Wolf, "Fast link control protection for surviving channels in multiwavelength optical networks," *Proceedings of the 22nd European Conference on Optical Communications*, Vol. 5, pp. 5.49–5.52 (1996).

[197] L. F. Mollenauer, J. P. Gordon, and M. N. Islam, *IEEE J. Quant. Elect.* **QE-22**, 157 (1986).

[198] L. F. Mollenauer, S. G. Evangelides, and H. A. Haus, *J. of Light. Tech.* **9**, 194 (1991).

[199] L. F. Mollenauer, K. Smith, J. P. Gordon, and C. R. Menyuk, *Opt. Lett.* **14**, 1219 (1989).

[200] J. P. Gordon and H. A. Haus, *Opt. Lett.* **11**, 665 (1986).

[201] L. F. Mollenauer, M. J. Neubelt, S. G. Evangelides, J. P. Gordon, J. R. Simpson, and L. G. Cohen, *Opt. Lett.* **15**, 1203 (1990).

[202] H. A. Haus, *Fiber and Int. Opt.* **12**, 187 (1996).

[203] A. Mecozzi, J. D. Moores, H. A. Haus, and Y. Lai, *Opt. Lett.* **16**, 1841 (1991).

[204] Y. Kodama and A. Hasegawa, *Opt. Lett.* **17**, 31 (1992).

[205] L. F. Mollenauer, J. P Gordon, and S. G. Evangelides, *Opt. Lett.* **17**, 1575 (1992).

[206] L. F. Mollenauer, E. Lichtman, M. J. Neubelt, and G. T. Harvey, *Elect. Lett.* **29**, 910 (1993).

[207] S. Kawai, K. Iwatsuki, and S. Nishi, *Elect. Lett.* **31**, 1463 (1995).

[208] L. F. Mollenauer, P. V. Mamyshev, and M. J. Neubelt, *Elect. Lett.* **32**, 471 (1996).

[209] H. Kubota and M. Nakazawa, *IEEE J. Quant. Elect.* **29**, 2189 (1993).

[210] M. Nakazawa, K. Suzuki, H. Kubota, E. Yamada, and Y. Kimura, *Elect. Lett.* **30**, 1331 (1994).

[211] G. Aubin, T. Montalant, J. Moulu, B. Nortier, F. Pirio, and J. B. Thomine, *Elect. Lett.* **30**, 1163 (1994).

[212] M. Nakazawa, E. Yamada, H. Kubota, and K. Suzuki, *Elect. Lett.* **27**, 1270 (1991).

[213] M. Nakazawa, K. Suzuki, H. Kubota, Y. Kimura, E. Yamada, K. Tamura, T. Komukai, and T. Imai, *Elect. Lett.* **32**, 828 (1996).

[214] M. Suzuki, N. Edagawa, H. Taga, H. Tanaka, S. Yamamoto, and S. Akiba, *Elect. Lett.* **30**, 1083 (1994).

[215] M. Suzuki, I. Morita, N. Edagawa, S. Yamamoto, and S. Akiba, *Elect. Lett.* **31**, 2027 (1995).

[216] S. Evangelides, L. F. Mollenauer, J. P. Gordon, and N. S. Bergano, *J. Light. Tech.* **10**, 28 (1992).

[217] T. Sugawa, K. Kurokawa, H. Kubota, and M. Nakazawa, "Polarization dependence of soliton interactions and soliton self-frequency shift in a femtosecond soliton transmission," in *Conference on Optical Fiber Communication*, Vol. 8, 1995 OSA Technical Digest Series (Optical Society of America, Washington, D.C., 1995), pp. 301–301.

[218] P. A. Andrekson, N. A. Olsson, P. C. Becker, J. R. Simpson, T. Tanbun-Ek, R. A. Logan, and K. W. Wecht, *Appl. Phys. Lett.* **57**, 1715 (1990).

[219] P. A. Andrekson, N. A. Olsson, J. R. Simpson, T. Tanbun-Ek, R. A. Logan, and K. W. Wecht, *Elect. Lett.* **26**, 1499 (1990).

[220] L. F. Mollenauer, S. G. Evangelides, and J. P. Gordon, *J. Light. Tech.* **9**, 362 (1991).

[221] Y. Kodama and S. Wabnitz, *Opt. Lett.* **18**, 1311 (1993).

[222] L. F. Mollenauer, *Opt. Lett.* **21**, 384 (1996).

[223] P. V. Mamyshev and L. F. Mollenauer, *Opt. Lett.* **21**, 396 (1996).

[224] L. F. Mollenauer, P. V. Mamyshev, and M. J. Neubelt, "Demonstration of soliton WDM transmission at up to 8 × Gbit/s, error-free over transoceanic distances," in *Conference on Optical Fiber Communication*, Vol. 2, 1996 OSA Technical Digest Series (Optical Society of America, Washington, D.C., 1996), pp. 415–418.

[225] L. F. Mollenauer, R. H. Stolen, and J. P. Gordon, *Phys. Rev. Lett.* **45**, 1095 (1980).

[226] L. F. Mollenauer and K. Smith, *Opt. Lett.* **13**, 675 (1988).

[227] L. F. Mollenauer, R. H. Stolen, and M. N. Islam, *Opt. Lett.* **10**, 229 (1985).

[228] K. Iwatsuki, K. Suzuki, S. Nishi, M. Saruwatari, and K. Nakagawa, *IEEE Phot. Tech. Lett.* **2**, 905 (1990).

[229] P. A. Andrekson, N. A. Olsson, P. C. Becker, J. R. Simpson, T. Tanbun-Ek, R. A. Logan, and K. W. Wecht, *Appl. Phys. Lett.* **57**, 1715 (1990).

[230] M. Nakazawa, K. Suzuki, and Y. Kimura, *IEEE Phot. Tech. Lett.* **2**, 216 (1990).

[231] M. Nakazawa, K. Suzuki, E. Yamada, and Y. Kimura, *Electron. Lett.* **26**, 1592 (1990).

[232] K. Suzuki, M. Nakazawa, E. Yamada, and Y. Kimura, *Elect. Lett.* **26**, 551 (1990).

[233] K. Suzuki and M. Nakazawa, *Elect. Lett.* **26**, 1032 (1990).

[234] E. Yamada, K. Suzuki, and M. Nakazawa, *Elect. Lett.* **27**, 1289 (1991).

[235] P. A. Andrekson, N. A. Olsson, M. Haner, J. R. Simpson, T. Tanbun-Ek, R. A. Logan, D. Coblentz, H. M. Presby, and K. W. Wecht, *IEEE Phot. Tech. Lett.* **4**, 76 (1992).

[236] H. Taga, M. Suzuki, N. Edagawa, Y. Yoshida, S. Yamamoto, S. Akiba, and H. Wakabayashi, *Elect. Lett.* **28**, 2247 (1992).

[237] M. Nakazawa, K. Suzuki, E. Yamada, H. Kubota, and Y. Kimura, "10 Gbit/s-1200 km single-pass soliton data transmission using erbium-doped fiber amplifiers," in *Conference on Optical Fiber Communication*, Vol. 5, 1992 OSA Technical Digest Series (Optical Society of America, Washington, D.C., 1992), pp. 351–354.

[238] M. Nakazawa, K. Suzuki, E. Yamada, H. Kubota, M. Takaya, and Y. Kimura, "20 Gbit/s-1850 km and 40 Gbit/s-750 km soliton data transmissions using erbium-doped fiber amplifiers," in *Conference on Optical Fiber Communication*, Vol. 4, 1993 OSA Technical Digest Series (Optical Society of America, Washington, D.C., 1993), pp. 307–310.

[239] M. Nakazawa, K. Suzuki, E. Yamada, H. Kubota, and Y. Kimura, *Elect. Lett.* **29**, 1474 (1993).

[240] B. Christensen, G. Jacobsen, E. Bødtker, J. Mark, and I. Mito, *IEEE Phot. Tech. Lett.* **6**, 101 (1994).

[241] B. Christensen, G. Jacobsen, E. Bødtker, and I. Mito, "Simple, bit-rate-flexible soliton generator and penalty-free 9-Gbit/s data transmission over 416 km of DSF," in *Conference on Optical Fiber Communication*, Vol. 8, 1995 OSA Technical Digest Series (Optical Society of America, Washington, D.C., 1995), pp. 304–305.

[242] M. Nakazawa, K. Suzuki, E. Yoshida, E. Yamada, T. Kitoh, and M. Kawachi, *Elect. Lett.* **31**, 565 (1995).

[243] A. Naka, T. Matsuda, S. Saito, and K. Sato, *Elect. Lett.* **31**, 1679 (1995).

[244] N. Edagawa, I. Morita, M. Suzuki, S. Yamamoto, H. Taga, and S. Akiba, "20 Gbit/s, 8100 km straight-line single-channel soliton-based RZ transmission experiment using periodic dispersion compensation," *Proceedings of the 21st European Conference on Optical Communication*, Vol. 3, pp. 983–986 (1995).

[245] F. Favre, D. Le Guen, M. L. Moulinard, M. Henry, T. Georges, and F. Devaux, *Elect. Lett.* **32**, 1115 (1996).

[246] K. Iwatsuki, K. Suzuki, and S. Kawai, "40 Gb/s adiabatic and phase-stationary soliton transmission with sliding-frequency filter over 4000 km reciprocating dispersion-managed fiber," in *OSA Trends in Optics and Photonics*, Vol. 5, Optical Amplifiers and Their Applications, 1996 OAA Program Committee, Eds. (Optical Society of America, Washington, DC, 1996), pp. 32–35.

[247] L. F. Mollenauer and S. G. Evangelides, "High bit rate, ultralong distance soliton transmission with erbium amplifiers," in *Conference on Lasers and Electro-Optics*, Vol. 7, 1990 OSA Technical Digest Series (Optical Society of America, Washington DC, 1990), pp. 512–514.

[248] N. A. Olsson, P. A. Andrekson, J. R. Simpson, T. Tanbun-Ek, R. A. Logan, and K. W. Wetch, *Elect. Lett.* **27**, 695 (1991).

[249] L. F. Mollenauer, M. J. Neubelt, M. Haner, E. Lichtman, S. G. Evangelides, and B. M. Nyman, *Elect. Lett.* **27**, 2055 (1991).

[250] L. F. Mollenauer, E. Lichtman, G. T. Harvey, M. J. Neubelt, and B. M. Nyman, *Elect. Lett.* **28**, 792 (1992).

[251] C. R. Giles, P. B. Hansen, S. G. Evangelides, G. Raybon, U. Koren, B. I. Miller, M. G. Young, M. A. Newkirk, J.-M. P. Delavaux, S. K. Korotky, J. J. Veselka, and C. A. Burrus, "Soliton transmission over 4200 km by using a mode-locked monolithic extended-cavity laser as a soliton source," in *Conference on Optical Fiber Communication*, Vol. 4, 1993 OSA Technical Digest Series (Optical Society of America, Washington, D.C., 1993), pp. 88–89.

[252] P. B. Hansen, C. R. Giles, G. Raybon, U. Koren, S. G. Evangelides, B. I. Miller, M. G. Young, M. A. Newkirk, J.-M. P. Delavaux, S. K. Korotky, J. J. Veselka, and C. A. Burrus, *IEEE Phot. Tech. Lett.* **5**, 1236 (1993).

[253] M. Suzuki, H. Taga, N. Edagawa, H. Tanaka, S. Yamamoto, and S. Akiba, *Elect. Lett.* **29**, 1643 (1993).

[254] L. F. Mollenauer, E. Lichtman, M. J. Neubelt, and G. T. Harvey, *Elect. Lett.* **29**, 910 (1993).

[255] T. Widdowson, A. Lord, and D. J. Malyon, *Elect. Lett.* **30**, 879 (1994).

[256] S. Kawai, K. Iwatsuki, K. Suzuki, S. Nishi, M. Saruwatari, K. Sato, and K. Wakita, *Elect. Lett.* **30**, 251 (1994).

[257] M. Suzuki, N. Edagawa, H. Taga, H. Tanaka, S. Yamamoto, and S. Akiba, *Elect. Lett.* **30**, 1083 (1994).

[258] T. Widdowson, D. J. Malyon, A. D. Ellis, K. Smith, and K. J. Blow, *Elect. Lett.* **30**, 990 (1994).

[259] L. F. Mollenauer, P. V. Mamyshev, and M. J. Neubelt, *Opt. Lett.* **19**, 704 (1994).

[260] M. Suzuki, I. Morita, S. Yamamoto, N. Edagawa, H. Taga, and S. Akiba, "Timing jitter reduction by periodic dispersion compensation in soliton transmission," in *Conference on Optical Fiber Communication*, Vol. 8, 1995 OSA Technical Digest Series (Optical Society of America, Washington, D.C., 1995), pp. 401–404.

[261] B. M. Nyman, S. G. Evangelides, G. T. Harvey, L. F. Mollenauer, P. V. Mamyshev, M. L. Saylors, S. K. Korotky, U. Koren, V. Mizhari, T. A. Strasser, J. J. Veselka, J. D. Evankow, A. Lucero, J. Nagel, J. Sulhoff, J. Zyskind, P. C. Corbett, M. A. Mills, and G. A. Ferguson, "Soliton WDM transmission of 8 × 2.5 Gb/s, error free over 10 Mm," in *Conference on Optical Fiber Communication*, Vol. 8, 1995 OSA Technical Digest Series (Optical Society of America, Washington, D.C., 1995), pp. 405–407.

[262] K. Iwatsuki, S. Saito, K. Suzuki,A. Naka, S. Kawai, T. Matsuda, and S. Nishi, "Field demonstration of 10 Gb/s - 2700 km soliton transmission through commercial submarine optical amplifier system with distributed fiber dispersion and 90 km amplifier spacing," *Proceedings of the 21st European Conference on Optical Communication*, Vol. 3, pp. 987–990 (1995).

[263] J.-P. Hamaide, F. Pitel, P. Nouchi, B. Biotteau, J. Von Wirth, P. Sansonetti, and J. Chesnoy, "Experimental 10 Gb/s sliding filter-guided soliton transmission up to 19 Mm with 63 km amplifier spacing using large effective-area fiber management," *Proceedings of the 21st European Conference on Optical Communication*, Vol. 3, pp. 991–994 (1995).

[264] M. Nakazawa, K. Suzuki, H. Kubota, and E. Yamada, "60 Gbit/s WDM (20 Gbit/s × 3 unequally spaced channels) soliton transmission over 10,000 km," in *OSA Trends in Optics and Photonics*, Vol. 5, Optical Amplifiers and Their Applications, 1996 OAA Program Committee, Eds. (Optical Society of America, Washington, DC, 1996), pp. 36–39.

[265] M. Suzuki, I. Morita, N. Edagawa, S. Yamamoto, and S. Akiba, "20 Gbit/s-based soliton WDM transmission over transoceanic distances using periodic compensation of dispersion and its slope," *Proceedings of the 22nd European Conference on Optical Communication*, Vol. 5, pp. 5.15–5.18 (1996).

[266] M. Nakazawa, Y. Kimura, K. Suzuki, and H. Kubota, *J. Appl. Phys.* **66**, 2803 (1989).

[267] M. Nakazawa, Y. Kimura, K. Suzuki, H. Kubota, T. Komukai, E. Yamada, T. Sugawa, E. Yoshida, T. Yamamoto, T. Imai, A. Sahara, H. Nakazawa, O. Yamauchi, and M. Umezawa, *Elect. Lett.* **31**, 992 (1995).

[268] N. Robinson, G. Davis, J. Fee, G. Grasso, P. Franco, A. Zuccala, A. Cavaciuti, M. Macchi, A. Schiffini, L. Bonato, and R. Corsini, "4× SONET OC-192 field installed dispersion managed soliton system over 450 km of standard fiber in the 1550 nm erbium band," in *Conference on Optical Fiber Communication*, Vol. 6, 1997 OSA Technical Digest Series (Optical Society of America, Washington, D.C., 1997), pp. 499–502.

Chapter 10

Four–Level Fiber Amplifiers for 1.3 μm Amplification

10.1 INTRODUCTION

In this chapter we will discuss optical fiber amplifiers for 1.3 μm signal wavelengths. To date, the rare earth ions that have been used for this purpose all operate on the four-level laser principle. The two principal candidates are Pr^{3+}, doped in a fluoride host, and Nd^{3+}, also in a nonsilica host. Of the two, Pr^{3+} has received the most attention and has found applications in systems demonstrations. The 1.3 μm fiber amplifiers have not yet found the same commercial success as their erbium cousins. One reason for this may be the lower efficiency of current 1.3 μm fiber amplifiers, which necessitates significantly higher pump levels than in the case of erbium-doped fiber amplifiers. Another factor is the need for fluoride fiber processing and fabrication technology, which is less widespread and more complex than that of silica fibers. Added to this list is the fact that 1.3 μm amplifiers are directed toward the older 1.3 μm fiber plant, while today's transmission networks are increasingly directed toward the 1.5 μm wavelength. Nevertheless, 1.3 μm amplifiers are still an area of current research and development interest.

10.1.1 Gain in a Four-Level System

There are some fundamental differences between the characteristics of Pr^{3+}-doped and Nd^{3+}-doped 1.3 μm amplifiers on the one hand, and Er^{3+}-doped amplifiers at 1.5 μm on the other. This results from the differences in the gain process between a four-level amplifier (Pr^{3+}, Nd^{3+}) and a three-level amplifier (Er^{3+}). Some elementary understanding of the gain in a four-level system can be obtained by deriving the gain and rate equations in an ideal four-level system.

We consider a four-level system as shown in Figure 10.1, with a ground state denoted by 0, an intermediate state, into which energy is pumped, labeled 3, the upper level of the amplifying transition, labeled state 2, and the lower level of the amplifying transition labeled state 1. The populations of these levels will be labeled N_i. The spontaneous transition rates between the various levels are denoted by τ_{32}, τ_{21}, τ_{10}, and τ_{20}. The total spontaneous transition rate out of level 2, τ_2, is obtained from the transition rates to levels 1 and 0

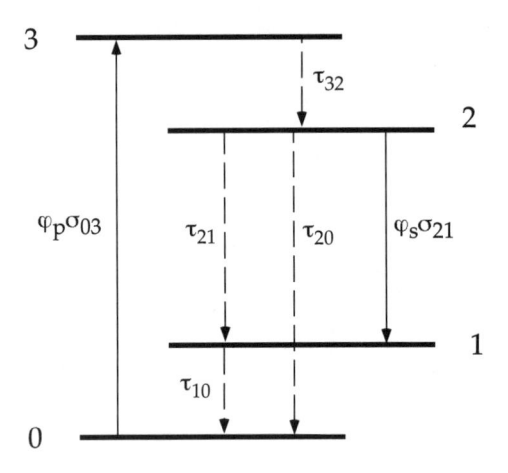

$$\frac{1}{\tau_2} = \frac{1}{\tau_{21}} + \frac{1}{\tau_{20}} \tag{10.1}$$

The transition rate τ_{20} has both a radiative component, $(\tau_{20})_r$, and a nonradiative component, $(\tau_{20})_{nr}$, such that

$$\frac{1}{\tau_{20}} = \frac{1}{(\tau_{20})_r} + \frac{1}{(\tau_{20})_{nr}} \tag{10.2}$$

We assume a fast relaxation rate from level 3 to level 2 so that $N_3 \simeq 0$. We also assume that level 1 empties into level 0 at a fast rate so that we can write in addition that $N_1 \simeq 0$. We then have

$$N_0 + N_2 = N \tag{10.3}$$

where N is the total population. The rate equation for population N_2 can then be written as

$$\frac{dN_2}{dt} = -\frac{N_2}{\tau_2} + (N - N_2)\sigma_{03}\phi_p - N_2\sigma_{21}\phi_s \tag{10.4}$$

where ϕ_s and ϕ_p are the signal and pump photon fluxes, respectively, and σ_{03} and σ_{21} are the absorption and emission cross sections, respectively, for the indicated transitions. In a steady-state situation, $dN_2/dt = 0$. This leads to an equation for N_2, which is also equal to that for the population inversion $N_2 - N_1$ (since we have assumed $N_1 \simeq 0$):

$$N_2 = N \frac{(\tau_2\sigma_{03}\phi_p) / (1 + \tau_2\sigma_{03}\phi_p)}{1 + \frac{\phi_s}{(1+\tau_2\sigma_{03}\phi_p) / \tau_2\sigma_{21}}} \tag{10.5}$$

10.1. INTRODUCTION

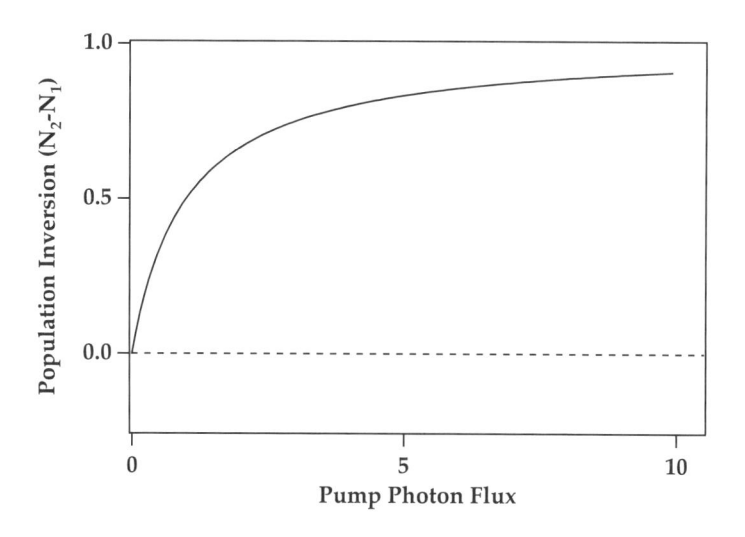

Figure 10.2: Population inversion in an ideal four-level system, from equation 10.8. The pump photon flux is in units of $(\tau_2 \sigma_{03})^{-1}$.

The population inversion saturates with the denominator $1 + (\phi_s/\phi_{sat})$, where

$$\phi_{sat} = \frac{1 + \tau_2 \sigma_{03} \phi_p}{\tau_2 \sigma_{21}} \tag{10.6}$$

For low pumping rates, such that ϕ_p is small compared to $(\tau_2 \sigma_{03})^{-1}$, the signal saturation photon flux is equal to

$$\phi_{sat} = \frac{1}{\tau_{21} \sigma_{21}} \tag{10.7}$$

For small signal levels, the population inversion is given by

$$N_2 \simeq N_2 - N_1 \simeq \frac{\tau_2 \sigma_{03} \phi_p}{1 + \tau_2 \sigma_{03} \phi_p} \tag{10.8}$$

In contrast to a three-level system, the population inversion in a four-level system is always positive, as shown in Figure 10.2.

In the case where the pump power is zero, the signal should suffer no attenuation and no gain. Thus, an advantage of a four-level fiber amplifier over an erbium-doped fiber amplifier is that in the event of a pump source failure the former becomes transparent whereas the latter becomes a strong absorber (at their respective signal wavelengths). As soon as the pump power becomes finite, the signal experiences gain. In practice, the situation is complicated by the fact that there is usually some nonzero background loss, such that even with zero pump power the signal experiences some absorption in a four-level amplifier ion doped fiber. Thus a small finite amount of pump power is needed to render the active medium transparent.

From equation 10.8, it is apparent that at low pump powers the population inversion, and thus the gain, is directly proportional to τ_2, the lifetime of the upper level of the amplifying transition. The search for a suitable ion thus involves finding a transition at 1.3 μm with a long τ_2. The τ_2 is directly affected by the nonradiative transition rates, and so the host glass is of paramount importance. The nonradiative transition rate decreases exponentially with the number of phonons required to bridge the energy gap between the levels, so that a smaller nonradiative transition rate results from lesser phonon energies (see section 4.4.2). In fluoride glasses the nonradiative transition rates are much lower than in silica, due to the lower phonon energy. In silica the phonon energy is on the order of 1100 cm^{-1} whereas in ZBLAN glass it is 500 cm^{-1}. Hence, the search for a suitable ion has centered mainly on the use of nonsilica hosts. A good candidate for a four-level amplification process at 1.3 μm has not yet presented itself in a silica host.

Analytical expressions for the overall gain in an ideal four-level fiber amplifier have been derived. Digonnet has expressed the small signal gain after the signal has traversed a section of pumped amplifier fiber as

$$g = \frac{\sigma_{em} \tau}{h\nu_p} \frac{P_{abs}}{A_{eff}} \frac{F}{\eta_p} \qquad (10.9)$$

whre σ_{em} is the emission cross section, τ is the upper-state lifetime, $h\nu_p$ is the pump photon energy, A is the fiber core area, P_{abs} is the absorbed pump power in the section of fiber considered, F is the overlap integral between the pump and signal fields in the transverse dimensions, and η_p is the fractional pump energy that propagates in the fiber core.[1] A more generalized version of this analytical solution for the gain has been presented, with separate expressions for the low pump power and high pump power regimes.[2]

For most four-level fiber amplifiers, an ideal four-level system such as the one just described cannot take into account the complex effects that can arise. Among these effects are

- a finite population in level 1 (nonzero τ_{10}).
- signal excited-state absorption.
- pump excited-state absorption.
- competing fluorescent transitions between other levels.

The two most actively studied ions for 1.3 μm amplification have been Pr^{3+} and Nd^{3+}. Both of these involve four-level transitions, with τ_2 lifetimes that are in the 100 μs range. The most promising candidate to date is Pr^{3+}, doped in a fluoride host. Excited-state effects, as mentioned above, play a non-trivial role in the gain process in Pr^{3+} at 1.3 μm. We will describe the details and results obtained so far with both of these ions.

10.2 PR^{3+}-DOPED FIBER AMPLIFIERS

10.2.1 Introduction

Figure 10.3 shows the energy level diagram of Pr^{3+}. The transition between levels $^1G_4$ and $^3H_5$ is at 1.3 μm. In silica, the lifetime of the $^1G_4$ level is very short and there is

10.2. PR³⁺-DOPED FIBER AMPLIFIERS

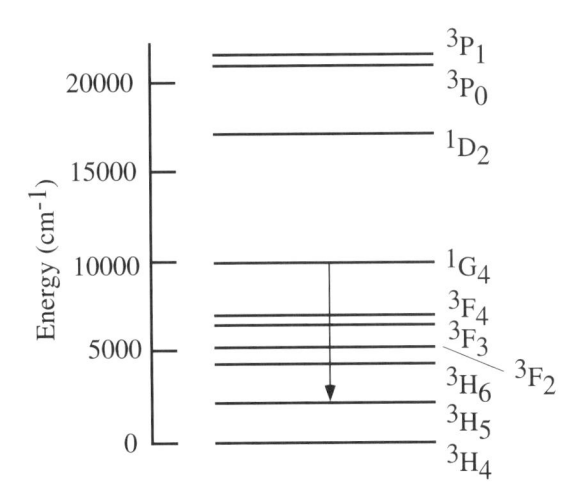

Figure 10.3: Energy level diagram of Pr^{3+}. The 1.3 μm transition is between levels $^1G_4$ and $^3H_5$.

no prospect of obtaining sufficient population inversion on the $^1G_4 - {}^3H_5$ transition. This is due to nonradiative relaxation from the $^1G_4$ level to the levels directly below it, i.e., the $^3F_{4,3,2}$ levels. This can be contrasted with the case of Er^{3+}, where there are no intermediate levels lying between the two levels of the amplifying transition, eliminating the possibility for a multitude of nonradiative decay processes. As mentioned previously, the fluoride glass hosts offer the opportunity to achieve greater lifetimes for levels deactivated by nonradiative transitions. The nonradiative relaxation rate results in an effective lifetime of 110 μs for the $^1G_4$ level, as measured in a fluoride host of ZBLAN composition.

The Pr^{3+}-doped fluoride fiber amplifier was first demonstrated by Ohishi and co-workers, and by other groups.[3, 4, 5, 6, 7] Gains in excess of 30 dB have been observed for pump powers on the order of several hundreds of mW. Laser diode pumping has been demonstrated. A gain of 28 dB was obtained when pumping with four 1.02 μm laser diodes, and a one laser diode pumped double-pass configuration module was shown to have a gain of 23 dB.[8, 9, 10] More recently, a gain of 40 dB was obtained with a two-stage Pr^{3+} fiber amplifier with each stage pumped by a solid state Nd:YLF laser. In this latter demonstration a saturated signal output power of 20 dBm was obtained, as well as a small signal noise figure of 5 dB, at 1.30 μm.[11] High-gain and high output power Pr^{3+} fiber amplifiers using high-power MOPA laser diodes as pumps have also been demonstrated.[12] Optimization of the host glass composition has yielded significant improvements in the performance of the amplifier.[13]

10.2.2 Spectroscopic Properties

The lifetime measured for the $^1G_4$ level of Pr^{3+} in ZBLAN is 110 μs, at low Pr^{3+} concentrations. At higher concentrations, quenching mechanisms set in and the lifetime is reduced. Figure 10.4 shows the lifetime of the $^1G_4$ level as a function of Pr^{3+} concen-

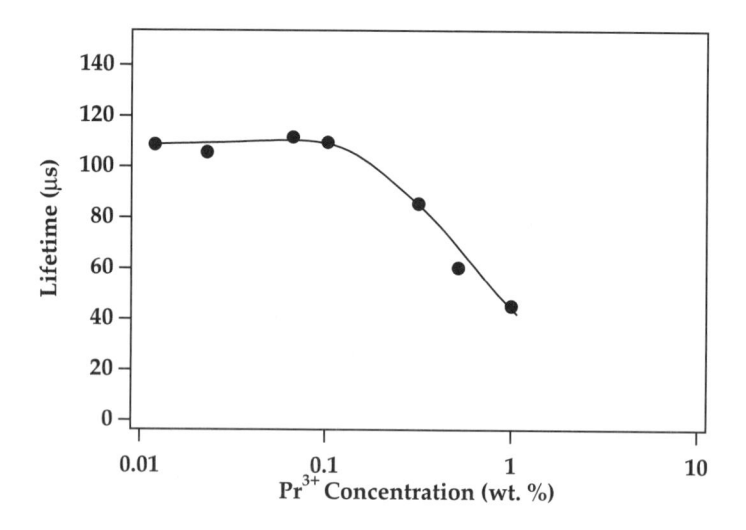

Figure 10.4: Concentration dependence of the Pr^{3+} $^1G_4$ level lifetime, in ZBLAN glass. From reference [15] (©1992 IEEE).

tration. This result suggests that concentrations below 1000 ppm are needed to prevent a reduction in lifetime due to cross relaxation induced concentration quenching.[15] The upper bound on the concentration determines the minimum fiber length required for a specific gain. In some cases, it can be advantageous to use a higher concentration of Pr^{3+} than 1000 ppm since the effects of excess background can negate the lifetime advantage of a lower concentration of Pr^{3+}.[16]

The fluorescence spectrum of Pr^{3+}, due to the $^1G_4$ to $^3H_5$ transition, has a peak at about 1.32 μm. The peak of the actual gain spectrum is shifted to lower wavelengths, by about 10 nm, due principally to the competing effect of the ESA transition from $^1G_4$ to $^1D_2$. The experimentally determined $^1G_4 - ^3H_5$ emission and $^1G_4 - ^1D_2$ absorption cross sections are shown in Figure 10.5.[17]

Interestingly, other glass hosts, namely, the mixed halides and the chalcogenides, have been shown to exhibit significantly longer upper-state lifetimes for the $^1G_4$ level of Pr^{3+} (on the order of 300 μs).[18, 19] These hosts offer the possibility of an order of magnitude increase in the product of the emission cross section and the upper state lifetime, and thus a significant increase in pump efficiency. Glass compositions based on the PbF$_2$-InF$_3$ fluoride host also offer an increase in the Pr^{3+} $^1G_4$ upper-state lifetime, to a value of 170 μs.[13]

10.2.3 *Gain Results for Pr^{3+}-doped Fiber Amplifiers*

In 1991, a Pr^{3+}-doped fluoride fiber amplifier was reported that achieved a net small signal gain of 5.1 dB at a signal wavelength of 1.31 μm, for 180 mW of launched pump power at 1.017 μm.[3] This result was rapidly bettered by improvements in the fiber fabrication process and overall fluoride glass quality, which resulted in a significant lowering of the background loss. Additionally, reductions in the core size, along with

10.2. PR^{3+}-DOPED FIBER AMPLIFIERS

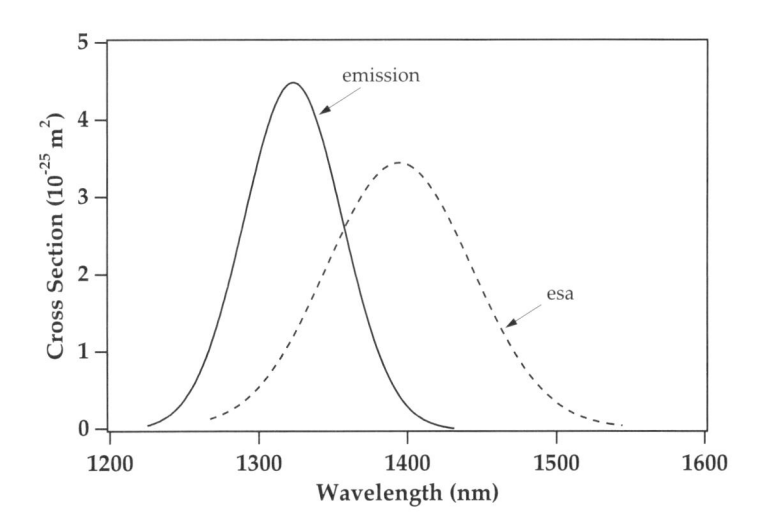

Figure 10.5: Cross sections of the $^1G_4 - {}^3H_5$ emission and $^1G_4 - {}^1D_2$ excited state absorption for Pr^{3+} in ZBLAN, from reference [17].

a commensurate increase in fiber NA, also improved the gain efficiency. In late 1991, a gain of 32 dB was obtained at 1.31 μm for 300 mW of pump power at 1.017 μm.[7] The 8 m long fiber used had a Pr^{3+} concentration of 2000 ppm, a core diameter of 2.3 μm, a relative refractive index difference of 3.8%, and a cutoff wavelength equal to 1.26 μm. A 3 dB saturation output power of 5 dBm was obtained for a pump power of 200 mW. The gain as a function of pump power is shown in Figure 10.6, and the variation of the gain with signal wavelength in this demonstration is shown in Figure 10.7. A 3 dB gain bandwidth of roughly 20 nm can be obtained. Studies have shown that the spectral dependence of the gain on signal wavelength is slightly dependent on the pump power, and that a slight shift of the gain peak to lower wavelengths can be observed as the pump power decreases.[8, 20] This is due to incomplete bleaching of the ground-state absorption of the Pr^{3+} ions.

A recent result reported a gain of 28 dB for a launched pump power of 280 mW.[8] The pump source in this case consisted of four diode lasers operating near 1.017 μm. Each laser diode was supplied with a current of 230 mA to achieve the high pump power operation. The overall configuration of the amplifier is depicted in Figure 10.8. The fluoride fiber is connected to silica fiber pigtails by placing each fiber in a V-groove and gluing them together. The ends of the fibers are polished at an angle prior to joining to minimize back-reflections. Such connections have been reported to have coupling losses as small as 0.2 dB and reflections less than 60 dB.[10, 20, 21] The silica fiber pigtails used have a high NA to ensure minimal coupling loss with the high NA fluoride fiber. Two pairs of laser diodes are used for both copropagating and counterpropagating pumping, with each pair being polarization multiplexed and then launched into the pump port of a wavelength division muliplexer. The net gain as a function of the drive current applied to the four laser diodes (plotted on a one-laser-diode basis) is shown in Figure 10.9. With the increase in power of laser diode pumps,

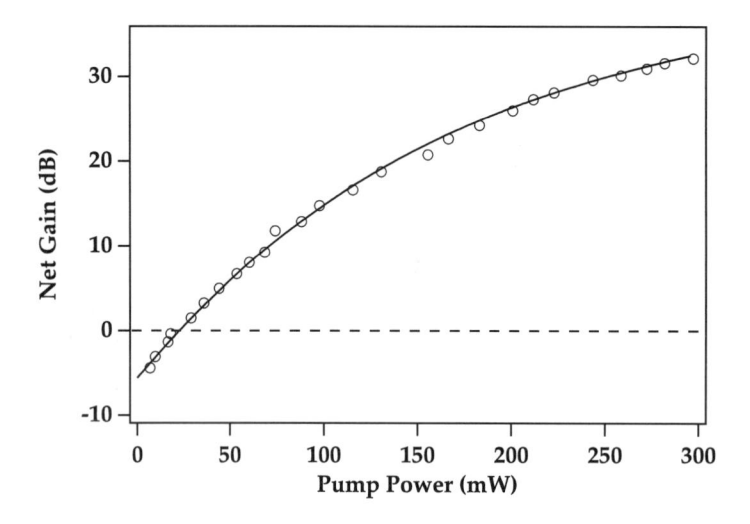

408 CHAPTER 10. FOUR−LEVEL FIBER AMPLIFIERS FOR 1.3 μM AMPLIFICATION

Figure 10.6: Net small signal gain at a signal wavelength of 1.31 μm, as a function of pump power, of a Pr^{3+}:ZBLAN amplifier pumped at 1.017 μm. From a 1991 experiment reported in reference [7]. The background loss of 6 dB has been included in the net gain.

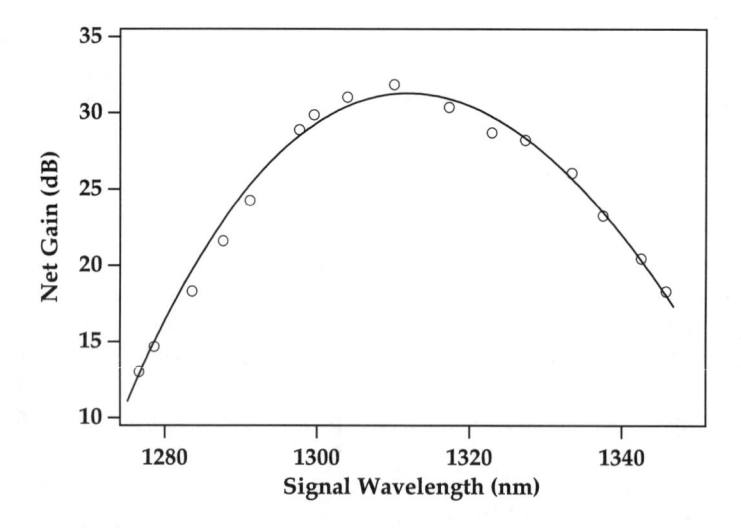

Figure 10.7: Net gain spectrum, as a function of signal wavelength, of a Pr^{3+}:ZBLAN amplifier pumped with 300 mW of power at 1.017 μm. From a 1991 experiment reported in reference [7]. The background loss of 6 dB has been included in the net gain.

10.2. PR^{3+}-DOPED FIBER AMPLIFIERS

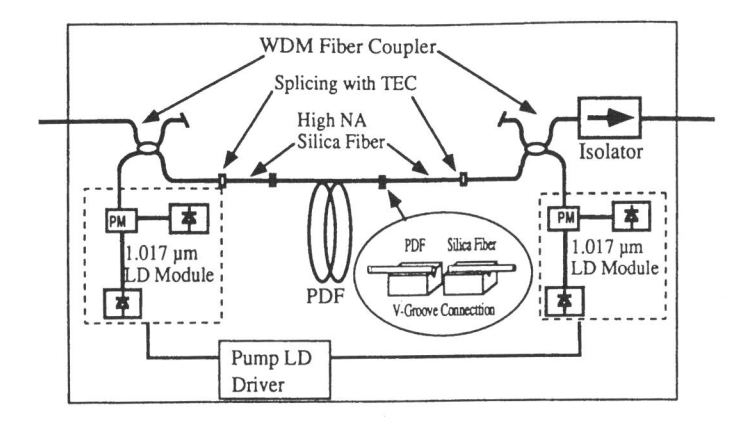

Figure 10.8: Configuration of a four-laser diode pumped the Pr^{3+}-doped fiber amplifier module. From reference [8] (©1993 IEEE).

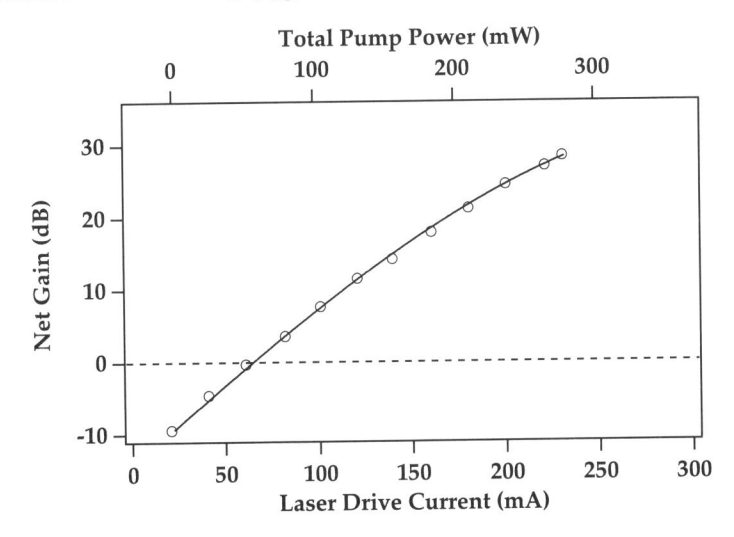

Figure 10.9: Small signal gain verses the drive current simultaneously applied to each laser diode, for a four-laser diode pumped Pr^{3+}-doped fiber amplifier module. From reference [8] (©1993 IEEE).

recent versions of such modules operate with only two laser diode pumps.[22]

A single laser diode pumped module has been demonstrated in which an optical circulator injects the signal into the amplifier, where it is double passed after reflection by an end mirror, and then exits the amplifier module.[9, 10] The end reflector also serves to reinject the leftover pump light after one pass through the fiber. The ZBLAN host fiber used had a Pr^{3+} concentration of 500 ppm, a core diameter of 1.6 μm, a relative refractive index difference of 3.7%, and a cutoff wavelength equal to

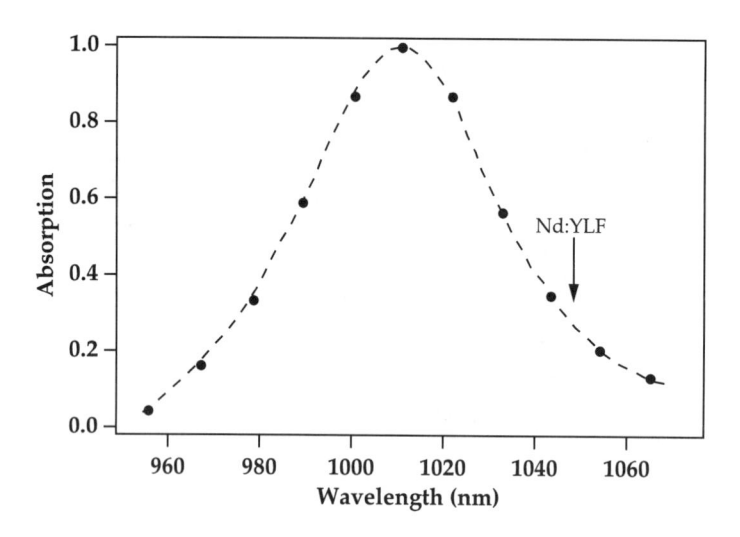

Figure 10.10: Absorption spectrum of a Pr^{3+}-doped fluoride glass in the 1 μm region. From reference [25] (©1993 IEEE). Also indicated is the pump wavelength corresponding to the Nd:YLF laser.

0.9 μm. The background scattering loss of the fiber was 0.05 dB/m at 1.30 μm. The combination of the small mode field diameters, which leads to high pump and signal intensities, coupled with the low background loss, allowed for the efficient operation of this amplifier.

Multistage Pr^{3+}-doped fiber amplifiers have also been demonstrated, leading to high gains and low noise figures. A two-stage fiber amplifier, where each stage is pumped by two laser diodes at 1017 nm and the stages are separated by an isolator and an optical filter, produced 42 dB of net gain with 250 mA of drive current for each of the four laser diodes (corresponding to 395 mW of launched pump power).[20]

An advance in pumping of Pr^{3+}-doped fiber amplifiers has been the demonstration that Nd:YLF solid state lasers, pumped by AlGaAs laser diode arrays, can be very attractive pump sources due to their high output powers.[11] This statement is also true for diode laser pumped cladding pumped Yb^{3+} fiber laser sources.[23] It was earlier shown that pump wavelengths that produce gain in Pr^{3+}-doped fluoride fibers range from at least 980 nm to 1030 nm.[7, 24] This can be understood from the absorption spectrum in the 1 μm pump wavelength region, corresponding to the $^3H_4 - {}^1G_4$ transition (see Figure 10.10). Interestingly, the transition can be pumped at 980 nm (one of the erbium pump wavelengths), at 1030 nm, the wavelength of operation of the Yb fiber laser, and also at 1047 nm, which corresponds to a high output power wavelength of the Nd:YLF laser. Since the 1.3 μm transition in Pr^{3+} is four-level in nature, it is feasible to pump in the wings of the absorption spectrum, in contrast to the case of erbium. For erbium, low-efficiency pumping results in incomplete inversion, which then produces signal absorption, due to the three-level nature of the erbium transition. For Pr^{3+}, incomplete inversion does not necessarily produce more signal absorption, except for the tail of the $^3H_4 - {}^3F_4$ transition that slightly overlaps the 1.3 μm region.

10.2. PR^{3+}-DOPED FIBER AMPLIFIERS

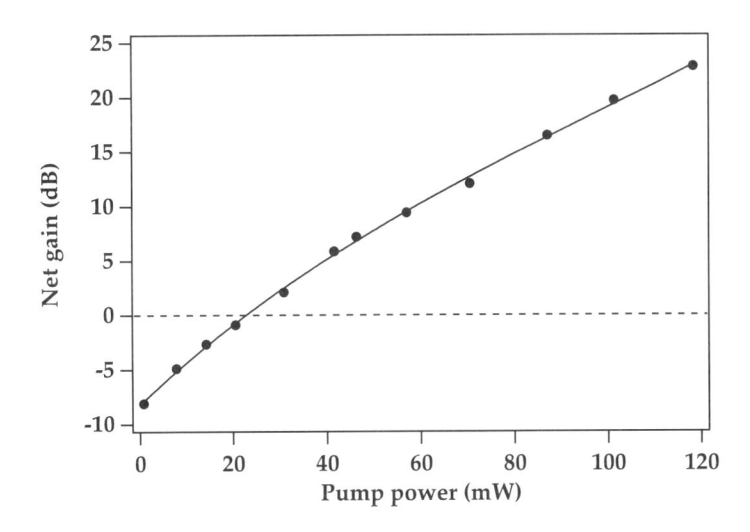

Figure 10.11: Net gain at 1.30 μm in a Pr^{3+}-doped indium fluoride fiber for a signal input power of -30 dBm. From reference [13] (©1997 IEEE).

The fiber used, however, must be made progressively longer as one moves away from the pump absorption peak, so as to obtain the same total number of excited Pr^{3+} ions. This can become an issue when the background scattering loss of the fiber is significant. The possibility of using alternative pump wavelengths for a Pr^{3+}-doped fiber opens the way for the use of existing high pump power lasers, such as a high-power diode laser pumped Nd:YLF laser.[25] Pumping with a cladding pumped Yb-doped fiber laser operating at 1030 nm has been demonstrated to necessitate less pump power than a 1047 nm pump to provide gain at 1300 nm, due to the closer proximity to the peak absorption at 1017 nm. Using such a fiber laser, an output power of 20 dBm was demonstrated for 800 mW of pump at 1030 nm.[23] High-power MOPA (master oscillator power amplifier) laser diodes have also been used successfully with Pr^{3+}-doped fluoride fiber amplifiers.[26]

The use of glass compositions based on PbF$_2$-InF$_3$ fluoride glasses have increased the efficiency of the amplifier through the increase in the upper-state lifetime to 170 μs.[13, 14] Using this glass host, a Pr^{3+}-doped single-mode fiber was fabricated. The gain performance of the amplifier at 1.30 μm, including the background loss of 8.2 dB of 19 m length of amplifying fiber, is shown in Figure 10.11. This particular result is the best achieved to date, with a net gain of 23 dB for a pump power of 120 mW at 1015 nm, using a MOPA laser diode pump source.

High output powers can also be obtained from Pr^{3+}-doped fluoride fiber amplifiers. One might expect that the conversion efficiencies, from pump photons to signal photons, would be low due to the low quantum efficiency of the 1.3 μm transition (i.e., its low radiative to nonradiative decay probability). In fact, in the saturated regime, it is possible to convert quite a large fraction of the pump energy into signal energy. Whitley and co-workers have demonstrated a Pr^{3+}-doped fluoride fiber power amplifier with a photon conversion efficiency of nearly 40%. Their measured output power, for a signal

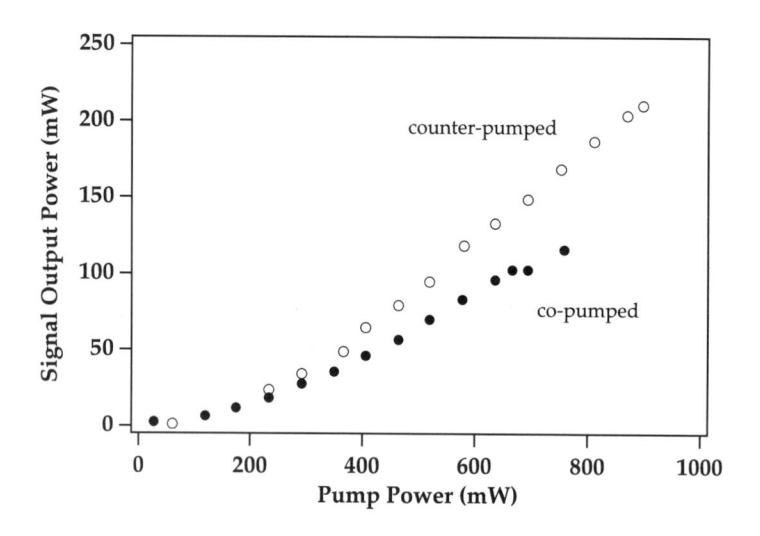

Figure 10.12: Signal output power versus launched pump power for a Pr^{3+}-doped fluoride fiber amplifier pumped in both copropagating and counterpropagating fashion. From reference [27] (©1993 IEEE).

input power of +2 dBm, is shown as a function of pump power in Figure 10.12.[27] An output power as high as a quarter of a watt was achieved using a diode pumped Nd:YLF laser as a pump source.[25]

10.2.4 Modeling of the Pr^{3+}-doped Fiber Amplifier Gain

Modeling of a Pr^{3+}-doped fiber amplifier is different than that of an Er^{3+}-doped fiber amplifier due to the four-level nature of the transition and the presence of additional effects, such as signal excited-state absorption and the $^3H_4 - {}^3F_4$ ground-state absorption transition. In practice, with certain assumptions, similar equations to those used for Er^{3+}-doped fiber amplifier modeling can be used.

Consider Figure 10.3 for the energy level diagram of Pr^{3+}. The lower level of the amplifier transition, $^3H_5$, lies some 2000 cm^{-1} above the $^3H_4$ ground state. This level needs to empty rapidly into the ground state to maintain a large inversion on the 1.3 μm transition. A long lifetime for the $^3H_5$ level will prevent such inversion and decrease the gain.[28] The lifetime of this level is not experimentally known. For reasons of simplicity, most modeling simulations assume it to be equal to zero, in which case the population that decays from the $^1G_4$ upper level instantaneously relaxes to the ground state and the $^3H_5$ level population is zero. Additionally, one can assume that ions excited to a pump level will instantaneously relax to the upper level of the amplifying transition. With this picture, the situation is very comparable to that in Er^{3+} for the 1.5 μm amplifier transition, and only two populations—those of the $^3H_4$ and $^1G_4$ levels—are nonzero. Signal ESA can be added similarly to the case of pump ESA in erbium. This is done by adding a term proportional to $(\sigma_{s-ESA}) \cdot P_s$ to the signal power propagation equation, where (σ_{s-ESA}) is the signal ESA cross section and P_s is

10.2. PR³⁺-DOPED FIBER AMPLIFIERS

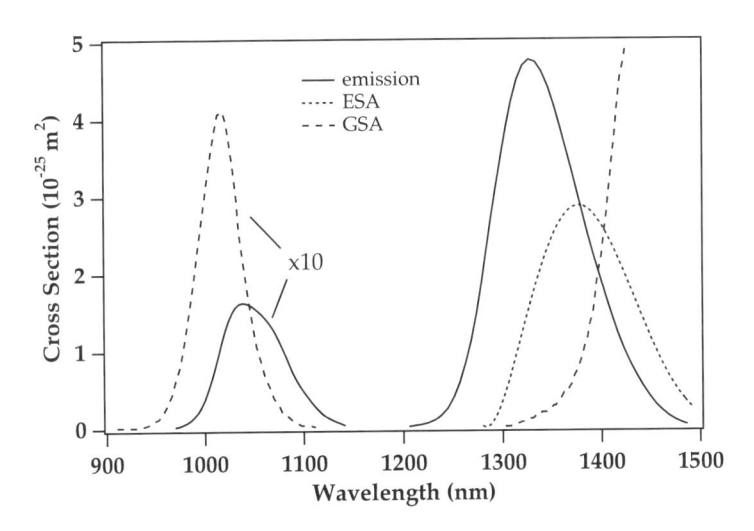

Figure 10.13: Absorption and emission cross sections for Pr^{3+}-doped ZBLAN, for pump and signal wavelengths, and signal excited-state absorption cross section. From reference [29] (©1992 IEEE).

the signal power. It is then assumed that the population excited to the $^1D_2$ level by the ESA process relaxes back instantaneously to the lower levels such that the population of the $^1D_2$ level is essentially zero. Similarly, signal ground-state absorption can be taken into account with a representative cross section σ_{s-GSA}, assuming that population excited to the $^3F_4$ level by the ground state absorption process relaxes instantaneously to the $^3H_4$ ground state.

Figure 10.13 shows the spectrally resolved pump and signal emission and ground-state absorption cross sections, as well as the signal excited-state absorption cross section.[29] These cross sections can be used in a full simulation of the Pr^{3+}:ZBLAN fiber amplifier, and good agreement has been found with experimental results.[29] The optimum cutoff wavelength for a step index fiber was calculated to be 0.8 μm, and increasing the fiber NA to values of 0.3–0.4 improves the pump efficiency significantly. For example, with a fiber NA of 0.35, the small signal gain is predicted to be more than 30 dB for 200 mW of pump power. The model can also be used to explore the effects of an increase in lifetime of the $^1G_4$ level of Pr^{3+}, through the use of a more advantageous host material. This is shown in Figure 10.14.[29] A tripling of the lifetime—as allowed by chalcogenide hosts—or even a simple doubling of the lifetime—as is the case in indium fluoride based hosts—yield significant improvements in the pumping efficiency of the amplifier. Power amplifiers can also be simulated. The quantum conversion efficiency (ratio of output signal photons to input pump photons) of a 1.3 μm power amplifier, calculated from the model just described, is shown in Figure 10.15, for different fiber NAs in a copropagating pump configuration.[30] The conversion efficiencies are expected to be higher for counterpumped or bidirectionally pumped amplifiers, as discussed below.

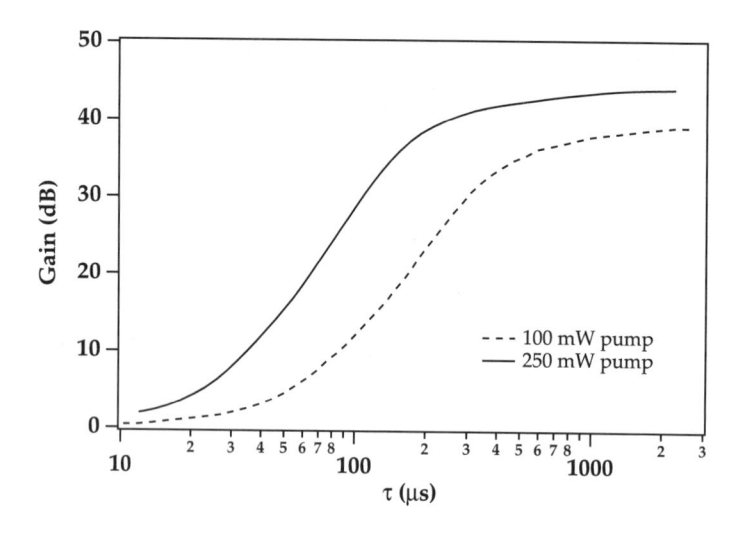

Figure 10.14: Small signal gain of a Pr^{3+}-doped fiber amplifier for two launched pump powers at 1017 nm as a function of τ, the $^1G_4$ upperstate lifetime. From the numerical simulations of reference [29] (©1992 IEEE). The fiber modeled has an NA of 0.3 and a cutoff wavelength of 800 nm.

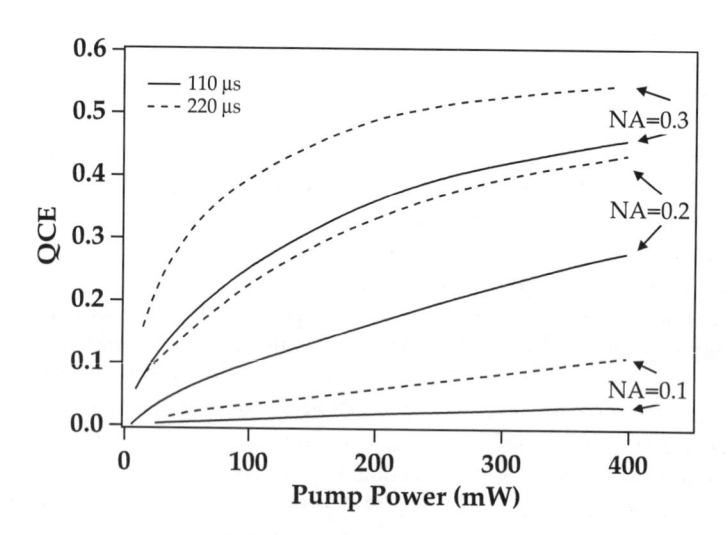

Figure 10.15: Calculated quantum conversion efficiencies for a Pr^{3+}-doped fiber amplifier for a signal input power of 5 mW at 1310 nm, for a variety of fiber NAs and for two different $^1G_4$ lifetimes. From the numerical simulations of reference [30].

10.2. PR^{3+}-DOPED FIBER AMPLIFIERS

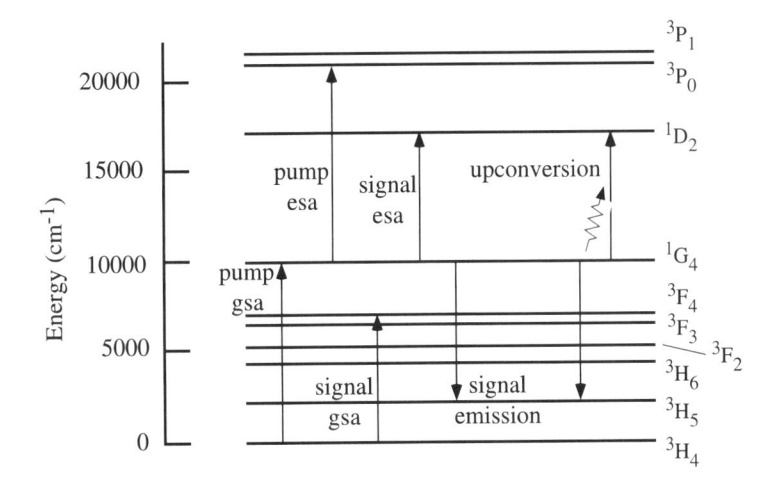

Figure 10.16: Radiative and energy transfer processes of Pr^{3+}, included in an extended model of amplification at 1.3 μm in Pr^{3+}-doped fiber amplifiers, after reference [15] (©1992 IEEE).

A more elaborate model, which takes into account additional effects and the finite population of levels aside from the $^1G_4$ and $^3H_4$ levels, has been developed by Ohishi and coworkers.[15, 31] The model includes signal emission, signal ground-state absorption, signal-excited state absorption, pump ground-state absorption, pump excited-state absorption, and energy transfer via upconversion between neighboring excited Pr^{3+} ion, as depicted in Figure 10.16. In the upconversion process an ion that deactivates from the $^1G_4$ level to the $^3H_5$ level simultaneously promotes a nearby ion from the $^1G_4$ level to the $^1D_2$ level. An important finding is that due to the excited-state absorptions and upconversion process, there is significant improvement to be obtained with bidirectional pumping as compared to copropagating pumping, for a given total pump power.[32, 33] This arises from the fact that when the pump power is strong, pump ESA, which is proportional to the square of the pump power, will waste pump photons by promoting ions from the $^1G_4$ level to the $^1P_0$ level. In addition, upconversion will reduce the $^1G_4$ population in regions of the fiber where the pump power, and the inversion, are high. As a result of these two effects, it is better to distribute the available pump power uniformly in the fiber rather than having it be strong in one section and weak in another. It has been shown that bidirectional pumping improves both the small signal gain (by about 5 dB when the gain is 15–20 dB) and the saturation properties.[32] Note that by the same token counterpropagating pumping can yield an improved signal output in the case of a strong signal amplifier, as discussed in sections 6.4.6 and 8.3.5. The strong signal can then deplete the inversion at the expense of pump ESA or upconversion. Figure 10.17 shows the significant improvement in output power obtained when pumping bidirectionally, as compared to copropagating pumping.

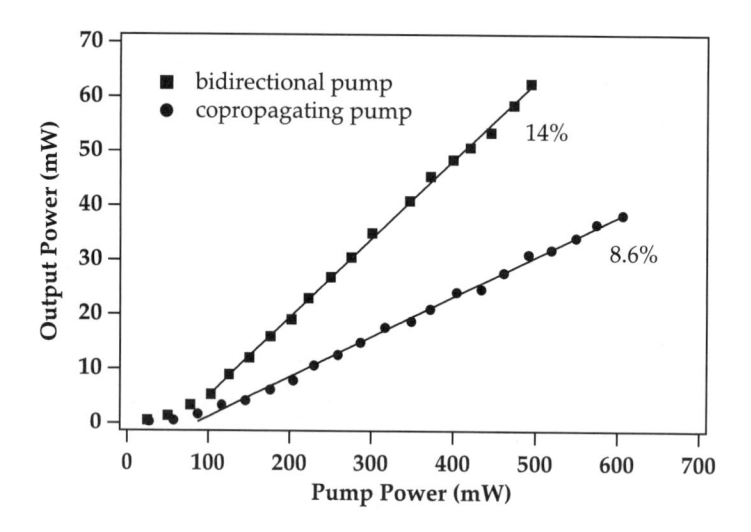

Figure 10.17: Signal output powers of a Pr^{3+}-doped fluoride fiber amplifier, for a signal input power of 1 mW at a signal wavelength of 1.31 μs, with both bi-directional and copropagating pumping at 1.017 μs. From reference [32] (©1994 IEEE). The slope efficiency improves from 8.6% (copropagating pump) to 14% (bidirectional pump).

10.2.5 System Results

The Pr^{3+}-doped fluoride fiber amplifier has been tested in a variety of system applications. It has demonstrated its ability to amplify digital and analog signals, while maintaining reasonably good noise characteristics. Systems experiments date back to 1991.

The noise figure of the Pr^{3+}-doped fluoride fiber amplifier has been measured to range from 3.8 to 8 dB, depending on the signal wavelength.[11, 34, 35] A measurement of the noise figure of an amplifier operating in the small signal regime is shown in Figure 10.18. The noise figure increases with wavelength. This is due to the reduction of the inversion attributed to both signal excited-state absorption and ground-state absorption, both of which increase with wavelength in the vicinity of 1.3 μs. Signal ESA is thought to provide the bulk of the noise figure degradation.[34] Additionally, it was found that the 5 dB noise figure at 1.3 μs is independent of signal output power, even with output powers as high as 15 dBm.[21, 34] Signal crosstalk experiments have also been performed, and they indicate that the 3 dB roll-off modulation frequency in a crosstalk experiment is 6 kHz, which is consistent with the 110 μs upper-state lifetime in Pr^{3+}.[34] This result implies that the Pr^{3+}-doped fluoride fiber amplifier will not suffer from signal crosstalk at high bit rate modulations.

The Pr^{3+}-doped fiber amplifier has been used in several digital transmission applications.[36, 37] A first system investigation at 2.5 Gb/s of the Pr^{3+}-doped fluoride fiber amplifier, pumped by a large frame Ti-sapphire laser, was performed with the amplifier used as a power amplifier, inline amplifier, and preamplifier.[36] The amplifier used had a small signal gain of 24 dB at 1.3 μm and a saturation output power of 18 dBm.

10.2. PR³⁺-DOPED FIBER AMPLIFIERS

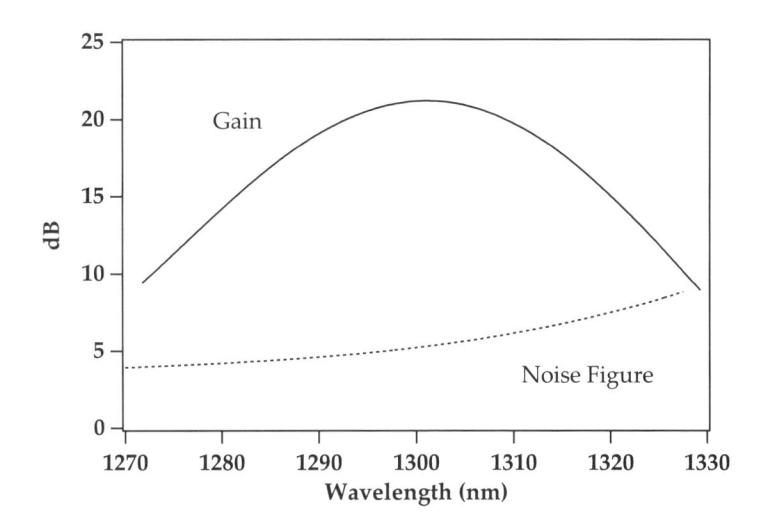

Figure 10.18: Noise figure and gain of a Pr^{3+}-doped fluoride fiber amplifier bi-directionally pumped at 1.047 μs, as a function of signal wavelength. From reference [34].

As a power amplifier, it produced an output power of 17.4 dBm with no degradation in receiver performance compared to back-to-back transmitter and receiver. As an in-line amplifier, it increased the system range by 19 dB over the unamplified system. As a preamplifier, the sensitivity of the receiver was improved to −37.3 dBm (490 photons per bit). In another power amplifier application at 100 Mb/s, the amplifier increased the transmission distance by 27 km.[37] The use of Nd:YLF pumped amplifiers has opened the way to additional digital systems demonstrations. With the additional pump power available, transmission over 110 km at 10 Gb/s with a total system power budget of 39 dB was achieved, using a single Pr^{3+}-doped fluoride fiber amplifier as a power amplifier.[11] More recently, a field demonstration was made over service-bearing routes in the UK. 5 Gb/s data was successfully transmitted over a 92 km route with a power amplifier and an inline amplifier, compensating for a total system loss of 36 dB.[21]

The Pr^{3+}-doped fluoride fiber amplifier has also been tested for analog transmission applications, using the AM-VSB and FM-SCM formats.[38, 39, 40] An early 40-channel experiment was done with a Pr^{3+}-doped fluoride fiber amplifier providing an output power of 14.3 dBm (net signal gain of 6.5 dB) at 1.3 μm. The loss budget was improved by 4.6 dB at a CNR of 51 dB, to 15.1 dB, as shown in Figure 10.19.[38] It was also observed that the composite triple beat was only slightly degraded by the presence of the amplifier, whereas the composite second-order distortion (CSO) and the cross modulation (XM) were significantly degraded (by 5–15 dB).[38] Some of this degradation was attributed to reflections at the silica to fluoride fiber interfaces. The transmitted picture was not of high quality, as measured by visual inspection. An improvement was achieved by using an externally modulated signal laser, rather than a directly modulated signal laser, in a 40-channel AM-SCM experiment.[41] The CSO

418 CHAPTER 10. FOUR–LEVEL FIBER AMPLIFIERS FOR 1.3 μM AMPLIFICATION

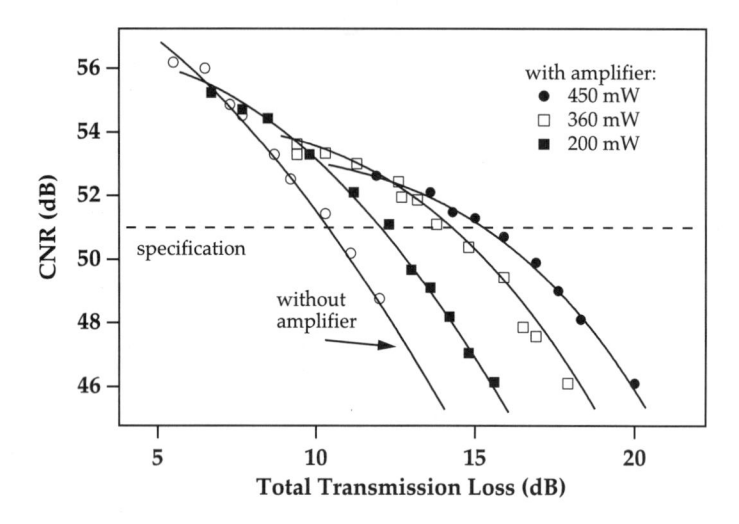

Figure 10.19: Measured CNR of a multichannel AM-VSB transmission system as a function of transmission loss, with and without a Pr^{3+}-doped fluoride fiber amplifier pumped with 200, 360, and 450 mW of power at 1.017 μm. From reference [38].

and XM problems were substantially removed compared to the previous experiments. Nevertheless, a low-frequency noise source was found to be present, which resulted in a poor quality picture. The conclusion was that backscatter in the amplifying fiber itself might be the source of the problem.

The first 40-channel AM-VSB experiment with a two-stage Nd:YLF pumped Pr^{3+} fiber amplifier resulted in CSO, CTB, and XM parameters that were almost within CATV specification. Actual TV transmission was excellent, as any degradation of the picture quality was imperceptible.[11] The use of wavelength stabilized 1.0 μm laser diode pumps has also yielded good picture quality for the transmission of 40 channels of AM-VSB television signals.[40]

Good quality FM-SCM (used for satellite TV) analog transmission has also been demonstrated.[21] Twenty-four channels were transmitted with high picture quality over 100 km, using a Pr^{3+}-doped fiber amplifier pumped by an Nd:YLF laser, for a 16 dB increase in power margin. Transmission of these 24 channels with high picture quality was also achieved over a 200 km distance, with one power amplifier and mid-point amplification with two inline amplifiers. The total power budget was 74 dB, representing the longest distance over which FM-SCM TV pictures have been transmitted.

10.3 ND^{3+}-DOPED FIBER AMPLIFIERS

10.3.1 Introduction

Nd^{3+}-doped fluoride fiber amplifiers were early candidates as amplifiers at 1.3 μm. Unfortunately, they suffer from a host of problems and have never yielded suitable

10.3. ND^{3+}-DOPED FIBER AMPLIFIERS

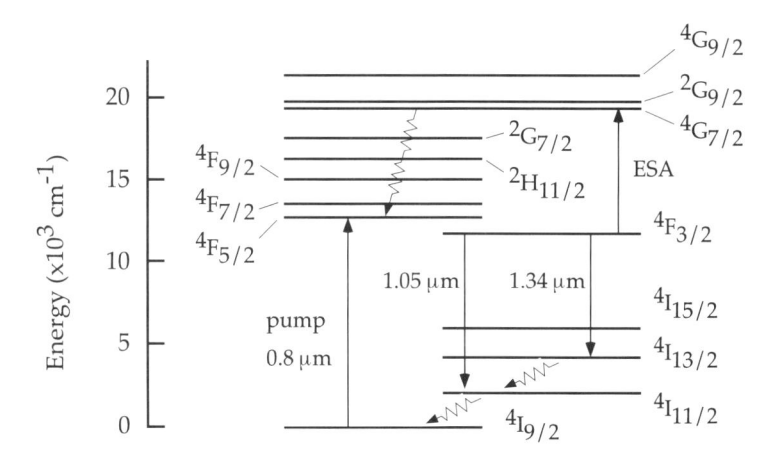

Figure 10.20: Energy level diagram of Nd^{3+} and relevant transitions for the 1.3 μm amplification process.

gains at the wavelength of interest for transmission applications (1.3 μm) despite the fact that they can be pumped in the 800 nm region where high-power laser diodes are available. Nevertheless, the study of the Nd^{3+}-based amplifiers has provided useful insights into the properties of rare earth doped fiber amplifiers. In addition to a very strong signal excited-state absorption effect, Nd^{3+} suffers from the fact that another transition, at a wavelength of 1.06 μm, originates from the same $^4F_{3/2}$ upper state as the 1.3 μm transition. The 1.06 μm transition has, in fact, an emission cross section about four times stronger than that at 1.3 μm, and thus the 1.06 μm transition tends to "steal" the gain from the 1.3 μm transition. Even without these deleterious effects, the product of the emission cross section and the lifetime for Nd^{3+} is an order of magnitude lower than that for Er^{3+}, indicating that the pump efficiencies for an Nd^{3+}-doped fluoride fiber amplifier will be significantly lower than what we are accustomed to for Er^{3+}-doped fiber amplifiers at 1.5 μm.[42, 43] Figure 10.20 shows the Nd^{3+} energy levels, along with the processes relevant to gain in an Nd^{3+}-doped amplifier.

10.3.2 Gain Results for Nd^{3+}-Doped Fiber Amplifiers

A silica host for Nd^{3+} is not suitable because the excited-state absorption in the 1.3 μm region is very strong in silica. The signal emission to signal ESA cross section ratio is calculated to be 0.92 for silica whereas it is close to 2.0 in a fluorozirconate host.[44]

A Nd^{3+}-doped fiber laser at 1.3 μm was demonstrated in a fluoride host.[44] The free-running laser was found to oscillate at 1.33 μm, when oscillation at 1050 nm was suppressed by a suitable choice of end mirror reflectivities. This result indicates that ESA must be reducing the gain at 1320 nm since it is the peak of the fluorescence spectrum and would be the natural oscillation wavelength in the absence of ESA. Early amplification experiments in a large core fiber also suggested that gain would be available only at wavelengths longer than 1320 nm in a fluoride fiber.[45] A gain of 4–5 dB was obtained with 150 mW of pump at 795 nm, for signal wavelengths between

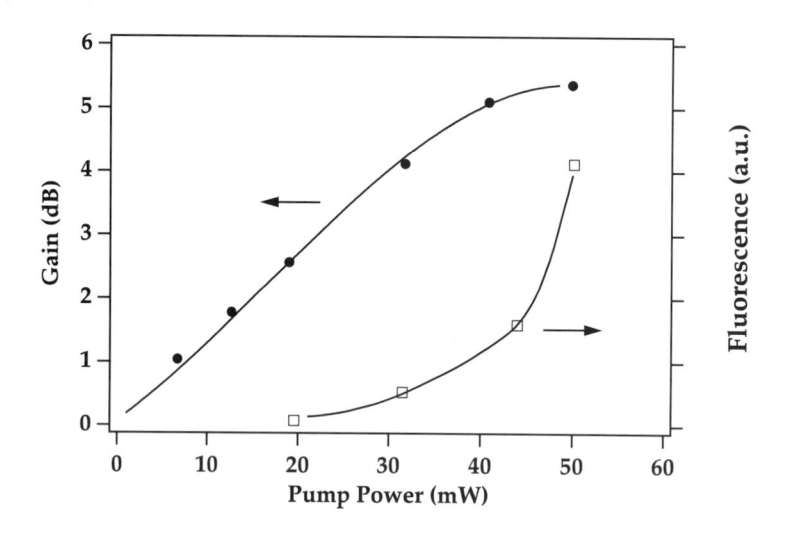

Figure 10.21: Gain at 1.3 μm and fluorescence at 1.05 μm as a function of pump power, in an Nd^{3+}-doped fluoride fiber amplifier. From reference [47].

1320 and 1350 nm.[46] The excited-state absorption resulted in net absorption for signal wavelengths below 1310 nm. A gain of 3.3 dB was also obtained at 1.32 μm for only 50 mW of pump power at 820 nm.[47] The gain was then flat for pump powers above 50 mW due to the increase in the 1050 nm ASE, as shown in Figure 10.21. Fluorophosphate glass hosts are also promising since the emission to excited-state cross section ratio is particularly favorable in this host. In addition, the ESA spectrum is shifted to longer wavelengths.[48] Amplifier investigations were carried out in an Nd^{3+}-doped fluorophosphate host.[49] It was found that positive gain can be obtained closer to 1.3 μm, as compared to Nd^{3+}-doped ZBLAN fiber amplifiers. Nevertheless, only 3.4 dB of gain is obtained at 1.32 μm and the gain saturates for pump powers above 80 mW due to the ASE buildup at 1050 nm. Since the ASE at 1050 nm is a major contributor to the low gain at 1.3 μm in Nd-doped fiber amplifiers, attempts have been made to reduce it by introducing filters that suppress its propagation along the fiber.[50, 51] Using mode coupling notch filters, a gain enhancement of up to 5 dB was achieved at a signal wavelength of 1328 nm, as shown in Figure 10.22.[50]

10.3.3 Modeling of the Nd^{3+}-Doped Fiber Amplifier Gain

The Nd^{3+}-doped fluoride fiber amplifier has been the subject of detailed modeling with respect to the impact of the excited-state absorption and its dependence on the host glass, and with respect to the effect of the 1050 nm ASE and its possible suppression by filters placed along the amplifier.

The simple analytical formula for the small signal gain in a four-level system can be modified for the presence of excited-state absorption.[42, 43] The gain in dB per

10.3. ND^{3+}-DOPED FIBER AMPLIFIERS

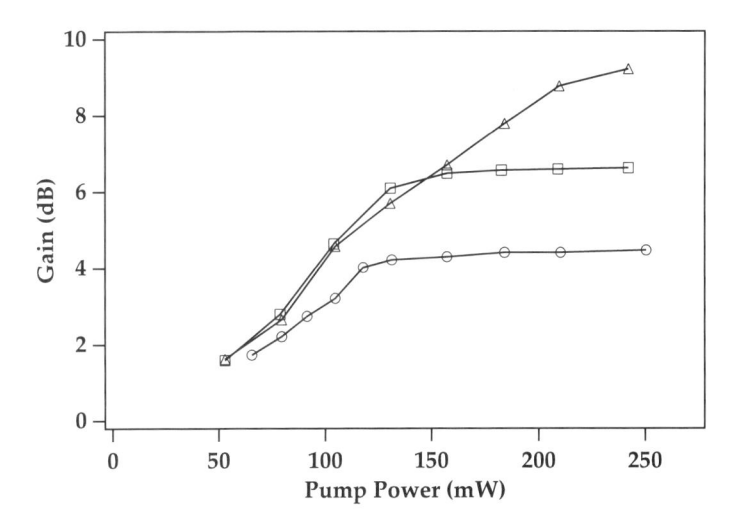

Figure 10.22: Gain at 1328 nm with and without a filter at 1050 nm in an Nd^{3+}-doped fluoride fiber amplifier. From the experiments of reference [50] (circles: no filter; squares: filter placed 25 cm from the fiber input; triangles: filter placed 39 cm from the fiber input).

unit absorbed pump power is then

$$\frac{4.34}{h\nu_p} \frac{\sigma_{em}\tau}{A_{eff}} \left(1 - \frac{\sigma_{esa}}{\sigma_{em}}\right) \qquad (10.10)$$

where $h\nu_p$ is the pump photon energy; σ_{em} and σ_{esa} are the signal emission and excited-state absorption cross sections, respectively; τ is the upper-state lifetime; and A_{eff} is the fiber effective area.[43] This equation, when $\sigma_{esa} = 0$, is seen to be a simplified version of equation 10.9. The presence of a large excited-state absorption cross section clearly reduces the small signal gain available. The excited-state absorption also reduces the gain in a power amplifier configuration. Figure 10.23 shows the effect of various cross section ratios $\epsilon = \sigma_{esa}/\sigma_{em}$ on the signal output power for an input power of 1 mW.[43] With $\epsilon = 0.5$, the photon conversion efficiency is reduced to 14%. Additionally, the ESA impairs the noise characteristics and increases the noise figure by reducing the population inversion.[42]

A full model that includes both signal excited-state absorption and parasitic ASE at 1050 nm has been investigated.[48] The effect of the 1050 nm ASE is to reduce the overall gain from 15–20 dB, in the 1320–1350 nm range, to 5 dB. At 1310 nm, the gain is very small due to the presence of signal ESA. The signal ESA effect is shown to be somewhat reduced in a fluorophosphate host, compared to ZBLAN glass.[48] Clearly, it is desirable to seek a host glass in which the effects of excited-state absorption are reduced at the signal wavelengths of interest (near 1310 nm). The most promising candidates appear to be a fluoroberyllate host.[52] In the fluoroberyllate host, Nd^{3+} has an emission spectrum that is shifted to lower wavelengths, and the ESA spectrum is not as detrimental at 1310 nm. The difference between the emission and ESA cross

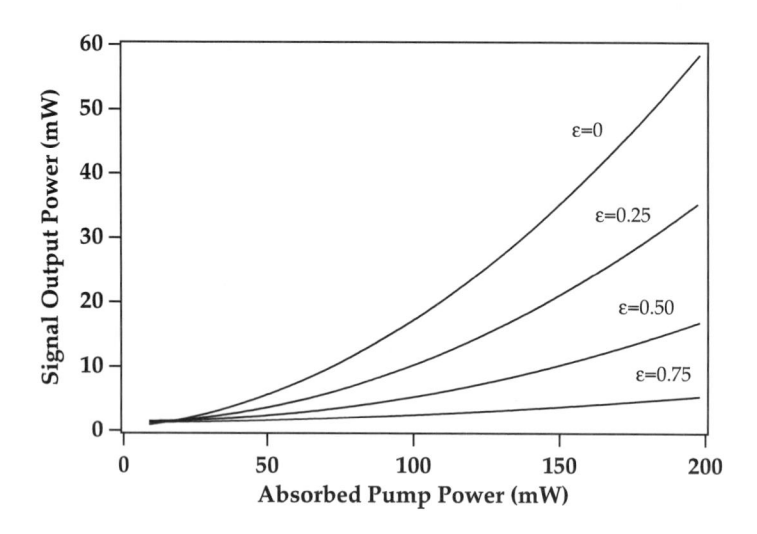

Figure 10.23: Signal output power at 1325 nm with a signal input power of 1 mW in an Nd^{3+}-doped fluoride fiber amplifier for various ESA to emission cross section ratios ϵ. From the numerical simulations of reference [43] (©1990 IEEE).

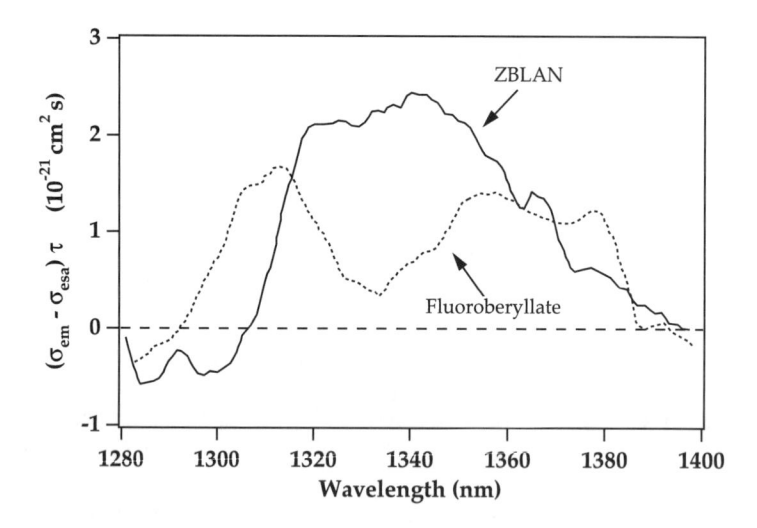

Figure 10.24: Effective cross section and lifetime product for Nd^{3+} in ZBLAN and fluoroberyllate glasses. From reference [52].

sections, multiplied by the upper-state lifetime, is plotted in Figure 10.24.[52] The upper state lifetime in the fluoroberyllate is 880 μs, as opposed to 450 μs in ZBLAN. The net result is a predicted gain of 25 dB at 1310 nm in the fluoroberyllate host, in contrast to 5 dB in ZBLAN, with a pump power of 100 mW at 800 nm (ignoring ASE at 1050 nm).[52] Unfortunately, fluoroberyllate glass is a very toxic material to fabricate

10.3. ND^{3+}-DOPED FIBER AMPLIFIERS

and fibers have not yet been obtained. To realize a practical Nd-doped fiber amplifier, a host with negligible signal ESA near 1310 nm will be need to be developed along with a suitable technique for near total suppression of the 1050 nm ASE.

Bibliography

[1] M. J. F. Digonnet, *IEEE J. Quant. Elect.* **26**, 1788 (1990).

[2] T. J. Whitley and R. Wyatt, *IEEE Phot. Tech. Lett.* **5**, 1325 (1993).

[3] Y. Ohishi, T. Kanamori, T. Kitagawa, S. Takahashi, E. Snitzer, and G. H. Sigel, "Pr^{3+}-doped fluoride fiber amplifier operating at 1.31 μm," in *Conference on Optical Fiber Communication*, Vol. 4, 1991 OSA Technical Digest Series (Optical Society of America, Washington, D.C., 1991), pp. 237–240.

[4] Y. Ohishi, T. Kanamori, T. Kitagawa, S. Takahashi, E. Snitzer, and G. H. Sigel, *Opt. Lett.* **16**, 1747 (1991).

[5] Y. Durteste, M. Monerie, J.Y. Allain, and H. Poignant, *Elect. Lett.* **27**, 626 (1991).

[6] S. F. Carter, D. Szebesta, S. T. Davey, R. Wyatt, M. C. Brierley, and P. W. France, *Elect. Lett.* **27**, 628 (1991).

[7] Y. Miyajima, T. Sugawa, and Y. Fukasaku, *Elect. Lett.* **27**, 1706 (1991). Note that to obtain the net gain for fiber A of this reference, 6 dB – the background loss – needs to be subtracted from the stated gains.

[8] M. Shimizu, T. Kanamori, J. Temmyo, M. Wada, M. Yamada, Y. Terunuma, Y. Ohishi, and S. Sudo, *IEEE Phot. Tech. Lett.* **5**, 654 (1993).

[9] M. Yamada, M. Shimizu, Y. Ohishi, J. Temmyo, M. Wada, T. Kanamori, and S. Sudo, "One-ld-pumped Pr^{3+}-doped fluoride fiber amplifier module with a signal gain of 23 dB," in *Optical Amplifiers and Their Applications*, Vol. 14, 1993 OSA Technical Digest Series (Optical Society of America, Washington, D.C., 1993), pp. 240–243.

[10] M. Yamada, M. Shimizu, Y. Ohishi, J. Temmyo, T. Kanamori, and S. Sudo, *Elect. Lett.* **29**, 1950 (1993).

[11] M. Yamada, M. Shimizu, T. Kanamori, Y. Ohishi, Y. Terunuma, K. Oikawa, H. Yoshinaga, K. Kikushima, Y. Miyamoto, and S. Sudo, *IEEE Phot. Tech. Lett.* **7**, 869 (1995).

[12] S. Sanders, K. Dzurko, R. Parke, S. O'Brien, D. F. Welch, S. G. Grubb, G. Nykolak, and P.C. Becker, *Elect. Lett.* **32**, 343 (1996).

[13] Y. Nishida, T. Kanamori, Y. Ohishi, M. Yamada, K. Kobayashi, and S. Sudo, *IEEE Phot. Tech. Lett.* **9**, 318 (1997).

[14] K. Isshiki, M. Kubota, Y. Kuze, S. Yamaguchi, H. Watanabe, and K. Kasahara, *IEEE Phot. Tech. Lett.* **10**, 1112 (1997).

[15] Y. Ohishi, T. Kanamori, T. Nishi, S. Takahasi, and E. Snitzer, *IEEE Phot. Tech. Lett.* **4**, 1338 (1992).

[16] C. Watanabe, K. Nakazato, and M. Onishi, "Effect of background loss on gain property of praseodymium-doped fluoride fiber," in *Optical Amplifiers and Their Applications*, Vol. 14, 1993 OSA Technical Digest Series (Optical Society of America, Washington, D.C., 1993), pp. 226–229.

[17] B. N. Samson, M. Naftaly, D. Hewak, S. Jordery, J. A. Mederios Neto, H. J. Tate, R. I. Laming, A. Jha, R. S. Deol, and D. N. Payne, "Pr^{3+}-doped mixed halide glasses for 1.3 μm fibre amplifiers," in *Optical Amplifiers and Their Applications*, Vol. 14, 1993 OSA Technical Digest Series (Optical Society of America, Washington, D.C., 1993), pp. 252–255.

[18] P. C. Becker, M. M. Broer, V. G. Lambrecht, A. J. Bruce, and G. Nykolak, "Pr^{3+}:La-Ga-S glass: a promising material for 1.3 μm fiber amplification," in *Optical Amplifiers and Their Applications*, Vol. 17, 1992 OSA Technical Digest Series (Optical Society of America, Washington, D.C., 1992), pp. 251–254.

[19] M. A. Newhouse, R. F. Bartholomew, B. G. Aitken, A. L. Sadd, and N. F. Borelli, "309 μs Pr^{3+} excited state lifetime observed in a mixed halide glass," in *Optical Amplifiers and Their Applications*, Vol. 17, 1992 OSA Technical Digest Series (Optical Society of America, Washington, D.C., 1992), pp. 296–299.

[20] M. Yamada, M. Shimizu, Y. Ohishi, J. Temmyo, T. Kanamori, and S. Sudo, *Optoelectronics* **10**, 109 (1995).

[21] T. J. Whitley, *J. Light. Tech.* **13**, 744 (1995).

[22] M. Shimizu, M. Yamada, J. Temmyo, M. Wada, T. Kanamori, Y. Terunuma, Y. Ohishi, and S. Sudo, *Micro. and Opt. Tech. Lett.* **7**, 159 (1994).

[23] V. Morin, E. Taufflieb, and I. Clarke, '+20 dBm praseodymium doped fiber amplifier single-pumped at 1030 nm'," in *OSA Trends in Optics and Photonics*, Vol. 16, Optical Amplifiers and Their Applications, M. N. Zervas, A. E. Willner, and S. Sasaki, Eds. (Optical Society of America, Washington, DC, 1997), pp. 104–107.

[24] T. Sugawa and Y. Miyajima, *IEEE Phot. Tech. Lett.* **3**, 616 (1991).

[25] T. Whitley, R. Wyatt, D. Szebesta, S. Davey, and J. R. Williams, *IEEE Phot. Tech. Lett.* **4**, 399 (1993).

[26] M. Yamada, T. Kanamori, Y. Ohishi, M. Shimizu, Y. Terunuma, S. Sato, and S. Sudo, *IEEE Phot. Tech. Lett.* **9**, 321 (1997).

[27] T. Whitley, R. Wyatt, D. Szebesta, S. Davey, and J. R. Williams, *IEEE Phot. Tech. Lett.* **4**, 401 (1993).

[28] P. Urquhart, *IEEE J. Quant. Elect.* **28**, 1962 (1992).

[29] B. Pedersen, W. J. Miniscalco, and R. S. Quimby, *IEEE Phot. Tech. Lett.* **4**, 446 (1992).

[30] B. Pedersen, W. J. Miniscalco, S. Zemon, and R. S. Quimby, "Neodymium- and praesodymium doped fiber power amplifiers," in *Optical Amplifiers and Their Applications*, Vol. 17, 1992 OSA Technical Digest Series (Optical Society of America, Washington, D.C., 1992), pp. 16–19.

[31] Y. Ohishi, T. Kanamori, Y. Terunuma, M. Shimizu, M. Yamada, and S. Sudo, "Investigation on efficient pump scheme of Pr^{3+}-doped fluoride fiber amplifiers," in *Optical Amplifiers and Their Applications*, Vol. 14, 1993 OSA Technical Digest Series (Optical Society of America, Washington, D.C., 1993), pp. 256–259.

[32] Y. Ohishi, T. Kanamori, Y. Terunuma, M. Shimizu, M. Yamada, and S. Sudo, *IEEE Phot. Tech. Lett.* **6**, 195 (1994).

[33] Y. Ohishi, M. Yamada, T. Kanamori, Y. Terunuma, and S. Sudo, *IEEE Phot. Tech. Lett.* **8**, 512 (1996).

[34] T. Whitley, R. Wyatt, S. Fleming, D. Szebesta, J. R. Williams, and S. T. Davey, "Noise and cross-talk characteristics of a praesodymium doped fluoride fibre amplifier," in *Optical Amplifiers and Their Applications*, Vol. 14, 1993 OSA Technical Digest Series (Optical Society of America, Washington, D.C., 1993), pp. 244–247.

[35] M. Yamada, M. Shimizu, H. Yoshinaga, K. Kikushima, T. Kanamori, Y. Ohishi, Y. Terunuma, K. Oikawa, and S. Sudo, *Elect. Lett.* **31**, 806 (1995).

[36] R. Lobbett, R. Wyatt, P. Eardley, T. J. Whitley, P. Smyth, D. Szebesta, S. F. Carter, S. T. Davey, C. A. Millar, and M. C. Brierley, *Elect. Lett.* **27**, 1472 (1993).

[37] N. Tomita, K. Kimura, H. Suda, M. Shimizu, M. Yamada, and Y. Ohishi, *IEEE Phot. Tech. Lett.* **6**, 258 (1994).

[38] K. Nakazato, C. Fukuda, M. Onishi, and M. Nishimura, *Elect. Lett.* **29**, 1600 (1993).

[39] H. Suda, N. Tomita, S. Furukawa, F. Yamamoto, K. Kimura, M. Shimizu, M. Yamada, and Y. Ohishi, "Application of LD-pumped Pr-doped fluoride fibre amplifier to digital and analog signal transmission," in *Optical Amplifiers and Their Applications*, Vol. 14, 1993 OSA Technical Digest Series (Optical Society of America, Washington, D.C., 1993), pp. 365–369.

[40] J. Temmyo, M. Sugo, T. Nishiya, T. Tamamura, M. Yamada, S. Sudo, F. Yamamoto, F. Bilodeau, and K.O. Hill, *Elect. Lett.* **32**, 1910 (1996).

[41] K. Kikushima, M. Yamada, M. Shimizu, and J. Temmyo, *IEEE Phot. Tech. Lett.* **6**, 440 (1994).

[42] M. L. Dakss and W. J. Miniscalco, "A large-signal model and signal/noise ratio analysis for Nd^{3+} doped fiber amplifiers at 1.3 μm," in *Fiber Laser Sources and Amplifiers II*, M. J. F. Digonnet, Ed. *Proc. SPIE* **1373**, pp. 111–124 (1990).

[43] M. L. Dakss and W. J. Miniscalco, *IEEE Phot. Tech. Lett.* **2**, 650 (1990).

[44] W. J. Miniscalco, L. J. Andrews, B. A. Thompson, R. S. Quimby, L. J. B. Vacha, and M. G. Drexhage, *Elect. Lett.* **24**, 28 (1988).

[45] M. C. Brierley and C. A. Millar, *Elect. Lett.* **24**, 438 (1988).

[46] M. Brierley, S. Carter, P. France, and J. E. Pedersen, *Elect. Lett.* **26**, 330 (1990).

[47] Y. Miyajima, T. Sugawa, and T. Komukai, *Elect. Lett.* **26**, 1398 (1990).

[48] S. Zemon, B. Pedersen, G. Lambert, W. J. Miniscalco, B. T. Hall, R. C. Folweiler, B. A. Thompson, and L. J. Andrews, *IEEE Phot. Tech. Lett.* **4**, 244 (1992).

[49] E. Ishikawa, H. Aoki, T. Yamashita, and Y. Asahara, *Elect. Lett.* **28**, 1497 (1992).

[50] M. Obro, J. E. Pedersen, and M. C. Brierley, *Elect. Lett.* **28**, 99 (1992).

[51] O. Lumholt, M. Obro, A. Bjarklev, T. Rasmussen, B. Pedersen, J. E. Pedersen, J. H. Povlsen, and K. Rotwitt, *Opt. Comm.* **89**, 30 (1992).

[52] S. Zemon, W. J. Miniscalco, G. Lambert, B. A. Thompson, M. A. Newhouse, P. A. Tick, L. J. Button, and D. W. Hall, "Nd^{3+}-doped fluoroberyllate glasses for fiber amplifiers at 1300 nm," in *Optical Amplifiers and Their Applications*, Vol. 14, 1992 OSA Technical Digest Series (Optical Society of America, Washington, D.C., 1992), pp. 12–15.

Appendix A

A.1 OASIX®AMPLIFIER SIMULATION SOFTWARE

The amplifier simulation software enclosed, known as OASIX®, has been provided by the Lucent Technologies Specialty Fiber Products Group. OASIX®software is copyright ©1995 Lucent Technologies. All rights reserved. OASIX® is a registered trademark of Lucent Technologies.

For technical information about OASIX®or the material in this appendix, please contact: Specialty Fiber Devices Group, Lucent Technologies, Franklin Township, NJ. Technical hotline: fax: (732) 748-7596.

A.2 INTRODUCTION

OASIX®is a user-friendly erbium-doped fiber amplifier (EDFA) simulation package capable of accurately predicting the performance of erbium-doped fibers when used as single-stage amplifiers, dual-stage amplifiers or ASE sources. OASIX®can simulate EDFA designs with a host of optical components such as filters, isolators, and reflectors. The educational version of OASIX®provided with this book, however, is only set up to simulate single stage amplifiers and ASE sources. To be able to make use of all the capabilities of OASIX®, the full package needs to be purchased from Lucent Technologies' Specialty Fiber group. Some of the commands which appear on the setup screen and which are operational in the full version will be either grayed out or non functional in the educational version presented here. This educational version will allow the user to simulate many of the amplifiers discussed in the text. In particular, many of the curves presented in Chapter 6 can be redrawn interactively by inputting the appropriate parameters to OASIX®.

With its graphical user interface, OASIX®makes the setup of even the most complex EDFA simulation quick and effortless. Because this version of OASIX®comes equipped with modeling parameters for some typical commercial erbium-doped fibers, all you need to do is load it on your PC and begin simulating.

A.2.1 System Requirements

To use OASIX®you need:

- The master OASIX®simulation disk.

- An IBM PC or compatible operating with a 386 or more recent microprocessor (a Pentium based machine is preferable).

- At least 2 MB of random access memory.

- Windows 3.1 or 95 (Windows 95 is preferable).

- At least 3 MB of hard-drive disk space.

A.2.2   Installing OASIX®

To install OASIX®, insert the master OASIX® simulation disk in the 3.5" drive and copy "Oascpp" to the desired directory on your hard drive. You should also copy the files which contain the parameters for 4 different kinds of erbium-doped fibers, to the same directory (these parameters correspond to the files Type1.bin, Type2.bin, Type3.bin, Type4.bin).

A.2.3   Starting OASIX®

OASIX®(Version 2.02) is a Windows compatible application. We describe here how to use it with Windows 95. OASIX® can be launched in Windows by changing to the folder which contains the executable file (the folder that Oascpp was transferred into during installation) and double clicking on the Oascpp icon. Since OASIX® Version 2.02 is not DOS executable, it can not be run from the DOS command line.

A.2.4   What to do next

Before proceeding it is recommended that you view the "readme.txt" file included with the OASIX® program for any recent developments and operating instructions, as well as the license declaration. Please read Section A.3: A Quick Overview and Tour, before attempting to run a simulation. Not all of OASIX®'s functions are immediately obvious and failure to set all parameters correctly will cause the simulation to generate useless data. The end of Section A.3 provides several step by step instructions for running sample simulations using OASIX®.

A.3   A QUICK OVERVIEW AND TOUR

A.3.1   Fibers and Modeling Parameters

OASIX® is intended for use with erbium-doped fibers (EDF) purchased from Lucent Technologies or its affiliates. Parameter sets for four fibers presently or previously available from Lucent Technologies are included with the software. These fibers are identified as Types 1, 2, 3, and 4 as indicated in Table 8.2 of chapter 8. Gains obtained with these fibers are shown in Figure 8.9 of Chapter 8. Type 1 is a high NA fiber suitable for low threshold and preamplifier applications; Type 2 is a medium to high NA fiber suitable for a preamplifier, inline amplifier, and power amplifier; Type 3 is a medium to low NA fiber suitable for power amplifiers; Type 4 is a low NA fiber suitable for high output power amplifiers. The parameter sets contain parameters based upon the erbium-doped fiber amplifier model set forth by Giles (see Journal Lightwave Technol.,

Vol. 9, No. 2, pp. 271, Feb. 1991). These include erbium ion gain, erbium ion loss, a saturation parameter, and a background loss. To improve high power amplifier simulation, Rayleigh reflection and erbium ion pairing have been added in a simple manner. The accuracy of this model depends on the case simulated, but it is typically within 0.5 dB for both gain and noise figure and within 5% for device efficiency.

To load a parameter set:

- Select the "Select Fiber" button on bottom of the amplifier first stage window (to reach this window, under menu "Setup" choose "Stage 1").

- Select a binary (.bin) file that corresponds to the desired fiber.

OASIX®simulates either 1480 nm band or 980 nm band pumping. For fibers with cutoff wavelengths below 980 nm, either band is available for simulation. For fiber cutoffs above 980 nm, only the 1480 nm band is available. Even though such fibers can be pumped at 980 nm, the result depends on the launching conditions. Since the fundamental and second-order modes are absorbed at different rates, no unique solution exists. OASIX®will not allow you to simulate 1480 nm fibers with 980 nm pumps. The fiber sets included here can all be simulated with either 980 nm or 1480 nm pumping.

A.3.2 Saving a Simulation Configuration

OASIX®will remember simulation configurations you use and recall them for later use. Upon entry into the simulation OASIX®is configured as it was the last time it was closed. OASIX®uses the file simul.cfg to remember your previous simulation. At any time during a session, if you wish to save the present configuration:

- Select "Save" under "File" in the menu bar of the main OASIX®screen.

- Enter a unique name for the present simulation. This is automatically assigned a type designation .sim.

- To recover the parameters for a given simulation, select "Open" under "File" in the menu bar of the main OASIX®screen.

- Select the name of the simulation to load (type .sim) and hit return. The setup for the simulation of your choosing should automatically appear in the appropriate locations.

A.3.3 Device Types Simulated

The full version of OASIX®is able to simulate three types of erbium-doped fiber devices, single-stage amplifiers, dual-stage amplifiers, and ASE sources. The educational version provided here will only model single stage amplificrs and ASE sources, without reflectors or feedback at the ends for either pump or signal. The type of simulation is selected in the main screen menu bar (described below). Based on the selected simulation type, different screens and dialogue boxes are activated. The single-stage amplifier can be pumped in both the co-propagating and counter-propagating directions, and can include one to four signals, each specified by a wavelength and launched input power.

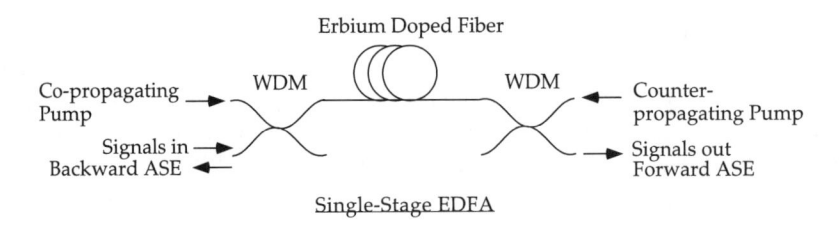

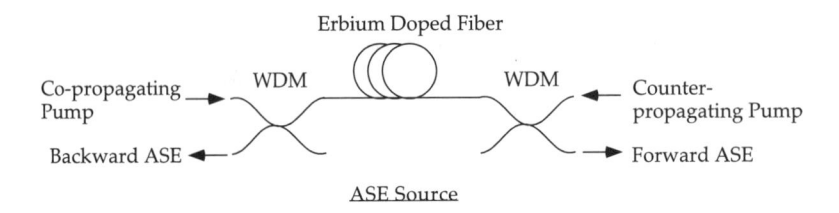

Figure A.1: Single-stage amplifier and ASE source configurations which may be modeled using the educational version of OASIX®described in this appendix.

The ASE source is a variation on the single-stage EDFA where no signals are present but all pump options are available. In all cases, OASIX®can simulate 10 or 31 ASE wavelengths across the gain spectrum of the erbium-doped fiber amplifier.

A.3.4 Data Entry and Device Conventions

OASIX®uses the following data conventions:

- All wavelengths are specified in nanometers.

- All powers are specified in mW. Feel free to use decimal or scientific notation.

- All lengths are specified in meters.

In all cases, power numbers specified upon setup represent the value launched into a particular EDF fiber end. Since OASIX®does not know what components you are using or how well you have coupled or spliced to the fiber, you must make these adjustments. Such losses are discussed in detail in section A.10 OASIX®treats certain components in an ideal manner. The WDM which combines pump and signal is considered to be lossless, as well as all connections to the amplifier.

A.3.5 Screens and Menus

The OASIX®simulation is controlled by information entered in the various screens, menus, and windows accessible for the two methods of simulation described above. Figure A.2 shows the main screen, which serves as the starting point for conducting the simulation. This screen is where the various menu selections can be made (i.e.,

A.3. A QUICK OVERVIEW AND TOUR 433

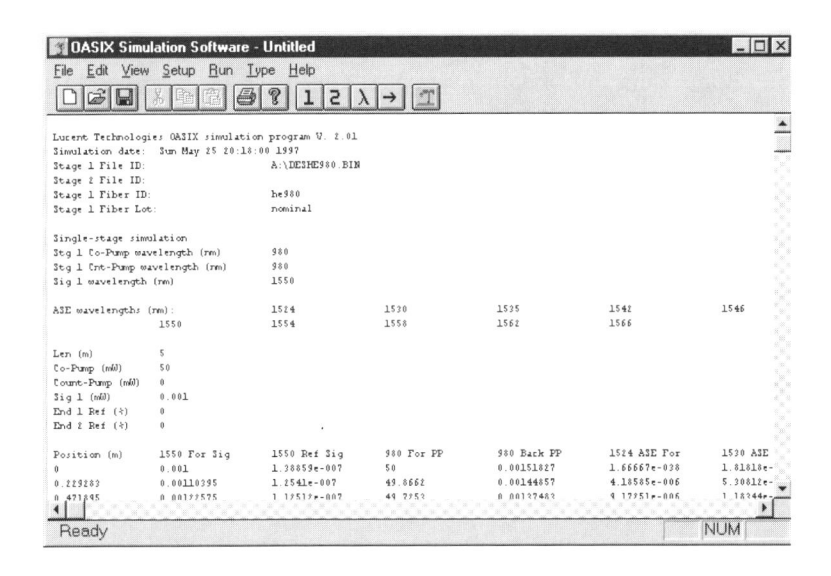

Figure A.2: OASIX®main screen (with results of a simulation). If OASIX®has just been launched, this screen is blank.

file opening, stage 1 setup). In Figure A.2, the window has been filled with the results of a previously conducted simulation and is ready for the next one. When OASIX®is launched, the screen is initially blank.

A.3.6 Simulation Looping and Output Modes

In the simplest simulation, OASIX®can be used to simulate one amplifier configuration and can output full details of the evolution of all signals, ASE waves, and pump power along the length of the amplifier. In its most complicated simulation mode, OASIX®will run three nested loops of simulations with design parameters such as fiber length, signal power and pump power allowed to vary in each loop. The number of points in each loop, the start point, the stop point and the mode of variation (linear or logarithmic) must be specified for each looping variable. All variables in the same loop should have the same number of points. OASIX®steps through the inner loop variables. After completing those simulations the middle loop is incremented, and the inner loop simulations are repeated. After stepping through all the middle loop values, the outer loop is incremented and the process is repeated until all combinations of all variables have been simulated. This structure is identical to nested FOR loops in programming. When variables are looping, the output file contains only the results at the fiber ends, not the detailed evolution of waves. In order to use the OASIX®looping structure correctly, it is important to consider the following conventions and details:

- The number of steps within each loop must be specified or the default of 1 step is used.

- If multiple variables vary in a given loop, the number of steps should be set to the desired value in front of all variables. Otherwise, the last variable in the list will determine the number of steps.

- Be careful to set the first and last point in each loop.

- Each loop can step in a linear or logarithmic fashion. Power levels should be entered in mW and lengths in meters. For example, a logarithmic variation of signal power from 1 mW to 1 mW (-30 dBm to 0 dBm) might be specified as 4 points from 1e-3 to 1.

- All variables that are not looping should have the first point set to the desired value.

- Do not vary a given variable in more than one loop.

The output file contents depends on the simulation order and the "Output Setup Screen" (described in section A.4.4). Anytime a loop box is selected, the simulation will output only the simulation results (power levels, gains, noise figures, etc.) at the fiber ends. This is true even if all loops are set to one step. This is a useful feature because the user can decide not to save full simulation details by just selecting a variable to loop and setting the number of steps to one. When full details of signal and pump evolution along the fiber are desired, simply remove all looping selections and enter values for all parameters in the starting point box.

A.4 SCREEN CONTENTS AND SIMULATION METHODOLOGY

A.4.1 Main/Entry Screen

After the opening logo, the first screen to appear is the main screen illustrated in Figure A.2. The menu bar performs these functions:

- The "Decode Fiber" selection under "File" is not active in this educational version and should be ignored.

- Saving and loading of simulation configurations under "File".

- Selection of EDF device type to simulate under "Type".

- Selection of First Stage Setup Window (see section A.4.2) for editing.

- Selection of Additional Signals Window (see section A.4.3) for editing.

- Selection of Output Setup Window for (see section A.4.4) editing.

- Launch of a simulation using the "Run" menu or the palm tree button. This action prompts the user for an output filename, displays the Simulation Status Box, and then starts the simulation. The simulation can be interrupted by pressing the STOP button in the Status Box. All cases previously solved will have been saved.

- Exit from program. Upon exit, the simulation setup is saved, this setup is then recalled when the program is launched.

A.4. SCREEN CONTENTS AND SIMULATION METHODOLOGY

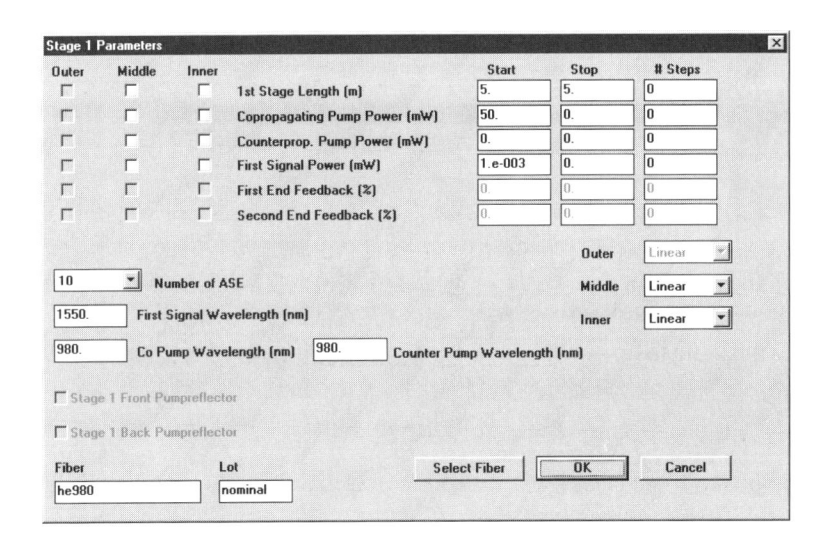

Figure A.3: Single-stage parameter setup window.

A.4.2 Single-Stage Setup Screen

The single-stage setup screen (shown in Figure A.3) is acessed from the main menu bar in the entry screen. It is used to setup the single-stage device. It controls the looping structure described in Section A.5.1. For any variable to be looped, select a start value, a stop value, a number of steps, the loop to be used and the step method (log or linear). For variables not to be looped, the proper value must be set in the start box for that variable. Also, a first signal wavelength (EDFAs only) and copump and counterpump wavelengths must be entered in nanometers. The available ranges for these wavelengths are:

- Signal band: 1520 nm – 1580 nm.
- First pump band: 1450 nm – 1500 nm
- Second pump band: 960 nm – 999 nm

The number of ASE wavelengths simulated can be set here at 10 or 31. The tradeoff is speed vs. spectral detail. No reflection can be specified in this educational version of OASIX®. The fiber type must also be selected. When leaving this setup window, hit "OK" to enter changes made or "Cancel" to revert to the previous values.

A.4.3 Additional Signals Screen

The additional signals screen of Figure A.4 is reached from the main screen. It is used to enter up to 3 additional signals in EDFA simulations. In the full version of OASIX®up to 10 additional signals can be entered. Set the number of signals, their power levels and wavelengths. The simulation uses the number of signals specified starting with the top left and moving down the column. Only the number of signals specified in the "Number of Signals" box will be used.

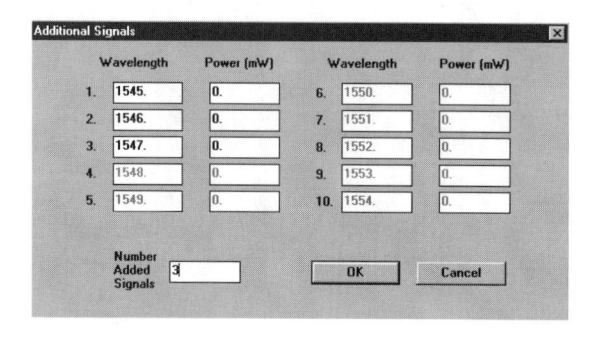

Figure A.4: OASIX®additional signals screen.

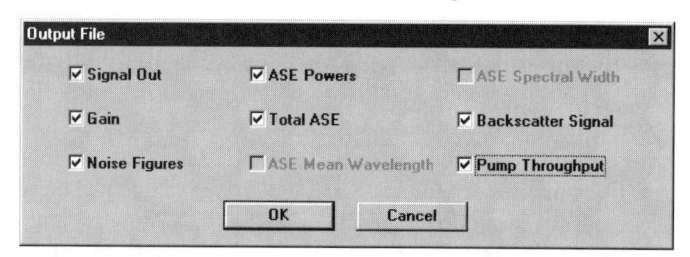

Figure A.5: OASIX®output setup screen.

A.4.4 Output Setup Screen

The output setup screen of Figure A.5 is entered from the main screen by choosing 'Output' under the 'setup' menu selections. It allows the choice of which wave power levels and computed values are stored in the output file. The output setup screen is shown in Figure A.5. It is reached by clicking on the 'Output Setup' button on the main screen. The output setup screen is used to customize the content of the result file OASIX®creates. All quantities which are selected in this box will appear in the output file. The possible selections include:

- The signal power in milliwatts.
- The gain for each signal in dB (defined in Section A.7 below).
- The noise figure in dB (defined in Section A.7 below).
- The pump power in milliwatts.
- The detailed ASE power spectral density in milliwatts/nm.

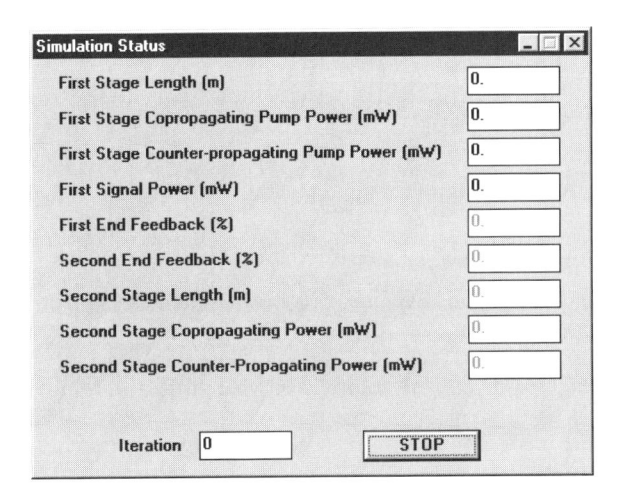

A.4. SCREEN CONTENTS AND SIMULATION METHODOLOGY 437

Figure A.6: OASIX® simulation status window.

- The total ASE power in milliwatts (defined in Section A.8 below).

- The mean ASE wavelength in nm (defined in Section A.8 below).

- The ASE spectral width in nm (defined in Section A.8 below).

It is recommended that only the desired values be saved. If all items are selected with 10 ASE wavelengths and 5 signals, the output file will contain more than 50 columns of data, which can be hard to manage. For many EDFA simulations, the output power, gain and noise figure are often adequate.

A.4.5 Simulation Status Box

The simulation status box, as shown in Figure A.6, provides the user with information to judge whether the simulation is converging and how long it will take to finish all simulations. However, the user must understand its operation before this information is useful. At the start of a particular simulation, OASIX® sets the values of all the possible looping variables and displays them in the status box. At this point, the iteration counter is set to 1. This indicates that OASIX® has made its first guess and is proceeding to propagate forward and then backward through the fiber. The number continues to count upward each time the simulation begins forward using better initial values for all waves. Eventually, if OASIX® solves a given case, the values in the displays change to the next simulation case, the iteration counter resets to 1, and OASIX® works on the next case. Usually, the greater the number of iterations and the longer the time OASIX® takes to solve a case, the more difficulty it had in convergence. If OASIX® fails to solve a case, it informs the user of this fact in the output file and in the result window. Causes of convergence failures and some methods for improving convergence are discussed in the next section.

A.5 SIMULATION LOOPING STRUCTURE

OASIX®has the ability to perform many simulations sequentially by allowing the user to specify up to three loops of design parameters to vary. OASIX®can vary within the following parameters: the length of the amplifier, the co-propagating and counter-propagating pump powers, and the signal power.

A.5.1 Specifying Loop Parameters

More than one parameter can vary in each loop. Set the looping parameters according to the following procedure:

- Decide which parameters to vary and in what order according to Section A.5.2.

- Activate the boxes adjacent to each parameter to be varied in the column corresponding to the desired loop.

- Enter the number of points to simulate for each parameter in each loop.

- For each parameter, enter the start point and stop point in the appropriate box.

- Decide whether to vary each loop in a linear or logarithmic fashion (Section A.5.3 below)

- Check that no single parameter varies in two loops.

- Check that all parameters varying in the same loop have the same number of points.

- Check that all static parameters are deselected and have proper constant values entered in their start point box.

Note that if no loops are selected, OASIX®will run one time and output the detailed information described in Section A.5.6.

A.5.2 Choosing Loop Order

The loops in OASIX®are labeled "outer loop," "middle loop," and "inner loop," to indicate the order in which simulations are performed. Rearranging the loops will generate the same total number of simulations but will change the order of looping and the output file format. In the output file, the column of the inner loop parameter and result columns specified in the output file configuration appear adjacent to each other. This makes graphing of the amplifier performance versus the inner loop parameter easy. However, the middle and outer loop parameters appear only in headers above each inner loop series. Consequently, graphing amplifier performance versus the middle or outer loop parameters requires substantial data manipulation. An example will clarify this distinction. Suppose a plot of signal gain versus signal input power is desired for a series of fiber lengths and pump powers. In this case, the inner loop should vary the signal input power while the length and pump power should appear in the middle and outer loops. Then signal gain and signal input power will appear in columns while the different lengths and pump powers will appear in headers only. If gain versus length is to be plotted, length is the better choice for the inner loop.

Another consideration for rearranging loops is the issue of simulation convergence. OASIX®uses the previously solved case within the inner loop to initially guess the solution for the next inner loop case. If a simulation set is run in a certain order, this produces a better guess and more solutions for difficult cases. If arrangement to produce easy graphing (as in the paragraph above) produces many nonconvergent cases, follow these rules:

- If long lengths are not converging, vary length in the inner loop and step from short lengths to long lengths.

- If low pump powers are not converging, vary pump power in the inner loop and step from high pump power to low pump power.

- If high or low signal powers are a problem, vary signal power in the inner loop and try different directions.

These rules can also be used to make a nonconvergent case converge when the problem parameter is a nonlooping parameter. At the expense of time, add an inner loop to solve an easier case first according to the above rules before proceeding to the tough case.

A.5.3 Linear or Logarithmic Looping

Each of the three loops in OASIX®can vary in a linear or a logarithmic fashion. All powers are specified in mW, and all lengths are in meters. A linear variation in pump power might step through the pump values 5 mW, 10 mW, 15 mW and 20 mW in that order while a linear variation in length might step through the length values 10 m, 20 m, 30 m, and 40 m in that order. These would each be 4 point linear variations with the appropriate start and stop points. However, some parameters used in EDFAs often vary over many orders of magnitude. In particular, signal power often varies from small-signal values of 1e$-$4 mW ($-$40 dBm) to power amp values of 1 mW (0 dBm). To perform such simulations, the logarithmic loop feature is used. For example, simulation of the values 1e$-$4 mW ($-$40 dBm), 1e$-$3 mW ($-$30 dBm), 1e$-$2 mW ($-$20 dBm), 1e$-$1 mW ($-$10 dBm) and 1 mW (0 dBm) would be performed as a 5 point logarithmic loop from 1e$-$4 to 1. Note that, even though orders of magnitude and a logarithmic variation are involved, these numbers are entered in mW not dBm. All number entry conventions must still be followed.

A.5.4 Multiple Parameters Varied in a Loop

In some complex simulations it is useful to be able to vary two or more parameters simultaneously in the same loop. This is especially true if these parameters are somehow physically related. For example, an amplifier could be designed to take one pump diode and split its power 50/50 between the co- and counter-propagating directions. In this case, turning up the pump diode increases both pump powers simultaneously. In OASIX®, both pumps can be selected for a given loop and can each be stepped through the same values.

It should be noted that, when varying multiple parameters in a loop, all numbers of points should be set the same for that loop. Otherwise, the last value read by the simulation routine will be used.

A.5.5 Influence on Output Format

If no loops are active, the output file will contain details of the evolution of all selected signals, pumps and ASE waves along the length of the fiber. Otherwise, only the results at the fiber end will appear. To eliminate the detail when running a single simulation, simply activate one loop and set the number of points to one.

OASIX®allows the user to control what data is saved to the output file and what values are computed using the results. The output files are all tab-delimited ASCII text files and can be read by any program which reads such data. The result files produced by OASIX®default to the .res ending. A number of conventions are observed in all output files:

- All columns of data are separated by tabs.
- All power values are recorded in milliwatts.
- All lengths are in meters.
- All gains and noise figures are in dB.
- All wavelengths, spectral widths and mean wavelength are in nanometers.

Two types of output formats are available, the detailed single-run output and the looping value output. These are described in the following section.

A.5.6 Output Modes

The output mode is directly controlled by the looping values selected during setup. If no looping boxes are selected on either stage screen, the detailed single-run output will result. This file contains the result for a single simulation in great detail. A sample single-run screen for a single-stage EDFA is shown in Figure A.2. The output file is shown in Figure A.7, obtained by opening the output file with a spreadsheet application. The following appear in the output in order:

- A header specifying the simulation date, fiber ID and lot, and simulation type.
- A block listing the pump wavelength, signal wavelengths, and ASE wavelengths.
- A block listing the values of fiber lengths, pump powers, and signal powers simulated.
- A column showing position along the fibers followed by columns showing the evolution of signals, pumps, ASE, etc., depending on what was selected in the output setup.
- Blocks of final results for the signal gain, noise figure, return loss, and pump power.

A.5. SIMULATION LOOPING STRUCTURE

Lucent Technologies OASIX simulation program V. 2.02					
Simulation date:	Sat Jul 05 21:57:41 1997				
Stage 1 File ID:	C:\NEWOAS~1.02\TYPE1.BIN				
Stage 2 File ID:					
Stage 1 Fiber ID:	Type 1				
Stage 1 Fiber Lot:	High NA				
Single-stage simulation					
Stg 1 Co-Pump wavelength (nm)	980				
Stg 1 Cnt-Pump wavelength (nm)	980				
Sig 1 wavelength (nm)	1550				
ASE wavelengths (nm):	1524	1530	1535	1542	1546
	1550	1554	1558	1562	1566
Len (m)	10				
Co-Pump (mW)	50				
Count-Pump (mW)	0				
Sig 1 (mW)	0.001				
End 1 Ref (%)	0				
End 2 Ref (%)	0				

Position (m)	1550 For Sig	1550 Ref Sig	980 For PP	980 Back PP	1524 ASE For
0.0000	0.0010	3.66E-03	50.0000	4.46E-03	1.67E-38
0.4484	0.0013	2.81E-03	45.7522	4.42E-03	1.30E-05
0.8900	0.0017	2.14E-03	42.5514	4.34E-03	3.06E-05
1.3321	0.0023	1.60E-03	40.1162	4.22E-03	5.53E-05
1.7744	0.0031	1.19E-03	38.2727	4.06E-03	9.05E-05
2.2168	0.0042	8.68E-04	36.8694	3.86E-03	1.41E-04
2.6594	0.0058	6.30E-04	35.7816	3.65E-03	2.15E-04
3.1023	0.0081	4.55E-04	34.9119	3.41E-03	3.23E-04
3.5457	0.0112	3.27E-04	34.1875	3.17E-03	4.79E-04
3.9897	0.0156	2.34E-04	33.5549	2.92E-03	7.07E-04
4.4347	0.0219	1.67E-04	32.9751	2.66E-03	1.04E-03
4.8811	0.0306	1.19E-04	32.4171	2.41E-03	1.52E-03
5.3298	0.0430	8.51E-05	31.8535	2.15E-03	2.22E-03
5.7819	0.0604	6.05E-05	31.2550	1.90E-03	3.25E-03
6.2394	0.0848	4.29E-05	30.5849	1.64E-03	4.74E-03
6.7056	0.1195	3.02E-05	29.7913	1.40E-03	6.90E-03
7.1859	0.1686	2.11E-05	28.7963	1.15E-03	1.01E-02
7.6893	0.2388	1.45E-05	27.4792	9.07E-04	1.46E-02
8.2322	0.3400	9.47E-06	25.6501	6.64E-04	2.12E-02
8.8459	0.4887	5.48E-06	23.0093	4.15E-04	3.07E-02
9.3267	0.6279	3.03E-06	20.5681	2.36E-04	3.91E-02
9.9253	0.8211	3.28E-07	17.2810	2.58E-05	4.97E-02
10.0000	0.8462	-1.34E-11	16.8661	-9.34E-10	5.10E-02

Sig wl (nm)	Input (mW)	Output (mW)	Ret Loss (dB)	Tot gain (dB)	Tot NF (dB)
1550	0.001	0.8462	-5.6385	29.2748	3.3681

Pump wl (nm)	End1 Stg1 In	End2 Stg1 In	End1 Stg1 Out	End2 Stg1 Out	
980	50	0	0.00446273	16.8661	

Tot For ASE (mW)	5.60363
Tot Bck ASE (mW)	6.3299

Figure A.7: OASIX®result of a single stage amplifier simulation for a fixed length and signal and pump powers. Only the leftmost portion of the output is shown, due to space limitations. Columns of data not shown are described in the text.

This mode is useful in order to understand the interaction between pump, signal and ASE waves along a specific amplifier or ASE source.

When any boxes are selected in the loop sections at the top of either stage setup screen (even if only 1 point is specified in the loops) the detailed single-run output is replaced by the less detailed multiple-run output. This format saves only results at the fiber ends. This file uses the following format:

- A header specifying the simulation date, fiber ID and lot, and the simulation type.

- A block listing the pump wavelength, signal wavelengths, and ASE wavelengths.

- A block listing the values of fiber lengths, pump powers, signal powers and reflections for all quantities not varying in the inner loop.
- A column specifying the value of the variable looping in the inner loop followed by columns showing the output signals, reflected signals, gains, noise figures, pump throughput, and ASE values (depending on what was selected in the output setup).

This mode is used whenever deciding what length, pump power, or signal power to use or what amplifier design to implement.

A.6 SAMPLE SIMULATIONS

The following sections contain some step-by-step simulation examples. The first example is a 10 m long amplifier with two signals. The second example varies the length of this amplifier in a loop.

A.6.1 Single-Run, Single-Stage EDFA

1. In main menu bar, select "single stage" under "Type."
2. Upon entry into OASIX®, select "stage 1" from the setup menu in the main screen menu bar and select Type 1 fiber for simulation using the "select fiber" button.
3. In number ASE wavelength box select 10.
4. Enter a pump wavelength of 980 nm in the "Pump Wavelength" box.
5. Enter a signal wavelength of 1550 nm in the "First Signal Wavelength" box.
6. Make sure no "Loop" boxes are selected.
7. In "Start" column near top of screen, enter a length of 10, a copropagating pump power of 0.001 mW, and a First Signal Power of 0.001 mW. Press "OK" to close this window.
8. Under "setup", choose "Additional Signals," and add a signal of 1553 nm and 0.0001 mW in the top left row in the screen which appears. Set the "number of added signals" to 1. Press "OK" to leave this screen.
9. Under "setup" choose "Output Setup". Select contents of output file (this can be anything for the demo). Press "OK" to return to the main screen.
10. Press "Simulate" (the palm tree button) and enter an output file name in dialogue box (after answering "Yes" to "store results in file?"). OASIX® will automatically add the extension ".res" to the filename if the extension is not specified.
11. Once simulation is done, exit the program and check results by importing the file into a spreadsheet or other program that recognizes tabs as column delimiters. The output is also visible on the main window screen, and the user can scroll through it to see all the results.

A.6.2 Multiple-Run, Single-Stage EDFA

(the above amplifier simulated for four fiber lengths between 10 m and 20 m).

1. Perform steps 1 through 7 in section A.6.1 above.

2. Select the "Inner" loop box for the "Stage One Length" (in the stage 1 setup screen). Enter 4 for "# steps". Enter 20 in the "Stop" column. This will run four simulations for lengths 10 m, 13.3 m, 16.6 m, and 20 m.

3. Perform steps 8 through 11 in section A.6.1 above.

A.6.3 Other simulations to try

The above sections only begin to illustrate the broad range of simulation sets OASIX®can perform. Experiment with the other two loops and look at the output files. Remember to follow the rules of Section A.3.4 and Section A.3.6.

A.7 COMPUTATION OF SIGNAL RELATED QUANTITIES

For each signal, the gain, noise figure and return loss are computed if these outputs are selected in the output setup screen. Computation of the gain is performed using the formula:

$$\text{Gain(dB)} = 10 \log_{10} \left(\frac{P_{signal-out}}{P_{signal-in}} \right) \tag{A.1}$$

where:

$P_{signal-out}$ = signal output power in mW at end of the amplifier
$P_{signal-in}$ = specified signal input power in mW at front of the amplifier

It should be noted that the output power is corrected for signal reflections within the amplifier from Rayleigh scattering. Because no particular electronic noise can be assumed, OASIX®computes only the optical noise figure of the EDFA. To do this, OASIX®adds an ASE wave at each signal wavelength that does not already coincide with a default ASE wavelength. Hence, the number of ASE wavelengths listed at the top of your output file may be larger than you specified on the setup screen. To compute the noise figure at a signal wavelength, OASIX®uses the following formula:

$$\text{NF(dB)} = 10 \log_{10} \left[\left(\frac{P_{ASE}}{h\nu\Delta\nu} + 1 \right) \left(\frac{P_{signal-in}}{P_{signal-out}} \right) \right] \tag{A.2}$$

where:

$P_{signal-out}$ = Output power (mW) in the ASE wave at the same wavelength as the signal
ν = frequency of signal and corresponding ASE wave
$\Delta\nu$ = spectral bandwidth represented by ASE wave

$P_{signal-out}$ = signal output power in mW at end of the amplifier
$P_{signal-in}$ = specified signal input power in mW at the front of the amplifier

The return loss of an EDFA is a measure of how much signal power is reflected back towards its source. This number can be valuable in deciding whether to isolate the amplifier input, a step which increases the noise figure. The reflected power from an EDFA can be due to discrete reflections at components, or fiber ends, or can be the accumulated Rayleigh backscatter of the fiber. In both cases, amplifier gain enhances the reflection. OASIX®, version 2.02, includes only the Rayleigh reflections specified by the background loss and a capture fraction particular to the fiber in use. The formula for the return loss is:

$$\text{RL(dB)} = -10 \log_{10} \left(\frac{P_{ref}}{P_{signal-in}} \right) \quad (A.3)$$

where:

$P_{signal-in}$ = specified signal input power in mW at front of the amplifier
P_{ref} = reflected signal power in mW at the front of the amplifier

The reflected power in equation A.3 is corrected for signal reflections at the front of the amplifier but not for splice losses. The negative sign makes equation A.3 consistent with the standard definition of return loss. It should be noted that return loss can be negative (return gain) in an EDFA because the gain of the device can enhance the reflected power far above the incident power.

A.8 COMPUTATION OF ASE RELATED QUANTITIES

The ASE simulated by OASIX® consists of a series of ASE waves at discrete wavelengths. Each wave represents the ASE in a bandwidth about the nominal wavelength. The equivalent input power in a given bandwidth at a given wavelength is assumed to be one photon per mode per bandwidth, which can be written:

$$P_0(\lambda, \Delta\lambda) = 2h\nu\Delta\nu = 2\frac{hc^2}{\lambda^3}\Delta\lambda \quad (A.4)$$

This quantity appeared in the noise figure equation, equation A.2 above. With this seed noise, the ASE accumulates and experiences gain in both directions in the EDF. If the number of ASE wavelengths simulated is, then the total ASE power is computed:

$$P_{ase,tot} = \sum_{1}^{nase} P_{ase,i} \quad (A.5)$$

where:

$P_{ase,i}$ = Power in ith ASE wave at location of computation

A.9. BASIC OPERATING PRINCIPLES

Two other quantities are of interest for ASE sources, the mean wavelength of the source and the spectral width of the source. Each ASE wave in OASIX® approximates the power within some bandwidth as occurring at the nominal wavelength of the wave. Hence, the mean wavelength of the ASE may be computed using the values calculated:

$$\overline{\lambda}_{ase} = \frac{\sum_{1}^{nase} \lambda_i P_{ase,i}}{P_{ase,tot}} \tag{A.6}$$

where:

$$\lambda_i = \text{wavelength of ith ASE wave.}$$

In some fiber sensor application, it is the stability of the mean frequency of an ASE source that matters for the sensor performance. OASIX® does not compute this quantity but the shift in the mean wavelength computed by equation A.6 is always nearly of the same magnitude and of opposite sign to the shift in mean frequency. If a precise mean frequency is desired, the ASE powers and wavelengths can be output in detail and proper computations can be performed to turn the wavelengths to frequencies. Then equation A.6 still applies but with frequencies replacing wavelengths. The ASE spectral bandwidth can be computed in a variety of ways depending on the application. The definition used by OASIX® is:

$$\Delta \lambda_{ase,tot} = \frac{\left[\sum_{i=1}^{nase} P_{ase,i} \times \Delta \lambda_{ase,i}\right]^2}{\sum_{i=1}^{nase} P_{ase,i}^2 \times \Delta \lambda_{ase,i}} \tag{A.7}$$

where:

$$\Delta \lambda_{ase,i} = \text{spectral width represented by ith ASE wave}$$

The width represented by the ith wave encompasses half the wavelength range from the (i-1)th wave to the (i+1)th wave. For the first point, this is taken as the whole range to the second point while, for the last point it becomes the entire range to the second to last point. Equation A.7 is the wavelength analog of the frequency width used most often when assessing noise characteristics of broadband sources. When a broadband ASE source produces beat noise upon detected on a square law detector, the relevant width is a frequency width weighting like equation A.7. OASIX® does not compute this value. The width represented by equation A.7 is often broader than a visual apparent spectral width. If desired, the user can allow the model to output all the ASE waves and can perform a computation of another width definition in frequency or wavelength units.

A.9 BASIC OPERATING PRINCIPLES

To predict the performance of an EDF device, OASIX® uses the basic governing equations for the inversion of the erbium ions and the evolution of signal, ASE and pump waves that can be found in many sources. Using these equations, OASIX®must meet the boundary conditions set at both ends of the device. Prior versions of OASIX®used what might be termed a "shooting method". In OASIX®Version 2.02, this technique has been replaced by a "back and forth method". In this method, all waves are propagated alternately forward and then backward, each time saving the values produced in each step. Using a Runge-Kutta technique, the waves are stepped rapidly to the far fiber end using the governing equations for erbium ions at each step. Boundary conditions are applied at the far fiber end and the waves are all stepped backwards using the governing equations. The result is compared at the front end to the result produced in the last iteration until successive iterations produce nearly identical results. Boundary conditions at the near fiber end and the previous solution are used to start forward in each iteration. This is a computationally intensive process so OASIX®uses as many tricks as possible to speed convergence and reduce the time to step along the length. In particular, the step size is self-adjusting to take the longest step possible without adding significant errors. The first guess is also made in such a way as to be close to the final solution.

A.9.1 Simulation Speed and the Number of Waves

The speed of simulation is partly determined by the number of waves OASIX®is simulating. We define the following:

$$\begin{aligned} \text{numase} &\equiv \text{number of ASE waves specified on the first} - \text{stage input screen} \\ \text{nsig} &\equiv \text{the total number of signals} \\ \text{npump} &\equiv \text{total number of pump wavelengths used} \end{aligned}$$

The total number of signals is the number of additional signals selected on the added-signal setup screen plus the 1 main signal assumed for all simulations. This number can be as high as 4. For ASE source simulations, 0 signals are present. When necessary, OASIX®automatically adds an ASE wave at each signal wavelength so that computation of the noise figure is possible. If a signal is at an already existent ASE wavelength, no ASE wave is added. This means that the true number of ASE waves simulated may be as high as nase = numase + nsig. The true number of simulated ASE waves appears at the top of the output file. The total number of waves simulated by OASIX®, including ASE waves, reflected signals, and pumps traveling in a given direction is represented by:

$$\text{numase} + \text{nsig} + \text{npump} \leq n_{wave} \leq \text{numase} + 2 \times \text{nsig} + \text{npump} \quad (A.8)$$

For example, for a simulation with 10 ASE waves selected, 2 signals, and 1 pump wavelength, the total number of simulated waves is between 13 and 15, depending on the specific wavelengths chosen. This number of waves propagates in each direction simultaneously. The speed of the mathematical portion of the simulation depends linearly on the number of waves. In the example above, if 31 ASE waves were selected (and the backward wave number was between 34 and 36), the mathematical portion of

A.9. BASIC OPERATING PRINCIPLES   447

the simulation would take approximately 3 times longer. To speed up simulations, try the following:

- Select 10 ASE wavelengths for simulations except when detailed ASE spectra are needed

- Eliminate all signals that you do not need for a particular simulation

- For 10 ASE wave simulations use signals at 1524, 1529, 1535, 1542, 1546, 1550, 1554, 1558, 1562, or 1566 nm work or any even wavelength for the 31 ASE case, if you are just doing exploratory work.

In some cases these steps make little difference. This is the case if the mathematical portion of the simulation takes less time than window monitoring, generating initial guesses and computing output.

A.9.2 Causes and Remedies for Convergence Failure

The convergence technique used is not guaranteed to solve all cases. The routines used in OASIX®check for indications of failure to converge. The number of iterations is limited because convergence is usually rapid, if successful. The number of steps taken along the fiber is limited because reasonable power levels rarely require too many steps unless unusual conditions have resulted. OASIX®also checks for other computational anomalies. If any of these conditions occur, OASIX®assumes it is not converging and moves on to the next case. OASIX®generally has difficulty solving the following cases:

- Very long fibers, especially when all signals are small or nonexistent (ASE source).

- Very low pump power cases.

To solve a difficult case, try the following:

- If the length is long, put length as the inner loop variable and allow OASIX®to solve a series of shorter lengths with the same operating conditions before solving the case of interest. The closer together the points are, the better OASIX®'s chances become.

- If the pump power is low, put pump power as the inner loop variable and allow OASIX®to solve a series of higher pump power cases with the same operating conditions before solving the case of interest.

- If an ASE source is the problem, try running this as an EDFA simulation. Include a signal as the inner loop variable and allow the signal to step (logarithmically) from a high power to an inconsequential power (like 1e-6 mW). The final simulation is your answer.

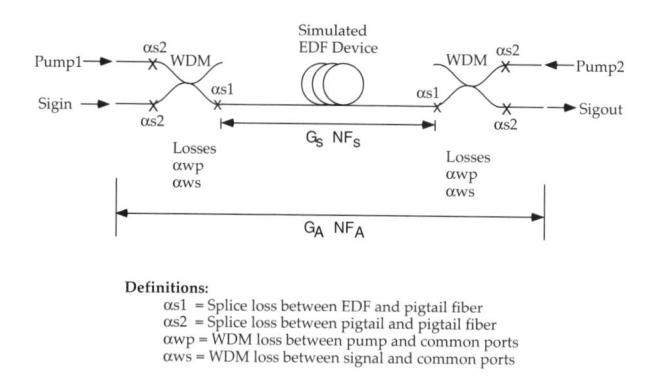

Figure A.8: Schematic of the amplifier and terminology discussed in section A.10, from consideration of losses. The labels are described in section A.10.

A.10 COMMENT ON THE TREATMENT OF LOSSES

Losses in an EDFA can be present before the first erbium-doped fiber, after the last erbium-doped fiber, or between stages. These losses affect the pumps, signals, and ASE. The treatment of such losses is best illustrated graphically. Figure A.8 depicts a typical EDF device with losses before and after the erbium-doped fiber. The region labeled "Simulated EDF Device" includes all components and fibers from the first EDF fiber end encountered to the last EDF fiber end. For a single-stage EDFA this includes only the EDF and lossless reflectors when desired. The simulation uses and generates the following quantities:

$$
\begin{aligned}
P_{1ins} &\equiv \text{Launched forward pump} \\
S_{ins} &\equiv \text{Launched signal power} \\
P_{2ins} &\equiv \text{Launched backward pump} \\
S_{outs} &\equiv \text{Exiting signal power} \\
G_s(\text{dB}) &= S_{outs}(\text{dB}) - S_{ins}(\text{dB}) \equiv \text{Signal gain of device} \\
NF_s(\text{dB}) &\equiv \text{Noise figure of device}
\end{aligned}
$$

Additionally, the simulation generates pump output powers, reflected signals, ASE powers, etc. All of these are internal values consistent with Figure A.8 and the above definitions.

Real EDF devices do not have lossless input and output paths. The losses considered in this discussion are those defined in Figure A.8. The losses in the pump, signal input and signal output paths can be written:

$$L_{pump}(\text{dB}) = \alpha_{s2}(\text{dB}) + \alpha_{wp}(\text{dB}) + \alpha_{s1}(\text{dB}) \tag{A.9}$$

$$L_{sigin}(\text{dB}) = \alpha_{s2}(\text{dB}) + \alpha_{ws}(\text{dB}) + \alpha_{s1}(\text{dB}) \tag{A.10}$$

A.10. COMMENT ON THE TREATMENT OF LOSSES

$$L_{sigout}(dB) = \alpha_{s1}(dB) + \alpha_{ws}(dB) + \alpha_{s2}(dB) \quad (A.11)$$

Clearly, all losses present in each path must be included in each of these quantities. If splice losses are different, if extra splices are present, or if more components (isolators, filters, etc.) are present, these must be included in these loss values.

The external pump and signal powers are the desired result of the simulations since these represent the performance of the overall device. These are defined:

$Pump_1 \equiv$ Total available forward pump power before losses
$Pump_2 \equiv$ Total available backward pump power before losses
$Sigin \equiv$ Total signal input power before all losses
$Sigout \equiv$ Resultant signal output power after all losses

These quantities *should not* be entered in the data entry boxes of OASIX®. The simulation values defined above should be used instead. These are related by:

$$P_{1ins}(dBm) = Pump_1(dBm) - L_{pump}(dB) \quad (A.12)$$
$$P_{2ins}(dBm) = Pump_2(dBm) - L_{pump}(dB) \quad (A.13)$$
$$S_{ins}(dBm) = Sigin(dBm) - L_{sigin}(dB) \quad (A.14)$$

The simulation reports signal power, gain and noise figure in the output files. These are the internal signal, gain and noise figures defined above. The desired external values required by the user are shown in Figure A.8. Assuming that the losses are known, the output results should be adjusted using the following formulae:

$$Sigout(dBm) = S_{outs}(dBm) - L_{sigout}(dB) \quad (A.15)$$
$$G_A(dB) = Sigout(dBm) - Sigin(dBm)$$
$$= G_s(dB) - L_{sigin}(dB) - L_{sigout}(dB) \quad (A.16)$$
$$NF_A(dB) = NF_s(dB) - L_{sigin}(dB) \quad (A.17)$$

Equations A.12, A.13, A.14 and Equations A.15, A.16, A.17 should be used to adjust input to OASIX® and to correct the results for losses. Similar equations apply for pump output powers, reflected signal powers, ASE powers, etc. All results reported by OASIX® are internal results and should be adjusted in a manner consistent with Figure A.8. The only adjustments OASIX® makes are to correct for user-specified reflections at the fiber ends. This is necessary to eliminate apparent violations of power conservation laws in the output file. Many losses are treated as wavelength-independent across the signal band including splice losses and losses in pump coupling optics.

OASIX® has the ability to perform many simulations sequentially by allowing the user to specify up to three loops of design parameters to vary. OASIX® can vary the following parameters: the lengths, the co-propagating and counter-propagating pump powers, and the signal power.

Index

1480 nm pumping, 67–70, 75, 155, 156, 161–163, 165, 167–170, 172–174, 176, 177, 185, 222, 273, 275, 282, 285, 289, 290, 292, 294, 298, 324, 335, 343, 347–349, 356, 357, 377
1580 nm signal band, 288, 294, 344
800 nm pump, 188–191
980 nm pumping, 58, 67, 68, 70–72, 135, 138–140, 149, 161–172, 174–177, 182, 184, 185, 187–189, 194–196, 235, 265, 267–274, 281, 282, 285, 287, 290, 292, 294, 295, 303, 304, 324, 326, 327, 343, 345, 377

absorption coefficient, 141–143
absorption cross section, *see* cross section, absorption
absorption spectrum
  Er^{3+}, 115
  Pr^{3+}, 410
active gain equalizer, 292
add/drop component, 58, 62–63
adding and dropping of channels, 339
Al co-doped glass, 21, 28, 29, 288, 294
Al-Ge silica, 114, 154, 156, 161, 164, 167, 189, 339
amplifier
  1.3 μm, 401–423
  analog, 351–354
  argon laser pumped, 326
  basic characteristics, 263
  bidirectional, 277–280
  bridge circuit configuration, 279

distributed, 28, 295–302
four-level, 401–404
gain, *see* gain
gain flatness, 263, 276, 280–282, 284–295, 339, 353
gain measurement, 251–257
gain measurement precision, 255
gain peaking, 336, 354–358
gain tilt, 245–247, 353, 354
inline, 263, 266, 267, 327–345
measurement techniques, 251–263
multi-arm, 279, 293
optimum fiber length, 268
polarization effects, 359–363
power, 175–178, 185, 187, 191, 226, 235, 257, 263, 268, 272, 273, 280–284, 291, 294, 304, 345, 347, 349, 351, 411, 413, 416–418, 421
Pr^{3+}, 404–418
preamplifier, 263, 267, 323–327
remote, 346–351
saturation, 138, 159, 169, 173–175, 226, 231, 232, 234, 235, 282, 287, 300, 355, 365, 407, 415, 416
self-filtering, 334, 354–358
self-regulation, 234, 336
small signal gain, 135–140, 154, 162, 169, 170, 173, 178–180, 183, 188, 190–192, 404
spectral dependence of the gain, 282, 284, 287
transient effects, 363–366

two-stage, 273–276, 285, 292, 293, 327, 410
waveguide, 302–304
WDM, *see* WDM
Yb-Er power amplifier, 282
amplifier chain, 226–235, 287
amplifier design, 263–265
   gain flatness, 284–295
   multistage amplifier, 273–276
   power amplifier, 280–284
   WDM issues, 256–257, 284–295
amplifier gain equations, 156–159
amplifier modeling
   absorption and emission cross section, 153
   analytic solution, 149–151
   ASE, 147–148, 155–197
   effective parameter model, 178
   excited state absorption, 188
   homogeneous broadening, 156–159
   optimum length, 139
   overlap factor, 140–144
   Pr^{3+}-doped fiber amplifier, 412–415
   rate equations, 146
   saturation, 138, 159, 169, 173–175
   small signal gain, 135–138
   three-level rate equations, 131–139
   two-level system, 144–147
amplifier noise, 171–175, 184, 188, 196, 202–212
   signal-spontaneous, 202, 206, 207, 216
   spontaneous-spontaneous, 202, 206, 208–210, 218
amplifier simulations, 161–197
   800 nm pumping, 188–191
   ASE spectral profile, 169–171
   erbium ion confinement, 180–186
   excited–state absorption, 186–191
   forward and backward ASE, 156–159, 161–171, 176
   pump absorption, 165, 168
   signal gain vs fiber length, 177, 266, 267
   transverse mode effects, 180–186

upper state population, 161, 163, 179, 268, 269, 284, 285
upper-state population, 167
amplifier spacing, in a long-haul system, 227–231
analog transmission, 240–247, 351–354
   1.3 μm, 417
analytic expression for gain
   four-level system, 404
   three-level system, 149–151
anomalous dispersion, 368
APD detector, 213, 324, 325
arrayed waveguide grating, 61
ASE, 55, 138, 147–148, 165–169, 259, 265, 267, 273, 324, 329, 334, 349, 354, 359, 370, 420
   backward, 52, 54, 258
   interaction with signal, 329
   power, 157, 222, 259
   spectral profile, 169–171
atomic Hamiltonian, 93
atomic structure of rare earth ions, 87–94
average inversion relationship, 158–159, 179, 357

backward ASE, 52, 54, 156–159, 161–172, 175, 176, 258
beat noise
   electrical power spectrum, 209
   signal-spontaneous, 202, 206, 207, 215
   spontaneous-spontaneous, 202, 206, 208–210, 215
bidirectional amplifier, 277–280
bidirectional pump, 175–177, 188, 191, 265–268, 415
bit error rate
   direct detection, 214–220
booster amplifier, *see* power amplifier
Bragg grating, 56–59, 71
   chirped, 58
bridge circuit amplifier configuration, 279
Brillouin amplifier, 211

INDEX

cascaded amplifiers, 226–235, 287, 327–345
casting and jacketing, 32
CATV transmission, 240, 241, 351
chalcogenide, 406, 413
chloride, 30
circulator, 53–54, 58, 276, 277, 279, 295, 303, 409
cladding pumping, 26, 73, 74, 282
clustering effect, 122
CMBH laser, 68
CNR, 243, 244, 351, 354, 417
coherent systems, 345
color center, 30
commercial erbium-doped fiber, 264
concentration quenching, 27, 113, 405
confinement of the erbium ion distribution, 180–186
connector loss
  angular misalignment, 45
  lateral misalignment, 44
conversion efficiency, 175–177, 185, 251, 257, 258, 264, 272–274, 280, 281, 349
cooperative upconversion, 120
copropagating pump, 157, 161, 175–177, 197, 265–268
counterpropagating pump, 157, 175–177, 188, 197, 265–268
counterpropagating signals, 279
CPFSK format, 345
cross saturation, 279, 290
cross section
  absorption, 102, 104, 105, 153–156, 159, 204, 287, 355, 402, 413
  Al-Ge erbium-doped fiber, 154
  as a function of inversion, 155
  definition, 99, 100
  emission, 102, 104, 105, 138, 139, 143, 153–156, 159, 204, 287, 355, 402, 404, 413, 419
  Er^{3+}, 116
  fundamental properties, 99–105
cross-phase modulation, 239, 373
crosstalk, 416

crystal fiber amplifiers, 5
crystal field, 91–94, 97
crystal field Hamiltonian, 93, 94, 96
CSO, 243, 245, 247, 351, 353, 417
CTB, 243, 351, 418
cutoff wavelength, 156, 183–185, 407, 409, 413, 414

defect, glass, 55
diode arrays, 67, 73, 74, 282
diode laser, 66–68, 70, 73, 74, 282
directly modulated laser, 353
dispersion, 336, 367
  post-compensation, 330, 338
  pre-compensation, 330, 338
  sawtooth dispersion map, 330
  zero dispersion transmission, 329
dispersion compensating fiber, 64, 244, 276, 338, 346, 349
dispersion compensation, 58, 63–66
dispersion limit to transmission distance, 63
dispersion management, 240, 280, 331, 338, 341, 345, 349
dispersion parameter, 368, 369
dispersion shifted fiber, 337, 338, 368
  non-zero, 338
distributed amplifier, 28, 295–302
dopant distribution, 30
double clad fiber, 26
double crucible method, 32–34
DWDM, 344

effective area, 136, 142, 236
effective parameter modeling, 178–179
Einstein coefficient, 100–102, 202
electrical filter, 205, 209
electrical noise figure measurement, 260–263
electrical spectrum analyzer, 262
emission cross section, *see* cross section, emission
emission spectrum
  Er^{3+}, 115
energy level diagram, rare earth ions, 95

energy level structure
  Er^{3+}, 111, 119
  Nd^{3+}, 419
energy transfer, 31, 120, 191–197, 415
Er-Yb-doped fiber amplifier, 73, 74, 282–284
Er/Yb-doped fiber amplifier, 107
erbium ion radial distribution, 140
erbium-doped fiber
  geometry, 165, 264, 270
  low concentration, 297
  short length, 267
erbium-doped fiber amplifier
  historical development, 5–9
  overview, 210
excited state absorption, 28, 31, 186–191, 265, 406, 415
  signal, 412, 416, 419
external modulation, 353
externally modulated laser, 247
eye opening, 337
eye pattern, 219–220

Faraday rotator, 52, 53
fiber connectors, 43–48
  insertion loss, 44
fiber fabrication
  casting and jacketing, 32
  chemistry, 14, 16
  chloride vapor pressures, 16
  double crucible method, 32–34
  fluoride, 31–33
  IMCVD, 14
  index-raising dopant, 16
  MCVD, 14, 16, 18, 19, 21, 22, 29
  OVD, 14, 16, 20, 21
  PCVD, 14
  rod and tube methods, 23–24
  rod-in-tube, 32
  seed fiber method, 24
  sol-gel, 15, 22
  solution doping, 21–23
  SPCVD, 14
  VAD, 14, 16, 20, 21
fiber grating, 55–60
  Bragg grating, 56–59, 71

chirped Bragg grating, 57, 58
  long period grating, 56, 59–60
fiber grating filters, 278
fiber laser, 6, 27, 67, 73–75, 282, 283, 348, 410, 419
fiber preform, 15, 16, 29
fiber reliability, 30
fiber strength, 30
filter, 55–56, 59, 341
fluorescence line narrowing, 115
fluoride, 14, 30, 31, 35, 114, 289, 294, 401, 404–415, 419
fluoride fiber fabrication, 31–33
fluoroberyllate, 421
fluorophosphate, 192, 420, 421
forbidden electric dipole transition, 92, 95
forward error correction, 345, 349
four wave mixing, 239, 277, 337, 344, 373
  efficiency, 242
  signal–ASE, 329
four-level system, 401–404, 420
four-wave mixing, 329, 330, 337, 338, 341, 344, 377
fundamental soliton, 369
fusion splicing, 13, 48–50

gain, 135–138, 169
  1.3 μm, 406–412, 419–420
  1580 nm signal band, 288, 289, 294–296, 344
  analytical solution, 149–151
  as a function of fiber length, *see* amplifier simulations, signal gain as a function of fiber length
  copropagating pump, 281
  counterpropagating pump, 281
  distributed, 295–302
  four-level system, 401–404
  inhomogeneous broadening, 159–161
  saturation, *see* amplifier, saturation, 109, 138, 173–175, 202, 232, 282, 287, 300, 355
  simulations, 161–197

spectral dependence, *see* amplifier, spectral dependence of the gain, 282, 284, 287
threshold, 265
tilt, 245–247, 353
transient effects, 363–366
waveguide amplifier, 302–304
gain coefficient, 138
gain equalization, 341
gain equation, 136, 137, 156–159, 404
gain flatness, 245, 284–295, 339, 353
gain measurement, 251–257
modulated signal method, 254
optical spectrum analyzer method, 252–253
precision, 255
reduced source method, 256
time domain extinction method, 257
WDM amplifier, 256–257
gain peak wavelength, 356
gain peaking, 287, 336, 354–358
Gaussian noise statistics, 214
germanate, 30
germano-silica fiber, 28
glass defect, 55
glass host, *see* host
Gordon-Haus effect, 370–373
grating, fiber, *see* fiber grating
GRIN lens, 52, 55
group velocity dispersion, 239, 329, 330, 338

halide, 14, 16, 31, 406
hermetic coating, 30
high concentration effect, 120–123, 191–197, 303
homogeneous linewidth, 108–110, 114–119, 291
Er^{3+}, 116
host
Al-Ge silica, *see* Al-Ge silica
alumino-silica, 339
chalcogenide, 406, 413
chloride, 30
composition, 27–29
fluoride, 14, 30, 31, 35, 114, 289, 294, 401, 404–415, 419
fluoroberyllate, 421
fluorophosphate, 420, 421
germanate, 30
halide, 31, 406
indium fluoride, 406, 411
iodide, 30
multicomponent glass, 28, 302
phosphate, 14, 111, 303
selenide, 30
soda-lime-silicate, 302
sulfide, 14, 30
tellurite, 14, 30, 35, 114, 290
ZBLAN, 14, 31, 413, 420, 421
hydrogen-induced loss, 30
hydrolysis, 14

IMCVD, 14
index-altering dopant, 29, 49
indium fluoride, 406, 411
inhomogeneous broadening, 290
gain in the presence of, 159–161
inhomogeneous linewidth, 108–110, 114–119
inline amplifier, 226–235, 263, 266, 267, 327–345
1.3 μm, 417
optimal spacing, 227–231
integrated components, 66
inversion, 163
inversion parameter, 202, 204, 222–224, 226, 253
iodide, 30
ion-ion interactions
pair induced quenching, 195–197
upconversion effects, 193–195, 266, 303
isolator, 52–53, 253, 273, 277, 326, 327, 349

Johnson noise, 213
Judd-Ofelt parameters, 96–99
Er^{3+}, 98
Judd-Ofelt theory, 96–99

Kerr effect, 368

Ladenburg-Fuchtbauer relation, 100–103
lanthanide contraction, 88
Laporte rule, 92
laser chirp, 244, 349, 353
laser damage, 67
lifetime
　concentration effect, 122, 123, 405
　Er^{3+} $^4I_{13/2}$ level, 112
　Er^{3+} energy levels, 111–113
　of an energy level, 105–107
lineshape, of a transition, 102
linewidth
　broadening, 114–119
　of a transition, 108–110
lithium niobate modulator, 361
long distance transmission, 323, 327–345
long period fiber grating, 56, 59–60, 292, 341
loop experiments, 332–334

Mach-Zehnder interferometer, 62
McCumber theory, 103–105, 154
MCVD, 6, 14, 16, 18, 19, 21, 22, 29
measurement techniques, 251–263
modal overlap, 140–144
mode field diameter, 46, 48, 264
modifier ion, 27
modulated pump, 363
MOPA laser, 67, 72, 405, 411
multi-arm amplifier, 279, 293
multicomponent glass, 28, 302
multipath interference, 261
multiplet, 145
multiplexer
　pump-signal, 50–51
　signal, 61–62
multistage amplifier, 273–276, 327

n_{sp}, 171, 172, 196, 202, 204, 222–224, 226, 253
Nd^{3+}, 6, 21, 25, 27, 28, 360, 401, 418–423
　laser, 282
　spectroscopy, 419
Nd^{3+}-doped fiber amplifier, 418–423

　gain results, 419–420
　modeling, 420–423
Nd:YLF laser, 405, 410–412, 417
noise
　current, 202
　device aspects, 202–212
　electronic, 324
　system aspects, 212–235
noise figure, 171–175, 220–225, 234, 253, 258–263, 268, 279, 291, 294, 304, 327, 416
　counterpropagating pump, 267
　electrical measurement, 260–263
　in an amplifier chain, 234
　measurement, 258–263
　multi stage amplifiers, 273
　Pr^{3+}-doped fiber amplifier, 416
noise power, 147, 204, 252, 259
noise variance, 214
non-radiative transition rate, 106–107, 402, 404
non-zero dispersion shifted fiber, 338, 344
nonlinear refractive index, 368
nonlinearities, 236–240, 298, 327, 334, 336, 344, 346, 367
　cross-phase modulation, 239
　effective length, 236
　four-wave mixing, 239, 329, 330, 337, 338, 341, 344, 373, 377
　self-phase modulation, 239
　stimulated Brillouin scattering, 237
　stimulated Raman scattering, 238
normal dispersion, 368
NRZ format, 327, 335, 370
numerical aperture, 27, 32, 184, 255, 264, 273, 413

optical amplifier
　Brillouin amplifier, 211
　device comparison, 210
　overview, 210
　Raman amplifier, 211
　semiconductor laser amplifier, 212
optical amplifier noise, 202–212
　system aspects, 212–235

optical mode, 140, 141
optical noise
    PIN diodes, 213
    receiver, 213–225
optical noise figure, *see* noise figure
optical preamplifier, *see* preamplifier
optical spectra of rare earth ions, 95–99
optical spectrum analyzer, 252, 256, 258, 259
optimum amplifier spacing, 227–231
optimum fiber length, 139, 268
oscillator strength of a transition, 98
OTDR, 255, 300
OVD, 14, 16, 20, 21
overlap factor, 136, 140–144

P co-doping, 22, 27, 294
pair induced quenching, 195–197
partition function, 104
passive gain equalizer, 291–293
PCVD, 14
phonon energy, 404
phosphate glass, 14, 28, 111, 303
phosphorus, 21, 282
photocurrent, 207
photon flux, 135, 187, 193, 226
PIN detector, 213, 324
polarization division multiplexing, 373
polarization effects, 359–363
polarization hole burning, 359–361
polarization mode dispersion, 64, 244, 362, 370
polarization nulling technique, 259
polarization-dependent gain, 359–361
polarization-dependent loss, 244, 359
population equation, 146, 150, 157, 160, 179
population inversion, 135, 163, 287, 402
    four-level system, 403
power amplifier, 235, 263, 280–284, 304, 321, 345, 411
    1.3 μm, 413, 417
    modeling, 175–178
    output power, 322
power conversion efficiency, 257, 281

effect of erbium concentration, 258
power self-regulation, 234, 290, 336
Pr^{3+}, 30, 32, 35, 72, 401, 404–418
    $^1G_4$ lifetime, 406
    energy level diagram, 405
    spectroscopy, 405–406
Pr^{3+}-doped fiber amplifier, 404–418
    gain results, 406–412
    modeling, 412–415
    system applications, 416–418
pre-emphasis method, 339
preamplifier, 220–225, 263, 267, 321, 323–327, 345
    1.3 μm, 417
    noise theory, 220–225
    schematic, 326
preform, *see* fiber preform
propagation equation, 137, 143, 146, 149, 157, 160, 179
    ASE, 148
pump
    bidirectional, *see* bidirectional pump
    copropagating, *see* copropagating pump
    counterpropagating, *see* counterpropagating pump
pump and signal multiplexing, 50–51
pump and signal overlap, 25
pump feedback, 365
pump laser, 66–75
    1.017 μm diode laser, 407
    1480 nm, *see* 1480 nm pumping
    980 nm, *see* 980 nm pumping
    argon laser, 7
    cladding pumped fiber laser, 73
    diode laser, 8, 67
    fiber laser, 67, 73–75
    MOPA laser, 67, 72, 411
    Nd:YLF laser, 410–412, 417
    Raman resonator laser, 75
    reflection of, 58
    reliability, 72
    solid state laser, 67, 73
    stabilization, 58
pump modulation, 363
pump polarization, 360

pump threshold, 137, 146

Q factor, 216–219, 359, 361
quantum conversion efficiency, 264, 280, 281, 414

radial distribution functions, electronic, 90
radiation effects, 30
radiative lifetime, 102, 105
Raman amplification, 211, 294, 296, 332, 347, 375
rare earth chloride, 18, 21
rare earth oxide, 27, 28
rare earth spectroscopy
    4f-4f transitions, 95
    absorption and emission cross sections, 99
    atomic configuration, 92
    energy level fitting, 94
    Er^{3+}, 110, 114
        absorption and emission spectra, 114
        homogeneous linewidth, 114
        upconversion, 121
    fluorescence line narrowing, 115
    high concentration effects, 120
    Judd-Ofelt theory, 96
    Ladenburg-Fuchtbauer relationship, 100
    lifetimes, 105
    linewidths and broadening, 108
    McCumber theory, 103
rate equations, 133, 146
Rayleigh scattering, 27, 31, 52, 244, 277, 349
receiver, 201, 202, 205, 206, 209, 210, 213–225, 227, 236, 239, 242, 243, 261, 284, 327
    APD, 202, 213, 323–325
    noise theory, 213–225
    PIN, 202, 213, 224, 324
    shot noise, 213, 215, 221
    thermal noise, 213, 215
receiver sensitivity, 322, 325, 327, 345, 347

quantum limit, 321, 327
remote pumping, 350
recirculating loop experiments, 332
reconfigured networks, 339
reduced source method, 256
reflection, 43, 52, 58, 254, 261, 277, 278, 417
refractive index profile, 17, 29
regenerators, 2
reliability, diode laser, 72
remote amplifier, 346–351
repeaterless transmission, 278, 345–346
RIN, 261
Russell-Saunders state, 93

saturation, *see* amplifier, saturation
saturation intensity, 137, 138
saturation power, 150, 151, 280, 363
sawtooth dispersion map, 330
seed fiber method, 24
selenide, 30
self phase modulation, 330, 338, 368
self-filtering, 334, 336, 354–358
self-phase modulation, 239
self-regulation, 234, 290, 336
semiconductor laser amplifier, 212
shot noise, 201, 208, 213, 215, 221
signal multiplexer, 61–62
signal polarization, 360
signal-spontaneous beat noise, 202, 206, 207, 215
single channel transmission, 356
single-channel transmission, 327–335
sliding guiding filters, 372
small signal gain, 135–138, 404
SNR, 216, 261, 336, 339, 351, 359
    analog system, 243
    electrical, 221, 226
    in an amplifier chain, 229–231
    optical, 218, 229, 231
    power amplifier, 235
soda-lime-silicate, 302
sol-gel, 15, 22
solid state laser, 67, 73, 282, 283, 351, 405, 410
soliton, 296, 367–377

INDEX 459

basic principles, 367–374
collision, 373
frequency filtering, 370
period, 369
sliding guiding filtering, 372
synchronous modulation, 372, 376
system results, 374–377
transmission experiments, 374
WDM, 374
WDM transmission, 377
solubility, 28
solution doping, 21–23
SPCVD, 14
spectral broadening, 239, 329
spectral hole burning, 117, 257
spectral inversion, 65
splice loss, 46, 48, 49, 64
spontaneous emission, 206, 402
spontaneous emission noise, 253
spontaneous-spontaneous beat noise, 202, 206, 208–210, 215
square-law detector, 205
stabilization, pump laser, 58, 71
Stark level, 91, 104, 109, 114, 119
stepwise upconversion, 120
stimulated Brillouin scattering, 237, 298, 345, 349
    threshold, 237
stimulated emission rate, 364
stimulated Raman scattering, 238, 298
    threshold, 238
Stokes shift, 211
submarine field trial, 8, 334
sulfide, 14, 30
synchronous modulation, 372, 376
systems demonstrations, 323–366

taper, fiber, 47
TAT-12, 3, 8, 67, 268
tellurite, 14, 30, 35, 114, 290
terrestrial transmission systems, 8, 323, 340, 344
thermal noise, 213, 215
thermal population, 104
three-level system, 131–139
threshold, 134, 137, 146

time domain extinction method, 257, 260
TPC-5, 3, 8
transient effects, 363–366
transition metal, 31
transmission
    long distance, see long distance transmission
    performance degradation, 330
    single channel results, 332
    single-channel, 356
    terrestrial, 8, 323, 340, 344
    transoceanic, 334
    undersea, 1, 3, 8, 323, 340
    WDM, 335–345
transoceanic transmission, 334, 376
transverse mode, 140, 141
transverse mode models, 180–186
TrueWave™ fiber, 338
twin core fiber, 55, 290
two-level system, 100, 103, 144–147
two-stage amplifier, 273–276, 293, 327

undersea transmission, 1, 3, 8, 323, 340
upconversion, 113, 120–122, 193–195, 303
    Pr^{3+}, 415
upper state lifetime, 404
upper-state population
    vs fiber length, 167

VAD, 14, 16, 20, 21

waveguide amplifier, 35, 302–304
wavelength division multiplexer, 50–51, 277, 279
WDM, 2, 110, 284–295, 329, 335–345, 374
    channel dropping, 365
    coupler, 268
    four wave mixing, 337
    gain measurement, 256–257
    inline amplifier, 335–345
    nonlinear interactions, 336
    pump and signal combiners, 50
    reconfigured networks, 339

single channel system upgrades, 336
  soliton transmission, 377
  transmission, 3, 295, 340–345

Yb^{3+}, 21, 22, 27, 66, 107, 282, 303
Yb^{3+} fiber laser, 74, 410
Yb-Er energy transfer, 282
Yb-Er power amplifier, 282

ZBLAN, 14, 31, 405–415, 420, 421

Optics and Photonics
(Formerly Quantum Electronics)

Editors: Govind Agrawal, University of Rochester, Rochester, New York
Paul L. Kelly, Tufts University, Medford, Massachusetts
Ivan Kaminow, AT&T Bell Laboratories, Holmdel, New Jersey

N. S. Kapany and J. J. Burke, *Optical Waveguies*
Dietrich Marcuse, *Theory of Dielectric Optical Waveguides*
Benjamin Chu, *Laser Light Scattering*
Bruno Crosignani, Paolo DiPorto and Mario Bertolotti, *Statistical Properties of Scattered Light*
John D. Anderson, Jr., *Gasdynamic Lasers: An Introduction*
W. W. Duly, *CO_2 Lasers: Effects and Applications*
Henry Kressel and J. K. Butler, *Semiconductor Lasers and Heterojunction LEDs*
H. C. Casey and M. B. Panish, *Heterostructure Lasers: Part A. Fundamental Principles; Part B. Materials and Operating Characteristics*
Robert K. Erf, Editor, *Speckle Metrology*
Marc D. Levenson, *Introduction to Nonlinear Laser Spectroscopy*
David S. Kilger, editor, *Ultrasensitive Laser Spectroscopy*
Robert A. Fisher, editor, *Optical Phase Conjugation*
John F. Reintjes, *Nonlinear Optical Parametric Processes in Liquids and Gases*
S. H. Lin, Y. Fujimura, H. J. Neusser and E. W. Schlag, *Multiphoton Spectroscopy of Molecules*
Hyatt M. Gibbs, *Optical Bistability: Controlling Light with Light*
D. S. Chemla and J. Zyss, editors, *Nonlinear Optical Properties of Organic Molecules and Crystals, Volume 1, Volume 2*
Marc D. Levenson and Saturo Kano, *Introduction to Nonlinear Laser Spectroscopy, Revised Edition*
Govind P. Agrawal, *Nonlinear Fiber Optics*
F. J. Duarte and Lloyd W. Hillman, editors, *Dye Laser Principles: With Applications*
Dietrich Marcuse, *Theory of Dielectric Optical Waveguides, 2nd Edition*
Govind P. Agrawal and Robert W. Boyd, editors, *Contemporary Nonlinear Optics*
Peter S. Zory, Jr. editor, *Quantum Well Lasers*
Gary A. Evans and Jacob M. Hammer, editors, *Surface Emitting Semiconductor Lasers and Arrays*
John E. Midwinter, editor, *Photonics in Switching, Volume I, Background and Components*
John E. Midwinter, editor, *Photonics in Switching, Volume II, Systems*
Joseph Zyss, editor, *Molecular Nonlinear Optics: Materials, Physics, and Devices*
Mario Dagenais, Robert F. Leheny and John Crow, *Integrated Optoelectronics*
Govind P. Agrawal, *Nonlinear Fiber Optics, Second Edition*
Jean-Claude Diels and Wolfgang Rudolph, *Ultrashort Laser Pulse Phenomena: Fundamentals, Techniques, and Applications on a Femtosecond Time Scale*
Eli Kapon, editor, *Semiconductor Lasers I: Fundamentals*
Eli Kapon, editor, *Semiconductor Lasers II: Materials and Structures*
P. C. Becker, N. A. Olsson, and J. R. Simpson, *Erbium-Doped Fiber Amplifiers: Fundamentals and Technology*
Raman Kashyap, *Fiber Bragg Gratings*

Yoh-Han Pao, Case Western Reserve University, Cleveland, Ohio, Founding Editor 1972-1979

LIMITED USE SOFTWARE EVALUATION LICENSE

YOU SHOULD READ THE TERMS AND CONDITIONS OF THIS AGREEMENT BEFORE YOU BREAK THE SEALS ON THE PACKAGES CONTAINING THE DISKETTE(S) AND THE DOCUMENTATION. ONCE YOU HAVE READ THIS LICENSE AGREEMENT AND AGREE TO ITS TERMS, YOU MAY BREAK THE SEAL. BY BREAKING SUCH SEAL YOU SHOW YOUR ACCEPTANCE OF THE TERMS OF THIS LICENSE AGREEMENT. THE AGREEMENT IS IN EFFECT FROM THEN UNTIL YOU RETURN ALL THE SOFTWARE TO THE LOCATION WHERE YOU OBTAINED IT OR TO ACADEMIC PRESS.

IN THE EVENT THAT YOU DO NOT AGREE WITH ANY OF THE TERMS OF THIS LICENSE AGREEMENT, RETURN YOUR INVOICE OR SALES RECEIPT, THE SOFTWARE AND THE BOOK WITH ALL SEALS UNBROKEN TO THE LOCATION WHERE YOU OBTAINED THE SOFTWARE, AND THE INVOICE WILL BE CANCELED OR ANY MONEY YOU PAID WILL BE REFUNDED. ALTERNATIVELY, YOU MAY RETURN THE BOOK AND SOFTWARE TO ACADEMIC PRESS FOR A FULL REFUND.

The terms and conditions of this Agreement will apply to the SOFTWARE supplied herewith and derivatives obtained therefrom, including any copy. The term SOFTWARE includes programs and related documentation and any derivative thereof including data generated by the SOFTWARE. If you have executed a separate Software Agreement covering the Software supplied herewith, such Software Agreement will govern.

1. TITLE AND LICENSE GRANT

The SOFTWARE is copyrighted and/or contains proprietary information protected by law. All SOFTWARE, and all copies thereof, are and will remain the sole property of LUCENT TECHNOLOGIES INC. or its suppliers. LUCENT TECHNOLOGIES INC. hereby grants you a personal, nontransferable and non-exclusive right to use, in all counties except those prohibited by U .S. Export laws and regulations, all SOFTWARE, in object code form, which is furnished to you under or in contemplation of this agreement. This grant is limited to use in accordance with paragraph 2.0 below on a single processor at a time. Any other use of this SOFTWARE or removal of the SOFTWARE from a country in which use is licensed shall automatically terminate this license. You agree to obtain prior LUCENT TECHNOLOGIES INC. approval for multi-processor usage. You agree to use your best efforts to see that any use of the SOFTWARE licensed hereunder complies with the terms and conditions of this License Agreement, and refrain from taking any steps, such as reverse assembly or reverse compilation, to derive a source code equivalent of the SOFTWARE.

2. PERMITTED USES OF THE SOFTWARE

You are permitted to use the SOFTWARE for internal educational and research purposes only. No commercial use of the SOFTWARE including use of the data generated therefrom for the manufacture of prototypes or product designs is permitted. A commercial version of this SOFTWARE may be obtained by contacting LUCENT TECHNOL-

OGY INC.'s Network Systems Specialty Fiber Devices Organization (800-364-6404). You may make a single archive copy, provided the SOFTWARE shall not be otherwise reproduced, copied, or, except for the documentation, disclosed to others in whole or in part. Any such copy shall contain the same copyright notice and proprietary marking, including diskette markings, appearing on the original SOFTWARE. Distribution of the SOFTWARE by electronic means is expressly prohibited. The SOFTWARE, together with any archive copy thereof, shall be destroyed when no longer used in accordance with this License Agreement or when the right to use the software is terminated, and further, shall not be removed from a country in which use is licensed.

3. LIMITED WARRANTY

A. LUCENT TECHNOLOGIES INC. warrants that the SOFTWARE will be in good working order. If the SOFTWARE is not in good working order, return the book and SOFTWARE to the location where you obtained it, within ninety (90) days from thepurchase date, for exchange with a new copy of the book and SOFTWARE disk. Alternatively, you may contact Academic Press directly for the exchange of the book and SOFTWARE.

B. LUCENT TECHNOLOGIES INC. does not warrant that the functions of the SOFTWARE will meet your requirements or that SOFTWARE operation will be error-free or uninterrupted. LUCENT TECHNOLOGIES INC. has used reasonable efforts to minimize defects or errors in the SOFTWARE. HOWEVER, YOU ASSUME THE RISK OF ANY AND ALL DAMAGE OR LOSS FROM USE, OR INABILITY TO USE THE SOFTWARE OR THE DATA GENERATED THEREBY.

C. Unless a separate agreement for software maintenance is entered into between you and LUCENT TECHNOLOGIES INC., LUCENT TECHNOLOGIES INC. bears no responsibility for supplying assistance for fixing or for communicating known errors to you pertaining to the SOFTWARE supplied hereunder.

D. YOU UNDERSTAND THAT, EXCEPT FOR THE 90 DAY LIMITED WARRANTY RECITED ABOVE, LUCENT TECHNOLOGIES INC., ITS AFFILIATES, CONTRACTORS, SUPPLIERS AND AGENTS MAKE NO WARRANTIES, EXPRESS OR IMPLIED, AND SPECIFICALLY DISCLAIM ANY WARRANTY OF MERCHANTABILITY OR FITNESS FOR A PARTICULAR PURPOSE.

Some states or other jurisdictions do not allow the exclusion of implied warranties or limitations on how long an implied warranty lasts, so the above limitations may not apply to you. This warranty gives you specific legal rights and you may also have other rights which vary from one state or jurisdiction to another.

4. EXCLUSIVE REMEDIES AND LIMITATIONS OF LIABILITIES

A. YOU AGREE THAT YOUR SOLE REMEDY AGAINST LUCENT TECHNOLOGIES INC., ITS AFFILIATES, CONTRACTORS, SUPPLIERS, AND AGENTS FOR LOSS OR DAMAGE CAUSED BY ANY DEFECT OR FAILURE IN THE SOFTWARE REGARDLESS OF THE FORM OF ACTION, WHETHER IN CONTRACT, TORT, .

INCLUDING NEGLIGENCE, STRICT LIABILITY OR OTHERWISE, SHALL BE THE REPLACEMENT OF LUCENT TECHNOLOGIES INC. FURNISHED SOFTWARE, PROVIDED SUCH SOFTWARE IS RETURNED TO THE LOCATION WHERE YOU OBTAINED THE BOOK AND SOFTWARE OR TO ACADEMIC PRESS WITH A COPY OF YOUR SALES RECEIPT. THIS SHALL BE EXCLUSIVE OF ALL OTHER REMEDIES AGAINST LUCENT TECHNOLOGIES INC., ITS AFFILIATES, CONTRACTORS, SUPPLIERS OR AGENTS, EXCEPT FOR YOUR RIGHT TO CLAIM DAMAGES FOR BODILY INJURY TO ANY PERSON.

B. Regardless of any other provisions of this Agreement, neither LUCENT TECHNOLOGIES INC. nor its affiliates, contractors, suppliers or agents shall be liable for any indirect, incidental, or consequential damages (including lost profits) sustained or incurred in connection with the use, operation, or inability to use the SOFTWARE or the data generated thereby or for damages due to causes beyond the reasonable control of LUCENT TECHNOLOGIES INC., its affiliates, contractors, suppliers and agents attributable to any service, products or action of any other person. This Agreement including all rights and obligations under it shall be construed in accordance with and governed by the laws of the State of New Jersey, United States of America.

YOU ACKNOWLEDGE THAT YOU HAVE READ THIS AGREEMENT AND UNDERSTAND IT, AND THAT BY OPENING ANY SEAL ON THE PACKAGE CONTAINING THE PROGRAM MEDIA YOU AGREE TO BE BOUND BY ITS TERMS AND CONDITIONS. YOU FURTHER AGREE THAT, EXCEPT FOR SEPARATE WRITTEN AGREEMENTS BETWEEN LUCENT TECHNOLOGIES INC. AND YOU, THIS AGREEMENT IS THE COMPLETE AND EXCLUSIVE STATEMENT OF THE RIGHTS AND LIABILITIES OF THE PARTIES. THIS AGREEMENT SUPERSEEDES ALL PRIOR ORAL AGREEMENTS, PROPOSALS OR UNDERSTANDINGS, AND ANY OTHER COMMUNICATIONS BETWEEN US RELATING TO THE SUBJECT MATTER OF THIS AGREEMENT.